2nd edition
microbiology

The National Medical Series for Independent Study

2nd edition
microbiology

David T. Kingsbury, Ph.D.

Professor, Department of Microbiology
George Washington University Medical Center
Washington, D.C.

Gerald E. Wagner, Ph.D.

Professor of Microbiology
St. George's University School of Medicine
St. George's, Grenada
West Indies

NMS

National Medical Series from Williams & Wilkins
Baltimore, Hong Kong, London, Sydney

Harwal Publishing Company, Malvern, Pennsylvania

Williams & Wilkins

Managing editor: Debra Dreger
Project editor: Terry Schutz
Production editor: Keith LaSala
Editorial assistant: Judy Johnson
Illustrator: Wieslawa Langenfeld
Composition and layout: TeleComposition, Inc.

Library of Congress Cataloging-in-Publication Data

Microbiology/editors, David T. Kingsbury, Gerald E.
  Wagner.—2nd ed.
    p.   cm.—(The National medical series for
  independent study)
(A Williams & Wilkins medical publication)
  ISBN 0-683-06234-4
  1. Medical microbiology—Examinations, questions,
  etc.
  2. Medical microbiology—Outlines, syllabi, etc.
  I. Kingsbury, David T. II. Wagner, Gerald E., 1950-
  III. Series. IV. Series: A Williams & Wilkins medical
  publication. [DNLM: 1. Microbiology—examination
  questions. 2. Microbiology—outlines.   QW 18 M623]
  QR46.M5385   1990
  616.01'076—dc20
  DNLM/DLC
  for Library of Congress   90-11991
                                                                  CIP

10 9 8 7 6 5 4 3

Contents

Preface

In the several years since the appearance of *Microbiology,* we have had the opportunity to discuss with students their response to the format and content of the volume. We discovered that the book is used equally for current classwork as an adjunct to the major textbooks and as a review volume for preparation for the National Boards. We also have detected that student expectations for each of these applications are slightly different.

While the goal of the second edition was to update the material presented, there also was pressure to expand the coverage. Expansion of the book has been primarily in the form of different, more easily referenced formatting and in the addition of new material in areas such as AIDS and Lyme disease. We hope we have resisted pressure to expand the volume so much that it loses its value as a study and review guide, and we feel that in its present form it serves both of its major uses well. Finally, we wish to say again that this is not a textbook of microbiology, but we hope it is the best medical microbiology study guide available.

David T. Kingsbury
Gerald E. Wagner

To the Reader

Since 1984, the *National Medical Series for Independent Study* has been helping medical students meet the challenge of education and clinical training. In this climate of burgeoning knowledge and complex clinical issues, a medical career is more demanding than ever. Increasingly, medical training must prepare physicians to seek and synthesize necessary information and to apply that information successfully.

The *National Medical Series* is designed to provide a logical framework for organizing, learning, reviewing, and applying the conceptual and factual information covered in basic and clinical studies. Each book includes a concise but comprehensive outline of the essential content of a discipline, with up to 500 study questions. The combination of distilled, outlined text and tools for self-evaluation allows easy retrieval and enhanced comprehension of salient information. Each question is accompanied by the correct answer, a paragraph-length explanation, and specific reference to the text where the topic is discussed. Study questions that follow each chapter use current National Board formats to reinforce the chapter content. Study questions appearing at the end of the text in the Challenge Exam vary in format depending on the book; the unifying goal of this exam, however, is to challenge the student to synthesize and expand on information presented throughout the book. Wherever possible, Challenge Exam questions are presented in the context of a clinical case or scenario intended to simulate real-life application of medical knowledge.

Each book in the *National Medical Series* is constantly being updated and revised to remain current with the discipline and with subtle changes in educational philosophy. The authors and editors devote considerable time and effort to ensuring that the information required by all medical school curricula is included and presented in the most logical, comprehensible manner. Strict editorial attention to accuracy, organization, and consistency also is maintained. Further shaping of the series occurs in response to biannual discussions held with a panel of medical student advisors drawn from schools throughout the United States. At these meetings, the editorial staff considers the complicated needs of medical students to learn how the *National Medical Series* can better serve them. In this regard, the staff at Harwal Publishing Company welcomes all comments and suggestions. Let us hear from you.

1
Overview of Medical Microbiology

David T. Kingsbury

I. MEDICAL MICROBIOLOGY is the study of interactions between humans and the microorganisms with which they coexist. The microorganisms involved are classified, according to the nature of their interactions with humans, on a spectrum that varies from beneficial to harmful.

A. Classification of microorganisms

1. **Commensals** are organisms that routinely colonize body surfaces without doing harm and are often referred to as the **normal microbial flora** (e.g., *Escherichia coli*, the alpha streptococci).

2. **Pathogens** are organisms that damage the human host either by direct invasion and injury (e.g., *Shigella* species) or by the production of harmful toxic products (e.g., *Clostridium* species).

B. Host-microbe interaction. The **pathogenic potential** of many organisms is variable. It is influenced by both the intrinsic properties of the microorganism and the state of health of the human host.

1. **Host defenses and natural immunity** refer to a multifactor system of protective mechanisms that prevent entry of microorganisms into normally sterile areas and limit the spread of those invaders that overcome the first line of defense.
    a. These mechanisms may be weakened by a variety of insults, including direct physical trauma, systemic disease, drugs, and toxins.
    b. When normal defenses are impaired, the person loses the ability to combat infection and the injury caused by pathogens, even those with low intrinsic virulence. In such cases, the **compromised host** often succumbs to infection.

2. **Microbial virulence** is the relative intrinsic ability of a microorganism to cause disease. Organisms of high virulence have evolved efficient mechanisms for circumventing normal host defenses. Virulent organisms are adept at gaining entry and doing damage even when the inoculum is small.

C. The human impact of infectious disease. The study of infectious diseases has had a dramatic impact upon human history. Much of modern scientific medicine is a direct result of the development of medical microbiology. In particular, the **germ theory of disease** (i.e., the recognition that a specific etiologic agent is responsible for a particular constellation of symptoms) is an outgrowth of enhanced understanding of infectious diseases.

1. The nineteenth century microbiologists Jakob Henle and Robert Koch formalized the concept that human disease is often caused by specific microorganisms. They developed a set of methodologic principles (**Henle-Koch postulates**) for establishing a specific microbial etiology. The following principles remain important guidelines in medical research.
    a. The organism occurs in every case of the disease in question and under circumstances that can account for the pathologic changes and clinical course of the disease.
    b. The organism occurs in no other disease as a fortuitous and nonpathogenic parasite.
    c. After being fully isolated from the body and repeatedly grown in pure culture, the organism can induce the same disease anew.

2. These postulates have provided an important conceptual model for studying the underlying pathology of infectious diseases. The model is not perfect, however; its conditions may not be met when a pathogen cannot be cultured in vitro or when the agent is not pathogenic for the available laboratory animals.

II. CLASSES OF PATHOGENIC MICROORGANISMS. Pathogens vary in size and biologic complexity. Some are able to extract sufficient nutrients from an inanimate environment and, hence, may be cultured on artificial media. Others are incapable of growth outside living host cells and are referred to as **obligate intracellular parasites**. Pathogens are divided into four major groups.

A. **Viruses** are the smallest intact organisms with demonstrated pathogenic potential. They are too small to be seen with a light microscope.

   1. Viruses are **obligate intracellular parasites** that depend entirely upon the host cell's synthetic machinery for reproduction.

   2. Viruses contain only one type of nucleic acid, **either DNA or RNA**, but not both.

   3. After entering a host cell, a virus sheds its coat and releases viral nucleic acid into the cell. Under the direction of the viral genes, the host cell diverts its activities toward producing new viral components, which are then assembled within the cell into new virus particles (**virions**). The virions are released, additional cells are infected, and the cycle is repeated.

B. **Bacteria** are both larger and more complex than viruses. Most bacteria are visible under the light microscope.

   1. Bacteria are termed **prokaryotes** because, unlike higher organisms, they lack a true cell nucleus. Since no nuclear membrane is present, the genetic material, in the form of a **nucleoid**, lies within the cytoplasm.

   2. Unlike viruses, bacteria possess **both DNA and RNA**.

   3. Bacteria reproduce by **binary fission**. Many pathogenic bacteria are capable of independent growth and, thus, may be cultured on artificial media. Some bacteria, however, lack the ability to produce important metabolites and are obligate intracellular parasites. These organisms must be grown in tissue culture if their recovery is necessary.

C. **Fungi** are larger than bacteria and have a more advanced cell structure.

   1. As **eukaryotic** organisms, their genetic material is separated from the cytoplasm by a nuclear membrane.

   2. Some fungi reproduce by **budding (yeasts)**, whereas others form growing **colonies of attached organisms (molds)**.

   3. Most pathogenic fungi exist in nature as **environmental saprophytes**, and human infection does not appear to be necessary for their life cycle.

D. **Parasite** is a general term often used in a narrow sense to refer to a variety of protozoan and multicellular eukaryotic organisms capable of causing disease. Many parasites undergo complex life cycles that may involve several host species, including humans. The parasitic diseases remain major health problems and sources of economic drain, especially in underdeveloped nations.

III. CLINICAL EVALUATION AND TREATMENT OF PATIENTS WITH INFECTIOUS DISEASES

A. **Clinical presentation.** Signs and symptoms of infection may develop acutely or gradually. The particular clinical picture varies with the etiologic agent of infection and with the patient's state of health. Because many important signs of infection are relatively nonspecific, workup must consider both infectious and noninfectious diseases. Important clinical findings that raise suspicion of infection include the following.

   1. **Fever** is perhaps the cardinal clue to the presence of infection; however, **not all infections cause fever and not all fevers result from infection**. The pattern of temperature variation may sometimes offer a clue to the diagnosis. **Noninfectious causes** of fever include:
      a. Neoplasm
      b. Central nervous system (CNS) disease
      c. Rheumatic disease
      d. Hypersensitivity reaction
      e. Metabolic derangement

2. **Pain** may be a reflection of inflammation and tissue destruction, or it may be the result of pressure on sensitive structures caused by an enlarging infected mass.

3. **Weight loss** may occur in the setting of chronic infection.

4. **Progressive debility** is a vague but important symptom. It may be a reflection of the catabolic effects of infection, or it may be traced to more specific problems such as malabsorption, CNS disease (e.g., chronic meningitis), or cardiovascular impairment (e.g., endocarditis).

5. **Change in physiologic function** of a particular organ system (e.g., progressive shortness of breath as a result of pneumonia, alteration of mental status due to meningitis, diarrhea caused by enteritis) may reflect infection.

6. **Abnormal laboratory test results** may suggest the presence of infection even in the asymptomatic patient. Common examples include abnormal chest x-rays, elevated erythrocyte sedimentation rates, elevated circulating leukocyte counts with abnormal differential cell counts, and the presence of white blood cells in normally acellular body fluids [e.g., urine, cerebrospinal fluid (CSF), joint fluid, peritoneal fluid].

B. Diagnosis

1. **The importance of accuracy.** Diagnostic procedures are aimed at identifying an individual etiologic agent responsible for a particular clinical picture. Accurate diagnosis allows implementation of a specific treatment regimen, which increases the likelihood of recovery. In the absence of an accurate diagnosis, the physician often is tempted to rely on shotgun empiric therapies, which should be employed only as a last resort. Empiric antibiotic regimens often create problems for reasons that include the following.
   a. The empiric regimen chosen may not affect a curable infection.
   b. Attention may be diverted from the need for further diagnostic studies.
   c. The patient is exposed to the toxicities of potentially unnecessary drugs.
   d. Empiric antibiotics can confuse interpretation of diagnostic tests such as bacterial cultures.

2. **Strategies.** When the possibility of infectious disease exists, several strategies may be employed to develop a diagnosis.
   a. An effort should be made to **define the anatomic location** of the infectious process. The physician must determine which organ systems are involved and must plan diagnostic tests to maximize the yield of information from these sites.
   b. The physician should consider **which groups of organisms** (i.e., viruses, bacteria, fungi, parasites) **are likely to cause the particular clinical picture**. Efforts to grow optimal cultures of the most likely organisms should be made.
   c. Data from several sources must be accumulated and interpreted in an orderly fashion. Since the **medical history** provides important information about infection, questioning should obtain the following data:
      (1) **Exposure history**, including exposure to diseases in family and friends, to both domestic and wild animals, and to insect vectors, as well as history of travel, hobbies, and occupation
      (2) **Related health history**, including history of immunosuppressive illnesses and drugs, heart and lung disease, trauma and surgeries, and serious infections
      (3) **Epidemiology**, including time of year the infection was acquired and a history of similar illness in the community
      (4) **Social history**, including sexual activities and illicit drug use
      (5) **Reliable indices of infection**, including documented fevers, shaking chills, weight loss, and rashes
   d. The **physical examination** may provide important information.
      (1) Documentation of temperature, weight, and blood pressure is needed.
      (2) The patient should be examined carefully for **localizing signs**, particularly localized tenderness, heat, swelling, and erythema. Other important localizing signs should be sought as well, including stiff neck, heart murmurs, lung conditions, skin lesions, abdominal tenderness, bony tenderness, and adenopathy.
   e. The **laboratory evaluation** is directed by elements in the medical history and physical examination.
      (1) **Gram stain and culture.** Material obtained from the suspected site of infection should be stained and cultured. Specimens must be collected carefully and delivered to the

laboratory in optimal condition. Culture results must always be interpreted in light of the normal flora of the sampled site. Gram's stain is especially important in evaluating specimens taken from a normally contaminated site: the presence of inflammatory cells and a predominant organism provide confirmation of infection. Because antibiotic therapy can interfere with isolation of an etiologic organism, every effort should be made to obtain all relevant culture material before antibiotics are started.

(2) **Biopsy and histologic examination** of potentially infected tissues may be especially important in the interpretation of culture results when specimens are obtained from normally unsterile sites. The pathologist should be informed of suspected etiologic agents so that appropriate tissue stains may be employed.

(3) **Serologic tests** are especially important when the suspected etiologic agent is known to be difficult to isolate or identify. A variety of methodologies (e.g., complement fixation, immunodiffusion, radioimmunoassay, enzyme-linked immunoassay, Western blotting, immunofluorescence) may be employed; the technique is not as important as an understanding of what is being detected.

  (a) **Antibody detection assays** [e.g., those for *Legionella* immunofluorescent antibody (IFA), *Mycoplasma* complement fixation, and human immunodeficiency virus ELISA and Western blot] detect an antibody response to the pathogen and, thus, may be positive even in the absence of a current active infection.

  (b) **Antigen detection assays** (e.g., the cryptococcal antigen test and hepatitis B surface antigen test) detect components produced by the pathogen itself and, thus, are more reliable indications of active disease.

(4) **Skin testing.** Delayed hypersensitivity testing may indicate past exposure to a pathogen but rarely indicates the nature of present activity.

(5) **Radiologic testing** (e.g., x-ray, nuclear scanning) may indicate the anatomic locus of infection but rarely offers an unequivocal etiologic diagnosis. On the other hand, radiographically directed percutaneous needle aspiration of abscess cavities is becoming important both as a diagnostic and a therapeutic modality.

STUDY QUESTIONS

Directions: Each question below contains five suggested answers. Choose the **one best** response to each question.

1. A newly identified pathogenic agent is isolated from the lung tissue of several patients. Which of the following characteristics proves conclusively that the organism is a virus?

(A) The agent is an obligate intracellular parasite

(B) The agent passes through a 1-μm filter

(C) The agent contains DNA but not RNA

(D) The agent is resistant to antibacterial treatment

(E) The agent lacks a nuclear membrane

2. Serologic diagnosis is an important tool in infectious disease treatment because

(A) it detects the presence of the specific antigen

(B) it is useful even if the etiologic agent is difficult or impossible to isolate

(C) it is a reflection of current infection

(D) it does not produce false-positive or false-negative results

(E) it is effective earlier in infection than any type of direct culture might be

Directions: Each question below contains four suggested answers of which **one or more** is correct. Choose the answer

A   if **1, 2, and 3** are correct
B   if **1 and 3** are correct
C   if **2 and 4** are correct
D   if **4** is correct
E   if **1, 2, 3, and 4** are correct

3. Diagnostic modalities that play an important role in the initial evaluation of a normal patient with a presumed infection include

(1) an empiric trial of antibiotics

(2) careful physical examination

(3) laparotomy

(4) review of the patient's sexual and occupational history

4. Principles that predicate the medical management of immunocompromised patients include

(1) relatively avirulent organisms often cause life-threatening infections

(2) antibiotics are required even in the absence of clinical infection

(3) antibiotic therapy often precedes a firm diagnosis of infection

(4) neutrophil function is depressed

ANSWERS AND EXPLANATIONS

1. The answer is C. [*II A*] Viruses are unique among microorganisms in that they contain only one type of nucleic acid: either DNA or RNA, but not both. Like viruses, bacteria also lack nuclear membranes, and some are able to pass through a 1-μm filter. Although all viruses are obligate intracellular parasites, agents such as *Chlamydia* also require a host cell for replication. Like viruses, fungi usually are resistant to the typical antibacterial antibiotics.

2. The answer is B. [*III B 2 e (3)*] Serologic diagnosis generally is based on the detection of antibody to a specific agent rather than of the agent itself. Because of this, positive results are expected not only in the case of current infection but also in the case of prior infection. This diagnostic tool is useful for organisms that are difficult or impossible to isolate, but the results must be interpreted in light of serologic cross-reactivity (which might give false-positive results) and in light of the fact that a false-negative result may occur early in infection, before a measurable antibody response has been developed.

3. The answer is C (2, 4). [*III B*] Careful history and physical examination provide the basis for the initial evaluation of a patient with a presumed infection. Empiric trials of antibiotics and invasive testing usually are reserved for situations in which all of the alternative diagnostic procedures have been attempted.

4. The answer is B (1, 3). [*I C 1*] The term "immunocompromised" refers to a variety of defects in host defenses, including neutrophil function as well as lymphocyte function (e.g., as occurs in patients with agammaglobulinemia) and breakdown of anatomic barriers to infection (e.g., as occurs in burn patients). Most physicians recommend antibiotic therapy at the earliest sign of infection; however, the use of antibiotics in the absence of such evidence (i.e., prophylaxis) in immunocompromised patients varies with the situation and often is controversial.

2
Bacterial Classification, Structure, and Physiology

David T. Kingsbury

I. BACTERIAL CLASSIFICATION. Identification of bacteria is of enormous practical significance in clinical microbiology.

   A. Nomenclature. A binomial nomenclature is applied to bacteria within a rigid framework of taxonomic rules. The most significant taxonomic subclasses are **genus** and **species**.

   B. Readily observable properties usually are the basis for bacterial classification and identification.

      1. Gram staining, the most important bacterial property for classification purposes, is a simple laboratory procedure that employs the aniline dye, **crystal violet**. The exact molecular basis for Gram's stain reaction is not understood completely, but cell wall structure appears to be the determining factor.
         a. Gram-positive bacteria are bacteria that take up and retain the crystal violet and resist alcohol decoloration. They appear blue to black.
         b. Gram-negative bacteria are bacteria that are decolorized completely by ethanol and take up safarin counterstain, appearing red.

      2. Morphologic features are the second most important factor in the classification of bacteria.
         a. Shape. Bacteria may exist in three basic shapes.
            (1) Rods (i.e., cylindrical cells) are referred to as **bacilli**.
            (2) Spheres. Spherical cells are referred to as **cocci**.
            (3) Spirals. The **spirilla**, **vibrios**, and **spirochetes** are examples of spiral bacteria.
         b. Arrangement. Bacteria are found as single cells or in regular groups of two or more cells.

   C. Metabolic reactions form the precise basis for bacterial **species identification**.

      1. Oxygen requirements
         a. Aerobic bacteria require oxygen for normal growth and development.
         b. Anaerobic bacteria cannot tolerate oxygen; they can grow only in the complete absence of oxygen.
         c. Facultative bacteria can grow in the presence or absence of oxygen.

      2. Carbohydrate utilization. Species often can be identified by the variety of carbohydrate substrates they use as energy sources. Some bacteria do not use glucose under any conditions.
         a. In oxidative metabolism, oxygen is required to degrade glucose.
         b. In fermentative metabolism, glucose is degraded anaerobically.

      3. Enzyme production
         a. Detoxifying or oxidative enzymes may be produced.
            (1) Catalase destroys hydrogen peroxide.
            (2) Oxidase transfers hydrogen directly from a substrate to oxygen.
         b. Proteolytic or toxic enzymes also may be produced. Gelatin liquefaction provides evidence of proteolytic enzymes.
            (1) Hemolysins lyse red blood cell membranes.
               (a) Alpha hemolysins lyse but do not dissolve red cell membranes.
               (b) Beta hemolysins both lyse and dissolve red cell membranes.
            (2) Coagulase causes clotting of blood plasma.

D. Genetic relatedness is based on similarities of nucleic acid sequence and provides a more phylogenetic scheme of bacterial classification than do the other methods. Genetic classification is based on the study of:

1. DNA homology among bacteria

2. Nucleotide base composition, which permits even broader classification of bacteria

3. Exchange of genetic information (i.e., transformation, conjugation) between related organisms

II. BACTERIAL STRUCTURE.

The basic components of bacterial cells (Figure 2-1) include the structures that form the cell envelope enclosing the cytoplasm, the cytoplasmic constituents, and the external structures that project through or cover the cell envelope.

A. Cell envelope. Most bacterial cells have a cell envelope consisting of a cell wall and an underlying cytoplasmic membrane.

1. Cell wall. Bacteria have a rigid cell wall that provides protection and imparts shape to the cell. The cell wall is entirely absent in a few unusual microorganisms (e.g., mycoplasmas).

　　a. The principal structural component of the cell wall is **peptidoglycan**, a mixed polymer (Figure 2-2).

　　　　(1) Peptidoglycan polymers, or glycan strands, consist of two hexose sugars, **N-acetylglucosamine** and **N-acetylmuramic acid**, which are linked to short peptides of identical chains of four amino acids.

　　　　(2) Peptidoglycan polymers contain several amino acids that are found only in bacterial cell walls; these acids are ***meso*-diaminopimelic acid** and the **D-isomers of glutamic acid and alanine**.

　　b. The cell wall of gram-positive bacteria differs in its structure and composition from that of gram-negative bacteria, although both contain peptidoglycan.

　　　　(1) Gram-positive bacteria have a somewhat simple, thick cell wall consisting primarily of a heavy, rigid layer of peptidoglycan as well as teichoic acids. The large amount of peptidoglycan make gram-positive bacteria susceptible to the enzyme **lysozyme** and to **penicillin**.

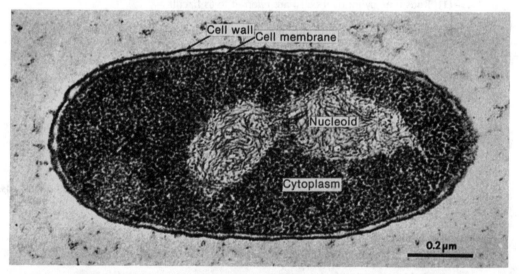

Figure 2-1. Electron micrograph showing the structural organization that is common to all bacterial cells. The nucleoid occupies the central area of the cell; the cytoplasmic membrane and cell wall surround the cell. This bacterium, *Alteromonas espejiana*, is a marine pseudomonad with a very simple structure. (Reprinted with permission from Cota-Robles EH, Ringo DL: The structure of the bacterial cell. In *Infectious Diseases and Medical Microbiology*, 2nd ed. Edited by Braude AI, et al. Philadelphia, WB Saunders, 1986, p 2.)

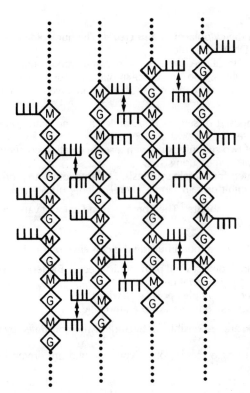

Figure 2-2. Schematic diagram of the structure of peptidoglycan. The glycan strands are composed of alternating units of *N*-acetylglucosamine (*G*) and *N*-acetylmuramic acid (*M*). Tetrapeptide side chains connect to the muramic acid residues. Some of the tetrapeptides are joined by other peptides (*short arrows*) to form cross-bridges between the glycan strands. (Reprinted with permission from Cota-Robles EH, Ringo DL: The structure of the bacterial cell. In *Infectious Diseases and Medical Microbiology*, 2nd ed. Edited by Braude AI, et al. Philadelphia, WB Saunders, 1986, p 5.)

     (a) Lysozyme hydrolyzes peptidoglycan by specific cleavage between *N*-acetylmuramic acid and the *N*-acetylglucosamine of the glycan strand.
     (b) Penicillin specifically inhibits peptidoglycan synthesis.
  (2) Gram-negative bacteria have a cell wall that is thinner than that of gram-positive bacteria, with far less peptidoglyan and no teichoic acids. A complex layer, called the **outer membrane**, covers the cell wall of gram-negative bacteria.
     (a) The outer membrane of gram-negative bacteria is thicker than the peptidoglycan layer and is composed of lipoprotein and lipopolysaccharide attached to the peptidoglycan.
     (b) The outer membrane contains the O antigen and the endotoxin that is characteristic of gram-negative bacteria.
  c. Removal of the cell wall, which protects the delicate underlying cytoplasmic membrane, results in **bacterial lysis**, because the membrane is unable to withstand the osmotic pressures found in nature.
     (1) **Protoplasts** are cells maintained in an osmotically protected environment following complete removal of the cell wall.
     (2) **Spheroplasts** also are osmotically fragile cells; they result from partial removal of the cell wall.

  2. Cytoplasmic membrane. The cytoplasmic membrane (also referred to as the **cell membrane** or **plasma membrane**) is the actual barrier between the interior and exterior of the bacterial cell.
  a. The cytoplasmic membrane exhibits a well-defined **selective permeability**.
  b. The bacterial **electron transport system**, the principal energy system, is located in the cytoplasmic membrane.
  c. **Mesosomes** are complex invaginations of the cytoplasmic membrane into the cytoplasm seen in many, but not all, bacteria. Their function is not well established.

B. Cytoplasmic components. The cytoplasm of most bacteria contains only DNA, ribosomes, and storage granules, because the cytoplasmic membrane performs many complex functions.

  1. DNA. The bacterial cell lacks a nuclear membrane (i.e., it is **prokaryotic**).

a. The cellular DNA is concentrated in the cytoplasm as a **nucleoid**. The nucleoid consists of one long, double-stranded, circular DNA molecule.

b. In some bacteria, a small portion of the DNA persists in an extrachromosomal state as molecules referred to as **plasmids**, which also are circular but are much smaller than bacterial chromosomes. Plasmids frequently carry genes that are involved in antibiotic resistance (**R factors**).

2. Ribosomes are complex structures composed of several RNA molecules and many proteins; they are approximately 20 nm in diameter.

a. Bacterial ribosomes are composed of two **subunits**, one with a sedimentation coefficient of 50 Svedberg units, or 50S, and the other measuring 30S.

b. Ribosomes function as the active centers for **protein synthesis**. The antibiotics streptomycin, tetracycline, and chloramphenicol inhibit protein synthesis on bacterial ribosomes.

3. Storage granules temporarily hold excess metabolites. Their presence and amount vary with the type of bacterium and its metabolic activity.

C. External structures. Capsules, flagella, and pili are found outside the cell envelope.

1. Capsules surround many bacterial cells, including several pathogenic species. Bacterial capsules lack the well-ordered structure found in the bacterial cell wall.

a. Most bacterial capsules are composed of complex **polysaccharides**.

b. Some bacteria have **polypeptide** capsules composed of D-amino acids.

2. Flagella are present on many bacteria and are responsible for the motility demonstrated by those species.

a. Bacterial flagella are composed of a single species of polypeptide and are driven by the rotary action of a swivel-like basal hook.

b. Some bacteria have many flagella distributed over their surface (**peritrichous flagella**), whereas other bacteria have a single flagellum (**monotrichous flagellum**) or a small bundle of flagella located at one end (**polar flagella**).

3. Pili are protein fibers that cover the entire surface of gram-negative bacteria.

a. Many gram-negative bacteria have long, slender **common pili** that originate at the cell membrane and extend through the cell wall.

(1) The common pili, or **fimbriae**, are composed of a single protein type, with the molecules in the form of a helical filament.

(2) Common pili appear to play a role in **bacterial adherence** to surfaces.

b. Sex pili occur less commonly and appear to be specifically involved in **bacterial conjugation**.

III. BACTERIAL PHYSIOLOGY

A. Growth

1. Nutritional requirements. Bacteria grow best in an environment that satisfies their nutritional requirements.

a. Oxygen. The proper oxygen tension is essential for balanced growth.

(1) The growth of aerobic bacteria in a nutrient-rich medium usually is restricted by limited availability of oxygen.

(2) Conversely, the growth of anaerobic bacteria may be inhibited by an oxygen tension as low as 10^{-5} atm.

b. Carbon dioxide is required by many bacteria for growth. Several species are stimulated by partial pressures (Pco_2) that are greater than those normally found in air.

c. Inorganic ions. Phosphate, potassium, magnesium, nitrogen, and sulfur as well as numerous trace metals are essential for bacterial growth.

d. Organic nutrients are required in different amounts by different species of bacteria.

(1) Carbohydrates are used as an energy source and as the initial substrate for many biosynthetic pathways.

(2) Amino acids and, in some cases, short-chain polypeptides are absolutely essential to many bacteria.

(3) Vitamins, **purines**, and **pyrimidines** also are commonly needed for growth.

2. Growth curve. Bacteria increase in mass and number in an exponential manner. Under optimal conditions, bacterial growth generally can be separated into at least four distinct phases (Figure 2-3).

   a. Lag phase. This is a period of physiologic adjustment involving the induction of new enzymes and the establishment of a proper intracellular environment.

   b. Logarithmic phase. This phase is characterized by maximal rates of cell division and increase in cell mass. The generation time during the logarithmic, or exponential, phase varies among bacterial species; it may be as brief as about 20 minutes or as long as several hours.

   c. Stationary phase. During this period, the essential nutrients begin to disappear, and there is a balance between cell growth and division and cell death.

   d. Death phase. During this phase, bacterial lysis and cell destruction cause a reduction in cell number.

3. Growth regulation

   a. As noted, bacterial growth is regulated by the **nutritional environment**.

   b. In addition, intracellular regulatory events (i.e., **feedback regulation**) occur.

   　(1) End-product inhibition occurs when the first enzyme in a pathway is inhibited by the end products of that pathway.

   　(2) Repression of enzyme synthesis through **catabolite repression** is another device for controlling the metabolic activity of a pathway.

   c. The rate of bacterial growth also is dependent on **temperature**. Most bacteria can grow in a wide range of temperatures but require a narrow temperature range for optimal growth.

B. Energy metabolism. Energy used by bacteria primarily is produced by fermentative, respiratory, and autotrophic metabolism.

1. Fermentative metabolism uses organic compounds as both the electron donors and the electron acceptors.

   a. The major fermentative pathway used by bacteria is the **glycolytic**, or **Embden-Meyerhof, pathway**, which converts glucose to pyruvate in a ratio of 1 mol to 2 mol. Glycolysis results in a net yield of 2 mol of adenosine triphosphate (ATP) per mol of glucose fermented.

   b. Many bacteria use pyruvate in **secondary fermentation processes** that oxidize the reduced form of nicotinamide-adenine dinucleotide (NADH) produced during the glycolytic process. Many secondary fermentation processes result in production of carbon dioxide, hydrogen, or methane gas during the breakdown of pyruvate.

   　(1) The simplest secondary fermentation process is **lactic fermentation**, which converts pyruvate to lactate.

   　(2) In **alcoholic fermentation**, pyruvate is converted to carbon dioxide and ethanol.

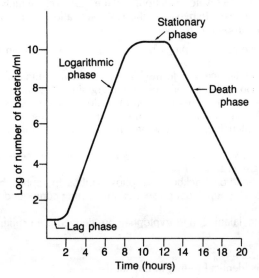

Figure 2-3. Bacterial growth curve. Initially there is a lag phase that permits the microbe to make metabolic adjustments to the new medium. When optimal conditions are met, the organisms double at a fixed rate to provide logarithmic growth. When nutrients decrease, a stationary phase occurs during which death and growth are balanced. When growth stops entirely, the death phase is seen.

      (3) In **propionic fermentation**, pyruvate is converted, first, to oxaloacetate with the addition of a carbon dioxide molecule and, eventually, to propionic acid.

      (4) **Mixed acid fermentation**, common among most members of the Enterobacteriaceae family, is a combination of lactate, acetate, and formate production via several similar pathways. In addition, carbon dioxide, hydrogen, and ethanol are produced.

  c. Many organisms use **alternate pathways** to metabolize glucose.

      (1) The major hexose-degrading pathway in pseudomonads is the **Entner-Doudoroff pathway**.

      (2) Some organisms use both glycolysis and the **hexose monophosphate shunt**.

2. **Respiratory metabolism** usually uses oxygen as the electron acceptor.

  a. In **aerobic respiration**, the most common form, pyruvate is converted to carbon dioxide and **acetylcoenzyme A** (acetyl-CoA); the acetate is degraded further, via the tricarboxylic acid (TCA) cycle, in a process known as **terminal respiration**.

      (1) **Complete respiration** results in 38 mol of ATP produced for 1 mol of glucose, a much higher energy yield than in fermentation. Respiratory metabolism also permits the use of energy sources that are so highly reduced that they cannot be fermented.

      (2) Many bacteria lack the enzymes needed for complete oxidation and, therefore, obtain energy from **incomplete oxidation**.

  b. The respiration process is facilitated by the **electron transport system**, a system whereby electrons move through a series of reversible electron carriers with successively higher oxidation-reduction potentials. These electron acceptors are embedded in the bacterial membrane and range from NADH to oxygen.

      (1) The major elements of the electron transport system are **cytochromes** (hemoproteins), **flavoproteins**, and **ubiquinones** (e.g., coenzyme Q).

      (2) In three stages of the electron transport sequence, energy is used to extrude a pair of protons (H^+) across an insulating membrane, thus creating an **electrochemical gradient** (**proton motive force**).

      (3) The return of the protons through the membrane reverses the activity of adenosine triphosphatase (ATPase) and is referred to as **oxidative phosphorylation**.

3. **Autotrophic metabolism** uses a variety of alternate inorganic sources for energy and reducing power.

  a. In **photosynthesis**, bacteria use energy from light to convert carbon dioxide to triose phosphate and, subsequently, to other cell constituents.

  b. In **anaerobic respiration**, bacteria use inorganic substrates as terminal electron acceptors in place of oxygen.

C. Biosynthetic pathways. The enormous diversity in the nutritional requirements of bacteria stems from the biosynthetic abilities of different groups of bacteria. Bacteria use various catabolic pathways for biosynthesis, including glycolysis, pyruvate oxidation, and the TCA cycle, which are bidirectional, or amphibolic, pathways. The **glyoxylate bypass** of the TCA cycle allows bacteria to use acetate as a precursor of a variety of 4-carbon dicarboxylic acids.

1. **Amino acid biosynthesis** falls into convenient groupings based on points of origin in the amphibolic pathways.

  a. **Glutamate family.** Glutamate and glutamine are formed from reductive amination of α-ketoglutarate and serve as the carbon skeletons for proline and arginine.

  b. **Aspartate family.** Aspartate is synthesized by transamination of oxaloacetate, gives rise to asparagine, and can be reduced to lysine, methionine, threonine, and isoleucine.

  c. **Pyruvate family.** Alanine, valine, isoleucine, and leucine are formed from the initial transamination of pyruvate.

  d. **Serine family.** Serine, glycine, and cysteine are derived from 3-P-glycerate; glycine results from the transfer of the 1-carbon fragment to tetrahydrofolate.

  e. **Histidine** is synthesized from the 5-carbon backbone of phosphoribosylpyrophosphate (PRPP) and is unique in that its carboxyl group is not present on the starting compound but is formed later.

  f. **Aromatic amino acids.** Tyrosine, phenylalanine, and tryptophan are derived from shikimic acid.

2. **Nucleotide biosynthesis**

  a. Nucleotide biosynthesis in bacteria is identical to that in animal tissue.

(1) **Purine nucleotides** are built on the ribose-P chain from PRPP, glycine, and carbon and nitrogen additions in a long sequence.

(2) **Pyrimidine nucleotides** are formed via a series of carboxyl-containing intermediates, starting with carbamyl phosphate with ribose-P added late in the sequence.

b. The principal **macromolecules**—DNA, RNA, and protein—are polymers of the nucleotide and amino acid building blocks and are synthesized via a highly regulated system in the cell cytoplasm. The sequence of information transfer for macromolecular synthesis involves **replication** (DNA to DNA), **transcription** (DNA to RNA), and **translation** (RNA to protein).

(1) **DNA** acts as the template necessary for synthesis of both DNA and RNA.

(2) **RNA** serves as the template for protein synthesis and provides the core of ribosomes on which the proteins are synthesized.

STUDY QUESTIONS

Directions: Each question below contains five suggested answers. Choose the **one best** response to each question.

1. The external barrier around the bacterial cell, the cell wall, has all of the following properties EXCEPT

(A) it consists of a mixed polymer called a pep-tidoglycan

(B) it is the structure principally responsible for the reaction to Gram staining

(C) it is a uniquely flexible plastic structure

(D) it contains D-isomers of amino acids

(E) the cell wall of gram-positive bacteria is more sensitive to lysozyme than that of gram-negative bacteria

2. Which of the following structures, found external to the bacterial cell wall, are involved in bacterial attachment to cell surfaces?

(A) Capsules
(B) Flagella
(C) Pili
(D) Mesosomes
(E) None of the above

3. Bacterial mixed acid fermentation yields all of the following organic compounds EXCEPT

(A) acetate
(B) butyrate
(C) formate
(D) lactate
(E) ethanol

Directions: Each question below contains four suggested answers of which **one or more** is correct. Choose the answer

A   if **1, 2, and 3** are correct
B   if **1 and 3** are correct
C   if **2 and 4** are correct
D   if **4** is correct
E   if **1, 2, 3, and 4** are correct

4. The classification of a microorganism reveals a great deal about the organism because

(1) it is based on the historic background of the organism

(2) it is based on readily observable properties of significance to the organism

(3) it groups pathogenic bacteria in different genera from nonpathogenic bacteria

(4) the primary criterion for classification, cell wall structure, is a significant factor clinically

5. The biosynthesis of amino acids by bacteria involves

(1) the reductive amination of α-ketoglutarate

(2) the complete biosynthesis of lysine from 1-carbon compounds

(3) the conversion of shikimic acid into tyrosine

(4) the transamination of oxaloacetate to produce alanine and valine

6. Bacteria have a variety of growth requirements. Balanced bacterial growth is influenced by

(1) oxygen tension
(2) inorganic ions
(3) partial pressure of carbon dioxide
(4) the nature of the organic compounds

7. True statements regarding protein synthesis in bacteria include which of the following?

(1) It requires mRNA and 50S ribosomes

(2) It is sensitive to chloramphenicol and streptomycin

(3) It usually is regulated at the level of transcription

(4) It is sensitive to penicillin

8. The bacterial cytoplasmic membrane has which of the following characteristics?

(1) It is freely permeable to all nutrient compounds

(2) It contains the cell's electron transport system

(3) It retains the shape of the bacterial cell if the cell wall is removed in an osmotically protected environment

(4) It is the actual barrier between the interior and exterior of the cell

Directions: The group of questions below consists of lettered choices followed by several numbered items. For each numbered item select the **one** lettered choice with which it is **most** closely associated. Each lettered choice may be used once, more than once, or not at all.

Questions 9–12

Match each description of bacterial structure or function with the most appropriate bacterial cell component.

(A) Nucleoid
(B) Cytoplasmic membrane
(C) Cell wall
(D) Common pilus
(E) Flagellum

9. A protein fiber covering the surface of gram-negative cells that plays a role in bacterial adherence

10. A protein fiber found on the surface of bacterial cells responsible for bacterial mobility

11. A structure in the interior of the cell not bounded by a membrane

12. Exhibits a well-defined selective permeability

ANSWERS AND EXPLANATIONS

1. The answer is C. [*II A*] The bacterial cell wall is a unique structure formed from a rigid complex of the peptidoglycan. The cell wall dictates the shape of the cell and provides an osmotic barrier for the cell interior. Digestion of the cell wall with lysozyme leads to lysis of the bacteria unless the surrounding medium is adjusted to increase the osmotic strength. In contrast, plasmolysis of the bacterial interior without digestion of the cell wall leaves isolated structures with the morphology and size of the original microorganisms. Cocci retain their spherical shape and bacilli their rod shape even though the cell interior has been disrupted.

2. The answer is C. [*II C 3*] Many bacteria have surface macromolecules that are essential to their adherence, and most of these surface-adhesive molecules are found on pili. It is now recognized that adherence of bacteria to cell surfaces is critical to the establishment of the normal flora as well as to invasion by pathogenic organisms.

3. The answer is B. [*II B 1 b (4)*] The usual mixed acid fermentation produces acetate, ethanol, formate, and lactate. The butyric acid fermentation involves a branch in the pathway that replaces the acetate and ethanol production with the production of acetoacetic acid and then butyric acid.

4. The answer is C (2, 4). [*I*] Organisms are identified and classified on the basis of readily observable properties such as their reaction to Gram staining, the presence of a capsule, and fermentation or metabolic activities. The nature of the cell wall of microorganisms is a significant factor in the infectious process of the organism as well as in its sensitivity to many antimicrobial drugs.

5. The answer is B (1, 3). [*III C 1*] The major biosynthetic pathways for amino acid synthesis involve the conversion of shikimic acid into aromatic amino acids and the use of α-ketoglutarate from the tricarboxylic acid cycle as a starting compound for the synthesis of glutamate and glutamine. Glutamate is the source of the α-amino group of all other amino acids; it also is the carbon skeleton of proline and arginine. Lysine and valine are produced by quite different pathways, neither of which involves 1-carbon compounds, which are used in the biosynthesis of the serine family of amino acids. Alanine, valine, and isoleucine all are derived as a result of the transamination of pyruvate

6. The answer is E (all). [*III A*] Balanced and optimal bacterial growth requires an environment that provides specific nutrients and the proper gaseous conditions. Bacteria vary considerably in their needs for oxygen and carbon dioxide. For example, an oxygen-rich environment often is perfect for the growth of aerobic bacteria but may inhibit the growth of anaerobic organisms. The proper inorganic elements are as critical to bacterial growth as is the proper source of carbon or nitrogen, which usually is supplied by the organic compound.

7. The answer is A (1, 2, 3). [*II B 2*] The ribosomes are the site of active protein synthesis within the cell. A 50S ribosome attaches to a messenger RNA (mRNA) strand and forms the active site of synthesis together with the required charged transfer RNA (tRNA) molecules. Antibiotics that interfere with protein synthesis (e.g., chloramphenicol, streptomycin) usually act on the ribosomes directly or on the combination of the mRNA and the ribosomes. Antibiotics that interfere with cell wall synthesis (e.g., penicillin) have no direct effect on protein synthesis; in fact, continued protein synthesis is required for penicillin to be effective. The major mechanism for regulating protein synthesis is not at the level of the mRNA-ribosome complex (translational control) but is far more commonly at the stage of mRNA synthesis.

8. The answer is C (2, 4). [*II A 2*] The cytoplasmic membrane forms the barrier between the interior and exterior of the bacterial cell. It is not a rigid structure and cannot withstand osmotic shock unless the cell wall remains intact to provide protection. However, if the external environment is osmotically protected, the cytoplasmic membrane will remain intact after removal of the cell wall. Since the cell wall gives the cell its shape, once the wall is removed the cytoplasmic membrane assumes a spherical structure. The cytoplasmic membrane in its role as barrier between the interior of the cell and the outside environment is not permeable to all nutrients but rather is highly selective in its permeability. The cytoplasmic membrane also contains the cell's electron transport system, which facilitates respiratory metabolism.

9–12. The answers are: 9-D, 10-E, 11-A, 12-B. [*II*] Many gram-negative cells are covered with hair-like structures known as common pili. These structures are made of a protein fiber that mediates interaction with structures in the cell's environment. The environmental structures might be cells in an infected individual, or they may be physical structures in an aqueous environment such as a lake or ocean.

Similarly, flagella are composed of a protein fiber organized around a swivel-like basal hook that provides a driving force to the flagellum and propels the microbe.

The DNA of bacteria condenses in the cytoplasm as a dense nucleoid that lacks any type of boundary membrane. The absence of a nuclear membrane is what makes bacteria prokaryotic.

The cytoplasmic membrane of bacteria serves many essential functions. It is the actual boundary between the interior and exterior of the cell. It contains many of the energy-yielding enzyme systems required by the cell. In addition, it acts as the barrier to the free flow of materials from the environment into the cytoplasm; however, this barrier is highly selective rather than absolute.

3
Bacterial Genetics
David T. Kingsbury

I. ORGANIZATION AND EXPRESSION OF GENETIC INFORMATION

A. DNA organization and replication

1. **Chromosomes** contain the genetic information that defines living bacterial cells.
    a. **Structure.** The bacterial chromosome is a **single, continuous strand of DNA**.
       (1) The "average" bacterial chromosome has a molecular weight of 3×10^9 and is a **closed, circular** structure.
       (2) The DNA is in a **native, double-stranded** form with the usual A-T and G-C base pairing.
    b. **Replication** of the bacterial chromosome is a precise process ensuring that each daughter cell receives an exact copy.
       (1) At the **origin of replication**, a specific site on the chromosome, the two DNA strands are locally denatured. Each strand serves as a template for a complete round of synthesis.
       (2) In a process called **semiconservative replication**, each daughter cell receives one parental strand and one newly synthesized strand of DNA.
       (3) Chromosome replication is carefully controlled to coordinate with **cell division**. Bacterial cell division usually is coupled with a complete round of DNA replication.

2. **Genes.** Genetic information encoded in the bacterial DNA is organized into a series of units known as genes. The normal bacterial chromosome encodes only one copy of each gene; therefore, bacteria are **haploid** organisms.
    a. **Structure.** A gene is a **chain of nucleotides** whose transcript is read without interruption and results in a polypeptide chain.
       (1) The genetic information in a gene is encoded in triple nucleotide groups (codons), each triplet coding for one specific amino acid.
       (2) Bacterial genes are contiguous DNA sequences lacking the intervening sequences (introns) that are characteristic of genes in higher organisms.
    b. **Phenotype.** Genes are defined functionally by observable characteristics (i.e., phenotype), and their boundaries are identified by genetic rather than biochemical tests.
       (1) **Fine-structure genetic mapping** determines the recombination frequency and may be used to locate mutations within a single gene.
       (2) By the construction of partial diploids with one well-defined gene copy, **complementation testing** may be used to define gene boundaries.

B. Regulation and expression of genetic information

1. **Processes affecting gene expression**
    a. **Transcription** is the process by which the genetic information carried in the bacterial DNA is transferred to **messenger RNA (mRNA)**.
       (1) **Mediators of transcription.** Transcription of bacterial DNA is mediated by **RNA polymerase**.
          (a) Purified RNA polymerase is found in two forms, **core enzyme** and **holoenzyme**.
          (b) **Sigma factor**—an initiation factor found in the holoenzyme—promotes the attachment of the enzyme to specific initiation sites.
       (2) **Initiation of transcription** involves an enzyme containing sigma factor interacting with a **promoter** at a well-defined **binding site** that is approximately 40 base pairs in length. After initiation, the **polypeptide chain is elongated** by the addition of single mononucleotides.

(3) Termination of transcription occurs at specific sites where the stability of the DNA helix is altered (e.g., a G-C–rich region).

b. Translation is the process by which mRNA, in conjunction with **transfer RNA (tRNA)** and the ribosomes, directs the synthesis of a specific protein.

2. Regulation of gene expression. Bacterial genes are precisely regulated to allow for balanced growth in a variety of nutritional conditions.

a. At the metabolic stage several regulatory mechanisms have little effect on the expression of genetic information.

(1) End-product inhibition (feedback inhibition) is a mechanism that causes the end product of a pathway to inhibit the first enzyme of the pathway directly.

(a) End-product inhibition is usually direct and immediate.

(b) Competitive antagonism between the pathway end product and the enzyme substrate is usually seen.

(i) This competition is usually not for a single binding site but involves alternative forms of the enzyme.

(ii) Allosteric enzymes demonstrate alternative forms as a function of interaction with a variety of regulatory molecules.

(2) Enzyme activation is a mechanism that causes positive effector molecules to stimulate the first enzyme of a pathway.

b. Initiation of transcription. Both positive and negative regulation occurs at this level. Negative control systems are mediated by an active repressor or aporepressor molecule.

(1) Repressor molecules are proteins that recognize specific DNA sequences (the operator); they also can bind the **inducer** or **corepressor.**

(a) The **lac repressor** is a 150,000 molecular weight polypeptide of four identical subunits; it binds to double-stranded but not single-stranded DNA.

(b) Because repressors bind DNA with extremely high affinity, only a few molecules are required per cell.

(2) In **enzyme induction**, a new mRNA and a new polypeptide are synthesized in response to a specific type of molecule (an **inducer**) that is not necessarily a substrate for the enzyme. The repressor is active in its native configuration.

(3) In **end-product repression**, the end product of a pathway decreases enzyme formation. In this case, the repressor alone is inactive but is converted to an active form after binding the end-product effector (a **corepressor**).

(4) An **operon** is a group of genes controlled by an **operator**, the genetic site of action of the repressor molcules.

(a) Operons begin with a **promoter**, which is the site of RNA polymerase binding, and are followed by an operator, which is followed by the structural genes. Many studies clearly show that the promoter and operator regions overlap.

(b) Mutations in the operator may lead to inability to bind the repressor and, thus, to a **nonrepressible phenotype** (i.e., the **operator constitutive** or **O^c mutation**).

c. Transcription termination and translation. Gene expression also is regulated at these stages.

(1) Two mechanisms control gene expression at the level of **transcription termination.**

(a) Polarity occurs in operons that are transcribed as a single mRNA species, and it works by transcription termination at weak terminator sites within the operon, leading to higher levels of mRNA for proximal genes of the operon than for distal genes.

(b) In some operons there is a variable, premature termination of transcription in the early mRNA (i.e., the "leader sequence") known as **attenuation.**

(2) Regulation of translation has been observed but is of minor significance in the overall regulation of gene expression.

II. TRANSFER OF GENETIC INFORMATION. Bacteria are haploid organisms that use a series of primitive mechanisms for gene transfer; all of these mechanisms produce partial diploids (**merozygotes**) as a prelude to a recombinational event. Transfer is most efficient between cells of the same species.

A. Transformation involves **uptake of fragments of free DNA** by competent cells. The active component is **naked DNA.**

1. **Competence** requires energy together with a convenient site of entry. In some species of bacteria that are not normally transformed, competence can be induced by treatment with calcium chloride and low temperature.

2. Some bacteria immediately convert the double-stranded DNA to single strands following uptake, whereas others leave the DNA in a double-stranded form until recombination.

3. Tranformation has been observed in cultures of a wide variety of both gram-positive and gram-negative organisms.

4. Transformation is a genetic tool that can be used for **gene mapping** and the quantitative determination of **gene dosages**.

B. **Transduction**, like transformation, involves the introduction of only fragments of the chromosome. It is mediated by **bacteriophages** (i.e., viruses that infect bacteria). Because the DNA is packaged in a bacteriophage coat, transduction is easier to perform than transformation. It also is more reproducible.

1. In **generalized transduction**, the phage packages host DNA at random and, therefore, may transfer any gene.
   a. The generalized transducing particle carries only a fragment of the bacterial chromosome and has no bacteriophage DNA.
   b. Because the generalized transducing particles are limited in the size of the DNA fragment they can package, they are valuable as genetic mapping tools.

2. In **specialized transduction**, phage DNA that has been integrated into the host chromosome incorporates a few specific genes upon excision and is limited to the transfer of those genes.
   a. Specialized transducing phages are only a small proportion of the progeny from a normal lysogenic culture and always contain both phage and bacterial genes.
      (1) Lysates from normal lysogenic cultures cause **low-frequency transduction**.
      (2) **High-frequency transduction** lysates are rich in transducing particles and can be produced by propagating the phage from cells transduced by low-frequency transduction.
         (a) With **nondefective transducing phages**, the particles in a single plaque are sufficient.
         (b) With **defective transducing phages**, the cells must be coinfected with a normal helper phage and, therefore, are doubly lysogenic.
   b. Specialized transduction is generally limited to genes near the attachment site of available temperate phases.

C. **Conjugation** accomplishes a **one-way transfer of genes** from donor cell to recipient cell by means of physical contact between the cells. Unlike transformation and transduction, conjugation demonstrates a marked **polarity**.

1. **Fertility (F) factor.** The F factor (also known as the **sex factor**) codes for a series of genes necessary for conjugal transfer. Donor cells are designated F^+ and recipient cells F^-. Donor cells will not act as recipients.
   a. The F factor genes are encoded by a plasmid (see section II D) often referred to as the **F plasmid**. The first known conjugative plasmid, the F plasmid transfers itself at high frequency, quickly converting an F^- population to F^+.
   b. The **F pilus**, which is encoded by the F plasmid, is the critical factor in recognition and establishing contact between donor and recipient. It has been proposed that the role of the F pilus is to overcome the mutual repulsion of the negative charge associated with the bacterial cell surface.

2. **F^+ strains.** The F^+ donor cell population only rarely transfers part of a chromosome to the recipient.
   a. The extent of transfer is dependent on duration of mating; however, regardless of time allowed, total transfer is rare. **The F factor must be incorporated into the bacterial chromosome** in order to mobilize transfer.
   b. Mapping by **interrupted mating** is one of the most powerful genetic tools of conjugation.

3. **High-frequency recombination (Hfr) strains** are bacterial cultures in which the F factor is stably integrated into the chromosome.
   a. Recombinants occur 1000 times more often in Hfr matings than in F^+ matings.
   b. Independently isolated Hfr strains differ in the sequence of gene transfer; however, the gene sequence determined from a series of transfers is circularly permuted.
      (1) Different Hfr isolates may transfer in opposite directions, depending on the orientation of the integrated F factor.
      (2) The genetic map of *Escherichia coli* is a closed circle. Its transfer takes 90–100 minutes, and the chromosome is conveniently represented as a closed circle of 100 equal parts (100 map units).

4. **F′ factors** are rare F factors excised from Hfr chromosomes along with a small segment of the chromosome. These factors are useful in achieving high-frequency transfer of selected genes.

D. **Plasmids** are extrachromosomal genetic elements in the bacterial cytoplasm. Although generally not essential to the survival of the cells, plasmids often carry genes for important traits (e.g., antimicrobial resistance).

   1. **Classification.** Plasmids are divided into two broad categories.
      a. **Conjugative plasmids** encode genes that are required for conjugal transfer of DNA.
         (1) **Structure.** Conjugative plasmids usually are large (i.e., about 60–120 kilobases) and are maintained at one or two copies per cell.
         (2) **Function.** In addition to conjugal functions (DNA transfer), these plasmids encode genes for autonomous replication as well as a variety of other properties. The conjugation and replication functions may be associated with other genes on the plasmid (e.g., those for bacteriocin or toxin production).
         (3) **Examples** of conjugative plasmids include the following.
            (a) **F factor** (see section II C 1) promotes conjugation and the transfer of plasmid DNA from one bacterium to another.
            (b) **R factors** are plasmids that mediate bacterial resistance to antimicrobial agents. Genes for multiple drug resistances may be located on the same R plasmid.
               (i) Frequently the drug-resistant (R) genes are carried on highly mobile **transposons**. Transposons are discrete genetic and physical entities able to move from one position to another in a cell or to a new genome. They assort in random fashion to produce a wide variety of R factors.
               (ii) R genes usually code for specific enzymes that modify or destroy specific antibiotics.
      b. **Nonconjugative plasmids** do not promote conjugal transfer of DNA.
         (1) **Structure.** Nonconjugative plasmids usually are small (i.e., 1.5–15 kilobases) and are present in multiple copies (i.e., 10–20 per cell).
         (2) **Function.** They lack the functions required to mobilize their own transfer, although they may be mobilized (transferred) by conjugative plasmids.

   2. **Recombination** between plasmids or between plasmids and the bacterial chromosome involves **insertion sequences**. Insertion sequences are specific nucleotide sequences ranging from 800 to 1400 base pairs in length, which are found in transposons [see section II D 1 a (3) (b) (i)].

III. GENETIC BASIS OF BACTERIAL DIVERSITY

A. **Mutations** are induced or spontaneous **heritable alterations in the DNA nucleotide sequence**. Mutations may produce changes in morphology, enzyme activity, nutritional requirements, susceptibility to antibiotics, or susceptibility to bacteriophage infection.

   1. **Types of sequence changes**
      a. **Nucleotide deletions** may be of single bases or of very large sequences.
         (1) **Microdeletions** of a single nucleotide shift the triplet code reading frame (**frameshift mutation**) and totally change the amino acid sequence of the protein downstream of the mutation.
         (2) **Large deletions** may or may not change the reading frame; however, they invariably lead to loss of the coding capacity for part of a gene or for as many as several genes.

    b. Nucleotide insertions usually are of a single base pair and result in a change of the reading frame similar to that caused by microdeletion.

    c. Nucleotide replacements are simple changes of one nucleotide base for another.

        (1) Transitions are replacements in which one purine is replaced by another or one pyrimidine is replaced by another (e.g., AT → GC).

        (2) Transversions are replacements in which a purine is replaced by a pyrimidine or a pyrimidine is replaced by a purine (e.g., AT → CG).

    d. Point mutations are changes in a single base; and they all give wild-type recombinants with each other. Point mutations are the most common biochemical alterations in genes.

        (1) Nonsense mutations are produced when a single nucleotide change results in the introduction of a termination codon in place of an amino acid codon.

        (2) Missense mutations are produced when a single nucleotide change results in the substitution of one amino acid codon for another.

2. Mutagens are agents that produce DNA alterations leading to mutation.

    a. Radiation, both ionizing and nonionizing, is a common source of mutagenesis arising during the **postreplication** repair of DNA damage.

    b. Chemical mutagens, the most common inducers of mutations, fall into several major classes.

        (1) Base analogs are incorporated in DNA during normal replication, but, because they tend to undergo transient internal rearrangements (**tautomerization**), they are copied incorrectly in the next round of replication.

        (2) Deaminating agents (e.g., nitrous acid, hydroxylamine) induce high-frequency **selective transition**.

        (3) Alkylating agents (e.g., nitrosoguanidine, ethyl ethane sulfonate) also produce **transition mutations**.

        (4) Acridine derivatives can intercalate successive base pairs of DNA and usually cause mutations due to **shifts in the reading frame**.

3. Reversions restore a mutant's original enzyme activity.

    a. True reversion is a restoration at the site of the mutation.

    b. Suppression of a mutation is the phenotypic restoration of gene function resulting from a change at a site different from the mutation.

B. Genetic recombination at the molecular level is the process by which DNA from a donor cell and DNA from a recipient cell combine to yield a new genome containing information from both sources.

1. Generalized recombination involves the alignment of similar base sequences from two different sources.

    a. In **symmetric assimilation**, single DNA strands from each source combine to form a DNA heteroduplex.

    b. In **asymmetric assimilation**, a single DNA strand from only one parent is transferred.

    c. In ***E. coli***, generalized recombination involves the full participation of at least three gene products; other genetic loci may also participate to a lesser extent.

        (1) The **rec A gene** is involved in both recombination and repair processes, and its absence leads to an almost complete loss of generalized recombination.

        (2) The **rec B gene** and **rec C gene** code for the two polypeptides of an adenosine triphosphate (ATP)–dependent nuclease (exonuclease V).

        (3) Loss of the rec B and rec C exonuclease decreases recombination less than does loss of the rec A function.

2. Site-specific and illegitimate recombination occur by a process independent of the rec A gene product.

    a. Site-specific recombination occurs between two identically matched sequences with the intervention of a gene product specific to that recombination. The most common examples of site-specific recombination are the integration of **bacteriophage** λ (at the att site) and the insertion of transposable genetic elements.

    b. Illegitimate recombination uses neither extensive homology nor specific sites. It occurs infrequently in bacteria, and its mechanism is unknown.

IV. GENETIC MANIPULATION OF BACTERIA. Novel techniques for rearranging genes within bacteria and for molecular recombination between DNA molecules of any source have revolutionized the science of genetics.

A. Operon fusion is one approach to the study of gene regulation and operation activity.

1. Fused operons are transcribed as a single unit under the control of the normal components that control the operator.

2. Operon fusions rely on the selection of **deletion mutations**, which remove the regulatory elements of one operon, thereby connecting it to the terminus of another.

B. Molecular cloning through recombinant DNA technology permits a segment of DNA from any source to be inserted into a foreign replicon (vector—i.e., a plasmid or bacteriophage) and replicated in the appropriate host. It is possible to clone DNA into bacteria, yeast, and tissue culture cells. The following steps are involved in molecular cloning.

1. **Restriction endonuclease cleavage.** A wide variety of restriction endonucleases cleave DNA at highly specific sites and leave overlapping complementary ends, which, after reannealing, can be closed with polynucleotide ligase.

2. When these cleaved DNA molecules are mixed with plasmid DNA similarly cleaved, occasionally a piece of foreign DNA reanneals with the plasmid to form a **chimeric (recombinant) plasmid**.

3. Chimeric plasmids may be introduced into a suitable host by transformation, and the plasmids will be propagated (cloned) in the cell's progeny.

4. For example, two types of bacterial replicons have been the basis for vectors used for cloning foreign DNA.
   a. Many different mutants of bacteriophage λ, which allow for rapid screening for specific DNA inserts or gene expression, have been derived.
   b. Varieties of small, nonconjugative plasmids that carry one or more antibiotic resistance markers are particularly valuable in the production of cloned gene translation products.

5. Despite the ready replication of eukaryotic DNA in *E. coli*, the organization of eukaryotic DNA, the nature of initiation signals, and the absence of proper glycosylation systems are clearly barriers to proper expression of eukaryotic genes in prokaryotes.

6. The use of yeast and tissue cultures has provided a highly effective mechanism for the production of cloned mammalian gene products.

STUDY QUESTIONS

Directions: Each question below contains five suggested answers. Choose the **one best** response to each question.

1. An F$^+$ strain and an Hfr strain of *Escherichia coli* share all of the following characteristics EXCEPT

(A) overall growth rate
(B) physical state of the F-factor DNA
(C) ability to mobilize nonconjugative plasmids
(D) sensitivity to male-specific bacteriophages
(E) electron microscopic appearance of the F pilus

2. All of the following regulatory mechanisms work at the metabolic stage and not at the gene expression stage EXCEPT

(A) end-product inhibition
(B) competitive antagonism
(C) enzyme activation
(D) allosteric activation
(E) end-product repression

3. All of the DNA elements carried within a bacterium—whether a plasmid, a cloned fragment of DNA carried by a vector, or the chromosome—must share which of the following?

(A) Common promoter sequences
(B) Common ribosome binding sites
(C) Bacterial origins for DNA replication
(D) Common repressor binding sequences
(E) None of the above

4. In bacterial genetics, researchers use all of the following procedures EXCEPT

(A) fine-structure genetic mapping
(B) complementation testing
(C) transformation
(D) meiotic segregation
(E) transduction

5. Which of the following genetic elements is capable of inserting into a variety of both chromosomal and nonchromosomal DNA sites?

(A) Prophage
(B) Transposon
(C) Conjugative plasmid
(D) Nonconjugative plasmid
(E) None of the above

6. Bacteria are simple genetic units with all of the following properties EXCEPT

(A) they are haploid
(B) their genetic material is organized into a single chromosome
(C) their DNA has intervening sequences (introns) in almost all genes
(D) they use the same genetic code as eukaryotic cells
(E) their genotypes and phenotypes are the same

Directions: Each question below contains four suggested answers of which **one or more** is correct. Choose the answer

A   if **1, 2, and 3** are correct
B   if **1 and 3** are correct
C   if **2 and 4** are correct
D   if **4** is correct
E   if **1, 2, 3, and 4** are correct

7. Statements that accurately describe DNA replication in bacteria include which of the following?

(1) It is coupled with cell division
(2) It begins at a single, unique site
(3) It requires active RNA synthesis
(4) It is independent of protein synthesis

8. True statements concerning the genetic unit that is defined as the operator include which of the following?

(1) The operator is the site of repressor binding
(2) Mutations occur in operator genes
(3) The promoter is a subregion of the operator
(4) The operator is a rigidly defined nucleotide sequence

Directions: The group of questions below consists of lettered choices followed by several numbered items. For each numbered item, select the **one** lettered choice with which it is **most** closely associated. Each lettered choice may be used once, more than once, or not at all. Choose the answer

    A if the item is associated with **(A) only**

    B if the item is associated with **(B) only**

    C if the item is associated with **both (A) and (B)**

    D if the item is associated with **neither (A) nor (B)**

Questions 9–12

Match each statement describing a characteristic of genetic information exchange with the transfer mechanism or mechanisms that demonstrate that feature.

(A) Transformation
(B) Transduction
(C) Both
(D) Neither

9. This process is sensitive to the presence of DNase in the suspension medium

10. This process has been observed in both gram-positive and gram-negative bacteria

11. This process requires that the bacteria be in a state of competence

12. This process involves packaging and transfer of DNA by a bacteriophage

ANSWERS AND EXPLANATIONS

1. The answer is B. [*II C 1–3*] F^+ and Hfr bacteria differ only in the physical state of the F factor DNA. In Hfr bacteria, the F factor (also referred to as the sex factor) is integrated into the bacterial chromosome and promotes the mobilization of chromosomal transfer. In the integrated state, the F factor DNA is not readily transferred from organism to organism; therefore, Hfr bacteria rarely convert recipients to F^+. Because Hfr bacteria transfer only chromosomal DNA, recombination for chromosomal markers occurs 1000 times more often in Hfr matings than in F^+ matings. The state of the F-factor DNA has no impact on any other characteristic of the organism, including growth, the appearance and composition of the F pilus, and the ability to mobilize the transfer of nonconjugative plasmids.

2. The answer is E. [*I B 2 a*] End-product repression reduces the formation of the enzymes in their own biosynthetic pathway rather than through alteration of their enzymatic activity. Although most commonly found in biosynthetic pathways, end-product repression is not limited to these pathways.

3. The answer is C. [*I A 1 b (1); IV B*] The source of the DNA that may be cloned in a bacterial cell is not restricted, and that DNA need not share any of its regulatory sequences with the bacterial host. The only requirement is that the foreign DNA be attached to a suitable vector that contains the appropriate origin of replication for the host organism. Without the appropriate promoters and ribosome-binding sequences, the cloned DNA cannot be properly expressed as a functional protein in the cell, but the DNA will be replicated as long as the origin of replication is held intact. It is possible to have more than one origin of replication on a given vector so that the vector may be used for more than one host.

4. The answer is D. [*I A 2 b; II A, B*] Bacterial genetics has proved to be an enormously effective tool for the study of basic gene organization and function. The study of bacterial genetics employs such techniques as fine-structure genetic mapping and complementation testing (for identification of gene boundaries) as well as transformation and transduction. Genetically, the normal bacterial cell is haploid. During no stage of its life does the cell undergo sexual reproduction; therefore, meiotic segregation does not occur. Owing to the haploid nature of the organisms, the genotype and phenotype almost always are the same, making genetic selection and screening much easier and faster.

5. The answer is B. [*II D 1 a (3) (b) (i)*] One source of variability within bacterial species is the widespread movement of genes, especially antibiotic resistance genes, between different species. Transposons are highly mobile genetic elements capable of inserting into a variety of DNA elements, including prophages, plasmids, and bacterial chromosomes. Many transposons encode genes for antibiotic resistance as well as genes to promote their own recombination. Transposons appear to have little species specificity; the same transposon can be found in a variety of different microorganisms. In contrast, bacteriophages are highly specific in both the species they infect and the site at which they integrate into other DNA. Conjugative plasmid can integrate at multiple sites within a bacterial chromosome but is limited to only a very few species and to only chromosomal DNA. Nonconjugative plasmid, except that carrying transposons, does not integrate into other DNA.

6. The answer is C. [*I A 1 a, 2*] The genetic material that defines bacteria is carried in a chromosome, which is a single continuous strand of DNA. Bacterial DNA is transcribed into both monocistronic and polycistronic messenger RNA (mRNA) in the cytoplasm without the need for any subsequent processing. This simple mechanism is possible because of the absence of introns (intervening sequences) and the absence of a nuclear membrane.

7. The answer is A (1, 2, 3). [*I A 1 b*] The bacterial chromosome is a single, continuous strand of DNA. Replication of the bacterial chromosome is a precise and carefully controlled process ensuring that each daughter cell receives an exact copy. The process begins at a specific site on the chromosome—the origin of replication—where the two DNA strands are denatured. In bacteria, DNA replication is coordinated with cell division and requires both protein and RNA synthesis.

8. The answer is E (all). [*I B 1 a (2), 2 b (1), (4)*] The operator is a rigidly defined DNA sequence that is involved in genetic regulation of bacteria. Repressor molecules recognize and bind to the operator, leading to reduction of transcription and expression of the structural genes (i.e., negative control). An operon is a group of genes controlled by an operator; operons begin with a promoter and are followed by an operator, which is followed by the structural genes. Operator mutations occur with regularity in many operons and lead to the nonrepressible phenotype referred to as the operator constitutive.

9–12. The answers are: 9-A, 10-C, 11-B, 12-A. [*II A, B*] Bacteria use a series of mechanisms for gene transfer; these mechanisms produce partial diploids as a prelude to a recombinational event. Transformation and transduction are two such mechanisms of genetic information transfer.

Transformation refers to the exchange of genetic information that involves uptake of fragments of naked DNA by competent bacterial cells. This exchange mechanism has been demonstrated in many bacterial groups, including both gram-positive and gram-negative bacteria. The process requires that the bacteria be in a well-defined metabolic state, which varies from species to species. A bacteriophage is not involved in transformation. Since the transforming DNA is in an unprotected (naked) state, it is sensitive to DNase in the medium.

Transduction refers to the exchange of genetic information between bacteria that is mediated by a bacteriophage carrier. Transduction may be either specialized or generalized, and it occurs in both gram-positive and gram-negative bacteria. Transduction does not require involvement of competent bacterial cells.

4
Antimicrobial Agents
Gerald E. Wagner

I. METHODS OF MICROBIAL CONTROL. The inhibition of growth and the destruction of pathogenic microorganisms are accomplished by either physical or chemical means. Nonselective methods of microbial control are applied only to inanimate objects (fomites); other methods display selective toxicity and may be used in vivo.

A. Physical methods. Potentially pathogenic microorganisms in the environment are reduced in number (in **disinfection**) or totally eliminated (in **sterilization**) usually with physical methods.

1. **Heat** is employed in a variety of methods to reduce or eliminate microorganisms from **heat-stable** materials.
   a. **Pasteurization** destroys pathogenic microorganisms by rapid heating of a substance to 71.7° C for 15 seconds followed by rapid cooling. Pasteurization is not sterilization, because not all microorganisms are susceptible to it. This technique has eliminated food-borne diseases such as gastrointestinal tuberculosis and Q fever.
   b. **Dry heat** of 160° C sterilizes a material exposed for 2 hours. Vegetative cells are destroyed within the first few minutes, but 2 hours are needed to kill all microbial spores. Dry-heat sterilization chars organic compounds and causes excessive evaporation of liquid materials.
   c. **Moist heat** (i.e., steam) of 121° C under a pressure of 15 psi is the most effective means of sterilizing heat-tolerant liquids. Heat-resistant spores are killed in less than 15 minutes in small volumes of liquid; volumes in excess of 500 ml require longer periods of time for equilibration to the sterilization conditions. An **autoclave**, which essentially is an industrial pressure cooker, is used for moist-heat sterilization.

2. **Radiation** of varying wavelengths in the electromagnetic spectrum is used for disinfection and sterilization of **heat-labile** materials.
   a. **Ultraviolet (UV) light** between the wavelengths of 250 and 270 nm is absorbed by nucleic acids.
      (1) UV light damages microbial cells by disrupting hydrogen bonds and causing thymine dimers to form in DNA. This structural alteration of the DNA often results in lethal frameshift mutations.
      (2) UV light has limited application for sterilization because of its poor penetrating energy and its absorption by glass and water.
   b. **Gamma radiation and x-rays** are forms of ionizing radiation that effectively sterilize many materials but must be used cautiously because of their potential danger to human cells.
      (1) These forms of radiation cause the formation of free radicals, which chemically react with proteins and nucleic acids to cause cell death.
      (2) Gamma radiation is used extensively for sterilization of plastic materials and is receiving renewed interest in the United States as a means of preserving foods.
   c. **Microwaves** have been used in the microbiology laboratory for rapid resterilization of media that have been stored for extended periods of time. Sterilization in this case, however, is the result of heat produced by the radiation rather than a direct effect of the microwaves.

3. **Filtration of liquids and gases** through natural or synthetic materials is an effective means of removing bacteria and eukaryotic microorganisms.
   a. Membrane filters with a pore size of 0.2 μm effectively remove all bacteria but do not sterilize the liquid or gas; they do not retain most viruses.
   b. A practical limitation of filtration is that flow rate decreases as viscosity of the liquid increases and pore size of the filters decreases.

B. Chemical methods of microbial control include disinfectants and antiseptics, which are nonspecific for the cells they affect, and antibiotics and synthetic antimicrobial agents, which have a selective toxicity.

1. **Nonselective chemicals** that control the growth of microorganisms on inanimate objects are referred to as **disinfectants**; those applied to human tissue are known as **antiseptics**. These chemicals ideally should effectively kill all microorganisms (including viruses), be soluble in water for ease of preparation and application, have a low toxicity for humans, and be reasonably economical.

   a. **Alcohols** (e.g., ethanol, isopropanol, benzyl alcohol) are effective **antiseptics** when used as 50%–70% aqueous solutions. Alcohols precipitate proteins and solubilize lipids present in cell membranes; as solvents, they also effectively clean human tissue. When properly used, alcohols kill the vegetative cells of many bacteria. They do not, however, affect microbial spores, fungi, and most viruses.

   b. **Halogens**, particularly iodine and chlorine, are widely employed as **antiseptics and disinfectants**.

      (1) **Iodine** solubilized in ethanol is a common antiseptic for cuts and abrasions. The iodine reacts with hydroxyl groups and inactivates proteins.

      (2) **Chlorine gas** reacts with water to form hypochlorous acid, which in turn reacts with water to form hydrochloric acid and hydrogen peroxide. Both of these compounds are strong oxidants that kill microbial cells. Household bleach (5.25% sodium hypochlorite) is another source of hypochlorous acid and an effective disinfectant.

   c. **Aldehydes** are alkylating agents that react with the amine, sulfhydryl, and carboxyl groups of proteins and small organic molecules to kill microorganisms. Formaldehyde (8%) and glutaraldehyde (2%) have limited use because of their noxious vapors.

   d. **Heavy metals** are effective as antimicrobial substances because of their ability to precipitate proteins and other organic molecules. Silver nitrate, copper sulfate, and merbromin (Mercurochrome) are widely used as antiseptics. Lead, arsenic, and inorganic mercury rarely are employed as disinfectants because they are concentrated by human tissues and cause cell death.

   e. **Phenols** and their substituted derivatives are highly effective disinfectants. Low concentrations are used as antiseptics.

      (1) Phenols function by denaturing proteins and disrupting cell membranes.

      (2) Phenol no longer is used widely, but phenolic derivatives such as carbolic acid and lysol are common disinfectants. Hexachlorophene is an excellent antistaphylococcal antiseptic available by prescription.

   f. **Cationic detergents** contain alkyl groups that interact with membrane lipids to disrupt the cytoplasmic membrane of bacteria. Quaternary ammonium compounds are bactericidal but have a low toxicity for mammalian cells and can be used as effective antiseptics. Cationic detergents are inactivated by low pH solutions, phospholipids, organic compounds, and metal ions.

   g. **Gases** of various types have been employed as disinfectants since ancient times. Sulfur dioxide is used as a food preservative. Ethylene oxide and propylene oxide used under pressure are effective sporicides for the sterilization of plastic materials.

2. **Selective chemicals** that inhibit the growth of or kill microorganisms include naturally occurring antibiotics, semisynthetic antibiotics, and some synthetic chemical compounds. Many of these agents are used in the treatment of infectious diseases because of their low toxicity for mammalian cells.

   a. **Agents**

      (1) **Antibiotics** are natural substances produced by an organism to kill or inhibit the growth of another organism. Their effectiveness as therapeutic agents is limited by their toxicity for human cells. Semisynthetic antibiotics generally are synthetic derivatives of naturally occurring antibiotics.

      (2) **Chemotherapeutic agents** are chemicals that are used for treating infectious diseases. Frequently these agents are analogs of microbial cell constituents or substrates.

   b. **Efficacy.** The efficacy of a drug is the product of qualities that make it useful in the treatment of infectious diseases. These qualities include stability in vivo, rate of absorption, rate of elimination, and the ability to penetrate the infected site.

   (1) The **therapeutic index** of a drug is the minimal dose causing toxicity to the host divided
        by the minimal dose required for antimicrobial activity. The higher the ratio, the greater
        the efficacy of the drug.
   (2) The **attainable serum level** of a drug is dependent on the dosage of drug administered,
        the host's body weight, the route and schedule of administration of the drug, and the
        rate of elimination.

II. CHARACTERISTICS OF THE MAJOR CLASSES OF ANTIMICROBIAL AGENTS. The major
groups of antimicrobial agents can be classified on the basis of their site of action on the microbial
cell. At normally achievable concentrations, most clinically employed antimicrobial agents are
inhibitory only; they do not kill the microbial cells. In addition, the inhibitory effect usually is exerted
only on actively growing microorganisms.

A. Inhibitors of cell wall synthesis

   1. Penicillins are produced either naturally, by species of the fungus *Penicillium*, or synthetically,
   by modifications of penicillanic acid. The penicillins are effective primarily against gram-
   positive and a limited number of gram-negative bacteria.

Penicillanic acid

   a. Mechanism of action. Penicillins may act as structural analogs of D-alanyl-D-alanine and
      inhibit the enzymatic reaction responsible for the terminal peptide linkage between alanine
      and glycine during the synthesis of the murein component of bacterial cell walls. The cells
      would be vulnerable to osmotic pressure and would lyse, although under special conditions
      the bacteria might survive as levo forms with incomplete cell walls.
   b. Penicillin G, the first antibiotic effective in systemic bacterial diseases, has the disadvantage
      of being inactivated by β-lactamase produced by bacteria and the acid pH of the stomach.
   c. Semisynthetic penicillins that overcome the disadvantages of natural penicillin to varying
      degrees have been produced by chemical modification. These agents include penicillin V,
      ampicillin, oxacillin, methicillin, and numerous others.
   d. The penicillins can be administered orally and have a low toxicity for mammalian cells,
      although a few people are allergic to them.

   2. Cephalosporins are produced by species of the fungus *Acremonium* (formerly called *Cephalo-
   sporium*). They are structurally similar to the penicillins in that they contain the β-lactam ring.

Cephalosporin

   a. The original, naturally occurring compounds have been chemically modified to produce a
      large number of cephalosporin antibiotics; so-called third-generation cephalosporins are
      currently available for clinical use. These agents inhibit cell wall synthesis in a broad
      spectrum of gram-positive and gram-negative bacteria.
   b. The **β-lactam ring** of the cephalosporins is more protected than it is in the penicillins and is
      less susceptible to the action of β-lactamase. Certain bacteria, however, produce a
      cephalosporinase and are resistant to the antibiotics.
   c. Cephalosporins can be given to people with allergies to penicillins.

3. **Bacitracin** is a peptide antibiotic produced by *Bacillus subtilis*. Just before sporulation, the bacilli produce bacitracin by an enzymatic process rather than by the normal process involving transfer of RNA and ribosomes. The antibiotic prevents peptidoglycan synthesis. It is relatively toxic for humans, but poor absorption makes it useful as a topical agent for the treatment of infected wounds.

4. **Vancomycin** is produced by species of *Streptomyces* and prevents the formation of the peptide portion of the peptidoglycan molecule. It is effective against gram-positive bacteria. The chemical structure of this high molecular weight antibiotic has not been confirmed.

5. **Cycloserine** also is produced by *Streptomyces* species. It inhibits cell wall synthesis by interfering with the production of alanine and its incorporation into the interpeptide bridges of peptidoglycan.

B. **Antibiotics acting on cell membranes**

1. **Polymyxin** is produced by *Bacillus polymyxa* and inhibits the normal function of the bacterial cell membrane. The polymyxins act as cationic detergents to disrupt the bacterial cell membrane. These antibiotics are too toxic for internal use, and only polymyxin B and polymyxin E (colistin) are used clinically as topical antibacterial agents.

2. **The polyene antibiotics nystatin and amphotericin B**, which are produced by *Streptomyces* species, are clinically useful as **antifungal compounds**. The polyenes bind to naturally occurring ergosterol molecules in the fungal cell membrane, causing small molecules to leak out of the cells.
   a. **Nystatin** currently is used as a topical antifungal agent, principally in treating mucocutaneous candidiasis.
   b. **Amphotericin B** is used to treat systemic mycoses, but it is highly nephrotoxic because it also binds to cholesterol in mammalian cell membranes.

Amphotericin B

3. **The imidazoles** also cause leakage of molecules through the cell membrane of fungi. It is believed that the imidazoles inhibit a metabolic step that is important in the synthesis of ergosterol. These compounds have a broad-spectrum antifungal action and also inhibit some bacteria and protozoa. **Clotrimazole** is limited to topical use in the United States, **miconazole** is administered parenterally, and **ketoconazole** is effective as an orally administered compound.

C. **Inhibitors of protein synthesis**

1. **The aminoglycosides** inhibit protein synthesis in bacteria by binding to the ribosome. The mode of action of streptomycin is known in more detail than that of other aminoglycosides.
   a. **Streptomycin** binds to the 30S subunit of the bacterial ribosome by irreversibly binding to a specific structural protein of the subunit. This binding produces three effects on protein synthesis.
      (1) It blocks normal function of the initiation complex. Streptomycin–ribosome complexes prematurely leave the messenger RNA (mRNA) assembly line. These complexes subsequently dissociate into their 30S and 50S subunits and reassemble at the normal

initiation site on mRNA. However, they remain irreversibly inactivated initiation complexes that cannot form peptide bonds. This is the most important mechanism of the bactericidal activity of streptomycin.

(2) It interferes with the attachment of transfer RNA (tRNA) to the ribosome–mRNA complex.

(3) It causes the production of faulty proteins by distorting the triplet code of mRNA and causing misreading. The misreading results in the insertion of the wrong amino acids into the polypeptide chain.

b. Amikacin, kanamycin, neomycin, gentamicin, and tobramycin probably have a mode of action similar to that of streptomycin.

2. **The tetracyclines**, broad-spectrum antibiotics produced by *Streptomyces* species, have a hydronaphthacene nucleus consisting of four rings. Chlortetracycline, oxytetracycline, doxycycline, and related compounds bind to the 30S subunit of bacterial ribosomes. Blockage of the attachment of tRNA to the ribosome–mRNA complex prevents the introduction of new amino acids into the polypeptide chain.

Tetracycline

3. **Chloramphenicol** is the only naturally occurring antibiotic that contains **nitrobenzene**, and it probably is this chemical structure that accounts for its toxicity to bacterial and mammalian cells. The drug combines with the 50S subunit of bacterial ribosomes. It prevents peptide bond formation by blocking the action of peptidyltransferase, which is located in the 50S subunit. Aplastic anemia is a serious side effect in patients receiving chloramphenicol therapy.

4. **Erythromycin**, produced by *Streptomyces erythreus*, is the only clinically important member of the macrolide antibiotics. The basic chemical structure is a large lactone ring to which unusual sugar molecules are attached. The mode of action is analogous to that of chloramphenicol; the molecules bind to the 50S subunits of ribosomes and interfere with peptidyltransferase activity.

Erythromycin

5. **Lincomycin** is similar to erythromycin in its antibacterial activity and its mode of action. Chemically it is dissimilar, however; it is an amino acid attached to a sulfur-containing amino sugar.

a. The drug prevents peptide bond formation by binding to the 50S subunit and blocking peptidyltransferase activity.

b. Interestingly, all *Escherichia coli* are resistant to the growth-inhibiting action of lincomycin.

c. Clindamycin (i.e., 7-chloro-7-deoxylincomycin) is a synthetic modification of lincomycin used in the treatment of infections caused by anaerobic bacteria.

6. **Emetine** is an ancient antimicrobial compound that is the principal agent of the antiprotozoan compound, **ipecac**. It inhibits the transfer of amino acids from tRNA to the polypeptide chain on the ribosome, preventing elongation. The drug is used in the treatment of amoebic dysentery and liver abscess, but it is relatively toxic to the patient, probably because emetine also inhibits protein synthesis in mammalian cells.

7. **Thiosemicarbazones** are synthetic compounds that appear to interfere with protein synthesis in smallpox virus by disrupting polyribosomes. This effect is produced by the ability of thiosemi-carbazones to break mRNA into small fragments that prevent polyribosome formation.

D. Inhibitors of transcription and nucleic acid synthesis

1. **The quinolones** are a new group of broad-spectrum antimicrobial agents. They inhibit the enzyme topoisomerase and prevent the coiling and supercoiling of the DNA molecule. A major clinical advantage of the quinolones is that they **can be given orally.** They are effective therapeutic agents in diseases caused by *Pseudomonas* and *Proteus* species.

2. **Rifamycin** antibiotics are products of fermentation reactions by *Streptomyces mediterranei*, although the currently most important compound, **rifampin**, is a semisynthetic derivative of rifamycin B.

   a. Rifampin consists of a double-ring structure with a long aliphatic bridge and a nitrogenous side chain. The drug is a potent inhibitor of DNA-dependent RNA polymerase in bacteria, but it shows no such activity in human cells. The synthesis of all forms of bacterial RNA is inhibited as the result of the binding of rifampin to RNA polymerase.

Rifampin

   b. Rifampin is administered orally, imparting a reddish orange color to urine and feces, and is effective in the treatment of tuberculosis, leprosy, and a variety of gram-negative and gram-positive bacterial infections.

3. **Chloroquine** is an important antiprotozoan drug employed in the treatment of some types of malaria as well as other protozoan infections. The drug interpolates between the stacked base pairs of DNA and interferes with its ability to act as a template. Chloroquine also inhibits nucleic acid synthesis in mammalian cells, but intracellular levels in protozoa, which concentrate the drug, are much higher than in human cells.

4. **Nitroimidazole** compounds such as metronidazole are selectively toxic for anaerobic bacteria and protozoa.

Metronidazole

  a. This selective toxicity probably involves the reduction of the nitro group to a nitrosohydroxyl amino group by a reduced electron transport protein similar to ferredoxin. This metabolic conversion results in an intracellular concentration of the drug that is 10–100 times greater than the environmental level.
  b. The metabolized drug is lethal to anaerobic bacteria and protozoa because it causes strand breakage in DNA. Mammalian cells are unaffected because they lack the enzyme needed to reduce the nitro group.

5. **5-Fluorocytosine (5-FC)** is a synthetic pyrimidine first synthesized for use as an antitumor agent. The drug shows limited antifungal activity and is primarily used in the treatment of systemic yeast infections.

5-Fluorocytosine

  a. Following uptake of the drug by the yeast cell, 5-FC is deaminated to 5-fluorouracil (5-FU). The primary action probably is inhibition of DNA synthesis by inhibition of the enzyme thymidylic acid synthetase.
  b. 5-FC is of low toxicity in patients because mammalian cells generally lack cytosine deaminase, the enzyme responsible for the conversion of 5-FC to the metabolically active 5-FU.

6. **5-Iodo-2′-deoxyuridine (IUDR)** is a nucleoside analog used in the treatment of herpes simplex infections of the cornea. The drug replaces thymidine in viral DNA by inhibiting the action of thymidylic acid synthetase, resulting in insufficient thymidylate for normal DNA synthesis. DNA strands containing the analog are more easily broken than normal DNA. Incorporated IUDR also may cause mispairing of bases, resulting in the synthesis of malfunctioning viral proteins.

7. **The sulfonamides** are a large group of synthetic antibacterial compounds that block the synthesis of thymidine and all purines. The parent compound is π-aminobenzenesulfonamide, and most of the hundreds of derivatives are made by substitutions on the sulfonamide group.
  a. Sulfonamides are structural analogs of para-aminobenzoic acid (PABA) and prevent the synthesis of folic acid by bacterial cells. Folic acid is required as a coenzyme in the transfer of 1-carbon units between molecules. Sulfonamides inhibit the synthesis of thymidine, purines, methionine, and serine.

Sulfonamides

  b. Human cells are unaffected because they cannot synthesize folic acid, which must be acquired through the diet.
  c. **Para-aminosalicylic acid (PAS) and the sulfones** also are competitive inhibitors of PABA in folic acid synthesis and are effective against certain mycobacteria.

8. **The diaminopyrimidines. Trimethoprim and pyrimethamine** are the only two diaminopyrimidines of any practical medical value. These compounds, originally synthesized as thymidine analogs, now are known to inhibit DNA synthesis by interfering with folic acid synthesis.

   a. The compounds structurally are similar to the pteridine portion of dihydrofolic acid reductase, which converts folic acid to tetrahydrofolic acid. Thus, as with the sulfonamides, DNA synthesis is blocked by inhibiting thymidine and purine synthesis in the bacterial cell. Both trimethoprim and pyrimethamine also inhibit DNA synthesis in some protozoa.

Trimethoprim

   b. Trimethoprim frequently is employed clinically in combination with a sulfonamide for the treatment of bacterial urinary tract infection.

III. MECHANISMS OF RESISTANCE TO ANTIMICROBIAL AGENTS.
Microbial resistance to drugs generally occurs either by structural alteration of the drug's target site or by the production of enzymes that inactivate the drug.

A. **Nongenetic drug resistance** involves metabolically inactive cells or loss of target sites.

   1. **Metabolic inactivity.** Most antimicrobial agents act effectively only on replicating cells. Mycobacteria survive in tissue for many years in a metabolically inactive state that creates resistance to drugs they are susceptible to when actively metabolizing and growing.

   2. **Loss of a target structure**, often induced by an antimicrobial agent, may result in drug resistance. Exposure of some gram-positive bacteria to penicillin results in the formation of cells lacking a cell wall (i.e., levo forms). These cells are penicillin resistant, having lost the cell walls that are the structural target of the drug.

B. **Genetic resistance** to antimicrobial agents may be encoded on bacterial plasmids or chromosomes.

   1. **Plasmids** are self-replicating extrachromosomal pieces of DNA that can impart drug resistance to bacteria.
      a. Plasmids may cause **epidemic resistance** among bacteria through either transduction or conjugation. They may transfer resistance in the absence of the antimicrobial agent.
      b. **Resistance (R) factors** are a class of plasmids in gram-negative bacteria that mediate resistance to one or more antimicrobial agents and to heavy metals.
      c. Plasmids frequently carry genes that code for the production of **enzymes that inactivate or destroy antimicrobial agents** (e.g., β-lactamase, which is effective against penicillins and cephalosporins, and acetyltransferase, which destroys chloramphenicol). They also may carry genes that code for enzymes responsible for the transport of some drugs (e.g., tetracycline) across bacterial cell membranes.

   2. **Chromosomal resistance** develops as the result of mutation in a gene locus that controls susceptibility to an antimicrobial agent.
      a. **Spontaneous mutation** in bacteria occurs at a frequency of 10^{-7}–10^{-12}, making chromosomal resistance a relatively rare event.
      b. By **selection pressure**, use of an antimicrobial agent may induce growth of a few bacteria with chromosomal resistance to the agent used, and these cells become the predominant type within the population. Some staphylococci become resistant to penicillin in this way.
      c. **Alteration of structural receptors** for an antimicrobial agent is the usual result of chromosomal mutation. For example, streptomycin resistance can result from a mutation in the chromosomal gene that controls the structure of the P12 protein of the 30S bacterial ribosome, which is the location of streptomycin attachment.

IV. ANTIMICROBIAL SUSCEPTIBILITY TESTS. Several standardized techniques have been developed for determining the susceptibility of bacteria to antimicrobial agents. The exact correlation between in vitro test results and the in vivo efficacy of a drug has yet to be established. In general, the attainable serum level of a drug should be two to four times greater than its minimal inhibitory concentration (MIC) for a specific bacterial isolate in order for the drug to be deemed an effective chemotherapeutic agent.

A. **The Kirby-Bauer disk diffusion method** (Figure 4-1) is a highly standardized test approved by the Food and Drug Administration for determining the susceptibility of bacteria to an antimicrobial agent.

   1. The standard medium is Mueller-Hinton agar, with or without sheep blood, poured to a depth of 4 mm in a Petri dish.

   2. The plate is inoculated by streaking the entire surface in three planes with a sterile cotton swab dipped into a standardized inoculum. The bacterial inoculum is prepared from an 18-hour broth culture of the microbe to be tested and is standardized with sterile physiologic saline (i.e., 0.85% sodium chloride) to contain 10^5 bacteria/ml.

   3. Standard commercial paper disks containing known amounts of the antimicrobial agents to be tested are placed on the surface of the agar. The plate is incubated in an inverted position at 35° C for 18 hours.

   4. The diameter of the zone of inhibition produced by the drug is measured for each disk. The zone diameter obtained experimentally is compared with the standard zone diameter provided by the disk manufacturer, and the bacterial isolate is designated susceptible, intermediately susceptible, or resistant.

   5. Zone diameters are designated susceptible or resistant by regression-line analysis of each zone diameter versus the MICs of thousands of bacterial isolates for each drug with the knowledge that serum levels of a drug must be 2–4 times the MIC to be effective. These data are accumulated by the manufacturer of the antimicrobial agent before its approval for clinical use.

B. **Broth dilution techniques** determine the MIC and the minimal bactericidal concentration (MBC) of an antimicrobial agent for a bacterial isolate.

   1. The liquid culture medium usually employed is Mueller-Hinton broth.

   2. Twofold serial dilutions of the antimicrobial agent are prepared in the broth dispensed in test tubes. Drug concentrations generally range from 128 to 0.06 μg/ml. A growth control consisting of drug-free broth is also prepared. Each test tube usually contains a final volume of 1.0 ml.

   3. Each test tube is inoculated with 0.05 ml of a standardized inoculum containing 10^6 bacteria/ml. The tubes are incubated at 35° C for 18 hours or until growth appears in the drug-free medium.

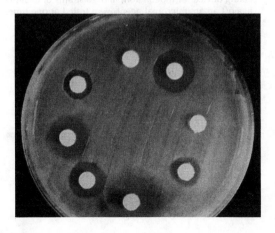

Figure 4-1. The Kirby-Bauer disk diffusion test is a standardized procedure for determining the susceptibility or resistance of a bacterial isolate to a variety of antimicrobial agents. Filter paper disks are impregnated with a known amount of an antimicrobial agent and are placed on an agar surface that has been seeded with the bacteria to be tested. Zones of inhibition of growth around the disks are measured and compared to a table of values that indicate susceptibility, intermediate susceptibility, and resistance to each agent.

4. The **MIC** is the lowest concentration of drug that inhibits bacterial growth as determined visually by the lack of turbidity.

5. The **MBC** is determined by inoculating drug-free medium with 0.01 ml of broth from each test tube that showed no growth in the MIC determination. The MBC is determined after incubation by the growth pattern seen on the drug-free medium. The MBC may be equal to or greater than the MIC.

C. Microtiter techniques for determining the MIC and MBC of an antimicrobial agent are simply modifications of the broth dilution method. The test is performed in a microtiter plate with a usual volume of 0.1 ml per well.

D. The agar dilution technique (Figure 4-2) is similar to the broth dilution method. The MBC cannot be determined easily with the agar dilution technique.

1. Twofold serial dilutions of the antimicrobial agent to be tested are prepared in molten 45° C Mueller-Hinton agar and then poured into separate Petri dishes.

2. The standardized inoculum usually is applied to the surface of the agar with an inoculating device that may place up to 36 bacterial inocula per plate.

3. Following incubation, the MIC is the lowest concentration of drug that inhibits colony formation as determined by visual inspection.

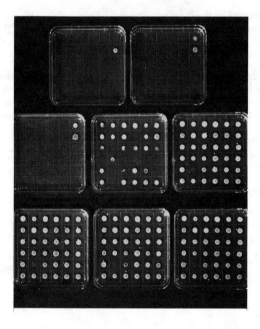

Figure 4-2. The agar dilution susceptibility test is used to quantitate the amount of an antimicrobial agent that inhibits the growth of a microorganism. The highest concentration of the antimicrobial agent is in the plate at the *upper left*, and the lowest concentration is in the plate at the *lower right*. The minimal inhibitory concentration (MIC) is the lowest concentration that inhibits the growth of the microorganism as determined by visual examination. The set of plates shown employed a replicating device for inoculation, and 36 strains of bacteria against an antimicrobial agent were tested at one time.

STUDY QUESTIONS

Directions: Each question below contains five suggested answers. Choose the **one best** response to each question.

1. What is the basic natural structure for the laboratory preparation of semisynthetic penicillins?

(A) The β-lactam ring
(B) Penicillin V
(C) Penicillanic acid
(D) Oxacillin
(E) Hydroxyethanbutol

2. Which of the following is an alkylating agent used as a disinfectant?

(A) Merbromin
(B) Iodouridine
(C) Silver nitrate
(D) Phenol
(E) Glutaraldehyde

3. What antibiotic is most commonly used to treat systemic mycoses?

(A) Nystatin
(B) 5-Fluorocytosine
(C) Clotrimazole
(D) Amphotericin B
(E) Rifampin

4. What antimicrobial agent displays a specificity for anaerobic microorganisms?

(A) Metronidazole
(B) Polymyxin
(C) Vancomycin
(D) Trimethoprim
(E) Emetine

Directions: Each question below contains four suggested answers of which **one or more** is correct. Choose the answer

A   if **1, 2, and 3** are correct
B   if **1 and 3** are correct
C   if **2 and 4** are correct
D   if **4** is correct
E   if **1, 2, 3, and 4** are correct

5. Sterilization may be accomplished by which of the following processes?

(1) Moist heat
(2) Filtration
(3) Radiation
(4) Pasteurization

6. Gas sterilization of heat-sensitive materials is accomplished with

(1) iodophor
(2) glutaraldehyde
(3) formaldehyde
(4) ethylene oxide

7. Microbial resistance to antimicrobial agents may occur as a result of

(1) the presence of a plasmid carrying genes that code for resistance
(2) alteration or absence of structural receptors that bind the antimicrobial agent
(3) selection pressure for a resistant population provided by exposure to an antimicrobial agent
(4) spontaneous mutation of a chromosomal gene

8. True statements about the Kirby-Bauer disk diffusion test include which of the following?

(1) It gives the serum concentration of antibiotic needed to inhibit the microbe
(2) It is a qualitative measurement of susceptibility or resistance
(3) It is a simple test of susceptibility used for all clinically important bacteria
(4) It is a highly standardized test involving medium, temperature, drug concentration, and inoculum

Directions: The group of questions below consists of lettered choices followed by several numbered items. For each numbered item select the **one** lettered choice with which it is **most** closely associated. Each lettered choice may be use once, more than once, or not at all.

Questions 9–13

Match each drug listed below with its appropriate mode of action.

(A) Inhibits cell wall synthesis
(B) Alters cell membrane function
(C) Inhibits protein synthesis by affecting the 30S ribosomal subunit
(D) Inhibits protein synthesis by affecting the 50S ribosomal subunit
(E) Alters nucleic acid synthesis or function

9. Quinolone

10. Bacitracin

11. Rifampin

12. Tetracycline

13. Amphotericin B

ANSWERS AND EXPLANATIONS

1. The answer is C. [*II A 1*] Semisynthetic penicillins are produced by modification of penicillanic acid. Many of these semisynthetic penicillins, such as oxacillin and penicillin V, are less susceptible than natural penicillin to β-lactamase and induce fewer allergic reactions. The β-lactam ring is a component of penicillanic acid, but the attachment of various aliphatic chains to the molecule can shield the ring from enzymatic degradation. Ethanbutol is a chemotherapeutic agent used in the treatment of tuberculosis.

2. The answer is E. [*I B 1 c*] Aldehydes, such as formaldehyde and glutaraldehyde, are alkylating agents that are used as disinfectants. Glutaraldehyde at a concentration of 2% reacts with the amine, sulfhydryl, and carboxyl groups of proteins and with small organic molecules to kill bacterial cells. It has limited use because of its noxious vapor. Merbromin, silver nitrate, and phenol are bactericidal by virtue of their precipitation of proteins. Iodouridine is a pyrimidine analog that affects nucleic acid synthesis and is used as an antiviral agent.

3. The answer is D. [*II B 2 b*] Amphotericin B traditionally is the most effective antibiotic in the treatment of systemic mycoses. Some of the imidazole compounds, such as miconazole and ketoconazole, are used to treat specific mycoses, but they have a high relapse rate. 5-Fluorocytosine is effective against yeast infections but has not been clinically effective against other agents of systemic mycoses. Nystatin and clotrimazole are used only as topical antifungal agents in the United States. Rifampin primarily is an antibacterial drug, although it has been used to treat some opportunistic fungal infections; its efficacy is questionable.

4. The answer is A. [*II D 4*] Metronidazole is a nitroimidazole compound effective in inhibiting DNA synthesis in anaerobic bacteria and protozoa. The active agent of metronidazole is a reduced metabolite with an intracellular concentration 10–100 times greater than the extracellular concentration. Vancomycin typically is used to prevent the overgrowth of *Clostridium difficile* in the colon during broad-spectrum antibiotic therapy, but it does not exclusively inhibit anaerobic bacteria. Polymyxin, trimethoprim, and emetine are antibacterial or antiprotozoal compounds that show no particular specificity for anaerobes.

5. The answer is B (1, 3). [*I A 1 c, 2*] A variety of physical methods are used to sterilize materials. One of the most common means of sterilizing liquids is moist heat (i.e., steam). Steam at 121° C and 15 psi will sterilize many liquids in 15 minutes. Also, some forms of radiation sterilize materials. Ionizing radiation has been used commercially to sterilize foods. Although filtration through a filter of pore size 0.2 μm removes most bacteria, it does not remove viruses. The process of pasteurization typically uses flash heating to kill pathogenic bacteria in liquids; it does not kill many other microorganisms.

6. The answer is D (4). [*I B 1 g*] Heat-sensitive materials such as plastics can be sterilized with the gases ethylene oxide and propylene oxide. Typically, the material is exposed in a gas autoclave to an atmosphere saturated with the gas for a period of hours. Glutaraldehyde, formaldehyde, and iodophor are useful disinfectants, and iodophor-impregnated gauze strip has been used as a wound antiseptic.

7. The answer is E (all). [*III B*] Microbial resistance to antimicrobial compounds frequently is associated with genetic factors. Plasmids carrying genes that code for antimicrobial resistance are present in many species; they can be transmitted among species and among different genera in some cases. Mutations affecting the function or physical structure of specific receptors for the antimicrobial agent, or simply the absence of specific receptors, also accounts for antimicrobial resistance. Selection pressures provided by exposure to antimicrobial agents do not alter the genetics of microorganisms, but they promote the growth of naturally resistant populations of the microbes. Finally, spontaneous mutations naturally occur at a frequency of about 10^{-7}–10^{-12} and rarely may account for antimicrobial resistance.

8. The answer is C (2, 4). [*IV A*] The Kirby-Bauer disk diffusion test is the most frequently used test to determine the susceptibility of a bacterial isolate to a variety of antimicrobial agents. The test is strictly a qualitative test based on a correlation between minimal inhibitory concentrations of the drug, attainable serum levels of the drug, and the diameter of the zone of inhibition produced by a disk impregnated with a known concentration of the drug. The test is highly standardized in reference to the type of medium used, thickness of the medium, incubation temperature, length of incubation, inoculum size, and concentration of drug in the disk. It is not quantitative; the only result obtainable is a judgment whether a bacterial isolate is susceptible to a given antimicrobial agent. Slow-growing, extremely fastidious, and filamentous microorganisms cannot be accurately tested by the Kirby-Bauer method.

9–13. The answers are: 9-E, 10-A, 11-E, 12-C, 13-B. [*II A 3, B 2, C 2, D 1, 2*] Quinolone inhibits the activity of nucleic acid topoisomerase, which is responsible for the coiling and supercoiling of DNA. This inhibition alters the function of the cellular DNA and results in cell death. The quinolones are broad-spectrum, orally administered antibiotics.

Bacitracin inhibits cell wall synthesis by preventing the synthesis of peptidoglycan. Peptidoglycan is a major component of gram-positive bacteria; it is present in smaller quantities in gram-negative cell walls. Because bacitracin is poorly absorbed by mammalian cells it can be used as a topical agent with few side effects.

Rifamycin antibiotics, such as rifampin, bind to DNA-dependent RNA polymerase and inhibit its activity. They inhibit the synthesis of types of RNA (i.e., messenger, transfer, and ribosomal). Rifampin is a broad-spectrum antibiotic capable of penetrating granulomata.

Tetracycline antibiotics inhibit protein synthesis by binding to the 30S ribosomal subunit. They prevent the attachment of tRNA to the ribosome–mRNA complex. This in turn prevents the attachment of additional amino acids to the polypeptide chain being synthesized.

Amphotericin B is one of several polyene antibiotics used to treat mycoses. Amphotericin B binds to ergosterol in the fungal cell membrane and causes formation of a pore that alters the permeability of the membrane. It must be administered intravenously for systemic infections, and because it also binds to cholesterol in mammalian cell membranes it can cause severe side effects.

5
Basic Immunology
David T. Kingsbury

I. THE HUMAN IMMUNE SYSTEM provides barriers against an enormous number of infectious agents in the environment. Some of these barriers are natural, or innate, and others are induced, or acquired.

  A. Innate immunity is nonspecific. It consists of natural barriers to infection that are a part of normal body function (see also Chapter 7, section III A, B).

    1. Physical barriers such as the skin and mucous membranes provide the first line of defense against infectious agents.

    2. Phagocytic cells such as neutrophils and macrophages are effective killers of microbes that penetrate the physical barriers. Phagocytes are the second innate line of defense against infectious agents.

  B. Acquired immunity is an inducible, specific immunologic response to exposure to a particular infectious agent. It may be humoral or cell mediated.

    1. Humoral immunity is an immune state resulting from the production of antibodies by **bone marrow-derived (B) lymphocytes.**

    2. Cell-mediated (cellular) immunity is an immune response primarily of **thymus-derived (T) lymphocytes.**

II. HUMORAL IMMUNITY. Antibodies, which form the basis of humoral immunity, are produced in response to antigens acquired through naturally occurring infection or through vaccination with specially prepared antigens.

  A. Antigens

    1. Properties
      a. All antigens share the properties of immunogenicity and antigenicity.
        (1) Immunogenicity is the ability of a substance to induce a specific cellular or humoral response.
        (2) Antigenicity is the ability of a substance to react with specific antibodies or immune lymphocytes.
      b. Immunogenic substances always are antigenic, whereas many antigenic substances are not immunogenic.
        (1) Haptens generally are small molecules that contain a limited number of antigenic sites (epitopes); they are antigenic but not immunogenic.
        (2) Haptens may become immunogenic by coupling with a carrier molecule that generally is much larger and immunogenic on its own.

    2. Epitopes are specific reactive sites on or within an antigen; they determine the specificity of the immune response and are responsible for its induction.

  B. Antibodies are serum glycoproteins found in the electrophoretically slow migrating fraction [the gamma (γ) fraction] of serum globulins (Table 5-1); hence, the term "gamma globulins" is sometimes used to refer to the antibody fraction of serum.

Table 5-1. Properties of Human Immunoglobulins (Ig)

Ig	Serum Concentration (mg/dl)	Total Ig (%)	Complement Fixation	Main Biologic Effect	Site of Main Action	Molecular Weight (kD)
IgG	800–1700	85	+	Defense—opsonin; secondary response	Serum	150
IgM	50–190	10	+	Defense—precipitin; primary response	Serum	900
IgA	140–420	5	–	Defense—prevents entry across mucous membranes	Secretions	160 (and dimer)
IgE	<0.001	<1	–	Anaphylaxis	Mast cells	200
IgD	0.3–40.0	<1	–	?	?; Receptor for B cell	185

1. **Basic structure of antibodies** (Figure 5-1). The antibody molecule consists of **two identical heavy and two identical light chains held together by interchain disulfide bonds**. The four chains can be separated following reduction of these bonds and acidification of the molecules.
   a. **Fragments.** Proteolytic digestion of intact antibody with the enzyme papain generates two identical antigen-binding fragments, each with a single antigen-binding site (**Fab**—fragment antigen binding), and a third fragment that lacks antigen-binding ability (**Fc**—fragment crystallizable).
   b. **Variable and constant regions.** Each chain has a variable and a constant region. The N terminal portions of both heavy and light chains show considerable variability, whereas the remainder of the molecule is fairly constant.
   c. **Hypervariable regions** are sequences in the variable region that show considerable sequence diversity among antibodies. Hypervariable regions have been localized to three segments on the light chain and three on the heavy chain (see Figure 5-1).
   d. **Domains.** Each immunoglobulin chain consists of globular domains with characteristic structures. These domains contain **intrachain disulfide links** that form loops in the peptide chain.

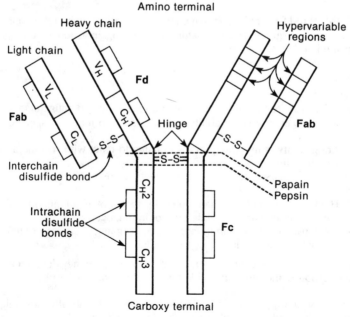

Figure 5-1. Basic unit (monomer) of IgG molecule consisting of four polypeptide chains linked covalently by disulfide bonds (S–S). *V* = variable region; *C* = constant region; *L* = light chain; *H* = heavy chain. (Reprinted with permission from Hyde RM, Patnode RA: *Immunology.* Media, PA, Harwal/Wiley Medical, 1987, p 32.)

(1) The **hypervariable** sequences appear at one end of the variable domain and are clustered near each other in 3-dimensional space.

(2) The **variable** domain is the **antigen-binding site**.

(3) **Constant region domains** determine secondary functions, such as the ability to fix complement and to bind to cell surface Fc receptors.

2. **Antibody classes.** On the basis of the structure of their heavy-chain constant regions, immunoglobulins are placed into major groups called classes, which are further divided into subclasses. These classes and subclasses are termed **isotypes**.

   a. **Light chains** are of two types: **kappa (κ)** and **lambda (λ)**. Both types have a molecular weight of approximately 23,000, and both types are common to all immunoglobulin isotypes.

   (1) In humans the proportion of κ to λ light chains is approximately 2:1.

   (2) A given immunoglobulin molecule may contain either identical κ or identical λ chains, but never both types.

   b. **Heavy chains** have a molecular weight of 50,000–75,000. Their composition is the basis for subdivision of immunoglobulin into five classes.

   (1) **IgG** contains the γ chain; there are four known subclasses.

   (2) **IgA** contains the α chain; there are two known subclasses.

   (3) **IgM** contains the μ chain; there are two known subclasses.

   (4) **IgD** contains the δ chain; there are no known subclasses.

   (5) **IgE** contains the ϵ chain; there are no known subclasses.

C. **Generation of antibody diversity.** The tools of recombinant DNA technology have rapidly advanced understanding of immunoglobulin gene structure.

1. **Allotype and idiotype.** In addition to class (isotypic) variation, antibody molecules also show allotypic and idiotypic variations.

   a. **Allotypic markers** are found on both heavy and light chains. These markers, which are localized to the **constant region**, are genetic in origin and segregate in a Mendelian pattern.

   b. **Idiotype** refers to those reactivities that define the individual determinants characteristic of each antibody. The idiotypic markers are associated with the **hypervariable regions and the antibody's antigen-binding site**. All antibody molecules produced by a single lymphocyte and its daughter cells express a single idiotype and, thus, are termed **monoclonal**.

2. **Gene organization.** In response to the need for millions of different antibodies, mammals have evolved a system whereby multiple gene segments code for antibody molecules. Antibody genes fall into **three clusters on three different chromosomes**, coding for the κ, λ, and heavy chains.

   a. **Mouse λ chain** appears to have a single variable region, and the gene is organized similar to normal eukaryotic genes, with variable insertions (introns) between the gene (exon) segments. The introns are removed by normal gene splicing (Figure 5-2).

   b. The **κ- and heavy-chain genes** use the same process for final maturation as the λ chain, but they have much **greater diversity**. The approximately 200 κ variable (V_κ) gene segments fall into five families, which arrange with one of five joint (J) region segments and a single constant region segment (Figure 5-3).

   c. The **heavy-chain gene cluster** shows, in addition, a D segment inserted between the V and J segments. D and J together encode the entire third hypervariable region.

D. **The T-cell receptor.** Like B cells, the T cell also has an antigen-specific receptor on the cell surface.

1. The T-cell receptor is a **heterodimer** composed of an α chain and a β chain of 40–50 kilodaltons (kD) each. These chains are not encoded by the immunoglobulin genes.

   a. Both α and β chains are required for antigen specificity.

   b. In immunocompetent cells the receptor is intimately connected to T3, a complex of three peptide chains responsible for signal transduction to the cell interior.

2. The generation of the T-cell receptor involves a similar rearrangement to V, D, J, and constant regions as occurs in the generation of functional immunoglobulin genes.

E. **The major histocompatibility complex (MHC).** The ability to mount an immune response is controlled by genes of the major histocompatibility complex.

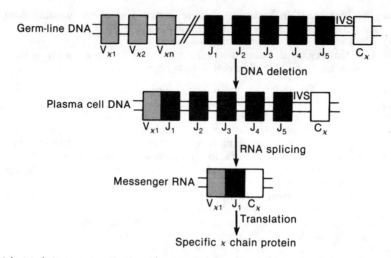

Figure 5-2. Light (κ) chain gene organization. The potential for a large variety of κ chains exists owing to somatic recombination in the DNA and to RNA splicing. As the germ-line DNA differentiates into a plasma cell, DNA deletion brings one of the variable (V_κ) genes next to one of the joining (*J*) genes—in this example, $V_{\kappa 1}$ and J_1. This unit and the remaining J genes are separated from the constant (C_κ) region by an intervening sequence (*IVS*) of DNA. The $V_{\kappa 1}$–J_1 unit codes for one of the numerous possible κ chain variable regions. The plasma cell DNA is transcribed into nuclear RNA, which is spliced to form messenger RNA, with $V_{\kappa 1}$, J_y, and C_κ joined and ready for translation into κ chain. (Adapted from David J: Antibodies, structure and function. In *Scientific American: Medicine*, vol 1, sect 6. New York, Scientific American, 1987, p 16.)

1. **Antigens of the MHC** were first recognized as the antigens responsible for rejection of tissue grafts.

   a. In **mice** the MHC is referred to as **H-2**, and the genes are located on chromosome 17.

   b. In **humans** the MHC complex encodes antigens referred to as **human leukocyte antigens (HLA)** and resides on chromosome 6.

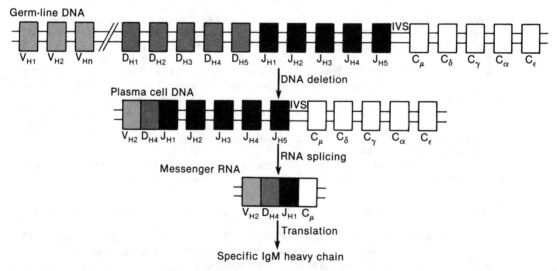

Figure 5-3. Heavy (μ) chain organization. The potential for variety in heavy chains, as in κ chains, is due to somatic recombination in the DNA and to RNA splicing. As the germ-line DNA differentiates into a plasma cell, DNA deletion brings one of the variable (V_H) genes, one of the diversity (D_H) genes, and one of the joining (J_H) genes together—in this example, V_{H2}, D_{H4}, and J_{H1}. This unit and the remaining J genes are separated from the constant (*C*) region by an intervening sequence (*IVS*). The plasma cell DNA is transcribed into nuclear RNA, which is spliced to form messenger RNA. In this process, the C_μ gene is selected and joined to the V_{H2}–D_{H4}–J_{H1} complex, and the entire unit is ready for translation into μ chain. (Adapted from David J: Antibodies, structure and function. In *Scientific American:Medicine*, vol 1, sect 6. New York, Scientific American, 1987, p 17.)

2. At least **two classes of MHC molecules** (termed class I and class II) are encoded by the MHC gene complex. Both classes of antigens are cell surface components, and both are heterodimers consisting of α and β chains (Figure 5-4).

   a. **Class I molecules** consist of a heavy chain of 43 kD covalently linked to a smaller peptide, β_2-microglobulin, of 11. Class I molecules are involved in the **effector phase of cell-mediated cytolysis**. Target cells can be killed only if the target cell and the T cell have the same class I membrane antigens.

   b. **Class II molecules** consist of an α chain of 34 and a β chain of 29 kD. They are found primarily on the surface of immunocompetent cells. Class II antigens are central to antigen presentation in the activation of helper T cells and B cells in the **initiation of the antibody response**.

F. The role of humoral immunity in host defense. Antibodies help prevent infections by several mechanisms.

   1. **Opsonization** is the process by which antibodies, through binding to the bacterial cell surface, make the microorganisms susceptible to phagocytosis.

   2. **Antitoxicity.** Antibodies can bind and thereby inactivate bacterial toxins.

   3. **Activation of complement.** After binding to a microorganism or a tumor cell, antibodies may activate the complement system. This system is a complex group of circulating enzymes, which, once activated, are capable of inducing lethal injury to the invading agent as a result of membrane damage.

   4. **Neutralization.** By binding to bacterial or viral receptors, antibodies can prevent microorganisms from attaching to, and thereby from gaining entry to, host cells.

G. Antibody production after immunization. The **clonal selection theory** explains some of the observed changes in serum antibody that occur after immunization.

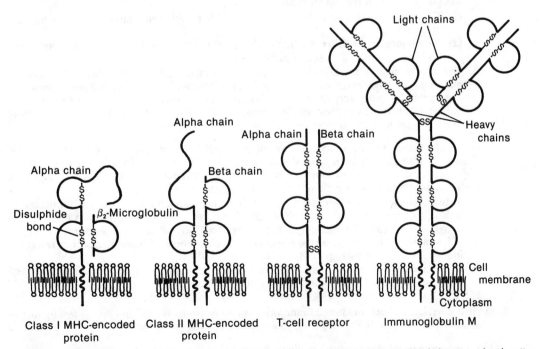

Figure 5-4. Molecular structures of the class I major histocompatibility complex (*MHC*) encoded protein, the class II protein, the T-cell receptor, and the immunoglobulin molecule. Note the structural similarity; the molecules also share similar sequences of amino acids. The molecules are characterized by loops made up of about 70 amino acids within each chain. (Adapted from The T cell and its receptor, by Philippa Marrack and John Kappler. Copyright © 1986 by Scientific American, Inc. All rights reserved.)

1. According to this theory, immunization causes the selective replication of lymphocytes producing antibody with high affinity for the inducing antigen. Over time, high-affinity lymphocyte clones comprise a progressively increasing proportion of the population of responding cells. Thus, the affinity of serum antibody for the inducing antigen increases after immunization.

2. Initially after immunization, serum antibody is characterized by a predominance of IgM; with time, however, IgM production is curtailed and IgG production ultimately predominates. Since the switch from IgM to IgG is an allotypic shift and does not influence idiotype, this change does not reflect clonal selection.

III. CELL-MEDIATED IMMUNITY. In some instances, the immune state depends upon the continued presence of immune cells for its expression. Transfer of this type of immunity is accomplished only by the transfer of immune lymphocytes.

A. **General description.** Cell-mediated immunity is best described in conjunction with particular phenomena, including delayed hypersensitivity, graft rejection, and cell-mediated killing of tumor cells. All of these phenomena depend upon lymphocytes first to recognize an antigen and then to recruit other cells to react to it. In the case of cell-mediated immunity, the **antigen receptor molecules remain attached to the lymphocyte cell surface**. Through interaction with the T-cell receptor, cell-mediated immune responses demonstrate antigen specificity similar to that seen with humoral immune responses.

1. **Delayed-type hypersensitivity (DTH)** is a localized cell-mediated immune reaction. DTH reactions may occur anywhere in the body but are classically described in conjunction with skin testing (e.g., tuberculin testing).
   a. Cutaneous DTH appears in an immune person approximately 24–48 hours after an antigen is introduced intradermally. The reaction is manifested by erythema and induration at the site of antigen introduction. Extreme reaction sometimes progresses to necrosis and sloughing of skin at the injection site.
      (1) DTH reactions occur **only in a person with established immunity** to the injected antigen.
      (2) Histologically, the DTH reaction is characterized by the accumulation of lymphocytes and monocytes in the area of induration.
   b. The development of the DTH reaction involves complex cellular interactions that result in **macrophage activation**. Stimulated by antigen, sensitized lymphocytes release chemotactic factors that promote entry of macrophages and other inflammatory cells into the area. The mediators released by the lymphocytes, termed **lymphokines**, increase the killing capacity of macrophages in a nonspecific fashion.

2. **Graft rejection and cytotoxicity** are forms of cell-mediated immunity best demonstrated when tissues are transplanted from one person to another. This mechanism of immunity may be important for defense against tumor cells that arise spontaneously; the cell surface antigens of the neoplastic cells often are altered and, thus, appear to be foreign.
   a. If surface antigenic differences exist between the recipient and donor cells (**histoincompatibility**), the new tissue is destroyed. Grafts that differ from recipient cells in their surface antigenic characteristics are referred to as **allografts**. Allograft killing depends on the actions of lymphocytes that recognize the presence of **foreign ("nonself") antigens** and, in response, induce injury to the foreign cells.
   b. Graft rejection is like DTH in that both depend upon specific recognition of foreign antigens by lymphocytes that react in a clonal manner to antigen exposure.

B. **In vitro correlates of cell-mediated immunity** frequently are employed in clinical testing just as measurements of antibody titer are used to evaluate humoral immunity.

1. **Lymphocyte proliferation assays** evaluate the ability of lymphocytes to recognize antigens and to proliferate in response to stimulation. Mixed lymphocyte reactions measure lymphocyte proliferation after stimulation with histoincompatible cells. The in vitro response correlates with the ability to reject an allograft in vivo.

2. Macrophage migration inhibition assays use macrophages as indicator cells to demonstrate the release of lymphokines from immune lymphocytes after exposure to antigen.

3. Cytotoxicity studies demonstrate killing of foreign target cells by immune lymphocytes.

C. Role of cell-mediated immunity in host defense. Cell-mediated immunity protects the host in several respects.

1. Lymphokines released by immune cells recruit phagocytes to a localized area of infection and, thus, help ensure **containment of the infecting agent**. Often the anatomic result of this interaction is a typical **granuloma**. This histologic pattern is frequently seen in a wide variety of diseases, such as tuberculosis, sarcoidosis, and coccidioidomycosis.

2. Immune lymphocytes activate macrophages and thereby improve their ability to kill ingested microorganisms. Because **macrophage activation** is a relatively nonspecific process, a cell-mediated immune response to one pathogen may potentiate host defense against a variety of other pathogens.

3. Cell-mediated cytotoxicity may be directed toward cells infected with virus, thus **destroying sites of viral replication**. Similarly, tumor cells with altered cell surface antigens may be detected and destroyed.

IV. ANATOMY AND PHYSIOLOGY OF IMMUNE RESPONSES

A. Lymphocytes. Both humoral and cell-mediated immune responses reflect the carefully regulated activities of lymphocytes.

1. Distribution. Lymphocytes reside in a number of sites throughout the body and often migrate from one location to another. Lymphocyte arrangements in solid lymphoid organs (e.g., lymph nodes, spleen, thymus, gastrointestinal tract) and their movement through the blood and lymphatic vessels are highly ordered processes that reflect the functions of the individual cells.

2. Heterogeneity. Although they appear similar by light microscopy, lymphocytes vary considerably with respect to structure, function, and ontogeny.

    a. All lymphocytes are derived from the same primordial stem cells; however, groups of lymphocytes develop under the influence of varying signals for differentiation. **Different development ultimately leads to differences in function**.

    b. Two distinct populations of lymphocytes can be identified by cell surface markers. These separate lymphocyte populations, designated T and B lymphocytes, play distinct roles in the complex processes underlying humoral and cell-mediated immune responses. T cells are distinguished by the presence of **differentiation cluster antigen 3 (CD3)**, and B cells are distinguished by surface immunoglobulin.

        (1) T cells develop under the regulation of hormone-like substances released by **thymic epithelial cells**.

            (a) Distinguishing features. During differentiation these cells develop **unique surface markers**, which can be identified by serologic reagents (Table 5-2). Alterations in the normal number of cells identifiable by these markers may be evident in certain immunodeficiency diseases (e.g., CD4 cells in AIDS).

            (b) Function. T cells are distinguished by their **ability to promote and regulate, but not carry out, antibody synthesis**.

            (c) Anatomic location. T lymphocytes are normally arranged in the periarteriolar lymphocyte sheath of the splenic white pulp and in the inner cortical areas of normal lymph nodes.

        (2) B lymphocytes probably develop their characteristic markers and functional capabilities under the influence of factors released from cells in either the **bone marrow or the gastrointestinal tract**.

            (a) Distinguishing features. B lymphocytes are distinguished by the presence of **immunoglobulin molecules on the cell surface**. All immunoglobulin molecules present on a single B lymphocyte display a single idiotype and are thus **monoclonal**. The antigen-binding site of this surface immunoglobulin serves as the cell antigen receptor.

Table 5-2. Surface Markers of Human Mononuclear Cells

Cells	Receptors for			Surface Markers*					
	Sheep RBC	Fc	Complement	CD4	CD8	CD2	Ig	HLA Class I	HLA Class II
Macrophages	−	+	+	−	−	−	+/−	+	+
Monocytes	−	+	+	−	−	−	+/−	+	+
T cells	+	+/−	−	...	...	+	−	+	−
Helper T cells	+	...	−	+	−	+	−	+	−
Suppressor T cells	+	...	−	−	+	+	−	+	−
B cells	−	+	+/−	−	−	−	+	+	+/−
Plasma cells	−	−	−	−	−	−	+/−	+	−

RBC = red blood cell; Fc = the crystallizable fragment of an immunoglobulin molecule, which binds complement; Ig = immunoglobulin; HLA = human leukocyte antigen; + = presence of the marker or binding activity; − = absence of the marker or binding activity; +/− = marker or binding activity present on some but not all cells in the population.

*Cell surface markers are identified through the use of specific monoclonal antibodies available as reference reagents. Through the use of these reagents it has been shown that the receptor for sheep red blood cells and CD2 are identical.

    (b) Function. B lymphocytes are unique in their **ability to carry out antibody synthesis**. With appropriate stimulation these cells differentiate into plasma cells.

    (c) Anatomic location. B lymphocytes are normally arranged in the peripheral pulp of the spleen and the outer region of the lymph node cortex, where they form germinal centers and secondary follicles.

B. Cellular interaction. A variety of cellular interactions modulate immune responses.

  1. Lymphocyte-macrophage interactions. The macrophage plays a significant role in regulating immune responses.

    a. Macrophages are derived from a hematopoietic stem cell that is distinct from the pre-lymphoid cell. Macrophages are phagocytic mononuclear cells that ingest microorganisms. Once ingested, the microorganisms are killed by internal lytic mechanisms that depend upon the action of enzymes and toxic molecules present in lysozomes. Such mechanisms are similar to the nonspecific microcidal activities of the polymorphonuclear cell. Macrophages may be selectively stained by histologic techniques that identify some of these lytic enzymes (e.g., esterase and peroxidase) in the cytoplasm.

    b. Macrophages migrate from the blood, where they are called **monocytes**, into tissues, where they are referred to as **histiocytes**.

    c. Macrophages nonspecifically ingest particulate antigens, digest them into smaller fragments, and present this processed antigen to lymphocytes. Antigen processing in this manner greatly stimulates the immune response.

    d. In addition to presenting antigen to lymphocytes, macrophages release other mediators, which further amplify the response of lymphocytes. Since antigen provides the initial signal for lymphocyte responses, the amplification factors released by macrophages belong to a class of molecules referred to as **second signals**. These second signals, produced by lymphoid cells (mainly T cells) and macrophages, are important in maintaining the immune response once it has been initiated.

  2. Lymphocyte-lymphocyte interactions also function in immune modulation.

    a. Humoral immune responses. Although T cells do not synthesize antibody, their participation is required for optimal antibody response.

      (1) Helper T cells. Some T lymphocytes recognize determinants on antigen molecules that are distinct from the determinants toward which antibody is directed. Termed **carrier recognition**, this form of antigen recognition results in the release of stimulating factors from the T cells that promote antibody production by B cells. These cells, termed helper

T cells, make up a distinct subpopulation of T cells and can be identified by the presence of **specific markers** on their cell surface.

(2) **Suppressor cells.** Another T lymphocyte subpopulation produces factors that **depress the response of B cells**. These T cells exert a negative effect upon the magnitude of immune responses and are, therefore, designated suppressor cells. Suppressor cells may be identified by the presence of **specific cell surface markers**, which are identified by specific serologic reagents.

(3) **Ratio of helper-to-suppressor cells.** A person's immune status can be evaluated by determining the ratio of T helper to T suppressor cells. **Normally this ratio is greater than 1**, reflecting a dominant influence of stimulatory factors upon immune responses. In some immunosuppressed states the ratio may be reversed, suggesting a pathologic predominance of suppressor cell effects underlying the immunosuppression. Such a reversed helper-to-suppressor cell ratio is observed in a variety of disease states, including AIDS.

b. **Cell-mediated immune responses**, like humoral immune responses, are influenced by cells that either amplify or suppress immune reactivity. The amplitude of any immune response is thus the result of ongoing activities of both helper and suppressor factors. This balance may be altered by genetic factors, disease states, drug therapies, and a variety of other poorly understood mechanisms.

V. TOLERANCE AND AUTOIMMUNITY. Normally, animals do not mount immune responses directed toward their own normal tissue antigens ("self" antigens).

A. **Tolerance** is a state of **specific unresponsiveness** to a potential antigen.

1. **Regulation of tolerance.** Like immunity, the tolerant state is elaborately regulated. In true tolerance, the response to the relevant antigen is muted, but responses to unrelated antigens are normal. Tolerance may be maintained by a variety of mechanisms involving suppressor T cells, genetic restrictions on immune responses, deletion of T and B cell clones bearing a particular receptor idiotype, and restrictions on lymphocyte–macrophage interactions.

2. **Tolerance toward one's own tissue antigens is appropriate.** Inappropriate tolerance toward foreign antigens can be induced by certain immunizing regimens, or it may develop during the course of some disease states.

B. **Autoimmunity.** Tolerance toward a self antigen may be circumvented by certain disease states that are termed autoimmune diseases (e.g., systemic lupus erythematosus, myasthenia gravis, rheumatoid arthritis). Several mechanisms may ablate normal tolerance. In most cases, the autoimmune process is ultimately more harmful than the original inciting disease (e.g., rheumatic fever after streptococcal infection).

1. Autoimmunity may result from the development of new carrier determinants or the alteration of existing carrier determinants of self antigens.

a. Tolerance may be circumvented when **self antigens are subtly altered**, as may occur when viral infection changes the normal surface antigen of tissue cells.

b. Tolerance may be circumvented when a person is exposed to **an antigen that cross-reacts with a normal self antigen** (e.g., cross-reactions between streptococcal proteins and myocardial tissue antigens in rheumatic fever).

c. Tolerance may be circumvented when **foreign antigenic groups become attached to cells**. Immune reactions mounted against the foreign antigens ultimately lead to destruction of self cells. Such a process accounts for the destruction of red blood cells in some drug-induced hemolytic anemias.

d. **Acute tissue injury** may lead to the exposure of tissue antigens that are normally isolated from the immune system. Under these circumstances the newly exposed self antigens may be recognized as foreign antigens. One example of such a loss of tolerance occurs in a condition known as **sympathetic ophthalmia**, in which a primary injury to the eye allows the immune system to process ocular antigens. Under these circumstances a new immune response often develops and can lead to the destruction of the eye even though the original injury resolves.

2. Autoimmunity may result from the polyclonal activation of B cells by the direct action of **mitogens** such as bacterial endotoxin lippopolysaccharide or from the secretion of factors by T cells that have been stimulated by a mitogen.

3. Autoimmunity may result from **regulatory bypass** within the immune system.
   a. One form of regulatory failure is the **failure of suppressor cells**.
   b. Regulatory failure may result from **inappropriate expression of class II MHC** molecules on the surface of cells not normally expressing these antigens.

STUDY QUESTIONS

Directions: Each question below contains five suggested answers. Choose the **one best** response to each question.

1. Which of the following statements best describes the properties of haptens?

(A) Haptens are immunogenic and reactive with antibody

(B) Haptens are immunogenic but not reactive with antibody

(C) Haptens are reactive with antibody but not immunogenic

(D) Haptens are neither immunogenic nor reactive with antibody

(E) Haptens are chemically complex, macromolecular structures

2. Daughter cells of the same antibody-producing clone are capable of producing antibodies with different

(A) isotypes

(B) idiotypes

(C) affinities

(D) antigen reactivities

(E) amino acid sequences in the hypervariable region

3. Two antibody clones that share an identical amino acid sequence at their antigen-combining site are said to share

(A) epitope

(B) isotype

(C) idiotype

(D) heavy chain

(E) Fc region

4. An immunodeficiency that results in increased susceptibility to viral and fungal infections is due primarily to a deficiency of

(A) macrophages

(B) neutrophils

(C) B cells

(D) T cells

(E) complement

5. The protein in the membrane of T cells that causes T cells to form rosettes with sheep erythrocytes is

(A) CD2

(B) CD3

(C) CD4

(D) CD8

(E) the Fc receptor

Directions: Each question below contains four suggested answers of which **one or more** is correct. Choose the answer

A   if **1, 2, and 3** are correct
B   if **1 and 3** are correct
C   if **2 and 4** are correct
D   if **4** is correct
E   if **1, 2, 3, and 4** are correct

6. An elderly man is found to have a high level of immunoglobulin in his serum. Data that would suggest a diagnosis of lymphoid malignancy include

(1) the finding of polyclonal immunoglobulin on electrophoretic studies

(2) anergy to skin test antigens

(3) the finding of similar amounts of κ and λ light chains

(4) the finding of mature plasma cells in the bone marrow

7. A patient is admitted to the hospital because of a suspected deficiency in T cell reactivity. Tests that would be useful for documenting this deficiency include

(1) tuberculin skin testing

(2) enumeration of blood cells bearing HLA determinants '

(3) enumeration of blood cells bearing receptors for sheep erythrocytes

(4) enumeration of lymphocytes bearing surface immunoglobulin

Directions: The group of questions below consists of lettered choices followed by several numbered items. For each numbered item select the **one** lettered choice with which it is **most** closely associated. Each lettered choice may be use once, more than once, or not at all.

Questions 8–12

For each therapeutic effect listed below, choose the substance that, when administered, induces the stated effect.

(A) Hepatitis B immune globulin antibodies
(B) Tetanus toxoid solutions
(C) Transfused neutrophils and monocytes
(D) Thymic epithelial factors
(E) Transfused bone marrow cells

8. Provide a transfer of passive immunity

9. Induce active, long-term immunity in host B cells

10. Provide short-term, nonspecific bactericidal activity

11. Provide precursors of a variety of host defense cells in order to induce long-term protection

12. Induce functional differentiation in lymphoid stem cells

ANSWERS AND EXPLANATIONS

1. The answer is C. [*II A 1 b (1), (2)*] Haptens generally are small molecules that have a limited number of antigenic sites. They are unable to stimulate an immune response (i.e., they are not immunogenic) unless they are attached to a larger carrier molecule. Haptens are capable of reacting with antibodies by themselves as well as after attaching to a carrier molecule.

2. The answer is A. [*II B, G 2*] The isotype refers to the heavy chain class present in an individual antibody molecule. Since the heavy chain produced by a clone may change with time, the isotype produced by different daughter cells may vary. The idiotype of the antibody product is preserved, however. Affinity (i.e., the strength of binding), antigen reactivity, and idiotype all are reflections of the amino acid sequence in the hypervariable region and, thus, are conserved within a clone. On very rare occasions, somatic mutation may modify the idiotype.

3. The answer is C. [*II C 1 b*] The idiotype of an antibody molecule is a reflection of the structure of the molecule at the antigen-combining site. The prime determinant of this structure is the amino acid sequence. The isotype is determined by the heavy chain class present in the antibody molecule. The antibody product of an individual cell may shift from one isotype to another. The Fc region is determined by the heavy chain class of the antibody molecule at the end distant from the antigen combining site. Epitopes are specific regions on an antigen molecule.

4. The answer is D. [*III C*] The primary line of defense against cellular antigens—whether fungi, virus-infected cells, or tumor cells—is the cell-mediated immune system, the central feature of which is the T cell. B cells do not play a significant role in this system; their primary role is antibody production. Macrophages may play some role in defending against these infections, but they are dependent upon T cell activation and are not the first line of defense. Neutrophils and complement are immune elements that are much more important to humoral immunity, which provides defense against bacterial infections but appears to play only a secondary role in fungal infection. Both humoral and cell-mediated immunity are important for full protection against viral infection.

5. The answer is A. [*IV A 2 b (1); Table 5-2*] CD2 and the sheep erythrocyte receptor are identical, each having been identified by different techniques but found to be the same structure. The Fc receptor binds only the Fc region of immunoglobulin and does not interact with sheep cells. CD3 and CD8 are developmental antigens found on cortical thymocytes and suppressor T cells; they do not interact with sheep red blood cells.

6. The answer is C (2, 4). [*II B 2 a, C 1 b*] A lymphoid malignancy implies that there is a neoplastic clone of lymphoid cells. Polyclonal immunoglobulin is heterogeneous and, therefore, is the product of multiple lymphocyte clones. Likewise, the presence of body κ and λ light chains implies that more than one immunoglobulin clone is present. Skin test anergy often is present in lymphoid malignancy. The presence of mature plasma cells in the marrow is suggestive of the histologic picture seen in multiple myeloma.

7. The answer is B (1, 3). [*III A 1; Table 5-2*] The tuberculin skin test reaction is a classic example of delayed-type hypersensitivity (DTH). The reaction depends upon intact function of T cells. T cells characteristically possess receptors for sheep erythrocytes. Surface immunoglobulin is present on B cells, and human leukocyte antigen (HLA) determinants are present on a variety of cells other than T cells.

8–12. The answers are: 8-A, 9-B, 10-C, 11-E, 12-D. [*IV A 2 b (1), (2), B 1 a*] Hepatitis B immune globulin is serum antibody administered to provide temporary protection against hepatitis virus. No host cell immunity is induced by this agent, and immunity lasts only as long as the transferred immune globulin escapes degradation. Thus, the transferred immunity is termed passive.

Tetanus toxoid is an immunogenic but harmless preparation made from the toxin produced by *Clostridium tetani*. It works by inducing the production of antitetanus antibody by the recipient's B cells. Immunity so induced lasts for years because of the persistence of immune memory cells.

Neutrophils and monocytes are able to phagocytize and kill bacteria in a nonspecific fashion. However, because these are not dividing cells, the protection so transferred lasts only as long as the transferred cells survive. Their activities do not depend upon prior immunization.

The bone marrow contains precursors for all of the blood-borne cells functioning in host defense. Stem cells so transferred eventually mature into granulocytes, macrophages, and lymphocytes. Because the precursor cells are self-perpetuating, long-term protection is provided.

Lymphoid stem cells are induced to differentiate into T cells by factors released by thymic epithelial cells. Transfused bone marrow provides a source of undifferentiated stem cells, which require additional stimuli in order to differentiate appropriately.

6

Clinical Immunology and Serology

Gerald E. Wagner

I. DETECTION OF ANTIGENS AND ANTIBODIES. In the infectious process, the human body is invaded by an antigen (the microorganism), and it responds by producing antibodies. Practical application of recent immunologic knowledge has led to the development of laboratory tests that detect antigens and antibodies. The qualitative and quantitative results of these tests help health care professionals diagnose and treat infectious diseases.

A. Immunodiffusion. The reaction of antigens and antibodies forms a complex that is visible as a precipitate in a medium such as agar when pH, temperature, amount of buffer electrolytes, and antigen-to-antibody ratios all are optimal.

   1. The **relative concentrations of antigen and antibody** are the most critical factor in the immunodiffusion technique.
      a. Maximal formation of precipitate occurs in the area of equivalence. Decreasing amounts of precipitate form when antigen or antibody is in excess.
      b. The **prozone phenomenon** is a suboptimal precipitate of antigen-antibody complex due to antigen or antibody excess. The phenomenon commonly leads to a misinterpretation of immunodiffusion results. Allowing various dilutions of the antibody to react with a fixed concentration of antigen permits the development of distinct lines of precipitate.

   2. In **single immunodiffusion**, one reactant remains fixed and the other is allowed to diffuse through a semisolid medium. Diffusion of the free reactant can be either linear or radial.
      a. A **line of precipitate** is formed at the point of maximal complex formation (i.e., equivalence).
      b. One **clinical application** of single immunodiffusion is the quantitation of serum immunoglobulins. Other serum proteins also can be accurately measured by single radial immunodiffusion when specific precipitating antibodies are fixed in agar medium. A standard curve is established with the use of known quantities of antigen.

   3. In **double immunodiffusion**—commonly referred to as the **Ouchterlony radial immunodiffusion technique**—both reactants (i.e., both antigen and antibody) diffuse toward each other. Double immunodiffusion also can be either linear or radial.
      a. As in single diffusion, a **line of precipitate** forms at the point of maximal complex formation.
      b. Clinical applications of double radial immunodiffusion are semiquantitation of antigens or antibodies and determination of the purity and relatedness of reactants.
         (1) Semiquantitation of antibody is accomplished by placing antibody in a central well and dilutions of antigen in surrounding wells on an agar plate. The formation of lines of precipitate gives a rough estimation of the antibody titer or, when specificity of the precipitate lines has been determined previously, of the antibody concentration.
         (2) Reactant purity or relatedness is determined by placing antibody in a central well and various antigens in surrounding wells. Three reactions are possible.
            (a) In **identity reactions** (Figure 6-1A) two lines of precipitate meet at a point and form a continuous line. This reaction indicates that the antigens being tested are serologically identical.
            (b) In **nonidentity reactions** (Figure 6-1B) two lines of precipitate cross or intersect at a point. This reaction indicates that the antigens being tested are serologically distinct.
            (c) In **partial identity reactions** (Figure 6-1C) one line of precipitate meets the other at a single point but does not cross it and does not form a continuous line with it. This reaction indicates that the antigens being tested are partially similar serologically.

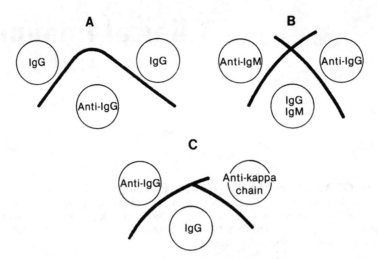

Figure 6-1. Diagrammatic representation of an example of the three characteristic results obtained by double diffusion. (*A*) The identity reaction occurs with anti-immunoglobulin G (*anti-IgG*) in the center well and immunoglobulin G (*IgG*) in both of the other wells. (*B*) Nonidentity occurs with IgG and immunoglobulin M (*IgM*) in the center well, anti-IgG in one peripheral well, and anti-IgM in the other well. (*C*) Partial identity occurs with IgG in the center well, anti-kappa chain in one peripheral well, and anti-IgG in the other well.

B. Electrophoresis is the separation of proteins in an electric field. It involves a variety of techniques that are useful for analyzing the heterogeneity of proteins. These methods also allow detection of specific antigens or antibodies within a complex mixture.

1. **Zone electrophoresis** separates proteins on the basis of surface charge. The technique is used in the clinical laboratory to detect abnormalities in the concentrations of immunoglobulins and other serum proteins. Multiple myeloma, Waldenström's macroglobulinemia, and hypogammaglobulinemia are some of the detectable diseases. Zone electrophoresis has been used recently to detect abnormalities in the cerebrospinal fluid (CSF) of multiple sclerosis patients.

2. **Immunoelectrophoresis** uses both electrophoretic separation and precipitation of proteins. Specific proteins in serum, urine, and other fluids can be identified and quantitated. Serum proteins are separated by electrophoresis, and lines of precipitate are formed by the diffusion of antiserum and the separated proteins. The technique is useful in identifying heavy- and light-chain paraproteins. A decrease or absence of immunoglobulins in immune deficiency diseases also can be determined, and the monoclonal nature of Bence Jones proteins in myeloma can be confirmed. Immunoelectrophoresis has been valuable in studying autoimmune and neurologic diseases.

3. **Radioimmunoelectrophoresis** is primarily a research tool that combines immunoelectrophoresis with the use of radiolabeled antigens. The radioactive antigens are prepared in tissue cultures of specific types of cells. When these antigens are reacted with antihuman antiserum and heavy- and light-chain antisera, the origin (i.e., specific tissue, organ, or cell population grown in culture) of a specific serum protein is confirmed.

C. Electroimmunodiffusion. Immunodiffusion techniques allow free diffusion of antigens and antibody toward each other with formation of a line of precipitate where they meet. Electroimmunodiffusion is the use of an electric field to drive the reactants toward each other at a relatively high speed and with great accuracy.

1. **One-dimensional double electroimmunodiffusion** (i.e., **counterimmunoelectrophoresis**) involves electrophoresis in a gel medium. The antigen and antibody diffuse toward each other from separate wells. A precipitate line forms on contact, as in double diffusion, but the reaction occurs in about 30 minutes rather than 24 hours and is more sensitive. This technique, however, is only semiquantitative. It is useful in detecting antigens in biologic fluids. Cryptococcosis,

meningococcal meningitis, *Haemophilus influenzae*-caused meningitis, and staphylococcal endocarditis can be diagnosed by this procedure.

2. **One-dimensional single electroimmunodiffusion** (i.e., **rocket electrophoresis**) employs electrophoresis of antigens from a well through a gel medium with fixed antibody. The resultant precipitates are shaped like spikes or rockets (Figure 6-2), and their lengths are proportional to the antigen concentration. The principal application of this technique is quantitation of antigens in biologic fluids.

D. Agglutination. The reaction of antibody with an insoluble antigen or antigen-coated particle results in the formation of clumps of antigen-antibody complexes that are visible to the eye. The technique is sensitive, semiquantitative, and applicable to a large number of antigens.

1. **Direct agglutination** tests involve the agglutination of antigenic particles such as red blood cells, bacteria, and fungi by serum antibodies. Antibody is serially diluted in a twofold dilution scheme, and a fixed amount of antigen is added to determine the relative amount of antibody present in the serum (i.e., the antibody titer). The reactants are incubated together long enough to allow the agglutination reaction to occur. The antibody titer is determined by visual examination of the reactions. A difference of two dilutions is considered significant when comparing titers.

2. **Indirect, or passive, agglutination** tests involve the attachment of soluble antigens to red blood cells or inert particles. A variety of absorption and chemical coupling techniques have been developed for antigen preparation. The tests are performed in the same manner as the direct agglutination tests.

3. **Hemagglutination inhibition** tests are performed by preventing the agglutination of antigen-coated red blood cells by homologous antigen. The test is sensitive and specific, and it allows detection of antigens that are soluble in serum and other biologic fluids. A known amount of antibody is incubated with serially diluted samples containing a soluble antigen. Antigen-coated red blood cells are added. Only antibody that is unbound can react and cause agglutination of the erythrocytes. This permits determination of the relative concentration (i.e., titer) of the antigen in the sample.

4. **The Coombs' test**, also known as the **antiglobulin test**, is a widely used technique for measuring nonagglutinating antibodies or amounts of antibody too small to agglutinate red blood cells effectively. Antiglobulin serum is produced in a heterologous species such as the rabbit and is added to the antigen-antibody reaction mixture. The antiglobulin forms the lattice that is necessary for agglutination of the red blood cells. The procedure is useful in clinical laboratory studies for determining human Rh factor as well as many other factors.

5. **The bentonite flocculation test** uses a passive carrier of antigen other than erythrocytes. Bentonite is a silicate that directly absorbs most proteins, carbohydrates, and nucleic acids. The absorbed antigens often are stable for several months, and the system has been used for commercial slide agglutination kits. Determination of antibodies to rheumatoid factor and *Trichinella* are two practical commercial applications of the technique.

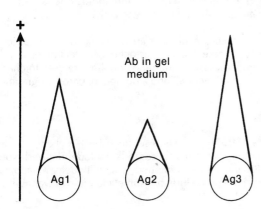

Figure 6-2. Diagrammatic representation of one-dimensional single electroimmunodiffusion (i.e., rocket electrophoresis). One reactant, usually the antibody (*Ab*), is placed in gel medium, and the antigen (*Ag*) is placed in wells at one end. An electric current attracts the protein antigen to the anode, and spikes (*rockets*) of precipitate are formed. The length of the spikes is proportional to the antigen concentration.

E. Complement fixation (CF). Reactions between antigens and antibodies fix complement; therefore, complement can be used to determine antigen and antibody concentrations. The test is widely employed in research and in the clinical laboratory. It is used in detecting hepatitis-associated antigens and *Coccidioides immitis* antigens; it also is used in the Wassermann antibody test for syphilis. The test has two stages.

1. In the **initial stage**, antigen and antibody are allowed to react in the presence of a known amount of complement. The reaction results in the consumption (i.e., fixation) of complement.

2. In the **second stage**, sensitized red blood cells are added to the reaction. Complement remaining after the initial stage reacts with the erythrocytes and causes lysis. The amount of complement fixed is proportional to the antigen or antibody content of the initial solution.

3. **Expression of results.** Antibody concentration is expressed as the highest dilution of serum showing CF. Antigen concentration is expressed as the titer of antigen that limits the hemolytic activity of complement.

F. Radioimmunoassay is a highly sensitive and specific method of quantitating any antigenic substance, such as hormones, drugs, immunoglobulins, and other compounds.

1. **Procedure.** Antibodies to the substance to be quantitated are prepared in a laboratory animal. The antigens are radiolabeled, usually with iodine 125 (^{125}I), and mixed with various known concentrations of unlabeled antigen. Each mixture of labeled and unlabeled antigen is combined with radiolabeled antibody. Labeled and unlabeled antigen compete for the antibody binding sites. Unbound reactants are removed.

2. **Application.** Measurements of radioactivity at different antigen concentrations are used to establish a standard curve that is employed in quantifying unknown amounts of antigen.

G. Enzyme-linked immunosorbent assay (ELISA) is used to detect either antigens or antibodies. It has approximately the same sensitivity as radioimmunoassay.

1. **Procedure.** For example, an antigen is attached to a solid-phase support such as a plastic surface or a paper disk and then bound to an enzyme such as horseradish peroxidase or alkaline phosphatase. It is essential that the antigen and the enzyme retain their activity after the binding process. Serum or other solution containing the antibody being assayed is applied and any excess antigen-enzyme complex is removed. Enzyme substrate is added.

2. **Interpretation.** The rate of substrate degradation determines the amount of antibody present. A substrate that yields a colored degradation product allows an easy spectrophotometric method of detecting the rate of reaction.

H. Tests using monoclonal antibodies. Hybridomas are formed when plasma cells or B lymphocytes from an immunized donor fuse with cells of a continuous cell line. The resultant cells have characteristics of both parent cells, and specific cells are selected for their continuous production of a specific antibody. These monoclonal antibodies are chemically, physically, and immunologically homologous. They are of high purity and are used in sensitive immunologic tests such as radioimmunoassay. Research into new applications of this technology is active.

I. Antigen and antibody characterization. A variety of other techniques are available for the detection and characterization of antigens and antibodies. These include histochemical techniques and immunofluorescence. Chemical and physical properties of the immunoglobulins are used to characterize antibodies. Ultracentrifugation, gradient centrifugation, and a number of chromatographic techniques are useful procedures.

II. DETECTION OF CELLULAR IMMUNE FUNCTION. Various tests have been developed for assaying the function of the cell-mediated immune system. These tests include delayed-type hypersensitivity (DTH) skin tests, assay of T and B lymphocytes, lymphocyte activity determinations, and tests to determine neutrophil function. Most of these procedures are subject to biologic variability

and to standardization difficulties, and some are complicated experiments. However, they provide important data on cell function in intracellular infections, tumor immunology, and transplant rejection.

A. DTH skin tests. Immunocompetence and exposure to certain infectious disease agents are measured by DTH skin tests. Performed and interpreted properly the procedures are simple, useful tools in medical practice.

1. Two **techniques** of skin testing are intradermal injection of antigen and direct contact testing. These tests generally can be interpreted within 24–72 hours of performance.
   a. **Intradermal injection of antigen** has been employed principally in the assessment of immunocompetence and of contact with etiologic agents of infections such as tuberculosis, mumps, and fungal diseases. The test also is useful in epidemiologic studies of these diseases.
   b. **Contact tests** are performed by placing antigen on the skin and covering it with a patch. The test is useful in determining hypersensitivity to chemical agents such as cosmetics and soaps.

2. Correct **interpretation** of the results of DTH skin tests is critical.
   a. **Induration** (i.e., hardness) around the site of injection results from infiltration of mononuclear cells and edema. This reaction occurs 24–48 hours after the test is performed. The diameter of the indurated area is an indication of the degree of cellular immunity against the antigen being tested.
   b. **Erythema**—the wheal and flare—at the site of injection is an indication of immediate-type hypersensitivity. It usually disappears in 12–18 hours but may persist for longer periods of time. Erythematous reactions should be disregarded in interpreting a DTH skin test.
   c. For **infants**, the tests are of little value and are not recommended, especially during the first year of life. Infants probably have not had exposure to antigens sufficient to elicit a response, and they may become sensitized to the injected antigen.
   d. **Immunosuppressed patients** such as organ transplant recipients usually are anergic and lack a response to the test.

3. **Adverse reactions** to DTH skin tests may occur in highly sensitive and immunized individuals. Severe local reactions with erythema, induration, and necrosis can occur. Systemic symptoms such as fever and anaphylaxis are rare.

B. Lymphocyte assays. Enumerating and evaluating T cell and B cell function in peripheral blood is important in detecting immune deficiency disease, autoimmune disease, infectious disease, and tumor immunity reactions. Density gradient centrifugation is used to separate T and B cells from peripheral blood. Although this procedure may selectively eliminate some lymphocyte subclasses, it isolates 70%–90% of peripheral mononuclear cells. Lymphocytes are distinguished from monocytes by morphology, phagocytic ability, and endogenous enzymatic activity.

1. **T cells.** The accepted identifying characteristic of T cells is that they tend to form rosettes with sheep erythrocytes.
   a. **The receptor sites** on the T cells probably are specific for glycoproteins or carbohydrates; immunoglobulin or complement receptors do not appear to be involved.
   b. **Subpopulations of T cells** are distinguished and assayed by the characteristic T-cell binding of immunoglobulin-coated ox erythrocytes in a rosette pattern. T cells have surface receptors for the Fc fragment of immunoglobulin G (IgG), IgM, and probably IgA.
      (1) **T gamma cells** are lymphocytes with IgG receptors. They compose about 20% of the lymphocytes.
      (2) **T mu cells** are lymphocytes with IgM receptors; they compose about 75% of the population.
      (3) **Alternative identification procedure.** T-cell subsets also can be detected by preparing antisera in animals against specific surface antigens of T cells.
      (4) **Clinical significance.** The ability to distinguish subsets has led to the discovery that patients with autoimmune diseases such as juvenile rheumatoid arthritis and systemic lupus erythematosus have naturally occurring antibodies to specific T cell subsets.

2. **B cells** have been shown to have up to 10^5 immunoglobulin molecules on their surfaces. These immunoglobulins are synthesized by the B cell on which they are found. Normally only one heavy- or light-chain type is found on each B cell. The immunoglobulins bind to the B cell by the Fc receptor.

   a. **Intracellular immunoglobulins** are detected by direct immunofluorescence using antiheavy- or antilight-chain antisera. Cytoplasmic immunoglobulins occur in the B cells of patients with certain malignancies, such as chronic lymphocytic leukemia and B-cell lymphoma. The intracellular immunoglobulins usually are identical to the ones expressed on the surface of the cell. Occasionally they may crystallize and form spindles or spicules within the B cells.

   b. **Complement protein 3 (C3) receptors**, referred to as **erythrocyte-amboceptor-complement (EAC) cells**, can be detected by rosette formation. The test is performed by coating ox erythrocytes with IgM antibodies against the erythrocytes and with complement without C5, thus preventing lysis. The EAC cells form rosettes around B cells with C3 receptors.

   c. **Fc receptors** on B cells are detected by heat aggregation of fluorescently labeled immunoglobulins. B cells with Fc receptors bind this aggregated immunoglobulin, but do not bind single immunoglobulin molecules; they are detected with a fluorescent microscope. The test is not specific for B cells; monocytes and neutrophils also bind the aggregated immunoglobulin.

   d. **Direct detection of B cells** can be accomplished with anti-B-cell serum. The antiserum is produced by immunizing laboratory animals with purified B cells. Large quantities of homologous B cells can be obtained from the blood of patients with chronic lymphocytic leukemia.

C. **Assays of neutrophils** are designed to measure the functional parameters of these cells, which play a major role in the body's immunity against infectious diseases. Effective neutrophil activity involves complex interactions with antibodies, complement, and chemotactic factors. **Neutrophil dysfunction in disease may have quantitative or qualitative causes.** In drug-induced neutropenia, the neutrophils are functionally normal, but counts often are less than 1000/μl. Chronic granulomatous disease is characterized by normal or elevated neutrophil counts, but the cells are nonfunctional.

   1. **Motility** is an important characteristic of neutrophils. Both random and directed movement occurs. Chemotaxis is the critical means of cell movement. Chemotaxins that attract neutrophils are produced by microorganisms (e.g., the endotoxin produced by gram-negative rods), by the activation of complement, and by other leukocytes.

      a. **The capillary tube method detects random motion.** Purified neutrophils are placed in a capillary tube, which is enclosed in a specially constructed chamber that can be placed on a microscope stage and observed with the oil immersion objective. The distance the cells move from the leading edge of packed cells is measured in millimeters at hourly intervals.

      b. **A Boyden chamber can measure chemotactic movement.** The chamber consists of two sections separated by a membrane filter of small pore size. Neutrophils are placed in one section, and the attractant is placed in the other. Migrating, attracted neutrophils are trapped in the filter and can be stained and quantitated microscopically. This technique has numerous inherent and performance difficulties.

      c. **A combined technique measures both random and directed movement at the same time.** An agarose medium plate with three wells is prepared. It is similar to the single radial immunodiffusion plate. Neutrophils are placed in the center well, a chemoattractant in the second well, and a nonattractant in the third well. The distance the cells migrate in each direction can be measured easily.

   2. **Receptors for opsonins can be detected** on the surface of neutrophils by rosette formation when the neutrophils are incubated with either antibody-coated or complement-coated erythrocytes. Opsonins (i.e., antibody and complement) on the surface of a pathogen enable phagocytic mononuclear cells and neutrophils to recognize the microorganism as a target. The detection of receptors for them has little clinical application to date, but development of these tests may aid in discovering defects in the recognition phase of phagocytosis.

   3. **The ingestion phase of phagocytosis is measured by several simple techniques** that, in general, use neutrophils in culture. The ingestion of microrganisms can be determined by fixing and staining the neutrophils for direct microbial counts using the light microscope. Some direct

count assays employ carbon particles or latex beads rather than microorganisms. Other techniques measure the ingestion of radiolabeled particles and the extraction and spectrophotometric quantitation of ingested lipids stained, for example, with oil red O.

4. **Clinical assays for measuring degranulation activity** have been used to study neutrophil dysfunction. Degranulation results in digestion of ingested microorganisms. The microbes are held in a phagosome after ingestion. Degranulation occurs when the intracellular lysosome fuses with the phagosome, forms a phagolysosome, and empties the lysosomal contents. The most commonly used assay measures the rate of release of lysosomal enzymes (e.g., β-glucuronidase and acid phosphatase) into a suspending medium after artificial stimulation of lysosome degranulation.

5. **Intracellular killing assays** measure the ability of neutrophils to destroy ingested microorganisms.
   a. **Direct counts of microorganisms** can be made using cultured neutrophils after ingestion and an incubation period. Counts usually are made by a direct plating method.
   b. **The nitroblue tetrazolium dye reduction test** measures the reduction of yellow nitroblue tetrazolium dye to a deep blue dye, formazan. The dye is reduced in neutrophils after ingestion of latex particles, indicating metabolic activity. The reduction of nitroblue tetrazolium dye can be quantitated by spectrophotometry.

III. CLINICAL IMMUNOLOGY AND SEROLOGY IN INFECTIOUS DISEASES.
Immunologic and serologic techniques currently are employed in the diagnosis of the entire spectrum of diseases, ranging from allergies to systemic and dermatologic infections. Since the recognition of antigen and antibody reactions, clinical immunology has been used to diagnose infectious diseases. Immunization as a means of controlling disease was practiced long before the role of microorganisms as etiologic agents and the body's immune response were understood.

A. **Extracellular infections involving opsonization** include a large number of bacterially caused diseases. These infections are controlled by phagocytosis of antibody-coated microbes.

1. *Streptococcus pneumoniae* is the cause of pneumococcal pneumonia. The disease is characterized by lobar consolidation, a single hard, shaking chill, pleuritic chest pain, and a cough producing bloody sputum. Bacteremia usually occurs and may result in systemic disease, including meningitis.
   a. **Immunologic diagnosis** is made with antisera to individual capsular types. Some clinical laboratories employ a polyvalent antiserum for rapid identification. The antisera produce capsular swelling known as the **quellung reaction**. The polysaccharide capsular antigen also may be detected in blood, CSF, and other body fluids by counterimmunoelectrophoresis and radioimmunoassay.
   b. **Immunization with capsular polysaccharide** is used to prevent pneumococcal pneumonia. A polyvalent vaccine containing the polysaccharides of the 14–23 most frequently occurring capsular types in human disease is available for patients at risk of developing disease.

2. *Streptococcus pyogenes* is the etiologic agent of severe pharyngotonsillitis. Untreated infections may induce the sequela of rheumatic fever. This group A β-hemolytic streptococcus produces more than 20 distinct extracellular antigens. A standardized test has been developed for the detection of serum antibodies to streptolysin O. A fourfold increase in antistreptolysin O titer appears in about 80% of children approximately 3 weeks after the onset of pharyngitis.

3. *Streptococcus agalactiae* is a group B streptococcus that is serologically divided into five types. **Type III** is responsible for 60% of the bacteremia and meningeal infections caused by group B streptococci in neonates and older infants. There is substantial evidence that type III capsular polysaccharide can be used as a safe vaccine.

4. *Staphylococcus aureus* produces infectious processes ranging from dermatologic infections to endocarditis. Detection of teichoic acid in serum by counterimmunoelectrophoresis has been used to differentiate endocarditis caused by *S. aureus* from that caused by *Staphylococcus epidermidis*. *S. aureus* synthesizes ribitol-teichoic acid, whereas *S. epidermidis* contains glycerol-teichoic acid in the cell wall.

5. *Haemophilus influenzae* is a major cause of neonatal meningitis and otitis media in children. It also may be involved in pneumonia and the exacerbation of chronic obstructive pulmonary disease in adults.

    a. Immunologic diagnosis is made by detection of *H. influenzae* type b capsular polyribitol phosphate antigen in the serum or CSF by a variety of techniques, including counterimmunoelectrophoresis.

    b. Immunization has been done with a purified type b polyribitol phosphate. The vaccine has not proven to be successful in children under 2 years of age (who are the most susceptible population), but it produces long-lasting immunity in adults.

6. *Neisseria meningitidis* causes a range of infectious processes including meningitis and fulminant septicemia with disseminated intravascular coagulation. The meningococci are grouped according to specific capsular polysaccharides. Serogroups A, B, and C commonly cause disease in humans.

    a. Immunologic diagnosis is made by detecting capsular polysaccharide in serum or CSF by counterimmunoelectrophoresis or by other methods. High levels of group C antigen are indicative of subsequent severe clinical courses.

    b. Immunization against meningococcal disease is achieved with a monovalent A, monovalent C, or a bivalent A and C capsular polysaccharide vaccine. Vaccination is highly effective and induces the formation of both opsonins and bactericidal antibodies. Group B capsular polysaccharide is not immunogenic in humans.

7. *Neisseria gonorrhoeae* is the etiologic agent of sexually transmitted gonorrhea and gonococcemia. Disseminated gonococcal infection is caused by cells that are distinct biochemically from those causing gonococcal urethritis. There currently is no reliable immunologic method for diagnosing gonorrhea. Attempts are being made to produce a vaccine from gonococcal pili.

8. Gram-negative rods contain an O (somatic) antigen as part of the lipopolysaccharide present on their surfaces. Motile forms also have an H (flagellar) antigen, and encapsulated strains have a K (capsular) antigen.

    a. Some strains of gram-negative rods can be typed with the use of antisera prepared in animals immunized with specific types of purified antigens. Although typing of gram-negative enteric rods is insignificant diagnostically, it is useful in epidemiologic studies.

    b. Gram-negative rods such as *Yersinia pestis* also have a variety of antigens that induce antibody formation in the host. A significant rise in the serum antibody titer indicates recent exposure to the pathogen.

    c. Vaccines have been prepared against *Y. pestis* and other gram-negative pathogens, but generally they are used only in individuals at a high risk of developing disease.

9. *Bacillus anthracis* is the etiologic agent of anthrax in humans and animals. It is a large gram-positive rod that produces a unique polypeptide capsule. The species also produces a complex toxin when growing in tissue. Anthrax vaccine must induce both anticapsular antibody and antitoxin to prevent disease in animals.

B. Exotoxin-produced diseases can be prevented, or their course can be altered, by antibodies against the toxins (i.e., antitoxins). The diseases are prevented through active immunization with toxoid vaccines. Passive administration of antitoxin may alter the course of disease substantially enough to prevent morbidity and mortality.

1. *Clostridium botulinum* produces a neurotoxin that is the most potent poison known. Infection produces no immunity; quantities of toxin large enough to induce antibodies kill the host. Immunization is achieved with a toxoid. Effective treatment of botulism requires early administration of equine-produced antitoxin.

2. *Clostridium tetani* is the etiologic agent of tetanus or lockjaw. The lethal dose of toxin is insufficient to induce protective antibodies. Immunization is accomplished by vaccination with tetanus toxoid. Passive immunity is produced by antitoxin obtained from humans and by transplacental passage of the antitoxin from an immune mother to the fetus.

3. *Clostridium perfringens* produces several exotoxins that, together, cause gas gangrene. Equine antitoxin has been used therapeutically, but its advantage is questionable. Antisera frequently neutralize only a few of the many toxins involved in the disease process.

4. *Clostridium difficile* produces an exotoxin that causes necrotic pseudomembranous colitis. There is no available vaccine, but antitoxin to both *C. difficile* and *Clostridium sordellii* neutralize the toxin.

5. *Corynebacterium diphtheriae* produces exotoxin causing pharyngeal and cutaneous diphtheria. Toxoid is used for active immunization. Passive immunization with equine antitoxin is critical in preventing toxin-induced paralysis and death.

6. *Vibrio cholerae* exotoxin causes massive diarrhea that can be fatal due to fluid and electrolyte loss. It is not clear whether infection induces immunity because detectable circulating antibodies probably have little effect on the localized toxicity in the intestine. Toxoid, killed vaccine, and live attenuated vaccine are available, but they appear to be only partially effective for short periods of time.

7. Enterotoxigenic *Escherichia coli* produces a toxin with the same mode of action as cholera toxin. Immunity is not clearly understood, and vaccines are not available at the current time. Research is underway to produce vaccines that will neutralize the toxin and inhibit attachment of the *E. coli* to intestinal epithelial cells.

8. *Shigella dysenteriae* **type 1** is the etiologic agent of dysentery. It probably produces an enterotoxin that can cause systemic nerve damage. There is no available vaccine. Toxoid produced from the neurotoxin induces high titers of serum antibodies, but it does not prevent dysentery.

9. *S. aureus* produces exotoxins that are responsible for scalded skin syndrome, food poisoning, and toxic shock syndrome (TSS). The immunity to these diseases conferred by infection is not understood. No effective vaccines are available.

C. **Infectious diseases in which humoral and cell-mediated immunity collaborate** in controlling the disease frequently can be diagnosed by immunologic techniques. There are no available vaccines for the diseases listed in this section.

1. **Syphilis** is caused by the spirochete *Treponema pallidum.* Two major types of serologic test are employed in the diagnosis.
   a. **Wassermann antibody detection.** Wassermann antibodies, erroneously called reaginic antibodies, are produced during the course of syphilis and other diseases; they react with a tissue-derived antigen known as cardiolipin. In a slide microflocculation test referred to as the **Veneral Disease Research Laboratory (VDRL) test**, the patient's serum is reacted with beef cardiolipin. A commercially available kit, the rapid plasma reagin card test, binds cardiolipin to carbon particles to facilitate visualization of the agglutination process.
   b. **Treponemal antibody detection** in patient serum is a more direct and specific test for syphilis.
      (1) The *T. pallidum* **immobilization test** is highly specific. However, it is cumbersome for routine laboratory diagnosis.
      (2) The **fluorescent treponemal antibody test** currently is employed in many clinical laboratories. Virulent *T. pallidum* from infected rabbits is incubated with the patient's serum. If antibodies are present, they bind to the treponemes and can be detected with fluorescein-labeled antihuman gamma-globulin antibodies.
      (3) **The specificity of the fluorescent treponemal antibody test is enhanced by the fluorescent treponemal antibody-absorbed serum test.** Patient serum is absorbed with nonpathogenic treponemal antigen before the antibody test is performed.

2. **Cryptococcosis** is a disseminated opportunistic infection caused by the encapsulated yeast *Cryptococcus neoformans.* Capsular antigen is detected in the CSF of patients with meningitis by the **latex agglutination method**. The latex beads are coated with anticapsular antibody produced in experimental animals. The test also has prognostic value; falling titers are associated with improvement. Rheumatoid factor in the CSF and serum may impair the diagnostic value of the test.

3. **Candidiasis** may be either a mucocutaneous or a systemic disease caused by the opportunistic species of *Candida*, particularly *Candida albicans.* Disseminated disease frequently is difficult to diagnose immunologically because it is necessary to differentiate between disease and colonization by *Candida* species. In addition, ill patients may be anergic, or the antibodies produced may be complexed with circulating antigen. Precipitin antibodies appear to be most indicative of disease. A **radioimmunoassay** method recently has been used to detect mannan, a candidal surface antigen. Detection of antibodies against mannan is of little value because of the incidence of saprobic colonization.

D. Granulomatous diseases without humoral immunity. Tuberculosis, leprosy, and the systemic mycoses (i.e., histoplasmosis, blastomycosis, and coccidioidomycosis) typically are chronic granulomatous diseases in which humoral immunity plays no protective role. **DTH skin tests** are useful epidemiologic tools when performed and interpreted correctly. A positive skin test merely indicates exposure to the etiologic agent and is of little diagnostic value.

E. Rickettsial and chlamydial diseases are caused by intracellular growth of these bacteria. The various diseases caused by these obligate intracellular parasites generally can be diagnosed by CF and immunofluorescence tests. New tests using radioimmunoassay techniques are being developed.

F. Viral infections usually are diagnosed by a significant rise in titer of serum antibodies. Immuno-fluorescence, agglutination, CF, and precipitin tests frequently are employed to detect viral antigens and virus-induced antibodies in biologic samples. Acquired immune deficiency syndrome (AIDS) is diagnosed with ELISA tests for antibodies induced by specific HIV antigens. A recently developed technique of DNA amplification called the polymerase chain reaction (PCR) technique may prove to be an important test for viral diseases such as AIDS and hepatitis.

STUDY QUESTIONS

Directions: Each question below contains five suggested answers. Choose the **one best** response to each question.

1. In which immunologic test is the use of a solid-phase support that allows retention of antigen activity important?

(A) Radioimmunoassay
(B) Enzyme-linked immunosorbent assay
(C) Single radial immunodiffusion
(D) Radioimmunoelectrophoresis
(E) One-dimensional electroimmunodiffusion

2. The distinguishing characteristic of a positive delayed-type hypersensitivity skin test is

(A) erythema
(B) necrosis
(C) induration
(D) vasculitis
(E) neuritis

3. The electrophoretic detection of ribitol teichoic acid in serum strongly suggests infection by

(A) *Streptococcus pyogenes*
(B) *Candida albicans*
(C) *Clostridium perfringens*
(D) *Streptococcus agalactiae*
(E) *Staphylococcus aureus*

Directions: Each question below contains four suggested answers of which **one or more** is correct. Choose the answer

A   if **1, 2, and 3** are correct
B   if **1 and 3** are correct
C   if **2 and 4** are correct
D   if **4** is correct
E   if **1, 2, 3, and 4** are correct

4. True statements regarding the prozone phenomenon of immunodiffusion include which of the following?

(1) It commonly causes a misinterpretation of results
(2) It occurs in areas of antigen or antibody excess
(3) It is a suboptimal precipitate of antigen-antibody complex
(4) It involves an area of rapidly diffusing antigen and antibody

5. Toxoids routinely are used prophylactically in the prevention of

(1) diphtheria
(2) shigellosis
(3) tetanus
(4) botulism

6. Immunogenic capsular polysaccharide is associated with

(1) *Streptococcus pneumoniae*
(2) *Neisseria meningitidis*
(3) *Haemophilus influenzae*
(4) *Bacillus anthracis*

7. Specific tests for the serodiagnosis of syphilis include the

(1) treponema immobilization test
(2) Wassermann antibody test
(3) fluorescent antibody-absorbed serum test
(4) latex bead agglutination test

Directions: The group of questions below consists of lettered choices followed by several numbered items. For each numbered item select the **one** lettered choice with which it is **most** closely associated. Each lettered choice may be used once, more than once, or not at all.

Questions 8–12

Match each description with the appropriate immunologic procedure.

(A) Double immunodiffusion
(B) Immunoelectrophoresis
(C) Indirect agglutination
(D) Complement fixation
(E) Radioimmunoassay

8. Antibody titer is determined by interaction with known quantities of soluble antigen attached to red blood cells

9. Freely moving reactants form a line of precipitate at the point of maximal complex formation

10. A highly sensitive and specific method of detecting antigens

11. Antigen concentrations are expressed as titers that limit the lysis of erythrocytes

12. Complex mixtures of antigens or antibodies are separated on the basis of charge prior to reaction

ANSWERS AND EXPLANATIONS

1. The answer is B. [*I G 1*] Enzyme-linked immunosorbent assay (ELISA) is a highly sensitive and specific means of detecting either antigens or antibodies. Critical to the assay is the binding of either antigen or antibody to a solid-phase support such as a plastic surface, and the reactant must retain its complexing activity after attachment to the support. Antigen-antibody complex formation is quantitated by the reaction of an enzyme bound to the complex with a substrate that yields a colored by-product. Other immunologic assay techniques, such as radioimmunoassay, single radial immunodiffusion, radioimmunoelectrophoresis, and one-dimensional electroimmunodiffusion, may immobilize one or both of the reactants in an agar, membrane filters, or paper medium; they do not bind the reactants to the support system.

2. The answer is C. [*II A 2 a*] Delayed-type hypersensitivity (DTH) skin tests are useful as epidemiologic screens for certain infectious diseases, especially tuberculosis and various fungal infections. The minimal indication of a positive DTH skin test is the presence of induration at the site of the intradermal injection of antigen. The induration usually occurs within 24–48 hours and is primarily the result of the infiltration of lymphocytes. Erythema frequently accompanies induration, but it is not indicative of a positive test. Individuals who display a strong positive reaction to the skin test may develop a necrotic lesion at the site of injection, and secondary bacterial infection is a potentially serious side effect. Vasculitis and neuritis are not associated with a positive DTH skin test.

3. The answer is E. [*III A 4*] Ribitol teichoic acid is a component of the cell wall of *Staphylococcus aureus.* Electrophoretic detection of this compound in serum is used to differentiate this pathogenic species from *Staphylococcus epidermidis.* The test typically is used for rapid identification of the specific etiologic agent of staphylococcal endocarditis. Detection of antistreptolysin O titers indicates current or past infection by *Streptococcus pyogenes.* Serotypes of *Streptococcus agalactiae* are determined on the basis of capsular polysaccharide, and *Candida albicans* is differentiated from other *Candida* species by radioimmunoassay.

4. The answer is A (1, 2, 3). [*I A 1 b*] Immunodiffusion depends upon the diffusion of one reactant toward another and the formation of a visible precipitate when maximal antigen and antibody concentrations occur. Maximal precipitate formation occurs in a zone of equivalency. The prozone phenomenon occurs in areas of antigen or antibody excess, and it is the suboptimal precipitation of antigen-antibody complexes. It commonly causes misinterpretation of immunodiffusion results. Immunodiffusion typically involves immobilization of one reactant in the agar medium and diffusion of the second reactant throughout the medium.

5. The answer is B (1, 3). [*III B 2, 5*] Diphtheria and tetanus routinely are prevented in humans by the prophylactic administration of toxoids. Toxoids produced by denaturation of the exotoxins of *Corynebacterium diphtheriae* and *Clostridium tetani* are administered as part of the DPT vaccine, which is mandated for infants in the United States. Toxoid produced from the neurotoxin of *Shigella dysenteriae* induces a high titer of antitoxin, but the antibodies do not protect against the neurotoxic symptomatology of shigellosis. Toxoid for botulinum toxin induces protective antitoxin, but it is not administered routinely because the incidence of botulism is low.

6. The answer is A (1, 2, 3). [*III A 1 b, 5 b, 6 b*] Capsular polysaccharide is a virulence factor that inhibits phagocytosis. In addition, the capsular polysaccharide of some bacteria is immunogenic and has been used successfully for immunizing humans against infection. Commercially available vaccines have been prepared from the polysaccharide capsule of *Streptococcus pneumoniae, Neisseria meningitidis* serogroups A and C, and *Haemophilus influenzae* type b. *Bacillus anthracis* produces a polypeptide capsule that induces antibody formation, but antibodies against the capsule alone do not protect against anthrax.

7. The answer is B (1, 3). [*III C 1 b*] Nonpathogenic species of spirochetes are members of the indigenous flora of humans, and for this reason specific tests are needed for the serodiagnosis of syphilis. The test most frequently used for a confirmed serodiagnosis of syphilis is the fluorescent treponemal antibody-absorbed serum test. Because it eliminates antibodies not specific for *Treponema pallidum* by absorbing them to nonpathogenic treponemes, this test is highly specific for *T. pallidum* antibodies. The treponema immobilization test also is specific, but it requires the use of live *T. pallidum* strains and is difficult to perform. Wassermann antibodies are produced in other infectious diseases as well as in syphilis. The Wassermann test lacks specificity; it is, however, used as a screening procedure. Latex bead agglutination is employed in the detection of cryptococcosis.

8–12. The answers are: 8-C, 9-A, 10-E, 11-D, 12-B. [*I A 3, B, D 2, E 3, F*] Indirect agglutination can be used to detect serum antibodies to soluble antigens. The soluble antigen is attached to red blood cells or other particles such as latex beads. The formation of a red cell C-antigen-antibody complex results in an agglutination reaction. The use of known quantities of antigen allows determination of antibody titer.

Immunodiffusion tests are characterized by a zone of equivalency, where maximal antigen-antibody complex formation results in a line of visible precipitate. In double immunodiffusion, both antigen and antibody freely diffuse toward each other in an agar medium.

One of the most sensitive and specific immunologic tests for the detection of antigens is the radioimmunoassay. Competition between radiolabeled and unlabeled antigen for radiolabeled antibody allows quantitation of the antigen concentration. The method is applicable for any type of antigen, such as hormones, drugs, and immunoglobulins.

Complement fixation (CF) employs the hemolytic activity of complement to estimate the concentration of antigens or antibodies. Antigen-antibody complexes bind complement and inhibit its hemolytic activity, but in test situations where there is an excess of unbound complement, erythrocytes are lysed. Antigen concentration, therefore, can be estimated and expressed as the titer of antigen that limits hemolysis.

Protein molecules possess either a positive or a negative electrostatic charge, or they are neutral, and a complex mixture of either antigens or antibodies can be separated by exposing them to an electric field. A variety of electrophoretic techniques and media are used in clinical serology to separate serum proteins and estimate their concentrations.

7

Indigenous Flora and Natural Barriers to Infection

Gerald E. Wagner

I. HOST-PARASITE INTERACTIONS. The normal, healthy human body is parasitized by a variety of microorganisms. These commensal parasites are referred to as the **indigenous flora,** and generally they are considered **nonpathogenic.** Challenge to the individual by known pathogenic microorganisms or a disturbance in the immune status causing overgrowth of indigenous flora may result in an infection and disease.

A. The infection process involves several distinct steps that are applicable to most pathogenic microorganisms.

1. **Adherence.** The initial step after the host is exposed to a microbe is the adherence (i.e., attachment or adhesion) of the microbe to the host cell. Both microbial and host cell characteristics are important in this process.
   a. The **recognition of specific receptors on the host cell surface** by the microbe appears to be a major mechanism of adherence.
   b. **Pili and other microbial cell structures** are known to assist adherence of bacteria to specific types of host cells.

2. **Colonization.** Once a microbe has adhered to tissue, it must colonize the host in order to survive. Colonization occurs if the microbe overcomes the natural resistance of the host. Indigenous flora colonize the human body without infecting it. It is important to recognize that **colonization does not necessarily lead to infection.** Sometimes highly pathogenic microorganisms colonize the human body without causing apparent disease.

3. **Multiplication.** The ability of the microbe to **reproduce within the host** in part determines its survival; this is true for colonizing microorganisms and for infecting microorganisms.
   a. The **ability to survive** (i.e., **to avoid phagocytosis and continue replicating**) is an important advantage to the microbe.
      (1) Extracellular pathogens avoid phagocytosis by producing capsules, fimbrae, pili, or other antiphagocytic virulence factors.
      (2) Facultative intracellular pathogens are phagocytized but not destroyed, usually because they prevent fusion of the phagosome and lysosome or inhibit the action of lytic enzymes.
   b. **Nutrition** is the primary need of the microbe. Extracellular enzymes such as nucleases, proteases, and kinases probably help provide nutrients; *Escherichia coli* produces a porphyrin-like compound that binds iron for growth stimulation.
   c. During multiplication, the microbe may produce **toxin,** which may cause a disease process. The toxin may be preformed and ingested, as in botulism, or it may be produced while the microbe is growing within the body, as in cholera.

4. **Penetration.** In order to cause infection as indicated by a pathologic process, most microbes must be able to invade tissue.
   a. The invading microorganism may be assisted by the host cell through the process of **pinocytosis.**
   b. Microbial spreading factors, such as **hyaluronidase** and **kinases,** are important in the penetration process.
   c. As is true with multiplication, **nutrition and replication** are critical adjuncts to the penetration and disease processes.

B. Acquisition of infectious microorganisms. The normal environment of most humans is contaminated with myriad microorganisms, generally in large numbers. Most of these microbes are considered **saprobes,** but under certain conditions most of them can cause disease. Humans and other animals, both homeotherms and poikilotherms, also may be reservoirs of infectious agents.

1. **Aerosols** (i.e., **droplet nuclei**) are a major means of transmitting respiratory pathogens from one person to another.
   a. Droplet nuclei smaller than 5 μm can be inhaled into the lungs. Pathogenic microbes transmitted by this mechanism avoid exposure to the rigors of the external environment. Microbes that can be acquired in this manner include *Streptococcus pneumoniae* and *Bordetella pertussis.*
   b. A few infectious diseases, particularly the opportunistic and systemic mycoses, are acquired by inhalation of dust particles containing the microorganisms.

2. **Contact with infected or colonized materials** may lead to infection in the susceptible host.
   a. **Human-to-human contact** is the principal means of acquiring the venereal diseases. In contrast to such intimate contact, casual contact with infected individuals also can result in acquisition of pathogenic microorganisms (e.g., anthropomorphic dermatophytic fungi). A few diseases are transmitted from mother to child in the uterus or at birth, such as congenital syphilis and neonatal listeriosis.
   b. **Animal-to-human transmission** of infectious agents occurs in the zoonotic diseases. Live animals can serve as reservoirs of pathogens, and commercial products from these animals can be contaminated. Anthrax, brucellosis, and tularemia are examples of zoonoses.
   c. **Ingestion of contaminated food and drink** is the major mechanism for acquiring gastrointestinal infections. Comestibles become contaminated in the first place by contact with colonized humans and animals, their waste products, and contaminated soil.
   d. **Fomites** are inanimate objects that become contaminated with microbes that can survive environmental conditions. Susceptible hosts become colonized or infected upon contact with a fomite. Tuberculosis, anthrax, and smallpox can be acquired from fomites.
   e. **Vectors** can transmit disease by transferring microorganisms from source to host (mechanical vectors) or by serving as reservoirs of pathogens (biologic vectors).
      (1) **Arthropods** are essential biologic vectors of most rickettsial diseases, such as Rocky Mountain spotted fever.
      (2) **Flies** serve as mechanical vectors of shigellosis under appropriate conditions.

C. The carrier state. Individuals with no signs or symptoms of disease may be colonized by pathogenic microorganisms and may transmit the pathogens to susceptible hosts who develop clinical disease. The mechanism of carriage of toxin-producing bacteria, such as *Corynebacterium diphtheriae*, by nonimmune, healthy individuals is not clear. There are several types of carriers.

1. **Persistent carriers** are colonized with the pathogen for long periods of time or even perpetually. The pathogen can be isolated every time a culture is taken from the individual. Treatment with an appropriate antimicrobial agent may appear to be effective, but the carrier state usually recurs when treatment is stopped.

2. **Transient carriers** occasionally are colonized with a pathogenic microorganism. Most individuals are transient carriers of some microbe at some time. *Staphylococcus aureus* and *S. pneumoniae* frequently can be isolated from a significant percentage of the population.

3. **Intermittent carriers** essentially are transient carriers with only one difference: The colonizing microorganism that is isolated from an intermittent carrier is always the same biotype or phage type (i.e., there is no variation in the microbe that is carried).

4. **Noncarriers** are colonized only by nonpathogenic indigenous flora.

D. Opportunistic infections. Individuals who are immunosuppressed by either natural or chemical means **are more susceptible to infections**. Opportunistic infectious processes often are caused by microorganisms that are considered to be nonpathogenic for healthy individuals. The immunosuppression leading to opportunistic infection may be transient, localized, or systemic.

E. Nosocomial infections are acquired in a hospital or health-care setting, usually as the result of debilitation and temporary immune incompetence. The hospitalized patient also has a higher incidence of exposure to virulent microorganisms that are resistant to antimicrobial therapy. Nosocomial infections that develop subsequent to mechanical manipulation are termed **iatrogenic infections**.

II. INDIGENOUS FLORA.
The human fetus is sterile under normal conditions but is exposed to a variety of microorganisms at birth. Microbial colonization of the neonate begins as the child passes through the birth canal and is exposed to the vaginal flora of the mother. The neonate then is exposed to microbes in the environment and to those that colonize hospital personnel and other individuals. Within a short time, the infant develops its own indigenous microbial flora. The indigenous flora (formerly called the normal flora) of any individual undergoes constant change in accordance with changes in life-style and environment.

A. **Skin.** The normal human skin generally is inhabited by a variety of microorganisms.

1. **Typical flora**
   a. The predominant **bacteria** are *Staphylococcus epidermidis, Micrococcus* species, sarcinae, and aerobic and anaerobic diphtheroids. Transient bacteria on the skin include *S. aureus*, α-hemolytic streptococci, and nonhemolytic streptococci. Saprobic mycobacteria occasionally colonize the outer auditory canal, the genital area, and the axillary areas.
   b. **Yeasts** that inhabit the skin are *Candida albicans, Torulopsis glabrata, Pityrosporum obiculare*, and *Pityrosporum ovale*.

2. **Habitat.** Most of the indigenous flora of the skin inhabit the **stratum corneum** and the upper portions of **hair follicles**. Deeper parts of the hair follicle and **sebaceous glands** usually serve as reservoirs of smaller numbers of microbes, which replace those lost from the skin in washing. The microbial flora of the **hair** is similar to that of the skin.

3. **Colonization.** The number of microbes on the skin generally is 10^3–10^4 organisms/cm^2 but may be as high as 10^6 organisms/cm^2 in moist areas such as the groin. Some of the fatty acids found on the skin probably are bacterial products that prevent colonization by other bacteria. Effective cleaning of the skin can reduce the bacterial count by 90%.

B. **Conjunctiva.** The microbial flora of the normal conjunctiva appears to be derived from the skin. The mechanical action of blinking and the chemical composition of secretions of the eye prevent infection. *S. epidermidis*, diphtheroids, and saprobic airborne fungi are the most commonly isolated microorganisms.

C. **The nares, nasopharynx, and accessory sinuses** are heavily colonized by bacteria removed from inhaled air by filtration through the nasopharynx.

1. **Typical flora.** The bacterial flora of the nasopharynx usually is a mixture of viridans streptococci, nonhemolytic streptococci, nonpathogenic *Neisseria* species, and *S. epidermidis*. Eradication of this typical flora by high doses of antibiotics usually results in colonization by gram-negative rods such as *E. coli, Klebsiella, Proteus*, and *Pseudomonas* species.

2. **Habitat.** The **nasopharynx** and associated structures frequently carry pathogenic bacteria. *Streptococcus pyogenes, S. pneumoniae, Neisseria meningitidis, S. aureus, B. pertussis*, and others are present in some of the population at all times.

3. **Colonization.** The nasopharynx of the neonate is sterile at birth. However, it becomes colonized with the indigenous flora of the mother and hospital staff within 2–3 days. The carriage rate of pathogens such as *S. pyogenes, S. pneumoniae*, and *Haemophilus influenzae* approaches 100%. Carriage rates among older children are lower than those for infants but higher than those for adults.

D. **Oral cavity.** The surfaces within the mouth are exposed to large numbers of bacteria. The indigenous oral flora can vary greatly in different individuals.

1. **Typical flora**
   a. **Viridans streptococci are the primary microbes in the oral cavity**; with other streptococci they comprise 30%–60% of the indigenous flora of the mouth. The species adheres specifically to buccal mucosa and tooth enamel. *Streptococcus mitior* is the species principally associated with buccal cells, *Streptococcus salivarius* is associated with the tongue, and *Streptococcus sanguis* and *Streptococcus mutans* with tooth enamel or plaque.

    b. Species of the **fungi** *Candida* and *Geotrichum* are found as indigenous flora in 10%–15% of the population.

  2. Habitat. Anaerobic bacteria inhabit **plaque** and **gingival crevices** where the oxygen level is lower than 0.5%. The anaerobic flora includes *Bacteroides melaninogenicus*, nonpathogenic treponemes, *Veillonella, Clostridium, Fusobacterium*, and *Peptostreptococcus* species. The potentially pathogenic obligate anaerobe, *Actinomyces israelii*, normally inhabits the gums.

  3. Colonization

    a. The neonate's mouth contains microorganisms found in the birth canal. These include lactobacilli, corynebacteria, staphylococci, micrococci, gram-negative enteric rods, yeast, and aerobic, microaerophilic, and anaerobic streptococci. This flora disappears in 2–5 days and is replaced by the oral flora of the mother and hospital personnel.

    b. Most of the anaerobic flora appears after the emergence of teeth and the formation of gingival crevices. Formation of plaque results in bacterial counts as high as 10^{11} organisms/g.

E. The urogenital tract normally is sterile in the upper portions (i.e., upper urethra, bladder, kidneys, testes, ovaries), but it may contain large numbers of microorganisms in the lower urethra, meatus, vagina, and vulva. *S. epidermidis*, nonhemolytic streptococci, and diphtheroids are predominant in the distal portions of the male and female urethras. Saprobic *Mycobacterium smegmatis* frequently colonizes urethral secretions of the female and the uncircumcised male.

  1. Typical flora of the adult female urogenital tract

    a. The adult female urogenital tract is colonized heavily by anaerobic lactobacilli, which probably assist in preventing overgrowth of other bacterial species. Diphtheroids, *S. epidermidis*, and *Ureaplasma* species also are prevalent. The vaginal flora of healthy women rarely contains enteric bacteria, despite the proximity of the anus. Anaerobic, microaerophilic, and aerobic streptococci also are common.

    b. Species of the fungi *Candida, Torulopsis*, and *Geotrichum* and the protozoan *Trichomonas vaginalis* may be present in small numbers without harm; overgrowth of these organisms can cause pathologic processes.

    c. About 15%–20% of pregnant women are colonized by *Streptococcus agalactiae* (i.e., group B streptococci), which potentially is pathogenic to the infant.

  2. Colonization

    a. The **urogenital tract of the neonate** is sterile, but the nonpathogenic flora of diphtheroids, staphylococci, micrococci, and nonhemolytic streptococci begins developing within the first 24 hours. The **neonatal vagina** is colonized by lactobacilli as long as vaginal glycogen deposits persist as the result of estrogen passively transferred from the mother.

    b. Alkaline conditions in **the vagina prior to puberty** enhance the growth of a different microbial flora. At puberty, estrogen causes deposition of glycogen, which is fermented by lactobacilli, producing an acidic environment and promoting growth of the flora typical of an adult vagina.

F. The gastrointestinal tract usually is the **most heavily colonized portion of the body**. The numbers and types of microorganisms vary with the microenvironments at different sites (Table 7-1).

  1. Typical flora in specific sites

    a. The **stomach** of the normal, healthy individual has no indigenous flora, although bacterial counts of less than 10^3 organisms/ml occasionally are detected. Ingested microorganisms usually are killed by the hydrochloric acid and enzymes of the stomach, or they pass into the intestine.

    b. The **ileum** also is sterile or contains relatively low numbers of microorganisms (i.e., less than 10^3/ml). The microbes that may be found in the duodenum and jejunum are lactobacilli, streptococci, and *C. albicans*, which can overgrow the ileum during abnormal conditions. Forward peristalsis in the ileum helps maintain sterility in the upper portion, but the terminal ileum usually is heavily colonized with bacteria.

    c. The **colon** is the **major reservoir of microorganisms in the human body**, with counts of 10^{12} bacteria/g wet weight of fecal material. The obligate anaerobes *Bacteroides* and *Bifidobacterium* species comprise greater than 90% of the bacterial flora of the colon. *E. coli* is the most numerous facultative anaerobe. A variety of factors can alter the flora of the colon and occasionally lead to overgrowth of pathogenic strains.

Table 7-1. Indigenous Microbial Species in the Normal Adult Gastrointestinal Tract

Location	Microbial Species	
Stomach*	*Candida albicans*	*Lactobacillus* species
Duodenum and jejunum	Enterococcus species	Diphtheroids
	Lactobacillus species	*Candida albicans*
Distal ileum	Enterobacteriaceae	*Bacteroides* species
Terminal ileum and colon	*Bacteroides fragilis*	*Clostridium perfringens*
	Bacteroides melaninogenicus	*Clostridium tetani*
	Bacteroides oralis	*Candida albicans*
	Fusobacterium necrophorum	*Clostridium septicum*
	Fusobacterium nucleatum	*Clostridium innocuum*
	Streptococcus faecalis	*Clostridium ramosum*
	Escherichia coli	*Streptococcus* species
	Klebsiella species	(groups A, B, C, F, and G)
	Enterobacter species	*Pseudomonas aeruginosa*
	Eubacterium limosum	*Salmonella enteritidis*
	Bifidobacterium bifidum	*Shigella* species
	Lactobacillus species	*Salmonella typhi*
	Proteus species	*Peptostreptococcus* species
	Staphylococcus aureus	*Peptococcus* species

*Usually sterile.

2. **Colonization.** The **gastrointestinal tract of the neonate** usually is sterile unless a few bacterial species have been acquired during delivery. Colonization occurs within the first 24 hours after birth; the flora generally is the same as that found in the adult intestine, although the relative numbers of individual species may vary.

   a. *Lactobacillus bifidus* is the primary bacterium in the stools of **breast-fed infants**. Other bacteria include enteric rods, enterococci, and staphylococci. The stools are soft, slightly acrid in odor, and yellowish brown.

   b. The stool of the **bottle-fed infant** is brown and is firmer and stronger in odor than is that of the breast-fed infant. The predominant bacteria are *Lactobacillus acidophilus*, gram-negative enteric rods, enterococci, and anaerobic rods such as *Clostridium* species. The addition of 12% lactose to cow's milk or other formulas may encourage growth of *L. bifidus*.

G. **Blood and internal tissues** are sterile under normal circumstances. Evidence now supports the contention that most individuals experience transient bacteremia nearly every time they brush their teeth or have a bowel movement. Medical manipulations (e.g., catheterization) also may induce transient bacteremia or the presence of organisms in other tissues.

III. **NATURAL BARRIERS TO INFECTION.** The healthy human body is resistant to infection except when it is exposed to a highly virulent microorganism. Colonization by the indigenous flora isbeneficial under the conditions and environment of a normal life. However, it has been shown that gnotobiotic (i.e., germ-free) animals survive under artificial conditions longer than do normal control animals. The natural barriers to infection generally are categorized as mechanical, chemical, and immunologic.

A. **Mechanical barriers** physically inhibit potentially infectious agents from attaching to and penetrating cells. Mechanical action of some specialized cells eliminates invading microbes from the body.

   1. **Skin** is the primary physical barrier to infection. Undamaged skin, consisting of the keratinized outer layer of dead cells and the successive layers of epidermis, is impenetrable to all but a few microorganisms.

2. **Hair** protects skin and mucous membranes against invading microbes. For example, hair in the nares is an initial filter that prevents large particles from entering the respiratory tract. Few particles larger than 10 μm pass this barrier.

3. **Mucous coating** of epithelial and mucosal cells prevents contact between many pathogens and areas that are not covered by skin. Microorganisms and other particles are trapped in the viscous mucus and removed by other mechanisms.

4. **Natural cleaning** is achieved by the **shedding of keratinized cells** that carry contaminating microbes; approximately 1 billion cells are shed each day. **Urine, saliva, perspiration**, and other body fluids assist in removing microbes by their mechanical cleansing action.

5. **Specialized ciliated epithelial cells** remove contaminating microorganisms from the upper respiratory tract. These cells usually are coated with mucus that entraps particles. The synchronized wave-like motion of the cilia transports the entrapped foreign material out of the respiratory tract.

6. **Other physiologic functions** also eliminate pathogenic microbes by mechanical action. **Sneezing, coughing, vomiting**, and **diarrhea** effectively eliminate large numbers of bacteria.

B. **Chemical inhibition** of microorganisms invading the body generally is associated with the mechanical barriers. The chemical and biochemical inhibitors that act as natural defenses against colonization and infection are found in body secretions.

1. **Perspiration and the products of sebaceous glands** help maintain a relatively acidic microenvironment on the skin. The salt content of perspiration and the fatty acid content of other secretions inhibit the growth of many microorganisms.

2. **Tears** contain lysozyme, a potent bactericidal compound. Lysozyme is effective particularly against gram-positive bacteria. The enzyme lyses the bacteria on contact by destroying the integrity of the cell wall.

3. **The acidic pH of most physiologic secretions** prevents colonization of tissues by pathogenic microorganisms. The hydrochloric acid in gastric secretions kills most microbes that are ingested in moderate numbers; when large numbers are ingested, some microbes may survive. Perspiration, as mentioned previously, also helps maintain acidic microenvironments, as do urine and vaginal secretions. Such an environment promotes the growth of nonpathogenic bacteria such as lactobacilli, which compete with less desirable bacteria for adherence sites.

C. **Immunologic barriers** to infection are humoral and cell-mediated responses. These two kinds of responses usually work together to eradicate an invading microorganism or to neutralize a toxin; one type of response may be more prominent than the other (for additional information, see Chapter 5).

1. **Humoral responses** primarily are specific to a particular microorganism. Upon initial exposure to a microbe, a host may **develop antibodies** against some specific antigenic component. Subsequent exposure to the microbe typically causes a rise in antibody titer. Neonatal immunity to certain infections can be due to the **transplacental passage of antibodies**. A nonspecific humoral response to invading microbes is **activation of the complement system**. Humoral responses often cause lysis of the invading microorganism.

2. **Cell-mediated responses** may be either nonspecific or specific. The major constituent of cell-mediated immunity is the phagocytic cell. Most microbes are destroyed by **phagocytosis**, although some microbes can survive and even replicate inside the phagocytic cell. Specific cell-mediated immunity usually is directed by antibodies.

STUDY QUESTIONS

Directions: Each question below contains five suggested answers. Choose the **one best** response to each question.

1. Disease processes that occur in hospitalized patients as the result of mechanical manipulation are known as

(A) nosocomial infections
(B) opportunistic infections
(C) fomite-induced infections
(D) iatrogenic infections
(E) hospital-acquired infections

2. Most bacteria in the normal adult gastrointestinal tract are members of the genus

(A) *Bacteroides*
(B) *Clostridium*
(C) *Streptococcus*
(D) *Lactobacillus*
(E) *Enterobacter*

3. The primary natural barrier against infection is

(A) the mechanical action of ciliated cells
(B) the presence of intact skin
(C) the acidic pH of body fluids and secretions
(D) lytic enzymes found in body fluids
(E) the cell-mediated immune system

Directions: Each question below contains four suggested answers of which **one or more** is correct. Choose the answer

A   if **1, 2, and 3** are correct
B   if **1 and 3** are correct
C   if **2 and 4** are correct
D   if **4** is correct
E   if **1, 2, 3, and 4** are correct

4. The indigenous flora of the normal, healthy adult human may include

(1) *Staphylococcus epidermidis*
(2) *Candida albicans*
(3) *Clostridium perfringens*
(4) *Bordetella pertussis*

5. The indigenous flora of the neonate is influenced by

(1) breast-feeding versus bottle-feeding
(2) the indigenous flora of the mother
(3) transplacental transfer of antibodies
(4) vaginal delivery versus delivery by cesarean section

6. Natural defenses against infection include

(1) lysozyme in tears
(2) mucous secretions of epithelial cells
(3) ciliated cells in the respiratory tract
(4) the acidic pH of skin

Directions: The group of questions below consists of lettered choices followed by several numbered items. For each numbered item select the **one** lettered choice with which it is **most** closely associated. Each lettered choice may be used once, more than once, or not at all.

Questions 7–11

Match each species of indigenous flora with its principal habitat.

(A) Skin
(B) Vagina
(C) Tooth enamel
(D) Adult gastrointestinal tract
(E) Infant gastrointestinal tract

7. *Clostridium septicum*

8. *Streptococcus mutans*

9. *Lactobacillus bifidus*

10. *Staphylococcus epidermidis*

11. *Torulopsis glabrata*

ANSWERS AND EXPLANATIONS

1. The answer is D. [*I E*] Iatrogenic infections are nosocomial infections (i.e., infections acquired in a hospital setting) that are the result of mechanical manipulation. Mechanical manipulation includes surgery, the insertion of an indwelling urinary or intravenous catheter, and intubation of the respiratory tract. Opportunistic infections occur under a natural or chemical immunosuppression, often by organisms that are nonpathogenic for healthy individuals. Susceptible hosts may become infected upon contact with fomites—inanimate objects contaminated with microbes. Iatrogenic infection does not preclude opportunistic pathogens as the etiologic agents or transmission of the infection by a fomite.

2. The answer is A. [*II F 1 c*] The anaerobic *Bacteroides* species are the most numerous bacteria in the normal adult gastrointestinal tract. Combined with the anaerobic *Bifidobacterium* species, the bacteroides comprise greater than 90% of the bacteria in the colon. The facultative anaerobe *Escherichia coli* is the most numerous bacterium in the gastrointestinal tract capable of aerobic growth. Species of *Streptococcus, Lactobacillus*, and *Enterobacter* are less numerous in the colon, although they exist in a large portion of the adult population. *Clostridium* species, including potentially pathogenic species (e.g., *Clostridium tetani*), are found in smaller numbers and exist in a smaller portion of the population.

3. The answer is B. [*III A 1*] Intact skin is the primary barrier to infection. Few microorganisms are capable of invading intact skin, although microscopic breaks may allow bacteria and viruses to enter into deeper tissues. The acidic pH of perspiration, fatty acids, indigenous flora, and salts are adjuncts to the mechanical barrier of the intact skin in preventing infection. Ciliated cells provide a mechanical barrier to microbes in the bronchi; lytic enzymes (e.g., lysozyme) are a chemical barrier found in some body fluids, such as tears; and the cell-mediated immune system is part of the natural human immunologic defense.

4. The answer is A (1, 2, 3). [*II A 1; Table 7-1*] A large number of bacterial species are considered indigenous flora of the healthy human. The distinction between indigenous flora and pathogens has become less well defined. *Staphylococcus epidermidis* usually is the most common bacterial species found on the skin. *Candida albicans* can be isolated from the skin, gastrointestinal tract, and urogenital tract of a significant percentage of healthy individuals. *Clostridium perfringens* and other clostridia, which can cause serious, life-threatening disease processes under certain situations, are considered indigenous flora of the gastrointestinal tract. *Bordetella pertussis* is not considered part of the indigenous flora, because infection with this bacterium always results in a disease process, although the severity of the disease can vary substantially. Carriage of *B. pertussis* is occasionally seen.

5. The answer is E (all). [*II C 3, D 3 a, E 2 a, F 2; III C 1*] Under normal circumstances, the healthy fetus is sterile. The neonate begins to be colonized with microorganisms during the birth process and within a few days has developed an indigenous flora. The skin, upper respiratory tract, oral mucosa, and urogenital tract are exposed to the mother's indigenous flora as the infant passes through the birth canal. If the baby is delivered by cesarean section, initial exposure is to the indigenous flora of the attending health-care professionals. Breast-feeding increases the incidence of *Lactobacillus bifidus* in the gastrointestinal tract as compared to bottle-feeding, and antibody-containing colostrum obtained during breast-feeding also affects the indigenous flora of the infant. The transplacental passage of immunoglobulin G antibodies, such as those induced by tetanus toxoid, can prevent colonization by certain microbes.

6. The answer is E (all). [*III A, B*] Natural defenses of healthy humans against infection include mechanical, chemical, and immunologic mechanisms. Lysozyme, found in tears and other secretions, is an enzyme capable of lysing the cell walls of gram-positive bacteria. Mucous secretions of epithelial and mucosal cells entrap microbes and prevent their adherence to tissue cells. The mechanical action of specialized ciliated cells in the respiratory tract, especially in combination with the mucous secretions of the respiratory epithelial cells, entraps and removes infecting microbes. In general, the pH of body fluids and tissues is acidic rather than alkaline. An acidic pH can predispose to the growth of nonpathogens (e.g., the lactobacilli) and can prevent the growth of pathogens with a narrow pH range for optimal growth.

7–11. The answers are: 7-D, 8-C, 9-E, 10-A, 11-B. [*II A 1 a, D 1 a, E 1 b, F 1 c, 2 a*] *Clostridium septicum* is an anaerobe found in the gastrointestinal tract of some healthy adult humans. Like many of the species comprising the indigenous flora, this microbe can cause serious disease if it overpopulates the colon as the result of broad-spectrum antibiotic treatment that reduces the numbers of other members of the indigenous flora.

Streptococcus mutans adheres to specific receptors on tooth enamel. It is a major constituent of plaque and is important in the etiology of dental caries.

Lactobacillus bifidus is the primary bacterium found in the stools of breast-fed infants. This bacterium in part accounts for the acrid odor, soft consistency, and yellowish brown color of the stool of a breast-fed infant.

Staphylococcus epidermidis is one of the principal members of the indigenous flora of the skin, although it also is found in the nares, oral cavity, pharynx, urogenital tract, and gastrointestinal tract. *S. epidermidis* can be isolated from the skin of 100% of healthy individuals.

The yeast *Torulopsis glabrata* can be found in small numbers in the vagina of healthy women. Metabolic disturbances and other conditions can predispose to the overgrowth of this yeast, resulting in a characteristic vaginitis or urinary tract infection.

8
Staphylococci, Streptococci, and Other Gram-Positive Cocci

Gerald E. Wagner

I. STAPHYLOCOCCI. The staphylococci are medically the most important group in the family Micrococcaceae. There are three species of clinical importance in the genus. They are ***Staphylococcus aureus***, which is responsible for most cases of staphylococcal infection in humans; ***Staphylococcus epidermidis***, which is considered to be indigenous flora but occasionally may cause infection; and ***Staphylococcus saprophyticus***, which may cause cystitis.

A. Morphology. Staphylococci are **gram-positive cocci** that classically grow in irregular clusters (Figure 8-1) because of their ability to divide along successive perpendicular planes.

   1. **Cellular morphology**
      a. **Size.** Individual cells range in diameter from 0.7 to 1.2 μm.
      b. **Capsules** are produced by some strains of *S. aureus*. Encapsulation probably is more frequent in vivo than after cultivation in vitro.
      c. **Cell wall.** Unique **pentaglycine bridges**, which link the tetrapeptides attached to the muramic acid residues, **characterize the peptidoglycan** of staphylococcal cell walls. The pentaglycine bridges are specifically sensitive to the enzyme **lysostaphin**, which may be used to identify the genus.

   2. **Colonies** of staphylococci are sharply defined, round and convex, and they measure about 4 mm in diameter.
      a. *S. aureus* usually produces a golden yellow pigment on solid media; however, the colonies may vary from white to orange. On blood agar, colonies usually are surrounded by a zone of complete hemolysis.

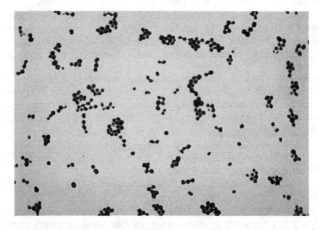

Figure 8-1. The staphylococci are gram-positive cocci that occur in clusters. The overall impression from the stained smear is that of clusters of bacteria, although single cells, pairs, and short chains occasionally are observed. It is impossible to identify the bacteria by microscopic morphology; therefore, the smear would be reported as containing gram-positive cocci in clusters that resemble staphylococci.

b. *S. epidermidis* colonies usually are white, and only a few strains produce hemolysis on blood agar.

c. *S. saprophyticus* colonies are white to grayish white. This species is nonhemolytic.

B. Antigenic composition. The staphylococci are antigenically complex. Antibodies are produced to many cellular and extracellular components of these bacteria.

1. **Species-specific antigens** of the staphylococci are the teichoic acids of the cell wall.
   a. **Ribitol teichoic acid** is specific for *S. aureus.* It is composed of a linear ribitol backbone linked by phosphodiester bridges. *N*-Acetylglucosamine is linked to the C-4 portion of ribitol in either α- or β-glucosidic linkages, and it is the glucosamine moiety that serves as the antigenic determinant.
      (1) *S. aureus* can be grouped on the basis of strain susceptibility to lysis by specific phages.
      (2) This susceptibility is due to the presence of specific receptors for the phage on the cell surface.
   b. **Glycerol teichoic acid** is the species-specific antigen for *S. epidermidis.* It is composed of glucose molecules, in either α- or β-glycosidic linkages, attached to a glycerol phosphate backbone and constituting the antigenic determinant.
   c. *S. saprophyticus* contains ribitol teichoic acid with either *N*-acetylglucosamine residues or glucose residues. **Both types of teichoic acid can occur in the same strain.**

2. **Cellular antigens of *S. aureus*** are complex and varied. Antibodies to some of these antigens are protective.
   a. **Capsules** appear to make *S. aureus* more virulent, on the basis of studies in animals with strains encapsulated in vitro; anticapsular antibodies protect against disease in laboratory animals. Different types of capsules have been noted in different groups of *S. aureus.* Capsules consisting of a polymer of glucosaminuronic acid are associated with one group of strains, a polymer of mannosaminuronic acid with another group, and a capsule with the characteristic components of peptidoglycan has been found in a third group.
   b. **Protein A** (agglutinogen A) is a surface protein that is covalently bound to the peptidoglycan of most strains of *S. aureus.* As much as one-third of the protein A of a cell may be released into the surrounding environment.
      (1) The most significant property of protein A is its nonspecific interaction with the Fc portion of certain subclasses of immunoglobulin G (IgG) of a variety of mammalian species (e.g., subclasses IgG_1, IgG_2, and IgG_4 of human immunoglobulin). This **nonspecific binding causes activation of complement** by both the classic and alternative pathways, resulting in such reactions as the local wheal and flare reaction, the Arthus phenomenon, local and systemic anaphylaxis, inhibition of phagocytosis of opsoninized particles, and induction of T cell and B cell proliferation in vitro.
      (2) Protein A also is a true antigen, inducing the **formation of specific antibodies**.
   c. **Clumping factor** (or bound coagulase) causes the clumping of nonencapsulated strains of *S. aureus* when they are mixed with a solution containing **fibrinogen**. Fibrinogen appears to have specific bacterial cell surface receptors, although they have not been characterized.

3. **Extracellular antigens and other products of *S. aureus*** also induce detectable antibodies.
   a. **Coagulase** is an extracellular product of *S. aureus* that exists in several antigenically different forms, all of which have the unique characteristic of clotting various mammalian plasmas. Coagulase does not react directly with fibrinogen but forms a thrombin-like substance by reacting with the coagulase-reacting factor of plasma, which probably is prothrombin.
      (1) In the past, coagulase-positive staphylococci by definition were identified as *S. aureus,* but recently coagulase-negative *S. aureus* organisms have been recognized that may be the etiologic agents of disease.
      (2) The **coagulase test** (Figure 8-2) remains the most frequently used laboratory test for the identification of virulent *S. aureus.*
   b. **Hemolysins** of four different antigenic types are produced by *S. aureus,* and all of them cause complete hemolysis. Many strains of *S. aureus* produce more than one type of hemolysin. The hemolysins have different spectra of activity with respect to various erythrocyte species, and they are cytotoxic to other types of cells.
      (1) **Alpha hemolysin** (i.e., alpha toxin) is the most frequently detected hemolysin in *S. aureus* isolated from clinical specimens. It has no effect on human erythrocytes but

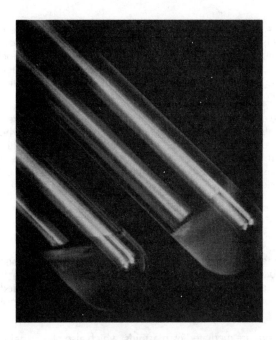

Figure 8-2. The coagulase test is the most widely used test to indicate virulent, coagulase-positive *Staphylococcus aureus.* The *upper tube* is coagulase positive, as indicated by the clotting of the plasma. The *lower tube* is coagulase negative, as indicated by the liquid character of the plasma.

rapidly lyses sheep erythrocytes. The hemolysin causes dermonecrosis at the site of injection in laboratory animals and is lethal when injected intravenously. The site and mode of action are unknown.

(2) Beta hemolysin is a sphingomyelinase that is only moderately toxic to human erythrocytes, and it is produced by less than 20% of clinical isolates of *S. aureus.* It is a hot-cold hemolysin, in that lysis is maximized when the hemolysin-erythrocyte mixture is held at low temperatures after incubation at 37° C.

(3) Gamma hemolysin is a two-component hemolysin with moderate activity against human erythrocytes. At least one component of the hemolysin is inhibited by sulfonated polymers such as agar, and its activity is not observed routinely on blood agar plates.

(4) Delta hemolysin is an aggregate of low molecular weight subunits with detergent-like activity, which gives this hemolysin a broad spectrum of lytic and cytotoxic activity. It is not known whether the hemolysin functions in vivo because it is inactivated by serum phospholipids.

c. Panton-Valentine (P-V) leukocidin is nonhemolytic and is cytotoxic only to human and rabbit neutrophils and macrophages. P-V leukocidin consists of two subunits, both of which are required for activity because specific antiserum against either subunit completely neutralizes the toxin.

d. Enterotoxins are produced by many strains of *S. aureus,* and in the United States they account for many cases of food poisoning associated with pastries, dairy items, and salted meats.

(1) There are five clinically and antigenically distinct enterotoxins (i.e., **enterotoxins A through E**) that are relatively heat stable and moderately resistant to trypsin digestion. The mechanism of action of the enterotoxins is not known completely, but they seem to act directly on the bowel itself as well as on the neural innervation.

(2) Toxic shock syndrome (TSS) apparently is caused by an enterotoxin-like substance (designated **enterotoxin F**) produced by some strains of *S. aureus* (section I E 2c).

e. Exfoliatin is an extracellular product of many strains of *S. aureus* belonging to phage group II. Exfoliatin cleaves the stratum granulosum layer of the epidermis and is responsible for several clinical syndromes; there probably are several antigenic types of the toxin.

f. Miscellaneous extracellular products such as staphylokinase, hyaluronidase, phospholipase, and deoxyribonuclease are produced by *S. aureus* as well as by *S. epidermidis* or *S. saprophyticus;* a few strains of *S. epidermidis* produce epsilon hemolysin, which is similar to the delta hemolysin of *S. aureus.*

C. Biochemical and metabolic characteristics of the staphylococci are useful in identifying these bacteria at the generic and species levels. The staphylococci possess heme-containing cytochromes, which, along with a group of quinones, produce a membrane-bound electron transport chain responsible for normal respiratory metabolism. The genus also produces catalase, which cleaves hydrogen peroxide and prevents the buildup of this highly toxic molecule. The staphylococci also are capable of growing in the presence of 7.5% sodium chloride.

1. **S. aureus and S. epidermidis** are capable of anaerobic growth, fermenting a variety of pentoses, hexoses, disaccharides, and sugar alcohols to produce large quantities of mostly lactic acid; the ability to ferment mannitol is a differential characteristic of *S. aureus*. Both species grow better under aerobic conditions than they do in an anaerobic environment, and they are capable of using a wide range of carbohydrates as energy sources. The products of aerobic glucose utilization are carbon dioxide and acetate.

2. **S. saprophyticus** grows very poorly under anaerobic conditions.

D. Genetic characteristics

1. Genetic information is exchanged among the staphylococci principally by **transduction** (see Chapter 3, section II B). Most staphylococcal strains are lysogenic, and serologic group B lysogenic bacteriophages are capable of generalized transduction, in which host DNA is carried by the phage. Under specialized conditions, transformation may occur, but conjugation is not seen among the staphylococci.

2. Many of the toxins and products of *S. aureus* are mediated by **plasmids**, which also play a major role clinically as mediators of antibiotic resistance.
   a. The **large plasmid of S. aureus** has a molecular weight of about 2×10^7 daltons, and it is covalently bonded at the ends, forming a circular piece of DNA. Replication of the plasmid is tied to that of the genome; thus, there usually is only one copy per cell. Multidrug resistance is not characteristic of this plasmid as it is of the resistance (R) plasmid factors of gram-negative bacteria; only β-**lactamase production** and sometimes **resistance to erythromycin** are associated with this staphylococcal plasmid.
   b. The **small plasmid of S. aureus**, with a molecular weight of about 3×10^6 daltons, carries **resistance determinants to tetracycline and chloramphenicol**. This plasmid replicates independently of cell division, and a single staphylococcal cell may contain many copies.
   c. In contrast to analogous resistance in gram-negative bacteria, β-lactamase and chloramphenicol acetyltransferase production in *S. aureus* are **inducible processes**.

E. Staphylococcal disease. The two forms of disease that are produced by *S. aureus* are invasive and toxigenic.

1. **Invasive disease is characterized by abscess formation**, generally in cutaneous tissue in otherwise healthy individuals. Patients who are debilitated or immunosuppressed or who have been extensively operated upon may develop deep abscesses in any organ system.
   a. **Pyoderma** occasionally is caused by *S. aureus* or *S. epidermidis*. It most frequently is seen in children who live in warm, moist climates. Gentle cleaning and the application of topical antibiotic ointment generally is effective treatment.
   b. **Folliculitis** is a localized infection of a single hair follicle with minimal involvement of surrounding tissue; it is caused by *S. aureus* or, occasionally, by *S. epidermidis*. Small abscesses form around the follicle. Treatment involves cleaning, application of topical antibiotics, and warm soaks to encourage spontaneous drainage.
   c. **Furunculosis** initially resembles folliculitis but spreads to involve a larger area. Typically, deep inflammatory nodules are surrounded by a large zone of intense inflammation. Treatment is the same as for folliculitis; occasionally, surgical drainage may be indicated.
   d. **Carbuncles** are burrowing, often multiple interconnected abscesses involving hair follicles, sebaceous glands, and the surrounding soft tissue. They are extremely painful and require surgical drainage of accumulated pus and *S. aureus*. Parenteral administration of antistaphylococcal antibiotics may be necessary if systemic signs of bacteremia are present.
   e. **Deep abscess formation** generally leads to **bacteremia** and **metastatic abscess formation** in other organs. Approximately two-thirds of affected individuals develop **endocarditis**, and many of these subsequently require prostheses because of valve perforation.

    f. Staphylococcal pneumonia results from inhalation of the bacteria. The disease usually occurs in debilitated patients secondary to viral (i.e., influenzal) pneumonia.

    g. Acute endocarditis is caused by *S. aureus*. It is characterized by a relatively short time course of less than 6 weeks.

      (1) The **infected heart valve typically is coated with numerous vegetations** composed of platelets, bacteria, and other amorphous cellular debris; perforation of the heart valve results from the many extracellular enzymes and products of *S. aureus*. Pieces of vegetation, known as **infective emboli,** may dislodge into the circulation and establish a focus of infection in any organ system.

      (2) **Emergency surgical replacement** of the heart valve usually is necessary to prevent death from cardiac insufficiency; antimicrobial therapy is of minimal value because the agents are unable to penetrate the vegetation and because the heart valve is avascular.

    h. Subacute endocarditis is caused by *S. epidermidis*. It has a time course of more than 6 weeks and typically follows thoracic surgery. Vegetation is not a prominent feature of subacute endocarditis, and **intravenous antimicrobial therapy** generally eliminates the infection.

  2. Toxigenic staphylococcal disease is caused by the production of extracellular substances by *S. aureus,* and it involves a variety of syndromes.

    a. Food poisoning is the result of ingestion of one of the five enterotoxins of *S. aureus,* usually through the ingestion of preformed toxin in improperly prepared food. The heat-stable enterotoxin may persist after cooking even though the bacteria are killed. Nausea, vomiting, and, occasionally, diarrhea occur 1–6 hours after ingestion of enterotoxin. Recovery generally is prompt without therapy, except in the young and elderly, who may require replacement of fluids and electrolytes.

    b. Exfoliatin, produced by phage group II strains of *S. aureus,* causes three syndromes, the occurrence of which probably depends upon the immune status of the individual. The presence of lesions is not necessary for initiation of the disease. Although the disease usually is not fatal, deaths have occurred in nursery epidemics, probably as the result of secondary infections.

      (1) Infants and young children develop **toxic epidermal necrolysis** (i.e., **Ritter's disease**), which is characterized by large areas of denuded skin and generalized bullae formation. The syndrome commonly is referred to as the **scalded baby syndrome** because of the moist red areas of denuded tissue.

      (2) Older children and adults develop localized bullae, which also are red and moist in appearance. The syndrome is known as **Lyell's disease**, or, more commonly, as **scalded skin syndrome**. Healing occurs without scarring in the absence of secondary bacterial infection.

      (3) Older children and adults also may develop a **scarlatina-like rash**. The syndrome mimics streptococcal scarlet fever except that the tongue and palate are not involved.

    c. TSS is a recently recognized disease attributed to a toxin (enterotoxin F) produced by *S. aureus*. The frequently fatal disease is most commonly associated with the improper, excessively prolonged use of an expandable tampon made of synthetic materials. **Wound TSS** has been recognized also in a variety of patients. The *S. aureus* probably is indigenous flora of the skin that contaminates the tampon or wound dressing during handling. The patient generally develops a variety of systemic symptoms including headache, nausea, vomiting, and shock, which is the most frequent cause of death.

F. Immunity. Neither humoral nor cell-mediated immunity appears to provide much protection against invasive staphylococcal disease, although antibodies to a variety of cellular components and products of *S. aureus* are found in normal human serum.

G. Laboratory diagnosis of staphylococcal diseases requires cultivation and identification of the bacteria, although the finding of clusters of gram-positive cocci in pathologic material is strong presumptive evidence.

  1. A **positive coagulase test and mannitol fermentation** confirm the identification of *S. aureus,* and further biochemical tests may provide additional definitive data.

  2. Rapid speciation of the staphylococci in systemic disease processes such as endocarditis can be made by the **electrophoretic analysis of serum**. The presence of antibodies against ribitol teichoic acid or the presence of antigen itself indicates infection by *S. aureus;* the presence of glycerol teichoic acid or antibodies against it indicates infection by *S. epidermidis.*

3. **Specific identification of strains** is of epidemiologic value, particularly in epidemic nosocomial outbreaks, and can be achieved by **phage typing.**

H. Epidemiology

1. The staphylococci are ubiquitous, and nearly all individuals are colonized by *S. epidermidis* on the skin. Approximately 30%–40% of the adult population are carriers of *S. aureus,* and the nares probably are the most important carriage site. Staphylococci are capable of extended survival on environmental surfaces and in the air, but **human-to-human transmission** probably is the most important means of spreading the bacteria.

2. It is not feasible to attempt to control the contact of healthy individuals with staphylococci, but patients at risk for infection, such as newborns and hospitalized patients with serious underlying diseases, should be isolated from contact with carriers.

II. STREPTOCOCCI.
The streptococci are a large group of bacteria belonging to the family Streptococcaceae. Some members of the genus are notable human pathogens, whereas others are indigenous flora of the oropharynx and gastrointestinal tract. Some species are responsible for the formation of dental plaque, and others produce nonsuppurative sequelae such as rheumatic fever, acute glomerulonephritis, and chorea. The most important species pathogenic for humans include *Streptococcus pyogenes, Streptococcus agalactiae, Streptococcus faecalis,* the viridans streptococci, and *Streptococcus pneumoniae.*

A. Genus characteristics as exemplified by *S. pyogenes.*
This section primarily deals with *S. pyogenes* (a group A streptococcus); details of other important streptococci are given in successive sections of this chapter.

1. **Morphology.** The streptococci are gram-positive cocci that occur in chains of varying length, often depending upon the species. Hemolytic streptococci such as *S. pyogenes* usually occur in long chains of 10–15 cells (Figure 8-3), whereas oral streptococci such as *Streptococcus mutans* frequently are seen in pairs.

   a. **Individual cells** range in **size** from 0.5 to 1.0 μm in diameter, depending upon growth conditions and the age of the culture.

   b. **Colonies** of most streptococci are small (0.5–2 mm in diameter), matte, and gray to grayish white on blood agar. In contrast, colonies of *S. faecalis* often are large and white, resembling colonies of *S. epidermidis.*

   c. **Hemolysis** on sheep blood agar is characteristic of specific groups of streptococci.

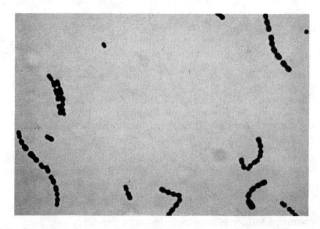

Figure 8-3. *Streptococcus pyogenes* typically appears as gram-positive cocci in chains. The chains of streptococci may be short or long and are formed by the division of the bacterial cells along a single plane.

(1) **Alpha hemolysis** is the incomplete destruction of the erythrocytes, resulting in a **green coloration** of the medium surrounding the colonies. The green pigment is due to biliverdin and other heme compounds. Many of the oral streptococci and *S. pneumoniae* are α hemolytic.

(2) **Beta hemolysis** is the complete lysis of erythrocytes and results in a **distinct clear zone** around the colonies. Beta hemolysis usually is visualized better under reduced oxygen tension (i.e., in stab areas in the agar) because of the oxygen lability of the hemolysin. Lancefield groups A, B, C, and G streptococci (see section II B) typically are β hemolytic.

(3) **Gamma hemolysis** is characterized by **no visible effect** on sheep blood agar; the designation indicates no hemolysis.

d. L forms, which are stable in hypertonic medium, are induced easily in streptococcus species by exposure to penicillin or by group C streptococcus phage-associated lysin. It has been suggested that L forms are responsible for the persistence of streptococci in tissue, but the exact role they play in disease is not known.

2. Antigenic composition of the streptococci is varied and complex. Cellular components induce both group-specific and type-specific antibodies.

a. Group-specific carbohydrates of the streptococcal cell wall are used to classify the genus serologically into at least **18 groups, designated A through R** by the Lancefield system. It now is known that the group-specific antigen of group D streptococci actually is a teichoic acid component of the cell membrane.

b. Type-specific M proteins are limited almost exclusively to **group A streptococci** and are responsible for the virulence of these bacteria by virtue of their antiphagocytic properties. The large M-protein complex contains fimbriae that are important in the attachment of group A streptococci to susceptible host cells. **More than 80 antigenic types** of M protein have been isolated from group A streptococci.

c. T antigens are proteins that are unrelated to virulence, and they do not induce protective antibodies in infected patients. Some T antigens are associated with a single M protein, whereas others are shared by several M proteins. T antigens are **useful in the laboratory identification** of group A streptococci that fail to synthesize M protein or for which there is no available antiserum.

d. R antigens are proteins that are associated with groups A, B, C, G, and L streptococci. The antigen is **nonprotective**, and its biologic significance is not known.

3. Metabolism. The streptococci are catalase-negative (i.e., they lack a cytochrome system) facultative anaerobes that ferment carbohydrates to form lactic acid. Most streptococci that are virulent for humans are fastidious in their nutritive requirements; the addition of whole blood or serum to culture media greatly enhances growth.

4. Pathogenesis of streptococcal disease involves both active infection with the production of extracellular virulence factors and nonsuppurative sequelae.

a. Virulence factors. *S. pyogenes* is the most important human pathogen, and, in addition to M protein, it **produces more than 25 extracellular antigenic substances.** The structure and biologic activity of only a few of these substances have been characterized.

(1) **M protein** probably is the **major virulence factor of group A streptococci.** It is antiphagocytic, and purified M protein has been shown to precipitate fibrinogen, clump platelets and leukocytes, lyse neutrophils, and inhibit the migration of leukocytes in capillaries. Serum factors, probably related to complement, are required for the cytotoxic activity of M protein. This antigen appears to be responsible for the adherence of group A streptococci to epithelial cells. It is highly immunogenic.

(2) **Streptolysin O** is an oxygen-labile protein responsible for β hemolysis around submerged colonies of groups A, C, and G streptococci. Its cytolytic and hemolytic activities occur only under reducing conditions because of a labile sulfhydryl group. It also is inhibited by cholesterol and other sterols. Streptolysin O is cardiotoxic in laboratory animals.

(3) **Streptolysin S** is an oxygen-stable nonantigenic peptide found on and released from the surface of streptococci. Release of streptolysin S involves a variety of nonspecific carrier proteins, which eventually may transfer the peptide to mammalian cells, where it reacts with and is inactivated by phospholipids. Streptolysin S is both hemolytic and cytotoxic.

(4) **Erythrogenic toxin**, more correctly known as **streptococcal pyrogenic toxin**, is produced by most strains of group A streptococci as one or more of three immunologically distinct substances. These toxins consist principally of protein and hyaluronic acid, and their production probably is mediated by a temperate prophage. Streptococcal pyrogenic toxin causes fever and rash, enhances susceptibility to endotoxin shock, alters antibody response, damages macrophages, and is mitogenic for lymphocytes. Its exact role in the pathogenesis of streptococcal diseases is not clear, although it is believed to play a role in the symptomatology of scarlet fever.

(5) Cardiohepatic toxin is an extracellular toxin of some group A streptococci; it causes lesions in the myocardium and diaphragm and giant cell granulomas in the liver when it is injected into a laboratory animal. Infection of the tonsils by a cardiohepatic toxin-producing strain of *S. pyogenes* causes the same effects within 24 hours.

(6) Streptokinase is a protein produced by streptococci in groups A and C; it activates the fibrinolytic system of human blood. Streptokinase acts as a virulence factor by lysing clots and fibrin near the streptococci. Purified streptokinase has been used clinically for debriding surface infections, treatment of fibrinous exudates, enhancement of wound healing, and lysis of intravascular thrombi.

(7) Spreading factors are extracellular enzymes that assist streptococcal growth in tissue by liquefying inflammatory exudates. Group A streptococci produce hyaluronidase, ribonuclease, and four varieties of deoxyribonuclease (DNase); all of these factors are immunogenic.

b. Sequelae of streptococcal infections are nonsuppurative complications of either overt or inapparent infection by group A streptococci. Because the sequelae have never been reproduced in an experimental animal model, there is no proven or totally accepted mechanism of pathogenesis.

(1) Rheumatic fever is thought to result only from upper respiratory tract infection and not from cutaneous infections by group A streptococci.

    (a) There are several absolute **prerequisites** for development of rheumatic fever.

        (i) Group A streptococci must have been present in the host at some time.

        (ii) The infection must be localized in the upper respiratory tract.

        (iii) The bacteria must persist in the tissue for a period of time in order for the complication to develop.

        (iv) The host must show an antibody response indicative of recent infection by group A streptococci.

    (b) A variety of **theories** have been proposed to explain the induction of sequelae by group A streptococci.

        (i) Rheumatic toxins of various types have been suggested as the cause of rheumatic fever. Some extracellular products of group A streptococci produce myocardial and endocardial lesions in laboratory animals; however, the lesions do not closely resemble the lesions that occur in humans.

        (ii) Autoimmunity, induced by the localization of complexes of extracellular streptococcal products and antibodies in tissue, has been proposed as the cause of tissue damage. Tissue injury could result either from the activation of complement or by the action of components of the cell-mediated immune system.

        (iii) Cross-reactivity has been suggested but probably does not play a role. Cross-reactivity is known to occur between mammalian and bacterial components, and group A streptococcal cell wall and protoplast membrane antigens cross-react with glycoproteins of heart valves and a number of other host tissues; however, there is no conclusive evidence for cross-reacting antibodies in rheumatic fever.

        (iv) Genetic and anatomic aberrations also have been suggested as causes of rheumatic fever. Some investigators believe that the rheumatic patient is predisposed genetically to develop adverse humoral and cellular responses to streptococcal antigens. Other investigators believe that there are direct lymphatic channels between the tonsils and the heart of rheumatic patients through which streptococci and their products pass, resulting in cardiac injury.

(2) Acute glomerulonephritis is one of the most common complications of upper respiratory tract and skin infections caused by group A streptococci; the disease never occurs concomitantly with rheumatic fever. Acute glomerulonephritis probably is caused by specific nephritogenic strains of streptococci, and certain M types have been associated most frequently with the disease. The causative substances elaborated by the nephritogenic strains and the pathogenesis of acute glomerulonephritis are not known.

(a) **Nephrotoxins**, which produce renal disease in laboratory animals, have been identified, but they do not produce the same spectrum of disease that is seen in human acute glomerulonephritis. Substances of nephritogenic streptococci that cause renal toxicity in laboratory animals include streptolysin, M protein, extracts of cell walls, lysates of whole cells, solubilized protoplast membranes, and uncharacterized diffusable substances released by the cells.

(b) **Immunologic cross-reactivity** occurs between antigens of protoplast membranes of nephritogenic streptococci and soluble components of glomerular basement membranes. Antibodies to protoplast membranes cause glomerular lesions.

(c) **Autoantibodies**, which were induced in animals immunized with a mixture of kidney extracts and streptococci, have been proposed as the cause of acute glomerulonephritis. This theory remains controversial, and the experiments have not been duplicated.

(d) **Immune complexes** have been observed in the glomeruli of kidneys from patients with acute glomerulonephritis by means of immunofluorescent and electron microscopic analyses. The relationship between the antigen in the complexes and group A streptococci is not clear.

(3) **Migratory arthritis**, probably caused by hypersensitivity and anaphylactic reactions in the joints, is part of the rheumatic syndrome caused by group A streptococci. In general, neither live streptococci, cellular components, nor extracellular products are found in the inflamed joints. The local activation of complement by immune complexes may explain the pathogenesis of this complication. Arthritic lesions can be induced in the joints of laboratory animals by the intra-articular injection of a variety of streptococcal substances.

5. **Immunity** to group A streptococcal infection is type specific with a humoral response induced by M protein. The anti-M antibodies are mainly IgG, and they act as opsonins. The antibody titer rises slowly over a period of 30–60 days following streptococcal infection. The production of a group A streptococcal vaccine must be individualized to any given population because there are more than 80 types of M protein. Antibodies induced by other group A streptococcal products are not protective and play no role in immunity.

6. **Laboratory diagnosis** of infection by group A streptococci requires detection of type-specific antibodies or the isolation and identification of *S. pyogenes*.
   a. More than 90% of gram-positive cocci that are in chains, that produce β hemolysis on blood agar, and that are inhibited by bacitracin are *S. pyogenes*. These criteria are used routinely in the clinical laboratory for the presumptive identification of these bacteria. Additional biochemical tests such as hippurate hydrolysis are used to confirm the identification.
   b. Antecedent group A streptococcal infections may be determined by detection of antibodies in the patient's serum. Antibodies against streptolysin O, hyaluronidase, streptokinase, and DNase B may be measured routinely. Serum titers of antistreptolysin O (ASO) are measured most frequently, and a twofold or greater rise in titer strongly indicates recent infection caused by group A streptococci.

7. **Epidemiology**
   a. Group A streptococci are the cause of pharyngitis and tonsillitis, most frequently in children between 5 and 15 years of age. The bacteria are spread by droplets from the upper respiratory tract and by fomites; isolates from dust and environmental surfaces usually are noninfectious.
   b. Asymptomatic carriers of group A streptococci in the upper respiratory tract and on the skin often are responsible for spread of infection. Mild and asymptomatic infections also occur in the general population, and most children have been exposed by 10 years of age.

B. **Beta-hemolytic streptococci** (exclusive of group A) that are pathogenic for humans include species in groups B, C, G, and D.

1. **Lancefield group B streptococci**, a classification based on the carbohydrate constituents of the cell wall, includes *S. agalactiae*.
   a. **Antigenic composition.** *S. agalactiae* also has type-specific capsular and protein antigens that are used to divide the bacteria into five **serotypes**: Ia, Ib, Ic, II, and III.

b. Group B streptococcal disease. *S. agalactiae* recently has been noted as a human pathogen. It has been found that type III frequently causes **neonatal meningitis**, and type Ia is the cause of **neonatal pneumonia**. Adult infections are sporadic and show no correlation with any particular serotype of bacteria.

c. Immunity to group B streptococci is induced by production of antibody against the type-specific antigens. Transplacental passage of maternal antibodies to capsules of type III protects the neonate against meningitis. Antibodies against type Ia antigens are opsonins and are not passed transplacentally.

d. Laboratory identification of group B streptococci is based on biochemical tests. *S. agalactiae* is differentiated from *S. pyogenes* on the basis of the CAMP test. This test identifies a protein produced by *S. agalactiae* that interacts with α-hemolysin of *S. aureus* to cause complete lysis of sheep erythrocytes.

e. Epidemiology. *S. agalactiae* frequently has been isolated from the nasopharynx of healthy individuals. Group B streptococci are part of the indigenous gastrointestinal flora of about 30% of the general population, and they are part of the indigenous urogenital flora of about 30% of women.

2. Lancefield group D streptococci include *S. faecalis* and *Streptococcus faecium* (i.e., the enterococci), *Streptococcus durans, Streptococcus bovis,* and *Streptococcus equinus.*

a. Antigenic composition. The **group-specific antigen** of group D is unique among the streptococci. It is a **glycerol teichoic acid** containing D-alanine and glucose. It probably is located in the region of the cell wall, with no primary link or bond to cell wall components.

b. Epidemiology. *S. faecalis* is a common inhabitant of the human gastrointestinal tract and is principally an opportunistic pathogen or a secondary invader of damaged tissue. The enterococci occasionally cause urinary tract infections, particularly in debilitated or immunosuppressed patients.

c. Metabolism. The **enterococci** differ from most other streptococci in that they **are not nutritionally fastidious** but grow on most routine bacteriologic media. The enterococci may produce either α or β hemolysis on blood agar. They hydrolyze esculin in the presence of 40% bile, and they can grow in 6.5% sodium chloride broth.

3. Groups C and G streptococci include a variety of species, such as *Streptococcus equisimilis, Streptococcus zooepidemicus, Streptococcus equi,* and *Streptococcus dysgalactiae.*

a. Virulence factors. Streptolysin O is produced by both group C and group G streptococci, but the significance of this hemolysin in pathogenesis is unknown. Group C streptococci produce a **hyaluronic acid capsule** that is antiphagocytic.

b. Epidemiology. Group C and group G streptococci are found in the nasopharynx of many apparently healthy individuals. Infections caused by these groups do not result in either rheumatic fever or acute glomerulonephritis.

C. Viridans streptococci are a poorly defined group of α-hemolytic organisms that do not have group-specific cell wall antigens and, therefore, do not belong to any Lancefield group. The most familiar species are *S. salivarius, Streptococcus sanguis, S. mitis,* and *S. mutans.*

1. Morphology and metabolism. The viridans streptococci usually occur in pairs or short chains. They are **fastidious** in their **nutritional requirements**, and growth media must be rich in proteins. **Colonies are small** and are surrounded by a narrow zone of α hemolysis on blood agar. Some species produce large quantities of extracellular carbohydrate polymers such as dextran. The viridans streptococci are **microaerophilic** and can be identified by their carbohydrate fermentation patterns.

2. Pathogenesis. Two major types of human disease are caused by the viridans streptococci. These pathogenic processes can be correlated with the natural habitat of these bacteria—the mouth and pharynx.

a. Dental caries has been associated with the presence of *S. mutans, S. salivarius, S. sanguis,* and *S. mitis* in humans, rats, and monkeys. *S. mutans* appears to be the most virulent species, and its pathogenicity seems to be due to its ability to adhere to the tooth surface. It produces large amounts of dextran from sucrose but not from other carbohydrates, resulting in large plaque formations.

b. Bacterial endocarditis results from the ability of most strains of viridans streptococci to adhere to human heart valves. Approximately 50% of all cases of bacterial endocarditis are caused by the viridans streptococci.

D. Pneumococci (i.e., *S. pneumoniae*) are α-hemolytic streptococci that cause more than 80% of all cases of **bacterial pneumonia** as well as numerous cases of meningitis, otitis media, and septicemia.

1. **Morphology and metabolism**
   a. **Cellular morphology.** *S. pneumoniae* classically is an ovoid or lancet-shaped, gram-positive coccus that occurs in pairs, although short chains may be present in clinical material or in culture (Figure 8-4).
   b. **Colonies usually are small** (i.e., about 1 mm in diameter), circular, raised, and smooth. As autolysis occurs in older cultures, the center of the colony sinks.
   c. Optimal **growth occurs on infusion media** supplemented with serum or blood under reducing conditions. A zone of incomplete hemolysis is present around colonies growing on blood agar.
   d. **Virulent pneumococci** possess a large **polysaccharide capsule**, which gives the colonies a mucoid appearance; these isolates are designated **smooth forms**. Nonencapsulated pneumococci form rough colonies and are avirulent.

2. **Antigenic composition.** The principal antigenic component of *S. pneumoniae* is the polysaccharide capsule. The cell wall consists of a peptidoglycan that is similar in structure to that of group A streptococci.
   a. **Antigenic typing** of the pneumococci is **based on the chemical variation of the capsule**. There are at least 82 recognized capsular types of pneumococci, and these capsules are chemically and immunologically distinct.
   b. **Capsular polysaccharides are immunogenic in humans** and mice but not in horses and rabbits. The injection of 30–60 μg of pneumococcal polysaccharide into adult humans induces the production of protective antibodies.
   c. **Immunologic cross-reactivity** occurs between some pneumococcal capsular surface polysaccharides and other α-hemolytic streptococci, the capsular polysaccharides of *Klebsiella* and *Salmonella* species, and blood group substances. Specific antibodies to pneumococcal polysaccharides cause capsular swelling (i.e., the quellung reaction).

3. **Pathogenesis.** The **capsule** of *S. pneumoniae* appears to be the **only significant virulence factor**; it is responsible for pneumococcal virulence in humans.
   a. The **capsule prevents phagocytosis** by both neutrophils and macrophages. Specific antibodies against the capsule (opsonins) enhance the phagocytosis of pneumococci. Complement may act as an opsonin for some capsular types.
   b. **Solubilized capsular polysaccharide** is resistant to degradation in the host and may overwhelm the mononuclear phagocyte system, resulting in **immune paralysis**. The solubilized polysaccharide also binds to circulating antibodies, nullifying host resistance.

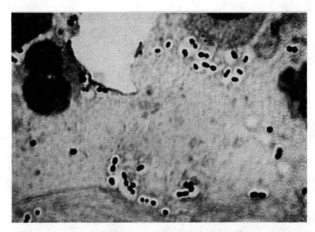

Figure 8-4. *Streptococcus pneumoniae* appears in pairs, and the individual cells are ovoid or lancet-shaped. The pneumococci are a classic example of an extracellular bacterial pathogen, and the typical morphology is seen best in clinical specimens. The clear area surrounding the pairs of cells is the large polysaccharide capsule, which is antiphagocytic.

    c. Surface phagocytosis occurs when pneumococci are trapped by phagocytes on a rough surface in the absence of opsonins. It is an in vitro phenomenon, and its role in the pathogenesis of pneumococcal infections is not clear.

    d. Neuraminidase and pneumolysin (i.e., α hemolysin) may play a role in the pathogenesis of pneumococcal disease.

        (1) Neuraminidase produced by *S. pneumoniae* cleaves terminal *N*-acetylmuramic acid from glycoproteins of the host cell membranes and may account for the neurotoxicity seen in pneumococcal meningitis.

        (2) Pneumolysin produces toxic effects on erythrocytes when it is injected into laboratory animals.

4. Immunity to pneumococcal infection is dependent upon the development of specific antibodies to the capsular polysaccharide. Although there are more than 80 capsular types, more than 80% of human infections are caused by 14 types, which have been incorporated into a polyvalent pneumococcal vaccine. The vaccine is recommended for the elderly and others who are at risk for developing pneumococcal infection. The vaccine is not effective in children under 2 years of age.

5. Laboratory diagnosis of *S. pneumoniae* is confirmed by **biochemical tests performed on a pure culture**. The nutritional requirements are complex but are met by infusion agar supplemented with blood.

    a. Isolation of pneumococci from clinical specimens is accomplished on blood agar; examination of colonies is facilitated by use of a dissecting microscope. The white mouse is susceptible to most types of pneumococci; injecting it intraperitoneally with sputum results in a septicemia and the isolation of a pure culture of *S. pneumoniae* from heart blood.

    b. Fermentation of inulin, producing acid, is considered diagnostic for *S. pneumoniae* in conjunction with microscopic and colonial morphology. Lactose, sucrose, raffinose, and trehalose also are fermented.

    c. Rapid presumptive identification is based on typical microscopic and colonial morphology, alpha hemolysis on blood agar, and susceptibility to optochin as indicated by any zone of growth inhibition surrounding an impregnated disk. An additional rapid test is based on the solubility of *S. pneumoniae* in a 40% solution of bile salt. The bile activates the autolytic system of the pneumococci. Bile solubility often is determined along with hydrolysis of esculin in bile-esculin agar.

    d. The quellung reaction, using specific antibodies (i.e., antiserum) against the capsule, can be used to identify and type pneumococci. Antiserum containing antibodies to all 82 capsular types is available for rapid identification, and monovalent antisera are used for typing individual isolates.

6. Epidemiology. Humans are the only known reservoir for *S. pneumoniae,* which is spread by droplets from infected individuals. Asymptomatic carriage of pneumococci is common among preschool-aged children, and there is a relatively high incidence of pneumococcal pneumonia among adults who routinely come into contact with children. Spread of pneumococcal infections by fomites has not been established.

III. ANAEROBIC GRAM-POSITIVE COCCI include *Peptococcus* species, *Peptostreptococcus* species, anaerobic *Streptococcus* species, and *Sarcina* species. The difficulty of classifying these anaerobic cocci taxonomically has led some investigators to doubt their pathogenicity for humans.

    A. Morphology. The anaerobic gram-positive cocci **decolorize easily** and commonly are **seen as gram negative.** Cells within a pure culture may be of different sizes, and sometimes the size differential is great enough to give the **impression of budding**.

    1. *Peptostreptococcus* species have a tendency to **elongate**, forming short rods. This feature is particularly common in *Peptostreptococcus productus.*

    2. *Sarcina ventriculis* is distinctive among the gram-positive anaerobic cocci. Cells measure 2 μm in diameter, have flattened adjacent sides, and are typically seen in packages of eight or more cells.

B. Pathogenesis. The isolation of gram-positive anaerobic cocci in pure culture from certain disease processes is the best evidence for their pathogenicity. The mechanism of pathogenicity of these bacteria is unknown.

1. Anaerobic gram-positive cocci have been isolated in pure culture from brain and liver abscesses, meningitis, extradural and subdural empyemas, osteomyelitis, breast abscesses, and other infectious processes.

2. Bones, joints, soft tissue, and vascular tissue are common sites of infection, and disease usually has been associated with surgery or prosthetic devices. The occurrence of infection at these sites indicates that natural defense mechanisms normally are sufficient to prevent disease caused by anaerobic cocci.

C. Epidemiology. Peptococci, peptostreptococci, and anaerobic streptococci are members of the human indigenous flora of the gastrointestinal tract, the oropharynx, the urogenital tract, and the skin. Sarcinae are found in soil and mud and are only transient members of the indigenous flora of humans.

STUDY QUESTIONS

Directions: Each question below contains five suggested answers. Choose the **one best** response to each question.

1. The single most important laboratory test for determining the virulence of staphylococci is

(A) mannitol fermentation
(B) hemolysis of sheep erythrocytes
(C) detection of coagulase
(D) the catalase test
(E) detection of penicillinase

2. Neonatal meningitis that is acquired in the birth canal is often caused by

(A) *Staphylococcus epidermidis*
(B) *Staphylococcus aureus*
(C) *Streptococcus pyogenes*
(D) *Streptococcus agalactiae*
(E) *Streptococcus pneumoniae*

3. Pneumococcal infections are cleared primarily by

(A) the humoral immune system
(B) the activation of complement
(C) the use of effective chemotherapy
(D) the cell-mediated immune system
(E) nonspecific immune mechanisms

Directions: Each question below contains four suggested answers of which **one or more** is correct. Choose the answer

A if **1, 2, and 3** are correct
B if **1 and 3** are correct
C if **2 and 4** are correct
D if **4** is correct
E if **1, 2, 3, and 4** are correct

4. Exfoliatin produced by some strains of *Staphylococcus aureus* causes

(1) a scarlatina-like rash
(2) Ritter's disease
(3) Lyell's disease
(4) scalded baby syndrome

5. True statements regarding acute endocarditis caused by *Staphylococcus aureus* include which of the following?

(1) The infection produces a positive ASO titer test
(2) Diagnosis is made by analyzing serum for ribitol teichoic acid
(3) The disease has a prolonged course lasting more than 6 weeks
(4) Acute endocarditis usually results in perforation of the infected heart valve

6. Laboratory test results used to identify *Streptococcus pyogenes* with more than 90% confidence include

(1) the presence of gram-positive cocci in chains
(2) susceptibility to bacitracin
(3) β hemolysis on blood agar
(4) a positive bile-esculin test

7. The enterococci differ from most other streptococci in that they

(1) grow in the presence of 6.5% sodium chloride
(2) are not nutritionally fastidious
(3) hydrolyze esculin in the presence of 40% bile
(4) are catalase positive

8. Effective human vaccines have been produced against

(1) *Staphylococcus aureus*
(2) *Streptococcus pyogenes*
(3) *Streptococcus agalactiae*
(4) *Streptococcus pneumoniae*

Directions: The group of questions below consists of lettered choices followed by several numbered items. For each numbered item, select the **one** lettered choice with which it is **most** closely associated. Each lettered choice may be used once, more than once, or not at all. Choose the answer

    A   if the item is associated with **(A) only**
    B   if the item is associated with **(B) only**
    C   if the item is associated with **both (A) and (B)**
    D   if the item is associated with **neither (A) nor (B)**

Questions 9–12

For each identifying characteristic listed below, select the correct bacterium or bacteria.

(A) *Staphylococcus aureus*
(B) *Staphylococcus epidermidis*
(C) Both
(D) Neither

9. Causes the complete hemolysis of sheep erythrocytes

10. Grows in 7.5% sodium chloride broth

11. Ferments mannitol

12. Has glycerol teichoic acid in its cell wall

ANSWERS AND EXPLANATIONS

1. The answer is C. [*I B 3 a*] Formerly, a coagulase-positive, gram-positive coccus in clusters was identified as *Staphylococcus aureus*. With the recognition of coagulase-negative *S. aureus*, the coagulase test is used to identify pathogenic staphylococci. The fermentation of mannitol is one of the biochemical tests employed in the differentiation of *S. aureus* from other staphylococci. Some strains of *S. aureus* are hemolytic, but hemolysis is a variable characteristic. All staphylococci are catalase positive. Penicillinase is an extracellular enzyme produced by some, but not all, virulent strains of staphylococci.

2. The answer is D. [*II B 1 a–c*] Group B β-hemolytic *Streptococcus agalactiae*, particularly type III, has been recognized as the etiologic agent of neonatal meningitis (neonatal pneumonia is caused by type Ia); the bacteria are passed from the mother to the child as it passes through the birth canal. Approximately 30% of healthy adult women have *S. agalactiae* in their urogenital tract. All species of staphylococci and streptococci are potential causes of meningitis in the neonate under the right circumstances. *Streptococcus pneumoniae* is an important etiologic agent of meningitis in older infants and children.

3. The answer is D. [*II D*] Pneumococcal disease is one of the classic examples of an extracellular infection. Once the bacteria are phagocytized, they are killed within the phagosome. Phagocytosis is enhanced by the presence of opsonins (i.e., antibodies against the capsule), and the use of an effective antimicrobial agent can reduce the number of pneumococci. However, even in the presence of effective antimicrobial therapy, the mortality rate in pneumococcal disease is approximately 10%, owing to insufficient phagocytic activity. Certain components of the complement system may act as opsonins against the polysaccharide capsule.

4. The answer is E (all). [*I B 3 e, E 2 b*] Many strains of *Staphylococcus aureus* that belong to phage group II produce exfoliatin, a toxin that can cleave the stratum granulosum layer of the epidermis. A scarlatina-like rash may develop in older children and adults as the result of exfoliatin. The disorder mimics scarlet fever caused by streptococci except that the tongue and palate are not involved. Infants exposed to exfoliatin-producing strains of *S. aureus* can develop a potentially serious disease known as Ritter's disease, or scalded baby syndrome. The loss of fluids and electrolytes or secondary bacterial infection can cause death. Older children and adults have a less serious form of the disease known as Lyell's disease, or scalded skin syndrome.

5. The answer is C (2, 4). [*I E 1 g, G 2; II A 6 b*] Acute endocarditis caused by *Staphylococcus aureus* is characterized by extensive vegetation on the heart valve and perforation by the many tissue-destroying enzymes and extracellular products of the bacteria. Acute endocarditis typically has a time course of less than 6 weeks, and surgical replacement of the heart valve is necessary. Rapid diagnosis of acute endocarditis caused by *S. aureus* can be made by the electrophoretic analysis of the patient's serum for ribitol teichoic acid, in contrast to glycerol teichoic acid from *S. epidermidis*, which causes subacute endocarditis. The ASO (antistreptolysin O) titer indicates prior infection with group A β-hemolytic streptococci.

6. The answer is A (1, 2, 3). [*II A 6 a*] In more than 90% of cases, the finding of gram-positive cocci that are in chains, that show complete (i.e., β) hemolysis on blood agar, and that are susceptible to bacitracin correlates with a positive identification of *Streptococcus pyogenes*. The lack of total assurance of the identification is due primarily to strains of group B *Streptococcus agalactiae*. Strains of *S. agalactiae* that are β-hemolytic and susceptible to bacitracin can be differentiated from *S. pyogenes* by the CAMP test; the bile-esculin test determines bile solubility and the hydrolysis of esculin. These two tests are used in the identification of *Streptococcus pneumoniae*.

7. The answer is A (1, 2, 3). [*II B 2 c*] The Lancefield group D enterococci include *Streptococcus faecalis* and *Streptococcus faecium*. These enterococci are normal inhabitants of the gastrointestinal tract, and they differ from the more pathogenic streptococci. The enterococci are not nutritionally fastidious, growing on routine bacteriologic media without the addition of blood or serum components. They can grow in the presence of 6.5% sodium chloride, and autolytic enzymes are not activated by bile. They grow in 40% bile and hydrolyze esculin. These characteristics are used in the laboratory to differentiate the enterococci from other streptococci and from staphylococci. Like all species of streptococci, the enterococci are catalase negative.

8. The answer is D (4). [*II D 4*] Among the gram-positive cocci, only the capsule of *Streptococcus pneumoniae* has yielded an effective vaccine that prevents human disease. There have been many unsuccessful attempts to produce protective vaccines against other staphylococci and streptococcal infections. Recently, with the advent of new biotechnology techniques, there has been a renewed impetus to produce a vaccine from the cellular components of *Streptococcus pyogenes* and *Streptococcus agalactiae*. The high degree of variability in *Staphylococcus aureus* and the lack of antigens that appear to produce protective antibodies have prevented production of a vaccine against *S. aureus* infection.

9–12. The answers are: 9-A, 10-C, 11-A, 12-B. [*I A 2 a, C, G 1, 2*] *Staphylococcus aureus* causes the complete hemolysis of sheep erythrocytes as the result of the production of α toxin, although hemolysis is a variable characteristic. *Staphylococcus epidermidis* does not cause hemolysis of sheep erythrocytes.

The ability to grow in 7.5% sodium chloride broth is a basic characteristic of the genus *Staphylococcus*.

Mannitol fermentation is one of the initial tests, along with coagulase, used in differentiating *S. aureus* from *S. epidermidis*.

Rapid differentiation of *S. aureus* and *S. epidermidis* can be accomplished by the electrophoretic analysis of cell wall teichoic acids. *S. epidermidis* contains glycerol teichoic acid, and *S. aureus* contains ribitol teichoic acid in its cell wall.

9

Gram-Negative Cocci:
Neisseria and Related
Species

Gerald E. Wagner

I. GENERAL CHARACTERISTICS OF NEISSERIAE. The genus *Neisseria* includes two species that are pathogenic for humans: *Neisseria gonorrhoeae* (the gonococci) and *Neisseria meningitidis* (the meningococci). Nonpathogenic species include *Neisseria sicca*, *Neisseria flavescens*, *Neisseria perflava*, *Neisseria mucosa*, and *Neisseria lactamica*; these species may be members of the indigenous flora, and they can be confused with the pathogenic species upon initial isolation.

A. Morphology

1. The neisseriae are gram-negative cocci ranging from 0.6 to 1.0 μm in diameter. The cell wall is similar in structure and composition to that of other gram-negative bacteria.

2. **Microscopic features**
   a. Neisseriae generally are seen as **pairs** with their adjacent sides somewhat flattened (Figure 9-1). Tetrads, short chains, and clusters of cells occasionally are observed, all showing characteristic pairing.
   b. **Capsules** are present on most freshly isolated strains of both *N. gonorrhoeae* and *N. meningitidis.*
   c. **Piliated cells** may demonstrate a **twitching** motility at certain times. The cells do not have flagella.

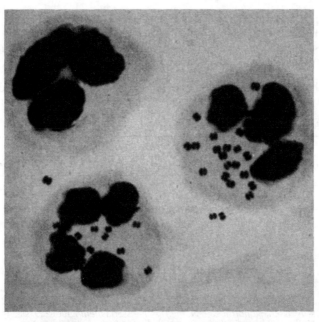

Figure 9-1. The microscopic morphology of *Neisseria gonorrhoeae* usually is observed as gram-negative, intracellular cocci in pairs. The individual cells typically are described as kidney shaped. The viability of phagocytized *Neisseria* species is not certain.

B. Metabolic activity. Pathogenic *Neisseria* species are nutritionally fastidious bacteria with specialized growth requirements, particularly upon initial isolation from clinical specimens. Nonpathogenic species are less fastidious.

    1. Oxygen. *Neisseria* species are aerobic and contain a highly active cytochrome c oxidase system. This characteristic is important in the initial differentiation of the genus. Although aerobic, most strains of *N. gonorrhoeae* require 4%–8% carbon dioxide in the atmosphere in order to grow when initially isolated. This requirement frequently is lost with continuous subculturing of the isolate.

    2. Carbohydrates are used in oxidative pathways to produce acetic acid in a pattern that is characteristic of each species (Table 9-1). *N. gonorrhoeae* and *N. meningitidis* use a combination of the Entner-Doudoroff pathway and the pentose-phosphate shunt to dissimilate glucose.

    3. The tricarboxylic acid cycle is functional in both gonococci and meningococci. The effectiveness of this pathway in producing biosynthetic precursors and adenosine triphosphate (ATP) is influenced greatly by growth conditions.

    4. Hemolyzed blood, increased atmospheric carbon dioxide, and solubilized starch enhance the growth of *N. gonorrhoeae* and *N. meningitidis*. Free fatty acids occurring in agar are inhibitory to the growth of these species, and the addition of soluble starch, serum, or charcoal to the medium detoxifies the fatty acids.

    5. Optimal growth temperature is 36°–39° C. The pathogenic neisseriae can grow at temperatures between 24° C and 41° C, and nonpathogenic species can grow at temperatures below 24° C. This difference is useful in separating pathogenic species from most nonpathogenic species.

    6. Optimal pH for growth varies from strain to strain, but many strains grow in medium that is buffered to a pH of 6–8.

    7. Commercial media for the isolation of clinically significant strains of *Neisseria* species have been formulated on the basis of the metabolic activity of the bacteria. Modified Thayer-Martin and TransGrow media are two of the most commonly used. In addition to chocolate agar with solubilized starch and an atmosphere of increased carbon dioxide, they include selected antibacterial and antifungal compounds to inhibit the growth of indigenous flora.

II. *N. GONORRHOEAE* is the etiologic agent of **gonorrhea**, the most frequently diagnosed venereal disease in the United States and western Europe. Gonococci are a frequent cause of **pelvic inflammatory disease (PID)** and sterility in women. Dissemination from simple urethritis may result in joint infection, meningitis, and septicemia, leading to infection of any organ system.

    A. Colony morphology. Four colonial types of gonococci have been recognized and correlated with virulence. Types 1 and 2 are virulent for humans; types 3 and 4 are less virulent.

Table 9-1. Acid Production from Carbohydrates by *Neisseria* Species

Carbohydrate	Neisseria Species						
	N. gon.	*N. men.*	*N. sic.*	*N. lac.*	*N. fla.*	*N. sub.*	*N. muc.*
Glucose	+	+	+	+	–	+	+
Maltose	–	+	+	+	–	+	+
Lactose	–	–	–	+	–	–	–
Sucrose	–	–	v	–	–	v	+
Fructose	–	–	+	–	–	v	+
Starch	–	–	+	–	–	v	+

Abbreviations—*N. gon.* = *Neisseria gonorrhoeae*; *N. men.* = *Neisseria meningitidis*; *N. sic.* = *Neisseria sicca*; *N. lac.* = *Neisseria lactamica*; *N. fla.* = *Neisseria flavescens*; *N. sub.* = *Neisseria subflava*; *N. muc.* = *Neisseria mucosa*.
+ = > 90% of strains are positive.
– = > 90% of strains are negative.
v = variability (i.e., inconsistency) exists within a strain.

1. Freshly isolated strains are either type 1 or type 2 and are piliated; their virulence can be maintained by selective, daily subcultures.

2. Nonselective transfers usually result in the replacement of types 1 and 2 by the less virulent, nonpiliated types, which form larger colonies.

B. Antigenic composition of *N. gonorrhoeae* **varies with continuous subculturing**. It is suspected that some gonococcal antigens are expressed only in response to certain environmental conditions, such as those that occur in vivo.

1. Capsules have been demonstrated on freshly isolated gonococci by reactivity with hyperimmune serum and by electron microscopy.
   a. The capsule is influenced by nutritive and environmental conditions.
   b. It is loosely associated with the cell wall, and its chemical structure is not known completely.
   c. The capsule is **antiphagocytic**, and opsonins produced against it promote phagocytosis.

2. Pili from *N. gonorrhoeae* have been purified and have been found to consist mainly of protein with some sugar and phosphate residues.
   a. Pili appear to play a critical role in the attachment of gonococci to host cells; purified pili bind to the greatest extent to cells resembling those found in naturally infected sites.
   b. Although avirulent strains of gonococci lack pili, other nonpathogenic species in the genus *Neisseria* are piliated.

3. Lipopolysaccharides of the gonococci appear to vary in composition depending upon the colonial type and the environmental conditions. The core region of gonococcal lipopolysaccharides is antigenic, and antibodies produced against it are bactericidal.

4. Outer membrane proteins are the basis for defining 16 gonococcal serotypes. The major outer membrane protein of a serotype accounts for about 66% of the total protein. Antibodies against these proteins are bactericidal in the presence of complement.

C. Genetic character of the gonococci is somewhat unique in that the species readily undergoes **transformation** and, unlike most bacteria, is competent for transformation throughout the growth cycle. Genetic transfer also is accomplished by **conjugation**, but transduction has not been described in the gonococci.

1. A **cryptic plasmid** with a molecular weight of 2.4×10^6 daltons is present in most strains of *N. gonorrhoeae*. No function has been associated with it.

2. A **plasmid that promotes** the **conjugal transfer** of genes and resistance (R) factors has been detected in some gonococcal strains. This sex-factor plasmid has a molecular weight of about 2.4×10^7 daltons.

3. A **plasmid associated with β-lactamase production** is either a 3.2×10^6- or a 4.4×10^6-dalton plasmid. Loss of this plasmid results in loss of penicillin resistance.

4. Competence for transformation at any time in the growth cycle is characteristic of the gonococci; it has been demonstrated in all strains of this bacterium that have been tested. **Transformation frequencies for types 1 and 2 colonies are about 1000 times higher than those for types 3 and 4 colonies.**

5. Conjugation has been described in the gonococci, but only those strains containing the 2.4×10^7-dalton plasmid serve as donors. Piliation bears no relationship to the ability of a strain to serve as a donor.

D. Pathogenesis of gonorrhea involves both bacterial and host factors, some of which are not fully understood. Exposure to *N. gonorrhoeae* does not always result in disease; whether the critical factor is variability in virulence, the size of the inoculum, specific host immunity, or nonspecific host resistance is unknown.

1. Bacterial factors
   a. An **extracellular enzyme** produced by some gonococci can cleave a proline-threonine bond in the heavy chain of immunoglobulin A (IgA). Cleavage inactivates the antimicrobial property of this secretory immunoglobulin.

b. It is debated whether gonococci are capable of **replication within neutrophils**. Gonococci are seen both inside and on neutrophils in smears of clinical material. It appears that most of the phagocytized bacteria are killed, but a few may survive and replicate.

c. **Pili** allow gonococci to attach to receptors on epithelial cells and, as such, are critical to the bacteria's ability to cause infection. In vitro studies indicate that piliated gonococci tend to remain extracellular, whereas nonpiliated cells are readily phagocytized and killed. Serum IgA and IgG antibodies to specific serotypes of gonococci enhance the association between the bacteria and neutrophils.

d. **β-Lactamase–producing gonococci contain an R plasmid**. A relatively low percentage of clinical isolates of *N. gonorrhoeae* produce β-lactamase.

e. **Strains causing disseminated disease** are very sensitive to penicillin, resistant to the bactericidal action of serum, and require arginine, uracil, and hypoxanthine for growth.

2. Host factors. Gender may be a factor in the pathogenesis of disseminated gonococcal disease, which can cause a variety of systemic syndromes (e.g., PID leading to sterility, ascending gonococcal arthritis, septicemia, meningitis).

   a. Women are more susceptible to disseminated disease because they often are asymptomatic (and, therefore, untreated) carriers of *N. gonorrhoeae*. Such carriers represent an important reservoir of gonococci.

   b. Men seldom have asymptomatic disease.

E. Immunity to gonorrhea involves local reactions at the mucosal surface of epithelial cells of the urogenital tract. It is not clear to what degree local immunity is stimulated by infection and whether local immune response protects against subsequent infection.

1. Isolates of *N. gonorrhoeae* from gonococcal urethritis frequently are resistant to the bactericidal action of the patient's serum.

   a. This resistance is lost on subculture unless the environmental conditions of the host's urethra are duplicated.

   b. Virulent colonial types are most resistant to serum bactericidal action and, when exposed to serum, can acquire total resistance.

2. Disseminated gonococcal infections induce antibody formation, but whether the antibodies are protective is unclear.

3. There currently is **no vaccine** to protect against gonorrhea or disseminated disease.

F. Laboratory diagnosis

1. Culture procedures differ depending on whether the isolated specimen is from a man or a woman.

   a. **Men.** Gram-stained smears of urethral exudate usually reveal gram-negative, intracellular cocci in pairs. This is strong presumptive evidence of gonorrhea.

   (1) Specimens for culture are obtained from the anterior urethra and are inoculated onto a selective medium such as modified Thayer-Martin agar. Cultures should be incubated in an atmosphere of increased carbon dioxide concentration.

   (2) Oxidase-positive colonies of gram-negative cocci in pairs are presumed to be *N. gonorrhoeae* in the appropriate clinical setting. Either staining by fluorescent antibody techniques or the production of acid from glucose but not maltose or sucrose confirms the identification.

   (3) Specimens for culture from homosexual men should be taken from the anal canal and pharynx as well as from the urethra. Identification of isolates proceeds in the same manner as that for heterosexual men.

   b. **Women.** Gram-stained smears of exudates are not useful for the diagnosis of gonorrhea in female patients.

   (1) Specimens for culture are taken from the cervix or cervical os and from the anal canal. (Asymptomatic carriage of *N. gonorrhoeae* frequently occurs in the anal canal of women and homosexual men.)

   (2) Specimens are plated on selective medium, and tests are performed on appropriate colonies.

2. Serologic tests have **not** proved **useful** for the diagnosis of gonorrhea in either sex.

III. *N. MENINGITIDIS* is the etiologic agent of meningococcal meningitis. This highly contagious disease usually occurs in epidemics and is associated with a high mortality rate (85%) when untreated.

A. **Antigenic composition.** The antigens of *N. meningitidis* are complex substances, some of which serve as virulence factors and some of which induce protective antibodies.

1. **Meningococci are grouped according to capsular polysaccharides.**
   a. The **serogroups** are designated A, B, C, X, Y, Z, 29e, L, and W-135 on the basis of highly specific anticapsular antibodies produced in rabbits. The capsular polysaccharides of some serogroups are immunogenic in humans.
   b. Serogroups B and C are subdivided into **serotypes** according to outer membrane proteins. Serotype 2 is the most common isolate from human meningococcal disease caused by either serogroup B or C. The type 2 antigens from both groups are chemically and serologically identical, and they induce antibodies that are bactericidal in the presence of complement.

2. **Lipopolysaccharide purified from meningococci is highly toxic.**
   a. Meningococcal lipopolysaccharide is analogous to lipopolysaccharide from enteric gram-negative rods in terms of lethality for laboratory animals; however, it elicits a dermal Schwartzman reaction at concentrations 10 times lower than those of lipopolysaccharide from enteric rods.
   b. Meningococcal lipopolysaccharide induces antibody formation, but the role of the antibodies in immunity to meningococcal disease is not established.

B. **Pathogenesis**

1. The ability of meningococci to cause disease appears to be due primarily to the antiphagocytic properties of the **capsule**. Specific antibodies produced against the capsular polysaccharides of the meningococci are bactericidal.

2. The toxic effects of meningococcal infection are attributed to **lipopolysaccharide**. A macular rash and petechial hemorrhages are common during bacteremic phases of the disease.

3. Meningococci are considered to be **extracellular parasites**. Although some organisms frequently are seen intracellularly in clinical specimens, they probably cannot multiply in neutrophils.

C. **Immunity**

1. Immunity is induced by the capsular polysaccharides of meningococci groups A and C. Highly protective **vaccines** are available for these two serogroups. Protection is provided by complement-mediated bactericidal antibodies, which are detectable for 5 years after vaccination.

2. Group B polysaccharide is only weakly immunogenic in humans and does not induce protective antibodies.

D. **Epidemiology.** Meningococcal disease occurs in epidemics among young adults, and outbreaks have occurred among military recruits. Overt disease is rare even though the bacteria may be prominent in the population.

1. **Reservoir.** The natural reservoir of meningococci is the human nasopharynx.

2. **Carriage.** About 3%–30% of normal, healthy persons are carriers at any given time.
   a. The carrier state provides the reservoir for infection as well as enhances the carrier's immunity.
   b. During epidemics, the carrier rate reaches 95%, but fewer than 1% of infected persons develop systemic disease.

3. **Incidence of disease**
   a. **Attack rates** and involved serogroups of meningococci vary with the age of the infected person, suggesting that host immunity may be an important factor in development of disseminated disease.
      (1) Carriers of meningococci usually are older than 21, yet the highest rate of disease is among children.

(2) Group B is the most common etiologic agent of systemic disease in children younger than 5 years, and group C is prevalent in children 4–14 years of age.

b. Sporadic meningococcal disease occurs in winter and spring. The frequency of disease is declining in the United States and Great Britain but is increasing in other countries, such as Finland and Brazil. Sporadic cases usually are caused by groups B, C, and Y; groups B and C appear to fluctuate in predominance.

c. Epidemics occur occasionally in adults and usually are caused by group A. When group A strains become predominant in carriers, cases of meningitis increase sharply. Groups B and C may cause small outbreaks.

E. Laboratory diagnosis of meningococcal disease is based upon cultural isolation and biochemical identification of *N. meningitidis*.

1. Specimens for gram-staining and culture include **cerebrospinal fluid (CSF), blood, and nasopharyngeal swabs** taken deep behind the soft palate. Specimens preferably are obtained before initiation of antimicrobial therapy.

a. Smears of blood or smears from the nasopharynx revealing **gram-negative cocci in pairs provide presumptive identification** of *N. meningitidis* in clinically suspected cases. Gram-negative cocci in pairs seen on smears of CSF usually are diagnostic of meningococcal meningitis; meningococci, however, often are difficult to find in CSF smears.

b. Specimens for culture are inoculated onto a selective medium such as modified Thayer-Martin agar. The cultures are incubated at 37° C in either a candle jar or a carbon dioxide incubator.

(1) Selected appropriate colonies with a **positive oxidase test** are identified presumptively as *Neisseria* species.

(2) The production of acid from the oxidation of glucose and maltose, but not lactose, sucrose, or fructose, confirms the identification of *N. meningitidis*.

(3) The serogroup to which the isolate belongs is determined by slide agglutination tests using specific antisera.

IV. *BRANHAMELLA CATARRHALIS*, formerly classified as *Neisseria catarrhalis*, is closely related morphologically and metabolically to the nonpathogenic neisseriae, but DNA base content studies have placed it in a separate genus. *B. catarrhalis* is a member of the human indigenous flora. However, it has been **a rare but significant cause of severe meningitis and endocarditis** and has been associated with cases of otitis media, maxillary sinusitis, pulmonary disease, and a urethritis indistinguishable from gonorrhea.

STUDY QUESTIONS

Directions: Each question below contains five suggested answers. Choose the **one best** response to each question.

1. Infection of urethral epithelium by *Neisseria gonorrhoeae* primarily is dependent upon

(A) production of an immunoglobulin-cleaving enzyme

(B) the antiphagocytic properties of a polysaccharide capsule

(C) the intracellular survival and multiplication of the bacteria

(D) the ability of piliated bacteria to attach to epithelial cell receptors

(E) resistance of the bacteria to the bactericidal action of serum

2. Genes that mediate the conjugal transfer of genetic information coding for the production of penicillinase by *Neisseria gonorrhoeae* are located on the

(A) genome

(B) cryptic plasmid

(C) sex-factor plasmid

(D) large plasmid

(E) β-prophage genome

3. The primary virulence factor of meningococci appears to be the

(A) lipopolysaccharide endotoxin

(B) antiphagocytic capsule

(C) outer membrane protein

(D) ability for intracellular survival

(E) pili responsible for attachment

4. The macular rash and petechial hemorrhages associated with meningococcal disease appear to be caused by

(A) capsular polysaccharide

(B) a protein exotoxin

(C) lipopolysaccharide

(D) outer membrane protein

(E) hyaluronidase

5. Meningococcal meningitis in children younger than 5 years most often is caused by strains belonging to

(A) serogroup A

(B) serogroup B

(C) serogroup C

(D) serogroup X

(E) serogroup W-135

6. Which of the following gram-negative cocci can cause urethritis that is indistinguishable from gonorrhea?

(A) *Branhamella catarrhalis*

(B) *Neisseria meningitidis* serotype 2

(C) *Neisseria sicca*

(D) *Branhamella flavescens*

(E) *Neisseria subflava*

Directions: Each question below contains four suggested answers of which **one or more** is correct. Choose the answer

A if **1, 2, and 3** are correct
B if **1 and 3** are correct
C if **2 and 4** are correct
D if **4** is correct
E if **1, 2, 3, and 4** are correct

7. Gonococci that cause disseminated disease are characterized by

(1) resistance to the bactericidal action of serum
(2) the potential to cause sterility in women
(3) the need for arginine, uracil, and hypoxanthine in order to grow
(4) the potential to cause ascending arthritis

8. Culture sites in a homosexual man suspected of having gonorrhea normally would include the

(1) urethra
(2) pharynx
(3) anal canal
(4) peripheral blood

9. A protective human vaccine available for prevention of meningitis is produced from the polysaccharide capsule of

(1) serogroup A meningococci
(2) serogroup B meningococci
(3) serogroup C meningococci
(4) serogroup Y meningococci

Directions: The group of questions below consists of lettered choices followed by several numbered items. For each numbered item, select the **one** lettered choice with which it is **most** closely associated. Each lettered choice may be used once, more than once, or not at all. Choose the answer

A if the item is associated with **(A) only**
B if the item is associated with **(B) only**
C if the item is associated with **both (A) and (B)**
D if the item is associated with **neither (A) nor (B)** ·

Questions 10–14

Match each property of microorganisms with the gram-negative cocci that demonstrate the feature.

(A) *Neisseria gonorrhoeae*
(B) *Neisseria meningitidis*
(C) Both
(D) Neither

10. Ferment glucose

11. Possess pili for twitching mobility

12. May be spread by asymptomatic carriers

13. Divided into serogroups on the basis of capsular polysaccharide

14. Engage in transformation throughout the growth cycle

ANSWERS AND EXPLANATIONS

1. The answer is D. [*II B 2, D 1 c*] Pili play a critical role in the virulence of *Neisseria gonorrhoeae* and are involved in the attachment of the gonococci to receptors on epithelial cells of the host. Attachment to epithelial cells is the initial step of the infectious process; without attachment the gonococci are relatively easily cleared by mechanical and immunologic means. Nonpiliated cells are phagocytized readily and killed. Avirulent strains of the gonococcus lack pili. Although other bacterial factors such as capsule, extracellular enzymes, intracellular survival, and resistance to serum bactericidal factors play a role in the pathogenesis of gonorrhea, attachment is a critically important process.

2. The answer is C. [*II C 1, 2, 3*] The sex-factor plasmid of *Neisseria gonorrhoeae*, with a molecular weight of approximately 2.4×10^7 daltons, promotes the conjugal transfer of genes that code for antibiotic resistance. β-Lactamase production has been associated with either a 3.2×10^6- or 4.4×10^6-dalton plasmid. The sex-factor plasmid mediates transfer of the resistance genes between conjugating gonococci. Penicillinase-producing strains of *N. gonorrhoeae* have been increasing in prevalence as the cause of disease in the United States. A smaller cryptic plasmid is found in most strains of *N. gonorrhoeae*, but no function has been associated with it. The β-prophage genome is present in lysogenic *Corynebacterium diphtheriae* and is responsible for production of diphtheria exotoxin.

3. The answer is B. [*III B 1*] The ability of *Neisseria meningitidis* to cause disease is dependent primarily upon the antiphagocytic polysaccharide capsule. Lipopolysaccharide endotoxin of *N. meningitidis* is relatively potent and appears to account for much of the clinical symptomatology of meningococcal disease, such as the rash and petechial hemorrhages. Outer membrane proteins are used as antigenic determinants of the serotype of serogroup B and C meningococci. The meningococci are considered extracellular pathogens and are rapidly killed once phagocytized. Pili are a critical factor in the infectious process of *Neisseria gonorrhoeae*.

4. The answer is C. [*III B 2*] The toxic effects of meningococcal disease appear to be due to the lipopolysaccharide endotoxin of *Neisseria meningitidis*. A macular rash and petechial hemorrhages are common during the bacteremic phase of disease. The capsular polysaccharide of *N. meningitidis* is relatively nontoxic to laboratory animals and apparently to humans; it is used as an effective vaccine against infection by serogroups A and C. The outer membrane proteins of serogroups B and C are used to serotype strains of *N. meningitidis*. No protein exotoxin or miscellaneous extracellular enzymes (e.g., hyaluronidase) have been identified as the cause of rash or petechial hemorrhages.

5. The answer is B. [*III D 3 a (2)*] In meningococcal disease, the attack rate and the serogroup of meningococci involved vary with the age of the patient. Serogroup B is the most common cause of systemic disease among children under 5 years of age. Serogroup C is prevalent among children between the ages of 4 and 14. The carriage of *N. meningitidis* is highest in young adults older than 21, and when serogroup A predominates, the number of cases of meningitis rises sharply. Other serogroups cause sporadic meningococcal disease.

6. The answer is A. [*IV*] *Branhamella catarrhalis* (formerly classified as *Neisseria catarrhalis*) is a member of the indigenous flora of humans, as are *Neisseria sicca*, *Neisseria flavescens*, and *Neisseria subflava*. *B. catarrhalis*, however, has been associated with rare but severe cases of meningitis and endocarditis. It also may cause a urethritis indistinguishable from gonorrhea and has resulted in misdiagnoses on the basis of clinical symptomatology and Gram stain.

7. The answer is E (all). [*II D 1 e, 2*] Disseminated gonococcal disease manifests as a variety of syndromes, including pelvic inflammatory disease in women, which may result in sterility, and a classic ascending arthritis. Women particularly are susceptible to disseminated disease because they frequently have asymptomatic and untreated urethritis. Gonococci causing disseminated disease are resistant to the natural bactericidal action of serum, and they require arginine, uracil, and hypoxanthine for growth.

8. The answer is A (1, 2, 3). [*II F 1 a (3)*] The regular sexual activity of a homosexual man indicates that culture specimens should be taken from other sites in addition to the urethra when gonorrhea is suspected. The pharynx and anal canal or anal mucosa are common sites of colonization. Unless disseminated gonococcal disease is included in the differential diagnosis, peripheral blood culture would not typically be performed.

9. The answer is B (1, 3). [*III C*] The polysaccharide capsules of meningococci serogroups A and C induce the formation of highly protective antibodies, and effective vaccines are available as prophylaxis for meningitis. Serogroup B, the most common cause of meningococcal disease in children under 5 years of age, does not induce the formation of protective antibodies against the capsule. Because serogroup Y is rare as a cause of meningitis, a vaccine is unnecessary for routine use.

10–14. The answers are: 10-D, 11-C, 12-C, 13-B, 14-A. [*I A 2 c, B 2; II B 4, D 2 a; III A 1 b, D 2 a; Table 9-1*] *Neisseria* species use carbohydrates by oxidative pathways to produce acetic acid. A combination of the Entner-Doudoroff pathway and the pentose-phosphate shunt is used by *Neisseria gonorrhoeae* and *Neisseria meningitidis* to dissimilate glucose.

Both *N. gonorrhoeae* and *N. meningitidis* have pili that provide a twitching motility at certain times. The neisseriae do not have flagella.

Both gonococcal and meningococcal disease can be spread by asymptomatic carriers of the bacteria. Women are considered an important reservoir of *N. gonorrhoeae*, because they may have asymptomatic infections of the cervix and anal mucosa. The carriage rate of *N. meningitidis* in asymptomatic, healthy persons ranges from 3% to 30% of the population at any given time.

N. meningitidis is divided into serogroups on the basis of capsular polysaccharide, and serogroup B and C are further subdivided into serotypes on the basis of outer membrane proteins. *N. gonorrhoeae* is genetically unique among bacteria in that it is competent for transformation at any time during the growth cycle.

10
Spore-Forming Gram-Positive Rods: Bacilli and Clostridia

Gerald E. Wagner

I. BACILLI. The genus *Bacillus* has 48 recognized species. They are large, gram-positive rods characterized by the ability to produce endospores. Some species are obligate aerobes, and others are facultative anaerobes. The heat- and chemical-resistant spores make the genus an important cause of spoilage and contamination of medical supplies. A few species are involved in human infections; *Bacillus anthracis* is the most important pathogen of this group.

A. *Bacillus anthracis* is the etiologic agent of anthrax in animals and humans. Anthrax primarily affects grazing animals (i.e., herbivores). Humans are accidental hosts.

   1. Morphology. *B. anthracis* is a large (3.0 μm long and 1.0 μm in diameter) rod. Clinically it is usually seen singly or in pairs. The ends of the bacilli are sharply defined and square. **Capsules** are seen in clinical specimens but are not produced in vitro except under special conditions. Centrally located **endospores** are produced in vitro, in soil, and in decaying animal tissues, but endospores are not formed when the bacilli are growing in living tissue.

   2. Cultural characteristics
   a. *B. anthracis* grows as large colonies (2–3 mm) on ordinary bacteriologic media; the addition of 5% blood to the media enhances growth. The colonies are grayish white, raised, and plumose with irregular edges. Tangled, hair-like masses of cells give a Medusa-head appearance to the colony.
   b. Optimal growth is obtained in aerobic culture at 37° C and near a neutral pH. Sparse growth can be obtained under anaerobic conditions and over a wide temperature range.

   3. Antigenic composition. Three major antigens of *B. anthracis* have been identified and partially characterized.
   a. Capsular antigen of *B. anthracis* is chemically unique among bacterial capsules. The antiphagocytic capsule consists of polypeptide composed solely of D-glutamic acid. Only one capsular type has been identified.
   b. Somatic antigen is a polysaccharide found in the cell wall of the bacillus. It consists of equimolar parts of D-galactose and N-acetylglucosamine. Antibodies induced by the somatic antigen do not protect against infection. The antigen cross-reacts with type 14 pneumococcal capsular polysaccharide and with human blood group A substances.
   c. Anthrax toxin is produced in vivo and can be isolated from thoracic and peritoneal exudates of infected animals. Specialized culture conditions using serum are required for in vitro production of the toxin.
      (1) The toxin is complex, consisting of a protective antigen, a lethal factor, and an edema factor. No single component of the toxin appears to be toxic, but a combination of the protective antigen and either the lethal factor or the edema factor produces a pathologic process in laboratory animals.
      (2) Chemically the components appear to be high molecular weight proteins or lipoproteins. Their molecules are heterogeneous, and they are heat labile, serologically active and distinct, and immunogenic.

   4. Pathogenic characteristics. The virulence of *B. anthracis* is dependent upon both presence of the capsule and production of the toxin. Isolates that produce neither are considered to be avirulent.
   a. The glutamic acid polypeptide capsule is antiphagocytic. It appears to be the major virulence factor in the early stages of disease. The capsular antigen induces antibody formation, but the antibodies are not protective.

b. The toxin appears to be responsible for the signs and symptoms characteristic of **anthrax**. Maximal accumulation of the toxin in tissues and its effect on the central nervous system (CNS) results in death from respiratory failure and anoxia. Toxoids have been produced, but they have not been studied fully.

5. **Epidemiology. Anthrax** occurs infrequently in the United States and western Europe, but epizootics occur in Africa.
   a. **Animals** become infected by ingestion of spores while grazing on contaminated pastures. Tissue is invaded probably through minute cuts and abrasions in the oral or gastrointestinal mucosa. Pastures contaminated with spores have been found to be a source of infection for longer than 30 years. The course of the disease is only a few days in duration; the animal remains asymptomatic until just before death. The mortality rate is about 80%.
   b. **Human disease** generally is an occupational hazard for people who come in contact with animal products (i.e., wool, hide, hair, bone, and skin) from Africa, the Middle East, and Asia. The spores gain access to tissue through cuts or abrasions, inhalation, and occasionally ingestion. The most common form of the disease in humans is cutaneous anthrax. There was only one case of anthrax in the United States in 1980 and no cases in 1982, although 25,000–100,000 cases occur worldwide each year.

6. **Clinical manifestations in humans.** Clinical cases of anthrax present in three different ways, depending upon the route of infection. All forms of anthrax can result in disseminated disease causing bacteremia and fatal meningitis.
   a. **Cutaneous anthrax** initially presents as a small papule at the site of inoculation. The papule generally appears 2–5 days after exposure. Within a few days it progresses to a vesicle filled with bluish black fluid. The vesicle sometimes is referred to as a malignant pustule because it contains a black eschar at its base surrounded by an inflammatory ring. Infection usually occurs on the hands, forearms, neck, scalp, and other exposed areas.
   b. **Pulmonary anthrax** occurs as a result of the inhalation of spores, usually in people who handle contaminated animal products. The disease frequently is referred to as **woolsorter's disease**. The initial symptomatology is that of any respiratory infection, with fever, a nonproductive cough, myalgia, and malaise. The disease worsens over a few days; severe respiratory distress and cyanosis develop. Death usually occurs within 24 hours of this abrupt change in the patient's condition.
   c. **Gastrointestinal anthrax** is a rare disease characterized by nausea, vomiting, and diarrhea. The presence of blood in vomitus or stools indicates severe disease and is associated with prostration and shock, resulting in death.

7. **Laboratory identification.** Depending upon the form of the suspected disease, specimens obtained are exudates from a malignant pustule, sputum, or blood. Gram-stains of smears and immunofluorescence are useful in the presumptive identification of *B. anthracis*.
   a. **Specimens** can be cultured on any routine bacteriologic medium. Confirmed identification is made by Gram stain, lack of motility, and a series of biochemical tests. The major problem in identifying *B. anthracis* is confusion with other bacilli, such as *Bacillus cereus*. In addition, laboratory personnel in the United States usually have never seen *B. anthracis* because of the rarity of clinical anthrax.
   b. **Serologic identification.** *B. anthracis* can be diagnosed serologically in individuals with acute disease and those in convalescence. The antibodies can be detected by a gel diffusion test, complement fixation (CF), and hemagglutination techniques. Serum usually is submitted to the Centers for Disease Control in Atlanta, Georgia, for these test procedures.

8. **Therapy.** The antimicrobial agent of choice in the treatment of anthrax is penicillin; tetracycline may be used in patients with an allergy to penicillin. Large intravenous doses of penicillin are curative if they are administered during the course of the disease. Treatment failure generally is the result of mistaken diagnosis. Excision and drainage of a pustule often result in disseminated disease and death.

9. **Immunization.** Currently, either a living spore vaccine obtained from a nonencapsulated strain of *B. anthracis* or alum-precipitated protective antigen is used to immunize animals. Combining the two vaccines provides the best protection. The protective antigen vaccine has been used successfully in humans who are at a high risk of infection and has produced no observable harmful side effects. Animals given only the living spore vaccine have developed localized disease.

B. *Bacillus cereus* is similar in morphology to *B. anthracis* except that it usually is motile and susceptible to the gammaphage of bacilli, and it completely hemolyzes sheep blood. *B. cereus* is often the unrecognized cause of food-borne disease in the United States.

1. **Two distinct clinical forms of food poisoning** are caused by *B. cereus*. Food poisoning is related to the presence of heat-resistant spores that survive cooking. If the cooked food is allowed to reach a low enough temperature, the spores germinate and enterotoxins are produced.
   a. **Short-incubation gastroenteritis.** One type of food poisoning caused by *B. cereus* has an incubation time of about 4 hours. Characteristic symptoms are severe nausea and vomiting. Epidemics have been associated with ingestion of food in which the bacillus has grown excessively. Fried rice is a frequent source of the disease, which often is attributed incorrectly to staphylococcal enterotoxin.
   b. **Long-incubation gastroenteritis.** The second *B. cereus* syndrome has an average incubation time of 17 hours, and a separate enterotoxin appears to be involved. The syndrome is characterized by abdominal cramps and diarrhea and frequently is misdiagnosed as clostridial food poisoning.
   c. **Pathogeneses.** The enterotoxins produced by *B. cereus* and their modes of action have not been completely identified, but stimulation of adenyl cyclase activity apparently is not involved.

2. **Predisposing factors.** *B. cereus* can cause serious infections when natural defenses are impaired.
   a. **Immunosuppression** for organ transplantation or as a result of debilitating disease such as acute leukemia can predispose to fatal *B. cereus* infections characterized by overwhelming bacteremia, endocarditis, and meningitis.
   b. **β-Lactam antibiotics**, which *B. cereus* inactivates, may selectively promote the growth of this bacterium.
   c. Foreign bodies, prosthetic devices, and restricted blood supply to tissues are other predisposing factors.

3. **Treatment.** Chloramphenicol, aminoglycosides, clindamycin, and vancomycin usually are effective antimicrobial agents.

C. Other *Bacillus* species principally are involved in spoilage of organic materials and contamination of sterile supplies and clinical laboratory reagents. Some species have been described as etiologic agents of human disease in severely immunocompromised patients; others are used in the biologic control of insects.

1. ***Bacillus subtilus*** is a ubiquitous microorganism in air, dust, vegetable matter, and brackish water. Its major importance is as a contaminant, but it can produce disease in the compromised host. *B. subtilus* has been isolated from street heroin and has caused eye infections and fatal bacteremia in addicts. Unlike *B. cereus*, this bacillus is susceptible to the β-lactam antibiotics.

2. ***Bacillus stearothermophilus*** is a member of the genus that is used for biologic control of insects. It is pathogenic for *Lepidoptera* larvae. The bacillus forms a crystalline protein granule during sporulation. The granule separates from the released spore and releases toxin by enzymatic action in the gut of the larva.

II. CLOSTRIDIA. The genus *Clostridium* is composed of gram-positive, spore-forming rods that are obligate anaerobes. Most species are saprobes, a few are commensal parasites residing in the intestines of humans and other animals, and many are pathogenic. Several species of clostridia are used for the industrial production of organic acids and alcohols.

A. *Clostridium perfringens* is one of the most important species of the genus. Although found in the colon of 25%–35% of healthy people, under certain conditions it can produce serious, life-threatening infections.

1. **Morphology**
   a. **Vegetative cells.** This strongly gram-positive rod is the only nonmotile species in the genus. Classically, it has blunt or square ends.
      (1) Microscopic appearance of individual cells varies from long rods to coccobacillary forms depending upon available nutrients and other growth conditions.
      (2) Capsules are formed in tissue.

    b. Spores. Large, ovoid, centrally located endospores often give the vegetative cell a swollen appearance. Spores usually are found in soil and intestinal specimens but only rarely are identified in laboratory cultures or cooked foods.

    c. Colonial morphology. Colonies can vary from round to irregular in shape but usually are low, convex, shiny, and semiopaque.

        (1) Hemolysis also is variable and may be complete or partially complete.

        (2) The bacterium is not a strict anaerobe; it can tolerate exposure to air for short periods of time.

2. Antigenic composition. Five different types of *C. perfringens*, designated A through E, are based on serologically distinct exotoxins. All strains produce alpha toxin (i.e., lecithinase). Type A strains are subdivided into many serotypes by agglutination tests. The differentiation of these serotypes aids in isolating the causes of food poisoning and gangrene.

3. Metabolic activity. *C. perfringens* is active biochemically, fermenting glucose, maltose, lactose, and sucrose. It produces hydrogen sulfide gas and the proteolytic enzyme, gelatinase. Nitrate reduction, lack of motility, sporulation, lactose fermentation, and lecithinase production differentiate *C. perfringens* from other clostridia.

4. Pathogenic characteristics. Type A is the most important type of *C. perfringens*; more than 80% of environmental isolates belong to this type. Types A, C, and D are etiologic agents of disease in humans. Type A causes food poisoning, necrotizing colitis, and gas gangrene; type C causes enteritis necroticans. It appears that types A through E cause disease in animals. A variety of exotoxins are produced and aid in producing gas in the tissues, blood-tinged edematous fluid, and septicemia.

5. Toxins. *C. perfringens* produces at least 12 identifiable toxins in addition to enterotoxin. The alpha, beta, epsilon, and iota toxins have proven to be lethal for laboratory animals.

    a. Alpha toxin demonstrates lecithinase activity, hydrolyzing lecithin to phosphorylcholine and a diglyceride. Intravenous injection of the toxin is lethal in mice, and intradermal injection in guinea pigs produces local necrosis. All types of *C. perfringens* produce alpha toxin, but type A produces the largest quantities.

    b. Beta toxin produces necrosis and is lethal for albino guinea pigs, but it does not lyse red blood cells. It is heat labile. Types B and C are known to produce beta toxin.

    c. Delta toxin is hemolytic for sheep red blood cells but not for horse or rabbit erythrocytes. The toxin is lethal for laboratory animals. It is produced by types B and C.

    d. Theta toxin is an oxygen-labile hemolysin. It lyses horse and sheep erythrocytes and shows minimal activity against mouse red blood cells. Type C produces large quantities of the toxin; types A, B, D, and E produce smaller amounts.

    e. Epsilon toxin and **iota toxin** are lethal; both produce necrosis in intradermal tests in guinea pigs. A protoxin is found in culture filtrates, and it is activated by treatment with trypsin. Types B and D produce epsilon toxin, and type E strains of *C. perfringens* produce iota toxin. In vivo the toxins are absorbed from the intestinal tract.

    f. Kappa toxin is a collagenase and gelatinase that is lethal and necrotizing. The toxin is produced by types A, C, E, and some strains of type D.

    g. Lambda toxin is a proteolytic enzyme with a relatively wide spectrum of activity. It breaks down gelatin, axocoll (i.e., commercially produced powder made of hide and coupled with a dye), casein, and hemoglobin. The toxin is produced by types B and E and by some strains of type D.

    h. Gamma toxin and **eta toxin** are lethal to laboratory animals. No other properties of the toxins are apparent.

    i. Mu toxin and **nu toxin** are a hyaluronidase and a deoxyribonuclease (DNase), respectively. They are detected in culture filtrates by biochemical tests of their activity.

    j. Miscellaneous toxins, including fibrinolysin and neuraminidase, have been detected in culture filtrates.

    k. Enterotoxin is produced by *C. perfringens* types A and C and is responsible for food poisoning. The toxin is a protein and is heat labile. It is produced in the large intestine and is released during sporulation of *C. perfringens*. It rarely is produced in laboratory cultures or in cooked food and is therefore unlikely to be detected. The toxin has both lethal and emetic characteristics and produces an erythematous skin rash in laboratory animals.

6. Epidemiology. *C. perfringens* is widely distributed in the environment; it is found in soil, water, and sewage. It also is found in the intestinal tract of humans and animals.

   a. Gangrene can result from contamination of a wound by *C. perfringens.* The bacteria grow and sporulate in devitalized tissue where the oxygen content is very low. Infecting cells or spores originate from the environment, from skin, or from the patient's own intestinal tract. The gangrene is characterized by necrosis and the production of foul-smelling gas.

   b. Food poisoning due to *C. perfringens* types A and C is an important form of gastrointestinal disease in the United States and the United Kingdom.

     (1) Mild food poisoning is caused by the enterotoxin of type A strains. It is characterized by nausea, abdominal pain, and diarrhea occurring 8–24 hours after eating heavily contaminated food. Symptoms last for 12–24 hours. Fatalities occur rarely, principally among debilitated elderly patients and malnourished children. Outbreaks have occurred in hospitals and other institutions.

     (2) Enteritis necroticans is a serious form of disease due to the beta toxin of type C strains. The disease is rare but has been reported from both Germany (after World War II) and New Guinea, where traditional cooking techniques and other factors allow survival of the bacteria. The intestinal lesions produced by the beta toxin frequently cause death.

     (3) Heat resistance of the spores of *C. perfringens* type A varies within a given strain and also from strain to strain.

       (a) In boiled meat some species survive only a few minutes, whereas others survive for more than 1 hour; the fat may be protective.

       (b) Germination is stimulated at 75° C, and optimal growth occurs at 43° C in cooked meat. The generation time of *C. perfringens* is 10–12 minutes under optimal growth conditions.

       (c) The vegetative cells and spores are common contaminants of raw meats.

7. Laboratory identification. It is important to assess the role of *C. perfringens* clinically because it is a member of the indigenous flora of humans. There are approximately 10^3 vegetative cells and 10 spores/g of fecal material from healthy people.

   a. Isolation of *C. perfringens* is accomplished on either blood agar or egg yolk agar containing neomycin sulfate under anaerobic conditions. Specimens also can be inoculated into cooked meat broth and then subcultured onto agar after 10–12 hours of incubation. Heating clinical stool specimens destroys vegetative cells; resistant spores then can be determined by culture on agar medium. In cases of clostridial food poisoning, cell counts are 10^6–10^7/g of feces; spores are found infrequently.

   b. Identification is made by growing suspected *C. perfringens* on agar medium containing lecithin (i.e., egg yolk agar). Colonies that are surrounded by an opalescent precipitate, which is produced by the action of alpha toxin (lecithinase) on the lecithin, probably are *C. perfringens.* The reaction is inhibited by antiserum, which can be placed on half of the plate at the time of inoculation. Other identifying tests include a negative motility test, nitrate reduction, and proteolysis.

B. *Clostridium tetani* is the etiologic agent of tetanus, sometimes referred to as **lockjaw**. The disease is the result of the neurotoxic activity of an exotoxin.

1. Morphology. This slender, gram-positive rod characteristically is motile and a strict anaerobe. The spores are spherical and terminal, giving the cells a drumstick or tennis-racket appearance; immature spores are oval. Cultured cells form fine filaments of growth, which are easily overlooked. Nonmotile mutants do not swarm, and they form discrete colonies. Complete or partial hemolysis is seen on horse blood agar.

2. Antigenic composition. The neurotoxins produced by all toxigenic strains of *C. tetani* serologically are identical. Ten types of antigenic variation among flagella have been identified.

3. Metabolic activity. Acetic, butyric, and propionic acids and ethanol are metabolic products. Glucose is fermented by a rare strain of *C. tetani*, but no other sugars are fermented. The bacteria are nonproteolytic, but they produce gelatinase and a rennin-like enzyme that causes a shady zone to form around colonies on milk agar. Indole is formed from tryptophan, and some strains produce DNase.

4. Pathogenic characteristics. The pathogenicity of *C. tetani* for humans and animals is caused by tetanospasmin, a neurotoxic exotoxin.

   a. Tetanospasmin blocks spontaneous and evoked transmitter release at both central and peripheral synapses.

    (1) Initial exposure to the toxin results in localized peripheral nerve and muscle spasms.
    (2) The potent toxin is produced in cultures after the logarithmic phase of growth, and it is heat labile.
    b. Tetanolysin, a second toxin, is hemolytic, but it is of minor importance in the pathogenesis of tetanus.

5. Immunity. Tetanus can be prevented by administration of tetanus toxoid. Immunization is required for children in the United States. Diseased patients can be treated successfully with hyperimmune human globulins. However, recovered patients still must be immunized because the amount of toxin producing serious disease is too small to induce protective antibodies. Booster immunizations generally are administered at times of risk. Immunization of pregnant women prevents tetanus in neonates.

6. Epidemiology. *C. tetani* is easily isolated from the soil. It also can be found in the intestinal tract of humans and animals. Human disease is produced when wounds and umbilical stumps become contaminated with soil or feces. Military action, automobile accidents, and mishaps among farm workers in particular are likely to result in wounds contaminated with soil and toxigenic *C. tetani.*

7. Laboratory identification. Clinical samples are placed on blood agar or in cooked meat broth and incubated under anaerobic conditions for 4 or 5 days. Inoculum is placed near the edge of the blood agar plate to allow swarming of the *C. tetani.* Pure cultures are obtained by subculturing the leading edge of growth. Nonmotile mutants can be isolated by adding neomycin sulfate to the medium or by carefully heating the specimen to kill non–spore-forming bacteria. Identification is confirmed by biochemical tests.

C. *Clostridium botulinum* produces the most potent poison known. It is the cause of **botulism**, a generally fatal form of food poisoning.

1. Morphology. The gram-positive rods are large and motile. The spores are oval and terminal or subterminal. Colonies are translucent and can be circular or irregularly shaped. The surface of colonies appears granular, and swarming cells may cover the entire culture plate. *C. botulinum* hemolyzes horse erythrocytes.

2. Antigenic composition. *C. botulinum* produces seven serologically distinct neurotoxins, designated A through G. The seven strains are placed in three groups because of some antigenic relationships.

3. Metabolic activity. All types of *C. botulinum* produce gelatinase and hydrogen sulfide gas. Types A, B, E, and F ferment glucose, maltose, and sucrose; types C and D ferment glucose and maltose; and type G does not ferment any sugars.

4. Pathogenic characteristics. The different types of *C. botulinum* show some degree of host specificity. Types A, B, E, and F cause botulism in humans; types C and D, and rarely types A and B, produce disease in animals and birds. Type G has not been isolated from diseased animals or humans.

    a. Clinical disease. The classic form of the disease is **food poisoning** produced by the ingestion of the neurotoxin in food. *C. botulinum* does not replicate easily within the body. **Wound botulism** can occur when devitalized tissue is contaminated with soil. **Intestinal botulism** in infants, known as **floppy baby syndrome**, is due to ingestion of spores, followed by germination and toxin production.

    b. Pharmacology. The seven different neurotoxins pharmacologically act in the same way; they are absorbed through the mucosal cells of the alimentary tract. The neurotoxin can be detected in the bloodstream, where it is carried to peripheral nerves. It acts at the motor nerve endings of the neuromuscular junction, where it interferes with the release of acetylcholine and is bound to the tissue. Neurotoxin production in *C. botulinum* is associated with bacteriophages.

    c. Symptomatology. The lapse of time between ingestion and paralysis usually is less than 24 hours. Vomiting, ocular paresis, and pharyngeal paresis occur, and death results from paralysis of the diaphragm and other respiratory muscles.

    d. Toxin. A fatal dose of botulinum toxin for humans has been estimated to be between 0.1 and 1.0 µg. The neurotoxin appears to be a simple protein, consisting of only 19 amino acids, with a molecular weight of 1.5×10^4 daltons. The toxin is destroyed completely by boiling.

5. **Immunity.** Polyvalent antitoxin is used to protect laboratory workers. Natural immunity is highly unlikely because the lethal dose of toxin is too small to induce protective antibodies.

6. **Epidemiology.** *C. botulinum* is found worldwide in soil. Types A and B are the most common isolates. Type G was isolated first from soil in Argentina. Type E occurs around shore waters where levels of these clostridia may become so high in mud that fish are affected. Drought appears to promote growth of the bacilli in these environments. Nitrites are required in the curing process of salted meats to prevent the germination of spores. Home-canned, low-acid vegetables and home-cured meats are a frequent source of intoxication.

7. **Laboratory identification.** The neurotoxin can be demonstrated in the blood and stools of affected individuals. It can be demonstrated in food by the injection of extracts into mice, and the type can be determined by mouse-protection studies using specific antisera. Reverse hemagglutination and immunofluorescence also are used to identify the toxin. Anaerobic cultures can be prepared on blood agar and egg yolk agar, but the procedure is time-consuming and may not produce viable bacteria.

D. ***Clostridium novyi*** is one of several etiologic agents of **gangrene**. It was the cause of about 42% of the gangrene in soldiers during World War II. It is widely distributed in soil and also can be recovered from the livers of healthy animals.

1. **Morphology.** This large, slender, gram-positive rod readily forms large subterminal spores. Young cultures resemble *C. perfringens*, but *C. novyi* is motile. Mature colonies are round or irregular and semitranslucent; they have a granular surface.

2. **Antigenic composition.** *C. novyi* has four serotypes, designated A through D. Types A, B, and D share two somatic antigens in differing proportions. Type B antiserum cross-reacts with the other three types.

3. **Metabolic activity.** All types of *C. novyi* produce gelatinase and hydrogen sulfide gas. Type D produces indole. Types A and C ferment glucose and maltose, whereas types C and D ferment only glucose. Large quantities of propionic and butyric acids, and smaller amounts of acetic and valeric acids, are produced.

4. **Pathogenic characteristics.** *C. novyi* types A and B cause gas gangrene in humans and animals; type A is the most frequently isolated etiologic agent. Eight soluble toxic antigens have been identified in culture filtrates. Types A and B produce alpha toxin, which is necrotizing and lethal; it causes increased capillary permeability, which produces the classic gelatinous edema of affected muscle. Lecithinase C enzyme also is produced, and it is necrotizing and hemolytic; type B is lethal.

5. **Laboratory identification.** As is true for other clostridia, specific antisera are used to identify and type *C. novyi*. *C. novyi*, however, is more difficult to grow than other species.

E. **Other *Clostridium* species.** A considerable number of other *Clostridium* species are distributed in soil and the mammalian intestine. Many of them can play a role in gas gangrene of mixed etiology. All are gram-positive, spore-forming rods, and they grow best under anaerobic conditions. Spores range in shape from spherical to ovoid to cylindrical. Some of them can tolerate oxygen for relatively long periods of time.

1. ***Clostridium histolyticum*** produces alpha, beta, gamma, and delta toxins that can be isolated from culture filtrates. Alpha toxin is lethal and necrotizing. Beta, gamma, and delta toxins are, respectively, a collagenase, a protease, and an elastase. *C. histolyticum* has been associated with other anaerobes in gas gangrene of mixed etiology.

2. ***Clostridium septicum*** causes gas gangrene in humans and animals. It produces three exotoxins. The alpha toxin is necrotizing, hemolytic, and lethal. The beta toxin is a DNase. The gamma toxin is hyaluronidase. It also produces neuraminidase and hemagglutinin.

3. ***Clostridium chauvoei*** is similar to *C. septicum* except that it has been reported as an etiologic agent of disease in animals only. It produces alpha, beta, and gamma toxins as well as an oxygen-labile hemolysin.

4. ***Clostridium sporogenes*** enhances the virulence of *C. perfringens* and *C. septicum* in mixed

cultures. It also causes putrefactive changes in muscle that has been inoculated with viable spores. No soluble antigens of clinical significance have been identified.

5. ***Clostridium bifermentans* and *Clostridium sordellii*** are related; they are differentiated on the basis of spore agglutinogens. Both produce a lecithinase C resembling that of *C. perfringens* and at least one additional soluble antigen. Infection causes proteolysis, hemorrhage, and gas production.

6. ***Clostridium difficile*** is associated with antibiotic-induced pseudomembranous colitis. It produces two toxins, one of which causes a cytopathic effect on tissue cells. The other is less cytopathic to cultured tissue cells. The toxins have been shown to be two distinct compounds. During broad-spectrum antibiotic therapy, resistant strains of *C. difficile* overgrow the intestine and secrete the exotoxins. Vancomycin therapy usually eliminates the toxigenic strains.

STUDY QUESTIONS

Directions: Each question below contains five suggested answers. Choose the **one best** response to each question.

1. What is the most common manifestation of anthrax in humans?

(A) Ulcerative gastroenteritis
(B) Pulmonary abscess
(C) Septicemia
(D) Cutaneous lesions
(E) Meningitis

2. What organism produces a frequently misdiagnosed food poisoning associated with fried rice?

(A) *Bacillus cereus*
(B) *Clostridium bifermentans*
(C) *Bacillus stearothermophilus*
(D) *Bacillus subtilus*
(E) *Clostridium perfringens*

3. Which bacterium characteristically produces large terminal endospores that give a drumstick appearance to individual cells?

(A) *Bacillus anthracis*
(B) *Clostridium botulinum*
(C) *Clostridium perfringens*
(D) *Bacillus subtilus*
(E) *Clostridium tetani*

4. In cultures of *Clostridium tetani*, the potent neurotoxin is produced during

(A) the lag phase
(B) the logarithmic phase
(C) the stationary phase
(D) the death phase
(E) the germination of spores

5. What condition may be produced by the germination of spores of *Clostridium botulinum* in the gastrointestinal tract?

(A) Pseudomembranous colitis
(B) Enteritis necroticans
(C) Lockjaw
(D) Floppy baby syndrome
(E) Rice water stools

6. Which strain of clostridia sometimes causes antibiotic-induced pseudomembranous colitis?

(A) *Clostridium bifermentans*
(B) *Clostridium difficile*
(C) *Clostridium sporogenes*
(D) *Clostridium sordellii*
(E) *Clostridium septicum*

Directions: Each question below contains four suggested answers of which **one or more** is correct. Choose the answer

  A   if **1, 2, and 3** are correct
  B   if **1 and 3** are correct
  C   if **2 and 4** are correct
  D   if **4** is correct
  E   if **1, 2, 3, and 4** are correct

7. True statements regarding the capsule of *Bacillus anthracis* include which of the following?

(1) It is a polymer of D-glutamic acid
(2) It is chemically unique among bacteria
(3) It is an antiphagocytic virulence factor
(4) It is the most effective anthrax vaccine

8. The toxic effects of anthrax are the result of

(1) edema factor
(2) protective antigen
(3) lethal factor
(4) alpha toxin

9. Extracellular toxins produced by *Clostridium perfringens* include

(1) lecithinase
(2) hemolysin
(3) collagenase
(4) enterotoxin

10. Tetanus can be prevented by

(1) administration of hyperimmune globulin
(2) transplacental passage of antitoxin
(3) injections of tetanus toxoid
(4) specific anti-clostridial chemotherapy

11. Clostridia that have been associated with gangrene in humans include

(1) *Clostridium difficile*
(2) *Clostridium septicum*
(3) *Clostridium chauvoei*
(4) *Clostridium novyi*

ANSWERS AND EXPLANATIONS

1. The answer is D. [*I A 5 b*] Cutaneous anthrax is the most common form of the disease in humans. It typically is acquired when a spore germinates in a small lesion containing an aerobic microenvironment. The disease is very rare now and is considered an occupational hazard. Pulmonary anthrax, also known as woolsorter's disease, occurs when spores are inhaled. It is a progressive, severe pneumonia that usually is fatal within a few days of onset. Gastrointestinal anthrax may occur following the ingestion of endospores of *Bacillus anthracis*. Each of these forms of anthrax may disseminate, resulting in a bacteremia. The meninges and other tissues may become involved in disseminated disease.

2. The answer is A. [*I B 1 a*] *Bacillus cereus* is the etiologic agent of food poisoning that resembles, and frequently is misdiagnosed as, staphylococcal food poisoning. The food poisoning occurs when heat-resistant spores of the bacillus survive cooking and germinate while the food is maintained at a lower temperature. Enterotoxin is produced; when ingested, it causes severe nausea and vomiting. Epidemics of this food poisoning have been associated with the consumption of fried rice. *Clostridium perfringens* also produces food poisoning that may resemble staphylococcal disease, but this intoxication typically causes diarrhea and occurs primarily among institutionalized people. *Bacillus stearothermophilus* and *Bacillus subtilus* are airborne contaminants that only rarely cause disease, usually in immunocompromised humans. *Clostridium bifermentans* is the etiologic agent of a myelonecrosis.

3. The answer is E. [*II B 1*] *Bacillus* and *Clostridium* species are medically important, gram-positive, spore-forming rods. The endospores formed by the different species sometimes are characteristic enough to be considered presumptive identification features. The mature endospores of *Clostridium tetani* are larger than the vegetative cells and are terminally located; they give the cells a drumstick or tennis-racket appearance. The endospores of *Bacillus anthracis*, *Bacillus subtilus*, *Clostridium perfringens*, and *Clostridium botulinum* range from central to subterminal to terminal; they are not as distinctly characteristic of their species as are the endospores of *C. tetani*.

4. The answer is C. [*II B 4*] Tetanospasmin, the potent neurotoxin produced by *Clostridium tetani*, can be isolated from culture media after the logarithmic phase of growth. The cells are in the stationary phase of the growth curve, indicating that toxin production occurs after cellular metabolism has slowed in mature bacterial cells. Toxin production may be associated with spore production.

5. The answer is D. [*II C 4 a*] The classic form of botulism is food poisoning, in which the neurotoxin is ingested and adsorbed through the mucosa of the alimentary canal. However, ingested spores of toxigenic *Clostridium botulinum* may germinate and produce toxin in the gastrointestinal tract when acidity is greatly decreased. This form of intestinal botulism has been described most frequently in infants; the disease is known as floppy baby syndrome. It is characterized by loss of muscular control due to the neurotoxic effects of the toxin. Pseudomembranous colitis, lockjaw, and enteritis necroticans are disease processes caused by other species of clostridia (i.e., *Clostridium difficile*, *Clostridium tetani*, and *Clostridium perfringens*, respectively). Rice water stools is a descriptive term of the diarrheal syndrome produced by *Vibrio cholerae*.

6. The answer is B. [*II E 6*] *Clostridium difficile* accounts for a small proportion of the indigenous flora in a percentage of healthy people. Broad-spectrum antibiotic therapy in these people may inhibit or kill a significant number of other colon bacteria, permitting *C. difficile* to proliferate. This clostridium produces two exotoxins that have cytopathic effects on cultured tissue cells and that appear to be responsible for pseudomembranous colitis. The condition is prevented by administering vancomycin to inhibit the growth of *C. difficile*. *Clostridium bifermentans* and *Clostridium sordellii* occasionally cause infections characterized by proteolysis, hemorrhage, and gas production. *Clostridium septicum* is one of the etiologic agents of gas gangrene in humans and animals. *Clostridium sporogenes* appears to enhance the virulence of *Clostridium perfringens* and *C. septicum* in vitro.

7. The answer is A (1, 2, 3). [*I A 3 a, 9*] The polypeptide capsule of *Bacillus anthracis* is chemically unique among bacteria; it consists of a polymer of D-glutamic acid. Like the polysaccharide capsules of other bacteria, the capsule of the anthrax bacillus is an aggressin that inhibits phagocytosis by neutrophils and other phagocytic cells. In humans, a protective antigen vaccine has been used successfully against anthrax; the best immunity in animals is achieved by administering a living spore vaccine in combination with alum-precipitated protective antigen.

8. The answer is A (1, 2, 3). [*I A 3 c*] The anthrax toxin consists of a protective antigen, a lethal factor, and an edema factor. None of these virulence factors alone produces the typical toxic effects on tissue seen in clinical cases of anthrax. However, either the edema factor or the lethal factor in combination with the protective antigen produces characteristic pathologic effects. *Bacillus anthracis* does not produce any extracellular substance known as alpha toxin.

9. The answer is E (all). [*II A 5 a, d, f, k*] *Clostridium perfringens* produces at least 12 extracellular toxins with known effects. Lecithinase activity is attributed to the alpha toxin, which is produced by all strains of this clostridium. Theta toxin is an oxygen-labile hemolysin produced in small quantities by *C. perfringens* type A and in large quantities by type C. It hemolyzes horse and sheep erythrocytes. Kappa toxin is a collagenase produced by type A and other strains. Kappa toxin also has gelatinase activity. Types A and C produce an enterotoxin that is responsible for clostridial food poisoning. The enterotoxin produces an emetic effect and is cytotoxic.

10. The answer is A (1, 2, 3). [*II B 5*] Tetanus is an intoxication; all prophylactic and therapeutic procedures target neutralization of the potent neurotoxin produced by *Clostridium tetani*. The most common means of preventing tetanus in humans is to administer tetanus toxoid periodically throughout life. The first administration normally is at 3 months of age in the form of the required DPT vaccine. Hyperimmune human globulin containing antibodies to tetanus toxin can be administered to nonimmune individuals who have been exposed to the neurotoxin. Protection against tetanus in the neonate can be acquired by transplacental passage of antibodies from an immunized mother. The administration of antimicrobial agents can eliminate the toxin-producing bacteria, but lethal doses of the neurotoxin are so small that chemotherapy is not a practical consideration for prevention of tetanus in humans.

11. The answer is C (2, 4). [*KII D, E 2, 3, 6*] Many species of clostridia are etiologic agents of gangrene in humans. *Clostridium seopticum* provides three exotoxins and is the etiologic agent of gas gangrene in humans and animals. *Clostridium novyi* produces several toxic products and also causes gas gangrene in humans and animals; *C. novyi* accounted for about 42% of the cases of gangrene during World War II. *Clostridium chauvoei* has been associated only with disease in animals. *Clostridium difficile* has not been identified as an etiologic agent of gangrene, although it produces clinical disease in humans.

11
Gram-Negative Rods: Facultative Enteric Rods, Bacteroides, and Pseudomonads

Gerald E. Wagner

I. ENTEROBACTERIACEAE. Most of the enteric gram-negative rods that are human pathogens or that inhabit the normal gastrointestinal tract belong to the family Enterobacteriaceae. The genera of this family are subdivided into five tribes: Escherichieae, Klebsiellae, Proteeae, Yersinieae, and Erwinieae. Yersinieae are zoonotic agents and are discussed in Chapter 12. **All species within the family are gram-negative, facultative anaerobes that ferment glucose.**

A. Morphology

1. **Cells.** Enterobacteriaceae are 2 μm in length and 0.4 μm in diameter; rapidly growing cells may appear coccobacillary. Many species are motile and encapsulated, particularly upon initial isolation from clinical specimens. **All species contain lipopolysaccharide endotoxin** as part of the outer cell membrane.

2. **Colonies.** Typical colonies on most bacteriologic media are circular, convex, and glistening or mucoid. The loss of capsules produces rough colonies that are flat, irregular, and granular in appearance. Highly motile species form a unique swarming pattern on agar cultures. Most Enterobacteriaceae are nonpigmented, although a few species include strains that produce red, pink, yellow, or blue pigments.

B. Antigenic composition.
All species in the family have the somatic (O) antigen, and most have the flagellar (H) antigen. The capsular (K) antigen is seen routinely in some species and is lacking in others. These antigens are used in the classification of species and may yield useful epidemiologic information.

1. **O antigens** are associated with the lipopolysaccharide of the outer membrane of all gram-negative bacteria.
   a. **Specificity.** More than 160 O antigens of *Escherichia coli* alone have been detected on the basis of specific sugars, α- or β-glycosidic linkages, and the presence or absence of substituted acetyl groups.
   b. **Loss of the O antigen transforms smooth colonies to rough ones.** It can be the result of mutation during long-term laboratory transfers. Rough colonies are avirulent.

2. **K antigens** consist of polysaccharides in capsules and in the outer envelope. K antigens cover the O antigens and can block agglutination by a specific O antiserum. The K antigens are destroyed by boiling in water.

3. **H antigens** are the flagellar protein antigens associated with motile species of the family Enterobacteriaceae. Many species are motile upon initial clinical isolation.
   a. *Klebsiella* **species** always are nonmotile.
   b. *Salmonella* **species** have two phases of H antigens—H1 and H2—depending upon the gene expression of the bacteria.

4. **Serologic strain identification** can be made with specific antisera for the three antigens. An example would be *E. coli* O111:K55:H3.

C. Laboratory identification.
Enterobacteriaceae are biochemically active and ferment a large number of carbohydrates. Many other biochemical tests are used to identify and differentiate species.

1. **Glucose fermentation** is characteristic of all of the enterobacteriaceae. Organic acids such as lactic, formic, and acetic acids are produced. Some species also produce hydrogen gas and carbon dioxide gas during glucose fermentation.

2. **Lactose fermentation** is used to separate the Enterobacteriaceae into two groups: the lactose fermenters and the lactose nonfermenters. Several differential and selective media have been developed for the separation of these groups, but eosin-methylene blue (EMB) agar and MacConkey agar probably are the most commonly employed.

   a. **EMB agar** is selective for gram-negative rods because the methylene blue inhibits the growth of gram-positive cocci. The lactose fermenters are identified by a deep purple to black coloration in isolated colonies. Some colonies of lactose fermenters (e.g., *E. coli*) display a characteristic dark green metallic sheen.

   b. **MacConkey agar** contains bile salts to inhibit the growth of gram-positive cocci. Lactose fermenters develop a pink to light red coloration in isolated colonies.

3. **Other biochemical tests** are employed for the identification of individual Enterobacteriaceae species. These tests include hydrogen sulfide production, indole formation from tryptophan, production of acetylmethyl carbinol from glucose (i.e., the Voges-Proskauer test), liquefaction of gelatin, decarboxylation of amino acids, use of citrate as a carbon source, and hydrolysis of urea.

D. Pathogenicity. All Enterobacteriaceae should be considered potentially pathogenic. Patients with underlying disease, immunosuppression, mechanical or medical manipulation, and other forms of debilitation are most susceptible to infection. As they die, all Enterobacteriaceae release **lipopolysaccaride (LPS)**, which can result in **endotoxin shock** in humans.

1. **Toxic LPS** consists of lipid A, the core polysaccharide, and the O antigen. The lipid A moiety of LPS is responsible for most of the symptomatology associated with endotoxin shock.

2. **Effects.** Small amounts of LPS in the bloodstream cause a variety of toxic effects. Large doses are lethal.

   a. **Fever.** A rise in body temperature usually occurs within 30 minutes of exposure to LPS. Tolerance occurs with repeated exposures; it is characterized by a diminishing height and duration of fever.

   b. **Hypotension.** Small doses of LPS cause a decrease in blood pressure in most people within 30 minutes of exposure. Hypotension is the end result of a series of hemodynamic effects caused by LPS. Large doses of LPS can cause permanent and fatal hypotension.

   c. **Intravascular coagulation.** Localized and generalized Shwartzman reactions occur with a second exposure to LPS. The generalized reaction can result in depletion of clotting factors, and serious bleeding may occur.

   d. **Alterations in circulating blood cells.** Neutropenia occurs within minutes of circulatory exposure to LPS. Leukocytosis follows; monoblasts and a variety of other immature cell types appear in the circulation.

   e. **Effects on cellular immunity.** LPS stimulates the proliferation of B lymphocytes, resulting in a decrease in nonspecific resistance to infection. Macrophages are stimulated to release interleukin 1 and lysosomal enzymes.

   f. **Increase in humoral immunity.** LPS activates complement by both classic and alternate pathways. It also stimulates interferon activity. Antibodies can be produced against each of the components of LPS, and they can neutralize some of the toxic effects.

   g. **Metabolism.** LPS can cause hypoglycemia and hypoferremia. Changes occur in carbohydrates, lipids, iron, and sensitivity to epinephrine.

   h. **Pregnancy.** LPS may cause placental hemorrhage because it stimulates the release of serotonin. LPS and lipid A alone have caused abortions in laboratory animals after intravenous injection.

E. Escherichieae. The tribe Escherichieae includes five genera: *Escherichia, Salmonella, Shigella, Edwardsiella,* and *Citrobacter.* The most important human pathogens in this group are *Escherichia coli* and the *Salmonella* and *Shigella* species. *E. coli,* like many members of the tribe, is among the indigenous flora of the human intestinal tract. The *Salmonella* and *Shigella* species are enteric pathogens.

1. ***E. coli***, a prominent member of the indigenous flora of the human intestinal tract, also is a major etiologic agent in urinary tract infection and gastroenteritis.

   a. Ascending urinary tract infection is caused most commonly by *E. coli* and frequently is seen in otherwise healthy people. Young women and elderly adults are most susceptible. The disease may progress from simple urethritis to serious pyelonephritis. It is characterized by dysuria, frequency of urination, pyuria, and, occasionally, hematuria.

   b. Gastroenteritis caused by *E. coli* ranges from simple diarrhea to a more severe form with a debilitating loss of fluids and electrolytes. Gastroenteritis is most serious and sometimes fatal in nutritionally deprived infants and elderly debilitated adults.

     (1) Enterotoxigenic *E. coli* produces one or both of two enterotoxins that are coded for by genes residing on a transmissible plasmid. One toxin is heat labile and the other is heat stable.

       (a) The heat-labile toxin causes diarrhea by the same mode of action as cholera enterotoxin.

         (i) The toxin binds to ganglioside receptors of intestinal mucosa and other cells, and it activates membrane-bound adenyl cyclase.

         (ii) Increased intracellular concentrations of cyclic adenosine 3', 5'-monophosphate (cAMP) cause the active excretion of electrolytes, which carry large amounts of water with them.

         (iii) The heat-labile toxin can be destroyed by heating at 65° C for 30 minutes.

       (b) The heat-stable toxin causes diarrhea by stimulating guanyl cyclase in intestinal mucosal cells only.

         (i) The receptor for this toxin is unknown, but it is obviously unique to the susceptible cell.

         (ii) The heat-stable toxin is a small, heterologous polypeptide molecule; it is not destroyed by heating at 100° C for 30 minutes.

     (2) Enteroinvasive *E. coli* causes diarrhea by invading the intestinal epithelial cells and eliciting an inflammatory response. The exact cause of the diarrhea is unknown, but it may be due to the cytotoxic effects of lipopolysaccharide.

2. ***Salmonella*** **species.** The three primary species are *Salmonella typhi*, *Salmonella choleraesuis*, and *Salmonella enteritidis*.

   a. Antigenic composition and biochemistry

     (1) Salmonellae have O and H antigens, and most virulent strains possess a capsular polysaccharide designated the **virulence (Vi) antigen**. There are more than 1700 serotypes of *S. enteritidis*.

     (2) *Salmonella* species ferment lactose slowly or not at all; *S. choleraesuis* and *S. enteritidis* produce hydrogen sulfide and gas from fermentation of carbohydrates, and *S. typhi* does not produce either.

   b. Epidemiology. The salmonellae are associated with the intestinal tract of humans and other animals, including reptiles, amphibians, fish, and birds.

     (1) Poultry products (i.e., flesh and eggs) currently are the leading source of salmonellosis in the United States. It is estimated that one in every four cases of food poisoning is salmonellosis. Improperly cooked commercial roast beef also has been a major source.

     (2) Pet turtles accounted for more than 300,000 cases of salmonellosis each year in the early 1970s in the United States. Today it is illegal to import turtles and turtle eggs and to transport turtles smaller than 4 inches in diameter across state lines.

     (3) Occupational salmonellosis affects veterinarians and workers in slaughterhouses.

   c. Pathogenicity. The salmonellae are capable of causing a variety of conditions, all referred to as salmonellosis. The three major categories of salmonellosis are enteric fever, septicemia, and gastroenteritis.

     (1) Enteric fever usually is caused by *S. typhi* and may be referred to as **typhoid fever**. Enteric fever, however, also may be caused by *S. enteritidis* serotype *paratyphi* and probably by other serotypes as well.

       (a) Pathology. Enteric fever is an invasive disease. Intestinal inflammation occurs 1–3 weeks after ingestion of the bacteria, which invade the mucosal epithelium. Regional lymph node infection is followed by bacteremia. Localized infections may occur in any organ of the body. The bacteria are facultative intracellular pathogens of macrophages, and the gallbladder may become a reservoir of infection.

       (b) Symptoms of enteric fever are systemic. Headache, loss of appetite, abdominal pain, weakness, stupor, and continuous fever are common. Diarrhea generally does not occur until late in the disease, after the onset of bacteremia. Splenomegaly almost always occurs, and multiplication of the bacteria in the skin may cause the appearance of rose-colored spots.

(2) **Septicemia** due to salmonellae is a fulminant, sometimes fatal, disease independent of intestinal symptoms. *S. choleraesuis* is a frequent agent and may cause focal infections throughout the body. Pneumonia, meningitis, and osteomyelitis may result from hematogenous spread of the bacteria. Vascular prostheses and graft sites particularly are susceptible to infection. People with sickle cell anemia are at high risk for developing osteomyelitis.

(3) **Gastroenteritis**, the most common form of salmonellosis, can be caused by any of the *S. enteritidis* serotypes. Symptoms appear 10–24 hours after ingestion of highly contaminated food or beverage. Nausea, vomiting, abdominal cramps, headache, and diarrhea may persist for 2–7 days.

d. Laboratory diagnosis. Salmonellosis is diagnosed by the isolation and biochemical identification of an etiologic agent. *S. typhi* can be isolated from the blood and stools, *S. choleraesuis* from the blood, and *S. enteritidis* from the stools. Serologic diagnosis may be made retrospectively by demonstrating a rise in serum titers of agglutinating antibodies.

e. Control and therapy. The prevalence of salmonellae, particularly *S. enteritidis*, among animals makes eradication impractical. The best means of controlling infection is proper cooking of foods derived from animal sources. Antimicrobial substances in animal feeds may be of some benefit. Antibiotic treatment is crucial in enteric fever and septicemia but may not be required in gastroenteritis except in severe cases. Ampicillin, because of its biliary circulation, currently is the drug of choice in eliminating the carrier state.

3. *Shigella* species are the etiologic agents of classically defined bacillary dysentery. The disease is characterized by the presence of inflammatory cells and, occasionally, blood in watery stools. The genus is divided into four species: *Shigella dysenteriae*, *Shigella sonnei*, *Shigella flexneri*, and *Shigella boydii*. Shigellae are related to *E. coli*; intergeneric conjugation results in exchange of genetic information between the two types of organisms.

a. Antigenic composition and biochemistry. The shigellae are nonmotile; there are no H antigens. They do not produce hydrogen sulfide and do not ferment lactose. (An exception is *S. sonnei*, which ferments lactose slowly.)

b. Epidemiology. Humans appear to be the only reservoir of the shigellae; they pass the bacteria to others by contaminating food or water. There is a high incidence of disease in institutionalized, debilitated people.

(1) The etiology of shigellosis in the United States is cyclic; sometimes *S. flexneri* predominates, and sometimes *S. sonnei*. Currently, *S. sonnei* is the most commonly isolated species.

(2) *S. boydii* is an infrequent cause of human disease.

c. Pathogenicity. The shigellae differ from the salmonellae by remaining localized in the intestinal tract. The shigellae invade the intestinal epithelium, causing an intense inflammatory response characterized by the presence of neutrophils and macrophages. The result is bloody, mucopurulent diarrhea (i.e., dysentery).

(1) *S. dysenteriae* type 1 produces both neurotoxic and enterotoxic effects, apparently due to a single exotoxin.

(2) *S. sonnei* and *S. flexneri* have been shown to produce small amounts of the enterotoxin.

d. Laboratory diagnosis. Isolation and biochemical identification of shigellae are necessary for a definitive diagnosis. These are accomplished best by directly swabbing an ulcerated intestinal lesion and culturing on a selective medium (e.g., selenite broth or EMB agar).

e. Control and therapy. Attempts to control the spread of shigellosis are directed toward improving sanitation because the bacteria are disseminated by the fecal-oral route. Severe cases of dysentery are treated by intravenous replacement of fluid and electrolytes and by administration of antimicrobial agents (e.g., ampicillin and a combination of sulfamethoxazole and trimethoprim).

4. *Edwardsiella* is a recently discovered genus of lactose-nonfermenting enteric rods resembling *Salmonella* species. The bacteria have been isolated from stools of patients with diarrhea syndrome, but their etiologic role is uncertain. They also are isolated from healthy people.

5. *Citrobacter* species also antigenically resemble the salmonellae. They are seen as opportunistic pathogens in immunocompromised individuals and have been reported to cause urinary tract infection, wound infection, osteomyelitis, and gastroenteritis in elderly hospitalized patients. Cases of neonatal meningitis also have been reported.

F. Klebsielleae. Several species of this tribe are indigenous flora of the intestinal and respiratory tracts of humans. They may cause serious disease.

    1. *Klebsiella pneumoniae* produces a large antiphagocytic capsule that partly accounts for its virulence. Capsular types 1 and 2 are the etiologic agents of a severe, destructive pneumonia. The bacterium frequently is a secondary invader in pulmonary and other diseases. It is a primary cause of nosocomial urinary tract infection, and it has also been linked to epidemic diarrhea of the newborn. *K. pneumoniae* can be isolated from the oropharynx or gastrointestinal tract of about 5% of healthy people.

    2. *Enterobacter* species occur as indigenous intestinal flora; they also can be found on plants and as free-living organisms. These bacteria rarely are seen as primary causes of infection. *Enterobacter cloacae* and *Enterobacter aerogenes* have been isolated in cases of nosocomial urinary tract infection.

    3. *Serratia* species, particularly *Serratia marcescens*, formerly were considered nonpathogenic. This species was used in a variety of aerosol dissemination studies because of its bright red pigment. Now *S. marcescens* is known to cause fatal disease in neonates, immunosuppressed people, and debilitated patients. It is an important cause of nosocomial urinary infection leading to bacteremia.

    4. The genus *Hafnia* includes only one species, but it has almost 200 serotypes; some taxonomists believe the species should be placed in the genus *Enterobacter*. The bacteria are opportunistic pathogens seen principally in nosocomial infections in debilitated patients.

G. Proteeae. *Proteus* **species and** *Providencia* **species** are related genetically and physiologically. They are highly motile bacteria; *Proteus* species are noted for their swarming characteristics. The bacteria are seen as opportunistic pathogens in urinary tract infections, bacteremia, osteomyelitis, and other pathologic processes.

H. Erwinieae. *Erwinia* **species** are plant pathogens and, unlike most of the other Enterobacteriaceae, are not found routinely in the human intestinal tract. *Erwinia herbicola*, however, caused several cases of bacteremia and some deaths due to endotoxin shock when it contaminated a series of glucose-saline intravenous solutions that were administered to hospitalized patients.

II. *VIBRIO* AND *CAMPYLOBACTER* SPECIES. Most *Vibrio* and *Campylobacter* species of medical importance are enteric pathogens of humans.

A. *Vibrio* species are small, curved, gram-negative rods that are motile by means of a single, polar flagellum. The genus includes two important human pathogens.

    1. *Vibrio cholerae* is the etiologic agent of human cholera, a diarrhea syndrome that can be fatal owing to severe loss of water and electrolytes.
        a. Antigenic composition and biochemistry
            (1) The bacterium is similar biochemically to the Enterobacteriaceae, except that it is oxidase-positive and grows most efficiently at a pH level of 9.0–9.5.
            (2) Serotyping is based on the O antigens, three major serotypes—the Ogawa, Inaba, and El Tor strains—have been placed in serogroup O type 1 (i.e., O:1). These three strains are capable of producing severe diarrhea.
        b. Epidemiology. Cholera is spread by the ingestion of fecally contaminated food and water. The nutritional status of infected people may play an important role in the disease; those with an alkaline pH in the stomach and intestines appear to be more easily infected. About 60% of affected people develop a severe diarrhea known as rice-water stools. The disease is prevalent in the Far East, India, the Middle East, and Africa. A few cases occur annually in the United States.
        c. Pathogenicity. *V. cholerae* colonizes the intestinal tract in very high numbers. The cells attach to but do not invade intestinal mucosa.
            (1) A potent enterotoxin is released and binds to ganglioside receptors on the mucosal cells.

        (2) After a lag period of 15–45 minutes, adenyl cyclase is activated and the cAMP concentration inside the intestinal cells increases.

        (3) Increased intracellular cAMP results in the excretion of electrolytes such as chloride and bicarbonate ions along with massive quantities of water.

    d. Symptomatology. The major symptom of cholera is diarrhea. The feces contain epithelial cells, mucus, and large numbers of *V. cholerae.* In severe cases, as much as 1 L of fluid may be lost each hour. The patient usually feels bloated and may have abdominal pain before the onset of diarrhea.

    e. Laboratory diagnosis. *V. cholerae* may be viewed directly in stool samples, particularly by dark-field microscopy. Fluorescently labeled antibodies can be used in identifying the observed cells as *V. cholerae.* Selective media for culture are based on the ability of vibrios to grow at an alkaline pH, and identification can be made by biochemical tests.

    f. Therapy

        (1) Intravenous administration of fluids and electrolytes is critical for recovery.

        (2) Oral administration of a solution containing glucose and electrolytes has been successful, but the patient must be capable of consuming the liquid by mouth. Severely ill patients generally are too weak to ingest fluids.

        (3) Antibiotic therapy does not affect the disease once the enterotoxin attaches to the intestinal cells but can prevent later attacks by reducing the number of toxin-producing *V. cholerae* in the intestines.

    g. Immunization with toxoid, live cells, or attenuated cells provides only limited protection for approximately 6 months.

  2. *Vibrio parahaemolyticus* is a marine vibrio that requires a relatively high salt concentration for growth.

    a. Epidemiology. *V. parahaemolyticus* is distributed worldwide in marine environments and now is recognized as a common etiologic agent of acute enteritis associated with the consumption of improperly cooked seafood. The bacterium accounts for about half of all cases of food poisoning in Japan. The best control measure is the consumption of only thoroughly cooked seafood.

    b. Pathogenicity. The enterotoxin of *V. parahaemolyticus* has not been characterized completely. All strains isolated from cases of gastroenteritis are capable of producing a heat-stable hemolysin that lyses human and rabbit erythrocytes. The hemolysis has been designated Kanawaga hemolysis, and the enterotoxigenic strains are referred to as **Kanawaga-positive *V. parahaemolyticus*.**

B. *Campylobacter* **species** are curved, spiral, gram-negative rods with a single, polar flagellum. Species that are important as human pathogens once were classified as *Vibrio fetus.* The major etiologic agents of human infection are *Campylobacter fetus* subspecies *jejunii* and *C. fetus* subspecies *intestinalis.*

  1. Epidemiology

    a. Most *Campylobacter* species are associated with infectious diseases, particularly venereal diseases, in animals. Some subspecies of Campylobacter *sputorum* appear to be indigenous flora of the urogenital tract of animals.

    b. Human infections apparently are transmitted by the fecal-oral route, especially by the ingestion of contaminated food or water.

    c. *C. fetus* subspecies *jejuni* and *C. fetus* subspecies *intestinalis* appear to be the cause of human infection more frequently than was previously suspected.

  2. Pathogenicity and symptomatology

    a. Gastroenteritis is the most common type of human infection caused by *Campylobacter* species, especially in children. It resembles dysentery and usually is self-limited but may last for several days. The heat-labile enterotoxin of *C. fetus* has not been totally characterized.

    b. These bacteria have been reported to cause bacteremia, meningitis, endocarditis, arthritis, spontaneous abortion, and, occasionally, urinary tract infection.

  3. Laboratory diagnosis and therapy

    a. *C. fetus* grows best under microaerophilic conditions (i.e., an environment of 6% oxygen and 10% carbon dioxide) on an enriched medium.

   b. Most cases of mild diarrhea resolve spontaneously, but intravenous fluid and electrolyte replacement may be necessary in more severe forms of the disease. Aminoglycosides, macrolides, and other antibiotics successfully treat disseminated disease.

III. BACTEROIDES. The genus *Bacteroides* is composed of obligately anaerobic, gram-negative rods. The habitat of these bacteria primarily is the human intestinal tract. *Bacteroides fragilis*, the most numerous bacterium in the normal human colon, may account for up to 99% of fecal flora. *Bacteroides* species also may be isolated from the normal oropharynx, vagina, and external genitalia.

A. Morphology

   1. Cells. The bacteroides are slender, nonmotile, gram-negative rods that do not form spores. These bacteria exhibit a relatively high degree of pleomorphism; coccobacillary and branching forms frequently are observed.

   2. Colonies. *B. fragilis* appears as a pearl-gray or white colony. Colonies of *Bacteroides melanin-ogenicus* are light brown to black in color. The colonies are smooth and glistening in appearance owing to capsular polysaccharide.

B. Growth. The bacteroides are strict anaerobes and grow best on a complex medium (e.g., brain-heart infusion agar) in an anaerobic atmosphere containing 10% carbon dioxide.

C. Epidemiology. The bacteroides routinely can be isolated from the intestinal tract and the oropharynx of healthy humans. *B. fragilis* is the most common species found in the intestinal tract and typically numbers 10^{12} bacteria/g of feces. *B. melaninogenicus* colonizes the mouth and pharynx.

D. Pathogenicity. The polysaccharide capsule of the bacteroides appears to be a major virulence factor. Antibodies against the capsule have been detected in the sera of patients with disease. *B. fragilis* produces a superoxide dismutase that allows it to remain viable in the presence of oxygen for several days. The bacteroides appear to lack the lipopolysaccharide endotoxin found in other enteric gram-negative bacteria.

   1. B. fragilis generally is linked with infections that occur below the diaphragm. Infections usually are associated with mixed flora that may include anaerobic streptococci, fusiform bacteria, and facultatively anaerobic rods. *B. fragilis* may cause postoperative peritonitis and gynecologic infections as well as occasional infections involving lung or brain abscesses.

   2. B. melaninogenicus commonly is isolated from anaerobic infections that occur above the diaphragm. This organism frequently is seen in a mixed etiology of disease, and it can cause lung, meningeal, and oropharyngeal infections.

   3. Abscess formation is common during bacteroidosis and may occur cutaneously (particularly at the site of a surgical wound), in mucocutaneous tissue, and in the deep organs (e.g., the lungs and brain). Bacteremic spread of these organisms may result in endocarditis.

E. Laboratory diagnosis. The bacteroides usually are suspected as the etiologic agent of a disease process on the basis of their anaerobic growth, microscopic morphology, and colonial appearance. Confirmation and speciation are made on the basis of biochemical tests or gas chromatography profiles of the fatty acid content of the cells. *B. fragilis* produces β-lactamase, which provides resistance to penicillin. *B. melaninogenicus* and other bacteroides, however, are susceptible to penicillin.

IV. PSEUDOMONADS. The genus *Pseudomonas* and other genera of the family Pseudomonadaceae are obligately aerobic, gram-negative rods. Many members are free-living species, many are plant pathogens, and a few are associated with human disease. The species causing human infection usually are referred to as the glucose-nonfermenting, gram-negative rods.

A. *Pseudomonas aeruginosa* is the most common etiologic agent of human disease. However, it can be isolated from the intestinal tract of approximately 5% of healthy people. Nearly 30% of hospitalized patients show colonization. Most strains of *P. aeruginosa* produce pyocyanin, a blue pigment, and fluorescein, a green fluorescent pigment. Although these pigments are lethal for other bacteria, there is no indication that they play any role in human disease.

1. **Pathogenicity.** *P. aeruginosa* produces several virulence factors, including exotoxin A, proteases, a leukocidin, and a phospholipase C. Exotoxin A has been characterized most extensively.

   a. **Exotoxin A** is a potent inhibitor of protein synthesis; in several mammals, it has a 50% lethal dose (LD_{50}) of less than 50 μg/kg body weight.

      (1) The toxin inhibits amino acid uptake at the cellular level and causes cell death.

      (2) The mode of action is the same as that of diphtheria toxin, in which elongation factor 2 is covalently bound and thus inactivated. Despite analogous modes of action, exotoxin A and diphtheria toxin are distinct antigenically.

   b. **Elastase** is a protease produced by *P. aeruginosa*; it hydrolyzes elastin in cells of arterial walls. This process allows *P. aeruginosa* to invade the tissue and become sequestered in arterial walls. The elastase also inactivates some of the protein components of complement.

   c. **Collagenase** is another protease produced by *P. aeruginosa*; it hydrolyzes the collagen in connective tissue. Collagenase appears to be a virulence factor in infectious processes that involve the cornea.

2. **Clinical disease.** Pseudomonads, though they are ubiquitous, rarely infect healthy people. These bacteria are not highly invasive; however surface contamination of wounds is a common problem. A variety of patient populations appear especially susceptible to infection.

   a. **Serious infection may occur** in patients receiving long-term antibiotic therapy for wounds, burns, and cystic fibrosis and other illnesses because of the resistance of *P. aeruginosa* to many antibiotics. Approximately 25% of burn victims develop *P. aeruginosa* infection, which frequently leads to fatal septicemia.

   b. **Catheterization** can result in urinary tract infection caused by *P. aeruginosa* introduced mechanically. Intravenous catheterization and lumbar puncture also may introduce the bacteria.

   c. **Respiratory therapy equipment** that is improperly maintained can lead to pulmonary infection with *P. aeruginosa*. The bacteria may contaminate and grow in distilled water and other fluids and may be inhaled in large numbers by the debilitated patient.

   d. **Natural and chemically induced immunosuppression** predisposes to *P. aeruginosa* infection. Leukemia patients and patients receiving immunosuppressive therapy for organ transplants are at high risk for developing septicemia. Septicemia, endocarditis, and septic arthritis due to *P. aeruginosa* also are seen in intravenous drug abusers.

3. **Therapy.** *P. aeruginosa* infections are difficult to control because these bacteria are resistant to many drugs. The quinolone and aminoglycoside antibiotics currently are being used with some success, but the mortality rate for septicemia remains at about 80%. Antiseptics containing active chloride ions are useful in disinfecting wound surfaces.

B. **Other *Pseudomonas* species** responsible for human disease include *Pseudomonas maltophilia*, *Pseudomonas mallei*, *Pseudomonas pseudomallei*, *Pseudomonas cepacia*, *Pseudomonas multivarans*, *Pseudomonas fluorescens*, *Pseudomonas putida*, and *Pseudomonas stutzeri*. Most of the human infections caused by these species are the result of mechanical introduction of the bacteria into debilitated or immunocompromised patients. Melioidosis has been reported in Vietnam veterans who apparently inhaled *P. pseudomallei* in dust and water aerosolized by helicopters.

C. ***Acinetobacter anitratus*.** *A. anitratus* belongs to the Neisseriaceae family but clinically is considered with the pseudomonads. This nonmotile organism is the second most common glucose-nonfermenting, gram-negative rod isolated from human clinical specimens. The disease processes it produces resemble those produced by the pseudomonads.

STUDY QUESTIONS

Directions: Each question below contains five suggested answers. Choose the **one best** response to each question.

1. What is the most common source of salmonellosis in the United States?

(A) Pet turtles
(B) Rare roast beef
(C) Eggs
(D) Poultry products
(E) Partially cooked seafood

2. What component of gram-negative lipopolysaccharide appears to be responsible for most of its effects?

(A) Core polysaccharide
(B) Lipid-glucose fragment
(C) Lipid A moiety
(D) O antigen
(E) Lipoprotein subunit

3. What is the single most frequent etiologic agent of ascending urinary tract infection?

(A) *Klebsiella pneumoniae*
(B) *Serratia marcescens*
(C) *Citrobacter freundii*
(D) *Enterobacter cloacae*
(E) *Escherichia coli*

4. Which bacteria are most numerous in the human gastrointestinal tract?

(A) *Escherichia* species
(B) *Bacteroides* species
(C) *Pseudomonas* species
(D) *Enterobacter* species
(E) *Proteus* species

Directions: Each question below contains four suggested answers of which **one or more** is correct. Choose the answer

A   if **1, 2, and 3** are correct
B   if **1 and 3** are correct
C   if **2 and 4** are correct
D   if **4** is correct
E   if **1, 2, 3, and 4** are correct

5. All virulent members of the Enterobacteriaceae family are characterized by

(1) glucose fermentation
(2) the O antigen
(3) lipopolysaccharide
(4) lactose fermentation

6. The fermentation of lactose is indicated by pink to red colonies on

(1) selenite medium
(2) eosin-methylene blue (EMB) agar
(3) triple sugar iron (TSI) agar
(4) MacConkey agar

7. Endotoxin shock is characterized by

(1) hypotension
(2) hypoglycemia
(3) intravascular coagulation
(4) hypothermia

8. Important virulence factors produced by *Pseudomonas aeruginosa* include

(1) collagenase
(2) exotoxin A
(3) elastase
(4) hyaluronidase

Directions: The group of questions below consists of lettered choices followed by several numbered items. For each numbered item, select the **one** lettered choice with which it is **most** closely associated. Each lettered choice may be used once, more than once, or not at all. Choose the answer

    A if the item is associated with **(A) only**

    B if the item is associated with **(B) only**

    C if the item is associated with **both (A) and (B)**

    D if the item is associated with **neither (A) nor (B)**

Questions 9–14

Match the identifying factors with the appropriate bacterial group.

(A) *Salmonella* species
(B) *Shigella* species
(C) Both
(D) Neither

  9. Typically cause bacteremia

10. Humans are the only natural host

11. Etiologic agents of gastroenteritis

12. Invade the intestinal mucosa

13. Optimal growth occurs at pH 9.0–9.5

14. Transmitted by the fecal-oral route

ANSWERS AND EXPLANATIONS

1. The answer is D. [*I E 2 b (1)*] Poultry products, flesh and eggs together, are the most common source of salmonellosis in the United States at the current time. Large numbers of the bacteria are ingested in improperly cooked or handled foods. Symptomatology generally occurs in 10–24 hours. Pet turtles were the cause of more than 300,000 cases of salmonellosis annually in the United States during the 1970s until their sale and transport were banned. Salmonellosis is increasingly associated with rare roast beef but not with seafood. It is estimated that one in four cases of food poisoning in the United States is salmonellosis.

2. The answer is C. [*I D 1*] Toxic lipopolysaccharide (LPS) from smooth colonial types of all Enterobacteriaceae consists of a core polysaccharide, lipid A, and the O antigen. The lipid A moiety appears to be responsible for the physiologic, metabolic, and cellular changes associated with gram-negative endotoxin shock. The LPS is a major cause of virulence among the Enterobacteriaceae; conversion of the bacteria to rough colony types lacking the O antigen component makes the cells avirulent.

3. The answer is E. [*I E 1 a*] *Escherichia coli* is the single most frequently isolated etiologic agent of ascending urinary tract infection in otherwise healthy people. Urinary tract infection is more common in females, particularly young girls, and in the elderly probably because of anatomic reasons in the female and underlying factors in the elderly. *Klebsiella pneumoniae*, *Serratia marcescens*, *Citrobacter freundii*, and *Enterobacter cloacae* have been isolated as etiologic agents of urinary tract infection at some time, particularly in people with underlying metabolic or immunosuppressive conditions. The presence of 10^5 or more Enterobacteriaceae in 1 ml of properly collected urine, by definition, indicates infection, although the absence of this finding in the presence of appropriate other symptomatology does not rule out a diagnosis of urinary tract infection. Smaller numbers of other groups of micro-organisms, such as gram-positive cocci and yeasts, generally indicate urinary tract infection.

4. The answer is B. [*III*] *Bacteroides* species, primarily *Bacteroides fragilis*, are the most numerous bacteria in the human gastrointestinal tract. They account for up to 99% of the fecal flora. *B. fragilis* is a strict anaerobe. *Escherichia coli* is the most common facultative anaerobe found in the colon. *Enterobacter* and *Proteus* species are Enterobacteriaceae found in the colon of nearly 100% of healthy people. *Pseudomonas* species also may be indigenous flora of the human gastrointestinal tract, although in a smaller percentage.

5. The answer is A (1, 2, 3). [*I A 1, B 1, C 1, D*] Members of the Enterobacteriaceae family have certain common characteristics. All are gram-negative rods that ferment glucose to form organic acids. The cell envelope of smooth colonial types contains lipopolysaccharide (LPS). LPS consists of three components: lipid A, the core polysaccharide, and the O antigen. O antigen may be lost on successive transfers in the laboratory, but the rough colonial types that result are avirulent. The fermentation of lactose is a differential characteristic for genera within the family.

6. The answer is D (4). [*I C 2 b*] The differentiation between lactose fermenters and lactose nonfermenters within the Enterobacteriaceae family is an important aid to identification of the bacteria. MacConkey agar contains a pH indicator that gives a pink to red coloration to isolated colonies that ferment lactose and thus produce organic acids. The medium also contains bile salts, which inhibit the growth of most gram-positive bacteria. Eosin-methylene blue (EMB) agar also is a differential medium for lactose fermenters, but the fermenting colonies turn deep blue to black or may develop a green metallic sheen. Triple sugar iron (TSI) agar detects glucose, lactose, and sucrose fermentation by a red to yellow color change in the medium. Selenite medium is a selective medium for *Salmonella* and *Shigella* species.

7. The answer is A (1, 2, 3). [*I D 2 a, b, c, g*] Endotoxin shock is characterized by a variety of physiologic, metabolic, and cellular effects. Hypotension occurs with small doses of LPS within 30 minutes of exposure in most people; larger doses can cause permanent hypotension and death. Hypoglycemia is a metabolic effect of endotoxin shock; it probably occurs as a result of changes in the utilization of carbohydrates and lipids and in sensitivity to epinephrine. Intravascular coagulation is the result of local and generalized Shwartzman reactions, usually upon a second exposure to LPS. Depletion of circulating clotting factors as a result of the generalized Shwartzman reaction may cause serious bleeding. Fever, rather than hypothermia, is characteristic of endotoxin shock. Fever typically occurs within 30 minutes of exposure to small amounts of LPS and probably is the result of the ability of LPS to stimulate the release of endogenous pyrogen.

8. The answer is A (1, 2, 3). *[IV A 1 a, b, c]* *Pseudomonas aeruginosa* is a member of the indigenous flora of some healthy persons and is ubiquitous in the environment. This gram-negative, glucose-nonfermenting aerobe is of low virulence as long as it is not introduced into the tissues. It essentially is noninvasive and lacks virulence factors such as hyaluronidase, which would allow it to invade tissues. Once it is introduced into tissue, for example by trauma, a variety of extracellular enzymes are produced that can break down the tissue. Exotoxin A is a potent inhibitor of protein synthesis; collagenase is a protease that hydrolyzes collagen in connective tissue; and elastase is a protease that hydrolyzes the elastin in arterial walls.

9–14. The answers are: 9-A, 10-B, 11-C, 12-C, 13-D, 14-C. *[I E 2 b, c; 3 b, c; II A 1 a (1)]* Salmonella species typically cause bacteremia following invasion of the intestinal mucosa. During the bacteremia, the salmonellae colonize the gallbladder and various lymphoid tissues, including the Peyer's patches of the intestinal wall. Usually, it is infection of the intestinal lymphoid tissue after the bacteremia that causes the diarrheal syndrome associated with salmonellosis.

Humans are the only known natural hosts of *Shigella* species, although primates can be infected experimentally. There is a high incidence of shigellosis in institutionalized persons, particularly where individual hygiene is difficult to control.

Both salmonellae and shigellae cause gastroenteritis, although the predominant form of the illness differs for the two genera. *Salmonella* species classically cause diarrhea and *Shigella* species classically cause dysentery, but these relationships are not mutually exclusive.

Both salmonellae and shigellae invade the intestinal mucosa after ingestion of the bacteria. The two genera differ, however, in their subsequent dissemination. *Salmonella* species disseminate via the lymphatics and bloodstream throughout the body, whereas *Shigella* species tend to remain localized in the gastrointestinal tract.

Neither salmonellae nor shigellae grow well at a pH of 9.0 or above. The optimal pH for growth is between 6.8 and 7.2. *Vibrio* species grow optimally at a pH of 9.0–9.5.

Both salmonellae and shigellae inhabit the gastrointestinal tract and are transmitted by the ingestion of contaminated food or beverages. Most cases of salmonellosis are caused by ingesting poultry products that are contaminated during the slaughtering of the poultry or, in the case of eggs, during the laying process. Shigellae can be transmitted from human to human by direct fecal contamination of food or beverages, or they may be transmitted by mechanical vectors such as flies.

12
Gram-Negative Rods: *Haemophilus, Bordetella,* and Zoonotic Agents

Gerald E. Wagner

I. *HAEMOPHILUS* SPECIES. The genus *Haemophilus* is characterized by an absolute requirement for blood components in its growth medium. The genus includes species that are members of the human indigenous flora and species that are human pathogens.

A. *Haemophilus influenzae* is the most important species in the genus because of its pathogenicity for humans. This bacterium is an important etiologic agent of **respiratory and meningeal infections** in children.

1. Morphology
   a. Cells. *H. influenzae* is a small, gram-negative coccobacillus with a tendency to be pleomorphic.
   b. Colonies. *H. influenzae* forms small, glistening, gray to white colonies on chocolate agar.

2. Growth requirements. *H. influenzae* is a facultative anaerobe that grows best under aerobic conditions.
   a. Fresh blood must be present in its growth medium, as it must be for all *Haemophilus* species. The blood provides two necessary growth factors.
      (1) X factor is a heat-stable substance that is now known to be **hematin**. *H. influenzae* requires hematin for the synthesis of a heme-containing cytochrome system and for the enzyme catalase.
      (2) V factor is a heat-labile compound that is now believed to be **nicotinamide-adenine dinucleotide (NAD)**. The addition of NAD, nicotinamide-adenine dinucleotide phosphate (NADP), or nicotinamide riboside to culture medium satisfies the need for V factor from blood.
   b. Red blood cells must be lysed by heat (80° C for 15 minutes) in agar medium to release the growth factors and to inactivate enzymes that can destroy the V factor. *H. influenzae* is not capable of lysing erythrocytes.

3. Antigenic composition
   a. Six serotypes are designated, a through f, based on the antigenic properties of the capsular material. Capsular types can be transferred between strains.
      (1) Protective antibodies are induced by the capsules. They enhance phagocytosis and stimulate a complement-requiring bactericidal action.
      (2) *H. influenzae* type b is the most common cause of infection in children. The capsular antigen consists of a polymer of ribose and ribitol phosphate.
   b. Nonencapsulated strains are antigenically heterogeneous.
   c. The M antigen is a protein common to all strains of *H. influenzae*.

4. Epidemiology. *H. influenzae* is an obligate human parasite that is transmitted by respiratory droplets.
   a. The carriage rate of nonencapsulated, avirulent strains is 30%–50% in children and much lower in adults.
   b. Meningitis due to *H. influenzae* occurs at an annual rate of about 1 in 2000. The incidence is highest in children between the ages of 3 months and 6 years.

5. Pathogenicity. *H. influenzae* acquired its name from the mistaken belief that it was the single etiologic agent of the influenza epidemics in the 1890s and early 1900s. The bacterium is a frequent secondary invader of the inflamed pharynx.

    a. Acute epiglottitis is a serious disease of children caused by *H. influenzae* type b. The child usually becomes suddenly ill and has a swollen, red, and edematous epiglotis. Respiratory distress is the characteristic symptom and usually has to be treated with a tracheotomy.

    b. Meningitis is the most serious disease caused by *H. influenzae*. It generally occurs when the bacteria in the nasopharynx of an asymptomatic carrier invade the regional lymph nodes. Hematogenous dissemination results in meningeal infection.

        (1) Permanent neurologic changes often are produced in children. Mental retardation, hydrocephaly, and blindness are common sequelae.

        (2) The risk of disease is inversely proportional to the titer of circulating antibodies against capsular polysaccharide.

        (3) The precipitating factors that result in infection and dissemination are not known.

6. Laboratory diagnosis. *H. influenzae* does not ferment sugars, and biochemical tests are of little value in identification. Several serologic tests have been developed for identification of the bacteria in clinical specimens.

    a. A **quellung reaction** obtained upon mixing sedimented cerebrospinal fluid (CSF) or conjunctival exudate with specific antiserum indicates *H. influenzae* infection. With appropriate symptomatology and history, the presence of gram-negative, pleomorphic rods in these specimens is presumptive evidence of *H. influenzae*.

    b. Fluorescently labeled antibodies against specific capsular polysaccharides are useful in detecting the antigen in the CSF. This test is highly specific for *H. influenzae*.

    c. Counterimmunoelectrophoresis of CSF can detect polysaccharide antigens. Formation of a precipitin band is strong evidence of *H. influenzae* meningitis.

    d. Assays of requirement for X and V factors. CSF, blood, or epiglotis material is cultured on chocolate agar at 37° C in a candle jar or carbon dioxide incubator. The isolated bacteria are assayed for their requirements for X and V factors.

        (1) Direct assay. Specimens of suspected *H. influenzae* are streaked on a nutrient medium for confluent growth on the surface. A paper strip impregnated with X and V factors is placed on the agar surface. The growth of bacteria around the strip but not on other portions of the medium is indicative of *H. influenzae*.

        (2) The satellite colony test also identifies a specimen as *H. influenzae*. The unidentified specimen is streaked on blood agar, and a small amount of *Staphylococcus aureus* is placed in a spot on the agar. Growing *S. aureus* hemolyzes the blood and releases X and V factors. Small satellite colonies around the *S. aureus* colony can be identified as *H. influenzae*.

7. Therapy and immunity

    a. *H. influenzae* type b infections usually are treated with ampicillin or with chloramphenicol. Drug-resistant strains of *H. influenzae* now have been reported for both agents.

    b. Passive immunity in neonates is provided by the mother, and natural immunity usually is acquired by 8 years of age. Vaccines employing capsular polysaccharide are protective but do not elicit a significant response in children under the age of 2 years, who comprise the high-risk group.

B. *Haemophilus aegyptius* currently is considered a variant of *H. influenzae*. It is the etiologic agent of **conjunctivitis**. The lay name for this disease is pink eye.

1. Pathogenicity

    a. Mild conjunctivitis is characterized by vesicular infection. A slight discharge may occur.

    b. Severe conjunctivitis usually is associated with a painful irritation. Lacrimation, swelling of the lids, photophobia, and a mucopurulent discharge are characteristic.

2. Laboratory diagnosis. *H. aegyptius* can be cultured on chocolate agar. Smears of the discharge reveal a gram-negative, pleomorphic rod.

3. Therapy. Tetracycline currently is the antibiotic of choice. It is applied as a topical ointment.

C. *Haemophilus ducreyi* is the etiologic agent of **chancroid**. The chancre is soft, in contrast to the hard chancre of syphilis.

1. Symptomatology. Chancroid is characterized by a ragged ulcer on the genitals; it is painful and surrounded by noticeable swelling. Regional lymphadenopathy occurs, and the lymph nodes may suppurate. Multiple lesions can occur by autoinoculation.

2. Diagnosis and therapy. Chancroid is transmitted sexually. The diagnosis is made by microscopic examination of smears and by culture on chocolate agar. Sulfonamides and streptomycin traditionally have been successful in treating the infection.

II. *BORDETELLA* SPECIES.

Several species of the genus *Bordetella* can cause human infections, and *Bordetella pertussis* is the etiologic agent of **whooping cough**. *Bordetella parapertussis* and *Bordetella bronchiseptica* cause mild forms of whooping cough in humans.

A. Morphology

1. Cells. *Bordetella* species are small, gram-negative coccobacilli. They stain poorly by Gram's method and are visualized more easily with toluidine blue stain.

2. Colonies. Upon initial isolation from clinical specimens, the cultures produce small, glistening, grayish, mercury-drop or pearl colonies.

B. Growth requirements.
B. pertussis grows best on Bordet-Gengou culture medium containing potato starch, blood, and glycerol.

1. Initial isolation of the species from clinical specimens usually requires Bordet-Gengou medium with 30%–50% blood in the agar. Increased carbon dioxide in the atmosphere enhances growth of the isolates.

2. Successive transfers on agar containing lesser amounts of blood eventually result in the ability of the bacteria to grow on nutrient agar.

C. Antigenic composition.
Successive cultures of *B. pertussis* isolated from a clinical specimen result in a shift from a phase I to a phase IV form. The molecular basis for the shift is not completely understood. The change is associated with decreasing virulence and with antigenic differences.

1. Phase I
   a. Phase I *B. pertussis* is virulent and usually categorized by colonial appearance. The antigens of phase I bacteria have not been completely characterized. The bacteria possess **pili** and an **O-specific antigen**.
   b. Other antigens of phase I *B. pertussis* include a **pertussis toxin** that is destroyed when it is exposed to a temperature of 80° C for 30 minutes. Phase I cells also excrete **two hemagglutinins**.

2. Phase II and phase III bacteria are less virulent than phase I. They have specific, intermediate colonial characteristics.

3. Phase IV bacteria lack pili and the O-specific antigen. They are avirulent. Electrophoretic analysis of proteins shows that there are at least two fewer proteins in phase IV cells than in phase I cells.

D. Epidemiology

1. *B. pertussis* apparently is an obligate human parasite.
   a. Transmission is by direct contact, by droplets from the respiratory tract, and by freshly contaminated fomites. Healthy carriers of the etiologic agent have not been detected.
   b. The incidence of whooping cough has declined dramatically since the institution of mandatory vaccination of children. However, the fatality rate per number of cases has remained relatively stationary.

2. *B. bronchiseptica* can be isolated from the oropharynx of dogs, and it is the etiologic agent of kennel cough in puppies.

E. Pathogenicity of *B. pertussis*

1. Virulence factors of *B. pertussis* phase I cells include histamine-sensitizing factor, lymphocytosis-promoting factor, heat-labile toxin, mouse protective factor, and islet-activating factor. These toxins and others play a role in the symptomatology of whooping cough. The major difference between virulent phase I cells and avirulent phase IV cells, however, is the presence of a heat-labile toxin, pertussis toxin, pili, and adenyl cyclase activity in the phase I cells.

    a. Pili (i.e., **fimbriae**) covering the cell surface are the agglutinogens of *B. pertussis* phase I cells. They permit the adherence of *B. pertussis* to the ciliated epithelium of the respiratory tract. The pili also are associated with hemagglutinin activity. Pili have many different antigenic types, some of which always are included in vaccines.

    b. Pertussis toxin probably is a group of substances that includes endotoxin. When injected into laboratory animals, it produces a variety of symptoms, including increased susceptibility to histamine and serotonin and, thus, to anaphylactic shock. The lymphocytosis-promoting factor alters the migration of lymphocytes. The islet-activating factor increases insulin production and inhibits epinephrine-induced hyperglycemia.

    c. Adenyl cyclase is synthesized and excreted into the surrounding medium by virulent phase I *B. pertussis*. Vaccines also contain this enzyme, but its role in the disease is unknown.

2. Symptomatology of whooping cough involves three stages after the incubation period, which usually lasts 7–10 days after exposure to the respiratory discharge of an infected person.

    a. The catarrhal stage is characterized by relatively mild, flulike symptoms. The cough is mild but persistent and irritating to most patients. It appears that *B. pertussis* is prominent in the respiratory tract at this time, and the disease is most contagious during this state, which lasts 1–2 weeks.

    b. The paroxysmal stage is the period when the characteristic whoop develops in the cough. Periods of violent coughing are followed by an inspiratory whoop. The coughing can be severe enough to cause cyanosis with vomiting and convulsions. Total exhaustion and prostration can be the result. The paroxysmal stage can persist for a period of weeks.

    c. The convalescent stage lasts for an additional 2–4 weeks. It is characterized by a gradual decrease in paroxysmal symptomatology.

F. Laboratory diagnosis

1. Specimen collection. A fine wire tipped with calcium alginate (cotton inhibits growth) is inserted into the posterior nares. The swab is left in place for 15–30 seconds, preferably while the patient coughs.

2. Cultures are grown on Bordet-Gengou medium, and the bacteria are identified by morphology and agglutination reactions with specific antiserum.

3. Fluorescently labeled antibodies may be used to identify bacteria from cultures or in smears made from clinical specimens.

G. Therapy. Most of the symptomatology of whooping cough is caused by the host's response to toxins and virulence factors released by *B. pertussis*. Antimicrobial therapy is not totally satisfactory because it does not alter the course of the disease. It does, however, prevent secondary bacterial infections. The tetracyclines, macrolides, and chloramphenicol are among the drugs used.

H. Immunization provides temporary protection from disease.

1. Effective vaccine is prepared from killed phase I bacteria. It must contain pertussis toxin, agglutinogens, and capsular material. The vaccine is generally given in conjunction with diphtheria and tetanus toxoids.

2. Administration. Because of the high mortality rate in children less than 1 year old, the vaccine is administered at 2 months. Booster shots are given at 4, 6, and 18 months and upon admission to school.

III. ZOONOTIC AGENTS. Several small, gram-negative rods are the etiologic agents of infectious diseases in animals (zoonoses); some of these diseases can be transmitted from animals to humans. The major genera involved in this type of infectious process are *Brucella*, *Yersinia*, *Francisella*, and *Pasteurella*.

A. *Brucella* species are the etiologic agents of **brucellosis** in livestock. Humans acquire the disease by direct contact with infected animals or by ingestion of contaminated livestock products. Species that are of medical importance include *Brucella abortus*, *Brucella melitensis*, and *Brucella suis*. Each species is subdivided into a number of biotypes.

1. **Morphology**
   a. **Cells.** The brucellae are small, gram-negative, nonmotile rods. Coccobacillary forms are predominant, but cocci and longer rods may be observed in young cultures.
   b. **Colonies.** The colonies of *Brucella* species are small, convex, and smooth; they generally require 2–5 days for growth.

2. **Growth requirements.** Enriched complex media are required for growth. *B. abortus* requires increased (i.e., 5%–10%) carbon dioxide for optimal growth.

3. **Antigenic composition**
   a. **Two major surface antigens**, A and M, vary in amount in the different species. Antisera can be used to identify the species.
   b. **L antigen** is an outer membrane antigen similar to the virulence (Vi) antigen of salmonellae.

4. **Biochemical characteristics.** The brucellae use carbohydrates but do not produce detectable amounts of either acid or gas. Some biotypes produce hydrogen sulfide gas, and some are susceptible to the growth-inhibiting effect of the dyes, basic fuchsin and thionin.

5. **Epidemiology.** The brucellae essentially are animal pathogens; humans are accidental hosts. The bacteria are transmitted from animal to animal and from animal to human by contact with contaminated feces, urine, milk, and tissue. Each species of *Brucella* generally is associated with a specific animal; *B. abortus* is found most frequently in cattle, *B. melitensis* in goats, and *B. suis* in swine. All three species of *Brucella* are capable of causing human disease.
   a. **In the United States**, brucellosis in humans primarily is an occupational hazard. The bacteria enter the body by way of small lesions in the skin. Inhalation of brucellae occasionally may cause disease.
   b. **In other countries**, where pasteurization of dairy products is not practiced routinely, brucellosis is more prevalent. The bacteria usually are ingested in contaminated milk and cheese. Nearly 100,000 cases of brucellosis occur annually in Spain, compared to fewer than 200 cases in the United States.

6. **Pathogenicity and symptomatology.** Human brucellosis typically is characterized by fever, weakness, and myalgia. These symptoms can occur in various patterns; the most common form is undulant fever.
   a. **Route of infection.** Ingested brucellae probably enter the body through small breaks in the mucosa of the oropharynx. Invasion of the body by brucellae results in lymphatic dissemination of the bacteria.
      (1) As regional lymph nodes become infected, the brucellae are phagocytized, but they are capable of replication inside the macrophages.
      (2) As macrophages die, brucellae are released into the bloodstream and establish localized infections in the liver, spleen, kidneys, and bone marrow.
   b. Mammary glands in **both humans and animals** can be infected. When they are, brucellae are shed into breast milk.
   c. Placental and fetal tissues are infected in animals, causing abortion, but this is not a human phenomenon. The difference between animals and humans appears to be due to erythritol in the animal placenta; the sugar alcohol enhances the growth of brucellae.
   d. **Course of human disease.** The incubation period in humans is 1–6 weeks. The classic symptoms are malaise, weakness, and an undulating diurnal fever. In rare cases, extreme pyrexia can cause death. Loss of appetite, backache, and headache are other frequent symptoms. Relapses are common over a 2- to 4-month convalescent period.
   e. **Chronic brucellosis.** The existence of chronic brucellosis is debatable because the bacteria have not been isolated from the rare cases of this disease. Chronic brucellosis is diagnosed on the basis of a history of brucellosis and the clinical symptoms of weakness, malaise, and emotional disturbances.

7. **Laboratory diagnosis.** Identification of the etiologic agent is required for a definitive diagnosis of brucellosis.
   a. **Culture.** Blood or biopsy material generally must be incubated in complex nutrient broth for longer than 5 days before it is discarded as negative. The isolated bacteria are identified by biochemical tests and can be biotyped by agglutination reactions.
   b. **Skin test.** A skin test antigen, **brucellin**, may be used for gathering epidemiologic data.

8. Therapy for brucellosis is difficult because of the intracellular growth of the bacteria. Rifampin is the drug of choice for effective treatment.

9. Control. Human brucellosis is controlled by controlling the disease in animals. In the United States, the incidence of disease in livestock has been reduced to less than 5%. Pasteurization of dairy products has helped to decrease human disease.

B. *Yersinia* species. The three species in the genus *Yersinia* are *Yersinia pestis, Yersinia enterocolitica,* and *Yersinia pseudotuberculosis.* Animals are the natural hosts of these species, all of which can cause human disease.

1. Morphology
   a. Cells. The *Yersinia* species generally appear as small, gram-negative, bipolar-staining coccobacilli. They have a tendency to display pleomorphism when grown in suboptimal conditions. With the exception of *Y. pseudotuberculosis,* they are nonmotile.
   b. Colonies of yersiniae derived from clinical specimens are grayish and mucoid, although irregular, rough colonies may be noted.

2. Growth requirements. The yersiniae are not fastidious; they can be grown on routine bacteriologic media.

3. Antigenic composition. All *Yersinia* species possess O antigens that are toxic for animals. The O antigens, which are lipopolysaccharide–protein complexes, can be subdivided on the basis of chemical and antigenic characteristics.
   a. *Y. pestis* has a variety of antigens that have not been completely characterized; the role of these antigens as virulence factors is unclear.
      (1) Fraction 1 is a capsular antigen that, chemically, is a glycoprotein. It is antiphagocytic, but it also stimulates other antibacterial reactions in the host.
      (2) The V/W antigen of *Y. pestis* consists of a protein, the V portion, and a lipoprotein, the W portion. The antigen is antiphagocytic and appears to promote the intracellular growth of the bacteria. *Y. pestis* strains that have only the V/W antigen remain virulent for mice.
      (3) Murine toxin is a proteinaceous intracellular toxin that has a 50% lethal dose (LD_{50}) of less than 1 μg in mice; it also is lethal for rats. The toxin causes shock and death by acting as an adrenergic antagonist.
      (4) Pesticin I and pesticin II are bacteriocins produced by *Y. pestis.* They are bactericidal for *Y. pseudotuberculosis* and some strains of *Escherichia coli.*
   b. *Y. pseudotuberculosis* is motile at 25° C and possesses an H antigen. At 37° C the H antigen is absent and the bacteria are nonmotile.

4. Epidemiology. The primary animal host for each of the *Yersinia* species is different. The mode of transmission from animal to human and the incidence of human disease also vary.
   a. *Y. pestis* is the etiologic agent of **plague.** The bacteria are associated with rodents, and the disease may exist as sporadic cases or as epidemics (i.e., epizootics).
      (1) Prairie dogs are believed to be the largest natural reservoir in the United States, and **rats** are the major source worldwide.
      (2) The rat flea is the primary vector in transmission of the disease to humans, although human fleas may become vectors and human-to-human transmission occasionally occurs.
      (3) Plague does not occur naturally in the eastern United States, but it is endemic west of the Mississippi River.
   b. *Y. enterocolitica* is the etiologic agent of **gastroenteritis** in humans. Rodents, dogs, cats, and domestic farm animals are the natural host of the bacteria. This species also has been isolated from lakes and well water. Humans probably become infected by the fecal-oral route.
   c. *Y. pseudotuberculosis* produces an **appendicitis-like syndrome** in humans. The species is a pathogen for both wild and domestic animals, in which it causes systemic disease. Humans apparently acquire the infection by the oral route after contact with infected animals.

5. Pathogenicity. Humans are accidental hosts of *Yersinia.*
   a. *Y. pestis* usually causes plague when the bacteria are introduced into the body by the bite of a flea. Rat fleas acquire the bacteria during a blood meal from an infected bacteremic rat or other rodent.

      (1) Transmission. *Y. pestis* produces an enzyme that, in conjunction with an enzyme in the flea's gut, clots the blood. The obstruction prevents the flea from feeding. The starving flea attempts to obtain a blood meal from a human and regurgitates the bacteria into the human body.

      (2) Human disease. At 28° C (i.e., normal flea temperature), *Y. pestis* does not possess either fraction 1 or V/W antigen. Cells that are phagocytized by neutrophils in humans usually are destroyed. Bacterial cells that survive synthesize both fraction 1 and V/W antigen.

         (a) Bubonic plague. *Y. pestis* spreads to the regional lymph nodes, usually in the groin or axillae, and buboes are formed. These buboes are swollen, tender lymph nodes that may suppurate and drain. The fatality rate associated with bubonic plague is about 75%.

         (b) Pneumonic plague. Lung involvement occasionally occurs. It can be spread from human to human by respiratory discharges. The fatality rate in untreated patients is almost 100%.

      (3) The complete pathogenesis of the disease remains unclear. Alone, none of the antigens or toxins produced by *Y. pestis* can produce plague.

   b. *Y. enterocolitica* causes enterocolitis characterized by abdominal pain, diarrhea, and fever.

      (1) Regional lymphadenopathy can occur; the syndrome mimics appendicitis.

      (2) Diarrhea is the result of a heat-stable enterotoxin that stimulates the synthesis of guanyl cyclase.

      (3) Complications of intestinal infection include septicemia with the development of lesions in internal organs and a severe arthritis that occurs 1–14 days after the acute attack.

   c. *Y. pseudotuberculosis* causes an enterocolitis similar to that produced by *Y. enterocolitica*. Swollen, tender lymph nodes are characteristic of the most frequent form of the disease, which resembles appendicitis. The bacteria rarely disseminate via the bloodstream or lymphatic system.

 6. Laboratory diagnosis. The best method of establishing the causative relationship of *Yersinia* species with disease is to isolate the bacteria from clinical specimens.

   a. Specimens. Biopsy material from lymph nodes provide a definitive source, but the etiologic agents also may be isolated from blood (in the case of *Y. pestis* and *Y. enterocolitica*) or from feces (*Y. pseudotuberculosis* and *Y. enterocolitica*).

   b. Identification is made by biochemical and serologic tests.

 7. Therapy. Disseminated systemic disease requires rapid treatment with appropriate antimicrobial agents. Antimicrobial therapy usually is not necessary in cases of enterocolitis, but supportive therapy such as administration of intravenous fluids may be of benefit in some patients.

C. *Francisella tularensis* is the etiologic agent of **tularemia**, sometimes called rabbit fever or deer fly fever. The disease is distributed worldwide.

 1. Morphology

   a. Cells. *F. tularensis* is a small, gram-negative rod. It shows a high degree of pleomorphism and may vary from coccoid to filamentous in shape. The bacteria are nonmotile.

   b. Colonies of *F. tularensis* are minute and droplike.

 2. Growth requirements. *F. tularensis* is a facultative anaerobe, although optimal growth occurs under aerobic conditions. Complex medium containing blood, tissue extracts, and cystine is required for growth. Small inocula must be used to minimize the addition of inhibitory factors.

 3. Epidemiology. *F. tularensis* has been isolated from numerous animals and also from insect vectors.

   a. In the United States, the bacteria have been isolated from deer flies in the Southwest and from wood ticks in the Northwest. Rabbits, the most common source of disease worldwide, account for 90% of cases in the United States. Flying squirrels are probably also a source of infection in the eastern United States.

   b. Human disease is acquired through insect bites and direct contact with the blood and tissues of infected animals.

 4. Pathogenicity. *F. tularensis* is an intracellular pathogen of phagocytic cells. It causes three major clinical syndromes. All are referred to as tularemia, and the specific form is largely determined by the route of infection.

a. **Ulceroglandular tularemia**, the most common form of the disease, is acquired by direct contact with infected animals.
   (1) **Primary lesions** develop at the site of inoculation, usually the fingers, and frequently become open ulcers in 7–8 days. Primary lesions also may occur on the eye and, in association with tick bites, on the lower extremities.
   (2) **Regional lymph nodes** swell, become tender, and occasionally break open and drain.
   (3) **Hematogenous spread** of the bacteria results in lesions in other organs, especially the liver and spleen.

b. **Pneumonic tularemia** results from hematogenous spread of the bacteria to the lung or from inhalation of aerosolized bacteria. Pneumonia develops and is fatal unless treated promptly. Pneumonic tularemia can be spread from person to person by respiratory droplets.

c. **Typhoidal tularemia** is acquired by the ingestion of contaminated food. Water contaminated by animals that have died of tularemia also is a source of infection. There are no primary lesions, and the high fever and gastrointestinal symptomatology resemble enteric fever caused by *Salmonella typhi*.

5. **Laboratory diagnosis**
   a. **Definitive diagnosis** of tularemia is made by culturing *F. tularensis*. The bacteria are difficult to grow, however, and pose a substantial threat to laboratory personnel.
   b. **Presumptive diagnosis** generally is based on appropriate history and clinical signs and symptoms, and on the application of fluorescently labeled specific antibodies to smears obtained from skin lesions. Retrospective diagnosis can be made by demonstrating a rise in serum titers of agglutinating antibodies.

6. **Therapy and control.** *F. tularensis* is a facultative intracellular parasite that is difficult to treat with antimicrobial agents.
   a. **Streptomycin, tetracyclines**, and **chloramphenicol** have been used successfully to treat tularemia.
   b. The disease is **impossible to control** because of the wide distribution of *F. tularensis* in nature.
   c. An effective attenuated **vaccine** is available for people at high risk for contracting the disease.

D. *Pasteurella multocida* is a member of the indigenous flora of the respiratory tract and oropharynx of many animals and birds. Infections in animals occur as pneumonia and septicemia during stress and are known as shipping fever and fowl cholera. Humans are accidental hosts.

1. **Morphology and growth requirements.** *P. multocida* is a small, gram-negative coccobacillus. The bacteria grow on routine bacteriologic media and form small grayish colonies.

2. **Epidemiology.** A common source of human infection is the bite of a dog or cat. Cats have a high degree of colonization in the nasopharynx, and most cases of human disease have followed cat bites.

3. **Pathogenicity.** Localized wound infection is common. Cleaning the wound and avoiding suturing generally prevent overt signs of infection. Septicemia can occur, and meningitis may develop. Chronic pulmonary infection probably is acquired by inhalation of the bacteria. The bacterial virulence factors are not clearly understood.

STUDY QUESTIONS

Directions: Each question below contains five suggested answers. Choose the **one best** response to each question.

1. In what disease stage is whooping cough most contagious?

(A) Catarrhal stage
(B) Paroxysmal stage
(C) Primary stage
(D) Recovery stage
(E) Convalescent stage

2. What is the etiologic agent of chancroid?

(A) *Yersinia pseudotuberculosis*
(B) *Haemophilus aegyptius*
(C) *Yersinia enterocolitica*
(D) *Haemophilus ducreyi*
(E) *Francisella tularensis*

3. Bordet-Gengou medium is used to isolate *Bordetella pertussis* from clinical specimens. What is the selective component of this medium?

(A) Bile salts
(B) Glycerol
(C) Potato starch
(D) Blood
(E) X factor

4. Which of the following microbes can cause an appendicitis-like syndrome?

(A) *Brucella abortus*
(B) *Francisella tularensis*
(C) *Brucella cholerasuis*
(D) *Pasteurella multocida*
(E) *Yersinia pseudotuberculosis*

5. What form of plague is most likely to be transmitted from human to human?

(A) Typhoidal
(B) Sylvatic
(C) Pneumonic
(D) Catarrhal
(E) Ulceroglandular

Directions: Each question below contains four suggested answers of which **one or more** is correct. Choose the answer

A   if **1, 2, and 3** are correct
B   if **1 and 3** are correct
C   if **2 and 4** are correct
D   if **4** is correct
E   if **1, 2, 3, and 4** are correct

6. Factors that appear to be important in the virulence of *Bordetella pertussis* include

(1) pertussis toxin
(2) pili
(3) histamine-sensitizing factor
(4) endotoxin

7. Etiologic agents of zoonoses that can cause human disease include

(1) *Brucella melitensis*
(2) *Pasteurella multocida*
(3) *Francisella tularensis*
(4) *Yersinia pestis*

8. Antiphagocytic virulence factors of *Yersinia pestis* include

(1) pesticin I
(2) V/W antigen
(3) A antigen
(4) fraction 1

9. Although humans are accidental hosts of the brucellae, *Brucella abortus* can cause

(1) fatal pyrexia
(2) infection of the mammary glands
(3) generalized muscular weakness
(4) septic abortion

Directions: The group of questions below consists of lettered choices followed by several numbered items. For each numbered item select the **one** lettered choice which it is **most** closely associated. Each lettered choice may be use once, more than once, or not at all.

Questions 10–14

Match each characteristic below with the appropriate microbe.

(A) *Yersinia pestis*
(B) *Francisella tularensis*
(C) *Pasteurella multocida*
(D) *Yersinia enterocolitica*
(E) *Brucella melitensis*

10. Vector is the wood tick

11. Indigenous flora of the nasopharynx of cats

12. Etiologic agent of an appendicitis-like syndrome

13. Can cause primary lesions on the eyes

14. Etiologic agent of undulant fever

ANSWERS AND EXPLANATIONS

1. The answer is A. [*II E 2*] Whooping cough, or pertussis, has three stages of clinical disease. The first stage, the catarrhal stage, is characterized by relatively mild, flulike symptomatology. It is during this mildest form of the syndrome, however, that *Bordetella pertussis* is most prevalent in the pharynx and the disease is most contagious. Symptomatology is most severe during the paroxysmal stage, but the clinical characteristics of this stage appear to be due primarily to the tissue reaction to bacterial virulence factors. It is difficult to isolate *B. pertussis* at this time. The third stage is the convalescent stage, which is characterized by a gradual lessening of symptomatology.

2. The answer is D. [*I C*] Chancroid is a sexually transmitted disease caused by *Haemophilus ducreyi*. The chancre is soft, painful, and surrounded by noticeable edema. *Haemophilus aegyptius* is the etiologic agent of conjunctivitis, commonly referred to as pink eye. *Yersinia pseudotuberculosis*, *Yersinia enterocolitica*, and *Francisella tularensis* are etiologic agents of zoonoses that can be transmitted to humans by contact with contaminated materials.

3. The answer is D. [*II B 1*] Bordet-Gengou medium normally is required for the initial isolation of *Bordetella pertussis* from clinical specimens. The medium contains potato starch, glycerol, and blood. The selectivity of the medium is the presence of blood in concentrations ranging from 30%–50%. The typical concentration of blood in standard bacteriologic blood agar is 3%–5%. The high concentration of blood in Bordet-Gengou medium inhibits the growth of most bacteria but has a detoxifying effect on the agar that allows *B. pertussis* to grow. Bile salts are selective for the growth of gram-negative enteric rods because they inhibit the growth of most gram-positive cocci. X factor (i.e., hematin) is one of the growth requirements for *Haemophilus* species.

4. The answer is E. [*III B 4 c, 5 c*] *Yersinia pseudotuberculosis* is the etiologic agent of human gastrointestinal disease that frequently is misdiagnosed as appendicitis. The bacteria cause disseminated disease in wild and domestic animals and appear to be acquired by humans through the oral route. The common feature in epidemic appendicitis-like outbreaks in children has been the consumption of chocolate milk. *Brucella* species cause disseminated disease in humans. *Pasteurella multocida* is the etiologic agent of wound infection following a cat or dog bite. *Francisella tularensis* can cause typhoidal tularemia characterized by high fever and gastrointestinal symptoms resembling enteric fever.

5. The answer is C. [*III B 5 a (4)*] Plague usually is transmitted to humans by the bite of the rat flea after it has taken a blood meal from a rodent infected with *Yersinia pestis*. Human to human transmission occasionally occurs from the bite of human fleas contaminated with the bacteria. Lung involvement with plague (i.e., pneumonic plague) is highly contagious and is spread among humans by respiratory droplets. Sylvatic plague is the disease in wild animal populations; more than 200 species of animals can develop plague. Ulceroglandular and typhoidal forms, as well as a pneumonic form, of disease are associated with tularemia. The catarrhal stage is the first clinical stage of whooping cough.

6. The answer is E (all). [*II E 1 a–c*] *Bordetella pertussis* produces a variety of virulence factors. Major differences between virulent phase I and avirulent phase IV forms is that the virulent form has pili, pertussis toxin, heat-labile toxin, and adenyl cyclase activity. Virulent strains also produce a histamine-sensitizing factor that appears to make tissue cells more susceptible to the tissue damaging effects of histamine. *B. pertussis* is a gram-negative coccobacillus, and like all gram-negative bacteria it possesses lipopolysaccharide endotoxin. The endotoxin is not as potent as that of the enteric gram-negative rods, but it does cause physiologic, biochemical, and metabolic effects.

7. The answer is E (all). [*III*] Zoonoses are infectious diseases that are endemic in animal populations. The etiologic agents of the classic zoonoses are the genera *Brucella*, *Yersinia*, *Francisella*, and *Pasteurella*. These infectious agents are transmitted to humans, usually through accidental contact with contaminated animal tissue, food, or fluid or by vectors. *Brucella melitensis*, the etiologic agent of brucellosis, is transmitted to humans by the ingestion of unpasteurized goat's milk or cheese. *Pasteurella multocida* is indigenous flora of the oropharynx of cats and dogs, but it causes serious respiratory and disseminated infections in other animals. Typically, it is transmitted to humans by the bite or scratch of a cat or dog. *Francisella tularensis* is the etiologic agent of tularemia. It causes disseminated disease in a variety of animals, and humans become infected by contact with the blood of infected animals or by vectors such as the wood tick or deer fly. *Yersinia pestis* is the cause of plague, and it is endemic in rodent populations such as prarie dogs in the United States and the rat worldwide. Human disease usually is contracted by the bite of an infected rat flea.

8. The answer is C (2, 4). [*III B 3 a (1), (2), (4), 4 a*] *Yersinia pestis* has two antiphagocytic virulence factors. Fraction 1 consists of a glycoprotein antigen of the capsule. V/W antigen of *Y. pestis* consists of a protein, the V portion, and a lipoprotein, the W portion. The V/W antigen is antiphagocytic, and this activity is retained even when the cell envelope is removed from the bacteria. Pesticin I and pesticin II are bacteriocins produced by *Y. pestis* that inhibit other species of bacteria. A antigen is a virulence factor of the brucellae, and it is most prominent in *Brucella abortus*.

9. The answer is A (1, 2, 3). [*III A 6 b–d*] The brucellae are major causes of zoonoses. Humans typically contract disease by ingesting contaminated milk or milk products. Brucellosis caused by *Brucella abortus* is characterized by fever, weakness, and myalgia. An undulating fever is most typical, but fatal pyrexia can occur. Generalized muscular weakness and myalgia are classic symptomatology of acute disease. The mammary glands of both humans and animals become infected, and the bacteria is shed in the milk. *B. abortus* is the etiologic agent of placental infection and septic abortion in animals, but these do not occur in humans. The specificity appears to be due to the large amount of erythritol in the placenta of animals; the human placenta does not contain erythritol.

10–14. The answers are: 10-B, 11-C, 12-D, 13-B, 14-E. [*III A 5, B 3 a, 5 b, c, C 3, 4 a, D 2*] Wood ticks are the major vector of *Francisella tularensis* in the northwestern United States. Human disease as a result of the tick bite typically is ulceroglandular. The principal vector of *Yersinia pestis* is the rat flea. *Yersinia enterocolitica*, *Pasteurella multocida*, and *Brucella melitensis* are not transmitted by vectors.

The nasopharynx of cats is highly colonized by *P. multocida*. The bacteria can be transferred to the claws by the cat's characteristic frequent licking and cleaning of them. Most human cases of pasteurellosis are the result of cat bites or scratches. The primary reservoirs of *Y. pestis* are the prairie dog in the United States and the rat worldwide. The major reservoir of *F. tularensis* worldwide is the rabbit. *B. melitensis* is associated primarily with goats. *Y. enterocolitica* has a variety of wild and domestic animal hosts.

Y. enterocolitica, like *Yersinia pseudotuberculosis*, is capable of causing a gastroenteritis that mimics appendicitis. Infection is acquired by ingesting food or beverages contaminated by infected animals, including dogs, cats, rodents, and domestic farm animals. The disease is characterized by abdominal pain, diarrhea, fever, and lymphadenopathy. *F. tularensis* can cause a typhoidal form of disease resembling enteric fever. *Y. pestis*, *P. multocida*, and *Brucella abortus* do not cause characteristic gastrointestinal infections.

F. tularensis causes primary lesions at the site of inoculation. When the bacteria are transmitted by a vector, the primary lesions occur at the site of the bite, frequently on the extremities. Handling of contaminated flesh can result in lesions on the fingers and hands, and contaminated droplets of blood can cause eye infection. *B. melitensis* and *Y. enterocolitica* cause disease following ingestion; *P. multocida* causes localized wound infection; and *Y. pestis* is transmitted by flea bites, but the lesion at the site of the bite is not characteristic.

B. melitensis, along with the other brucellae, causes human disease characterized by fever, weakness, and myalgia. The most common syndrome of brucellosis is undulant fever with a diurnal pattern. *Y. pestis* is the etiologic agent of plague; *Y. enterocolitica* causes gastroenteritis; *F. tularensis* causes tularemia; and *P. multocida* causes localized cat and dog wound infection.

13
Spirochetes
Gerald E. Wagner

I. *TREPONEMA* SPECIES. The genus *Treponema* includes three species that are pathogenic for humans: *Treponema pallidum, Treponema pertenue,* and *Treponema carateum.* The genus is one of three major genera in the family **Spirochaetaceae.**

A. *T. pallidum* is the most important human pathogen in the genus. It is the etiologic agent of **syphilis,** a venereal disease that can produce serious complications.

1. **Cellular morphology.** *T. pallidum* is a slender, straight spiral measuring about 0.2 μm in diameter and 5–15 μm in length. The coils in the spiral are evenly spaced about 1 μm apart.
   a. The bacteria are actively **motile** by means of an axial filament. The treponemes **replicate** by transverse fission, and, at times, the cells may remain joined to each other.
   b. **Microscopy.** *Treponema* species do not stain with aniline dyes; dark-field illumination, fluorescently labeled antibody, or silver stain is needed to demonstrate the spiral morphology.

2. **Growth characteristics.** *T. pallidum* has not been successfully cultured with any consistency on artificial media or in tissue culture.
   a. Pathogenic *T. pallidum* remains viable for 3–6 days at a temperature of 25° C and in an appropriate suspending medium that contains reducing agents. It also remains viable for 24 hours in blood stored at 4° C. The bacterium is very susceptible to environmental conditions such as heat, dryness, and sunlight.
   b. A nonpathogenic saprobic treponeme, the Reiter strain, has been grown anaerobically on a complex medium that includes amino acids, vitamins, and serum albumin.
   c. Some laboratory animals (e.g., the rabbit) can be infected artificially with strains that are pathogenic for humans.

3. **Antigenic composition.** The antigens of *T. pallidum* are unknown, but certain immunologic responses are elicited by infection.
   a. **Induced antibodies** are capable of immobilizing and killing the spirochete; such antibodies may be employed in staining by indirect immunofluorescence.
   b. A nonspecific, reagin-like substance (i.e., the **Wassermann antibody**) that reacts with lipid extracted from mammalian cells also is produced during *T. pallidum* infection.

4. **Epidemiology.** Except in the case of congenital syphilis, the treponeme in syphilis usually is transmitted during sexual contact (venereal syphilis). The bacteria also may be transmitted by contact with a lesion on mucous membranes, such as the lips. Clinical laboratory workers, hospital personnel, and blood transfusion recipients occasionally are infected accidentally by nonvenereal means.

5. **Pathogenicity**
   a. **Venereal syphilis** develops in several stages with different characteristics that appear over a long period of time.
      (1) **Primary syphilis** develops after inoculation during sexual contact.
         (a) The bacteria probably enter the body through tiny breaks in the skin and mucous membranes. The treponemes enter the lymphatics, the regional lymph nodes become involved, and bacteremia occurs.
         (b) A **chancre** appears at the site of inoculation about 3 weeks after initial contact, although it can appear at any time within 1–12 weeks. The lesion contains large numbers of *T. pallidum*; it is not painful, and it heals spontaneously.

(2) Secondary syphilis generally occurs at about the time the primary chancre heals. The regional lymph nodes often remain swollen, and disseminated lesions appear.

   (a) The **lesions** are apparent on the skin and mucous membranes, but they occur throughout the body. Body hair (e.g., the eyelashes) frequently is lost because of hair follicle involvement.

   (b) The lesions contain many treponemes, and the disease is **transmissible** in this stage. The lesions heal spontaneously.

   (c) The disease enters a latent period characterized by occasional lesions. **Latency** is due to suppression of the host's cell-mediated immunity.

   (d) Prognosis. Approximately 25% of patients are cured, 25% remain in latency for life, and the remaining 50% progress to the tertiary stage.

(3) Tertiary syphilis principally is a cellular immune response to *T. pallidum* and its metabolic products. Lesions, referred to as **gummas,** can appear in any organ 5–40 years after initial infection. Few treponemes are found in the gummas, which frequently rupture and ulcerate. Central nervous system (CNS) involvement can result in paralysis. Cardiovascular lesions causing aortic aneurysms are common.

b. Congenital syphilis is a separate syndrome with distinct signs and symptoms. It develops by transplacental passage of *T. pallidum* from an infected mother. Often, the fetus dies and is aborted.

   (1) Survivors appear to have **common cold symptoms** at birth. A **maculopapular rash** develops. The skin on the palms and soles of the feet frequently desquamates, and liver damage can cause jaundice.

   (2) In untreated cases, the disease enters a **latent period** after about 1 year.

   (3) Later manifestations include severe bone damage, liver damage, blindness, deafness, and a variety of neurologic problems.

6. Laboratory diagnosis. Direct observation of *T. pallidum* in conjunction with appropriate signs and symptoms of the disease is the best means of diagnosis.

a. Primary and secondary stages. Direct observation of *T. pallidum* by dark-field microscopy or indirect immunofluorescence is possible only during the active primary and secondary stages.

b. Diagnosis during **other stages** of syphilis depends upon serologic techniques.

   (1) Nonspecific serologic tests for syphilis generally employ techniques to detect Wassermann antibodies. The tests are sensitive but are not specific because other diseases elicit Wassermann antibodies. The **Venereal Disease Research Laboratory (VDRL) test** and the **rapid plasma reagin card test** are employed for routine screening purposes, and patients with positive results are tested with one of the specific tests for confirmation of the diagnosis. The VDRL test and the rapid plasma reagin card test are the two most widely used nonspecific methods.

     (a) The VDRL test is a slide flocculation test that mixes cardiolipin (i.e., the tissue antigen), lecithin, and cholesterol with the patient's serum.

     (b) The rapid plasma reagin card test is an agglutination test that employs cardiolipin-coated carbon particles.

   (2) Specific serologic tests employ the patient's serum in reactions with live *T. pallidum* or with the Reiter strain. The Reiter strain is thought to be an avirulent strain of *T. pallidum.*

     (a) The fluorescent treponemal antibody absorbed serum (FTA-ABS) test is the most widely used specific serologic test for syphilis.

       (i) The patient's serum is absorbed with an extract of antigens from the Reiter strain. This removes from the serum antibodies that are due to saprobic treponemes.

       (ii) The serum is applied to a smear of lyophilized, testes-grown *T. pallidum* (i.e., the Nichols' strain) on a slide. After 30 minutes' incubation at 37° C, the slide is washed.

       (iii) Fluorescently labeled antihuman gamma globulin is added to the smear. The slide again is incubated for 30 minutes, rinsed, and examined for fluorescently stained *T. pallidum.*

     (b) The *T. pallidum* immobilization test mixes the patient's serum with live *T. pallidum* grown in the testes of rabbits.

       (i) The bacteria are suspended in a supportive medium under anaerobic conditions. If specific antibodies are present in the serum, the treponemes are immobilized within 18 hours, as observed by dark-field microscopy.

(ii) This test is costly and cumbersome to perform in the routine clinical laboratory.
(c) Other specific tests occasionally used for the diagnosis of syphilis include the hemagglutination test for *T. pallidum* and the *T. pallidum* agglutination test.

7. **Therapy and control**
   a. **Primary and secondary syphilis** respond to penicillin and other antimicrobial agents. Congenital syphilis is treated with the same drugs and can be prevented when appropriate therapy is given to the mother before the fourth month of pregnancy.
   b. **Tertiary syphilis** cannot be cured. Some of the symptomatology of this form of disease can be alleviated with supportive therapy.
   c. **No vaccine** has been produced, and the inability to grow *T. pallidum* in culture is a major inhibitory factor to the development of an effective immunization program.

B. *T. pallidum* **var. endemic syphilis** is the name given to the etiologic agent of **bejel**. The bacterium may be identical to the etiologic agent of syphilis, but the disease is not transmitted venereally.

1. **Epidemiology.** Bejel occurs primarily in Africa and the Middle East in children living in rural areas. It appears to be transmitted from person to person by the shared use of eating utensils and drinking vessels.

2. **Pathogenicity.** Primary lesions of bejel occur in the oral mucosa. After a period of weeks, secondary lesions generally appear at the corners of the mouth as well as on other oral mucosa. Tertiary gummas appear in any organ, including the nasopharynx, skin, and bone.

3. **Therapy.** The disease process in humans can be eliminated by a single injection of penicillin.

C. *T. pertenue* is the etiologic agent of the tropical disease **yaws**. *T. pertenue* is closely related to *T. pallidum*, and the two cannot be distinguished morphologically. Yaws also is known as **frambesia**.

1. **Epidemiology.** Yaws is endemic in children living in tropical regions. The disease appears to be disseminated either by direct contact or by vectors such as flies.

2. **Pathogenicity.** Yaws occurs in three stages.
   a. **Primary lesions** appear at the site of inoculation about 3–4 weeks after exposure. The lesions ulcerate and then spontaneously heal.
   b. **Secondary lesions** appear several months after the primary lesions appear. These lesions also ulcerate and heal but may recur in crops over a period of several years.
   c. Following the secondary stage, either the disease enters a **latent period or tertiary lesions** appear in the skin and bones. The tertiary stage of the disease frequently is characterized by gross disfigurement of the face.

3. **Laboratory diagnosis.** In endemic areas, yaws generally is diagnosed by characteristic clinical signs and symptoms. It can be confirmed by tests analogous to those used in the diagnosis of syphilis.

4. **Therapy.** *T. pertenue* is highly susceptible to penicillin. Eradication programs in some endemic areas have been effective, especially in the Western Hemisphere.

D. *T. carateum* is the etiologic agent of **pinta**. The disease occurs primarily in Central and South America and appears to require person-to-person contact for transmission.

1. Unlike other treponemal diseases, the lesions of pinta remain **localized in the skin**. The lesions are flat, with a red and blue discoloration of the affected area.

2. *T. carateum* is susceptible to penicillin, and eradication programs in Mexico and Colombia have met with some success.

E. *Treponema vincentii* has been suggested as the etiologic agent of **Vincent's angina** when it grows in combination with *Bacteroides melaninogenicus*. The disease medically is known as **acute necrotizing ulcerative gingivitis** and is referred to as **trench mouth** in lay terms. The disease actually may be caused by a herpesvirus that simply promotes the overgrowth of the two indigenous bacteria.

II. *LEPTOSPIRA* SPECIES can cause a variety of syndromes. All are designated **leptospirosis**, although several common names are used for the disease.

A. Cellular morphology

1. *Leptospira* species are characterized by the morphologic appearance of a thin, tightly coiled spiral with a hook at one or both ends.

2. Individual cells are 0.1 μm in diameter and 10–20 μm in length. They actively move, apparently by rhythmic contractions or by axial fibrils (i.e., flagella) at the hooked ends.

B. Growth requirements. *Leptospira* species are **obligate anaerobes**. They grow on artificial media supplemented with bovine or rabbit serum, and some strains have been cultured in chemically defined media. Optimum growth occurs at a temperature of 30° C in a medium with a pH near 7.

C. Antigenic composition and taxonomy. Two species currently are recognized, although the taxonomy is in a state of flux. The species are the pathogenic *Leptospira interrogans* and the saprobic *Leptospira biflexa*.

1. *L. interrogans* has been divided into several serotypes based on microscopic agglutination tests using specific antiserum produced in rabbits. The current names generally employ earlier designations, such as *L. interrogans* **serovar** *icterohaemorrhagiae.*

2. *L. biflexa* contains all members of the genus that are nonpathogenic for humans.

D. Epidemiology. *Leptospira* species are parasites of animals; they occasionally infect humans. Rodents, dogs, cats, pigs, and other animals can carry the bacteria asymptomatically in their kidneys throughout their lives.

1. Some serotypes are associated with specific animal species (e.g., *L. interrogans* serovar *icterohaemorrhagiae* with rats; *L. interrogans* serovar *autumnalis* with mice). Dogs have become the most common source of human leptospirosis in the United States.

2. The bacteria frequently are shed in the urine, and humans can be infected through contact with any of the serotypes of *L. interrogans*. The disease is an occasional hazard for workers in slaughterhouses and sewage disposal facilities. Leptospirosis also may be seen in campers who come into contact with contaminated soil or water.

E. Pathogenicity. Humans become infected with *L. interrogans* serotypes by direct skin contact with the urine of infected animals or with recently contaminated soil or water. It is unclear whether the bacteria can invade unbroken skin, but minute breaks are sufficient for entry.

1. **Tissues involved.** Bacteremia occurs following inoculation, and the bacteria establish infectious processes in the kidneys, liver, spleen, meninges, conjunctiva, and, occasionally, other tissues.

2. **Symptoms** associated with leptospirosis frequently tend to be nonspecific. Headache, fever, chills, myalgia, and photophobia are common. Jaundice and meningitis are more severe manifestations of the disease.

3. **Weil's disease** is a serious form of leptospirosis caused by *L. interrogans* serovar *icterohaemorrhagiae*. The disease is characterized by severe jaundice and has a fatality rate that approaches 25%.

4. **Pretibial fever** (also known as Fort Bragg fever) is characterized by a skin rash on the shins. The disease is caused by *L. interrogans* serovar *Fort Bragg*. Other serotypes of *L. interrogans* may cause a similar skin rash.

F. Laboratory diagnosis. A definitive diagnosis of leptospirosis requires isolation of the bacteria from the patient's blood. Observation of representative bacteria in blood by dark-field microscopy is strong evidence of leptospirosis, but confirmation requires identification of the etiologic agent.

1. **Isolation and identification.** *L. interrogans* can be grown on artificial culture media under anaerobic conditions. The inoculation of hamsters or guinea pigs with clinical specimens

usually enhances growth and increases the probability of isolating the leptospirae. The identification of individual serotypes requires specialized procedures that are not routinely performed in the clinical laboratory.

2. **Serologic diagnosis** can be made with convalescent serum from a patient suspected of having the disease. Agglutinins in the serum lyse *L. interrogans* when complement is present. Complement fixation (CF) tests also may be employed.

G. **Therapy.** The administration of penicillin, erythromycin, or tetracycline early in leptospirosis appears to be effective. Later in the disease, antimicrobial therapy apparently is of minimal efficacy. No vaccine for human use is available, but domestic animals can be vaccinated as a control measure.

III. *BORRELIA* SPECIES are large spirochetes, when compared to *Treponema* and *Leptospira* species.

A. ***Borrelia recurrentis*** causes **relapsing fever** in humans.

1. **Cellular morphology.** *B. recurrentis* is an irregular spiral that measures 0.3–0.5 μm in diameter and 20–30 μm in length. The coils are irregularly spaced, 2–4 μm apart. Motility is accomplished by both a rotating and a twisting motion. The borreliae stain well with aniline dyes and can be observed with the ordinary light microscope.

2. **Growth characteristics.** *B. recurrentis* can survive for several months at 4° C in either blood or culture medium. The bacteria also may be passed transovarily in ticks; they survive long periods of time in these arthropods. Little is known about the growth requirements and metabolism of the borreliae.

3. **Antigenic composition.** The borreliae are antigenically **variable**. *B. recurrentis* demonstrates antigenic variation in vivo.
   a. **Classification.** Instability of the antigenic structure has made classification of the borreliae difficult. Generally, borreliae that are isolated from different sources are given new species names or are designated as strains of *B. recurrentis*.
   b. **Variability.** The antigenic composition of *B. recurrentis* apparently can change during the course of an infection. It is possible that the production of antibodies induces antigenic variation.
   c. **High titers of serum antibodies** are produced in the infected host. Lytic antibodies, agglutinins, and complement-fixing antibodies are induced during the course of a single infection.

4. **Epidemiology.** Relapsing fever caused by *B. recurrentis* occurs worldwide. Endemic areas usually have large populations of rats.
   a. **Ticks** in the genus *Ornithodoros* usually are involved in infection. Infected rats are the source of infection for ticks, which can pass *B. recurrentis* transovarily.
      (1) Infected ticks are most commonly found in the mountainous regions of the western United States.
      (2) **Human disease** is acquired either by a tick bite or by crushing a tick and inoculating minor skin lesions. **Tick-borne relapsing fever is not epidemic**.
   b. **Human lice** become infected by a blood meal on an infected person. In 3–5 days, the infected louse can spread the disease to other humans by bites or by being crushed and inoculating minor lesions. **Louse-borne relapsing fever can be epidemic**, and crowded living conditions can be a predisposing factor.
   c. **Infected rodents** occasionally pass the disease to humans who come in contact with their blood. This method of transmission usually is seen in endemic areas.
   d. The **mortality rate** from relapsing fever is low in endemic areas. During epidemics, however, it frequently approaches 30%.

5. **Pathogenicity and clinical manifestations.** Relapsing fever in humans is characterized by a febrile bacteremia. The disease derives its name from the fact that **three to ten recurrences** are common.
   a. **Onset** of illness is sudden, with chills and a sharp rise in body temperature after a 3- to 10-day incubation period. Many borreliae are present in the blood at this stage, and the fever

persists for 3–5 days. A **second attack** can occur in 4–10 days, and chills, fever, and severe headache are typical symptoms. **Successive recurrences** follow the same pattern.
  b. Antibodies against *B. recurrentis* appear in a patient's serum during the initial febrile period. Agglutinating and lytic antibodies may end the attack and also may select for antigenic variants of *B. recurrentis*. However, different antigenic variants can be isolated from recurrent episodes in a single patient, even one who has been inoculated experimentally with a single antigenic type.
  c. Tissues affected. Large numbers of *B. recurrentis* can be found in the spleen and liver of persons who have died of relapsing fever. Hemorrhagic lesions typically are found in the gastrointestinal tract and kidneys. Borreliae occasionally are found in the cerebrospinal fluid and brain of patients with meningitis.
  d. Acquired immunity following clinical disease appears to be short-lived. Currently, no vaccine is available for human use.

  6. Laboratory diagnosis. Relapsing fever can be suspected in patients with relevant clinical signs and symptoms and a blood smear showing typical spirochetes. If handled properly, blood cultures can yield borreliae, and inoculation of white mice or guinea pigs with specimens can enhance the yield. The preparation of antigens for serologic testing is difficult, but patients may have positive results with the VDRL test.

  7. Therapy. The variability in the disease makes it difficult to evaluate the efficacy of antimicrobial agents. A single day of therapy with erythromycin, tetracycline, or penicillin may be effective in eradicating the borreliae from the body. No vaccine is available.

B. *Borrelia burgdorferi* is the etiologic agent of **Lyme disease**.

  1. Epidemiology. Lyme disease was recognized in 1975 as the result of an epidemiologic survey in Lyme, Connecticut. The survey sought the cause of a cluster of arthritis cases frequently preceded by expanding annular skin lesions.
  a. Geographic distribution. Although initially recognized along the Atlantic seaboard of the United States, Lyme disease now has been reported from 35 states. Cases also have been reported in Australia and in many countries of Europe and Asia.
  b. Ecology. The ecology of *B. burgdorferi* is not completely known, but a cycle involving insect vectors appears to be involved. **The bacteria are transmitted to humans by a minute tick** belonging to the genus *Ixodes*.
    (1) Microscopic and cultural examination of *Ixodes* ticks from endemic areas has revealed that 20%–80% harbor the spirochete.
    (2) *B. burgdorferi* also has been isolated from 80% of the **mice and deer** in endemic areas.
  c. Seasonality. Lyme disease typically is seen between May 1 and November 30, with 80% of clinical cases occurring in June and July.

  2. Pathogenicity and clinical manifestations
  a. The **pathogenicity** of *B. burgdorferi* is difficult to demonstrate. There appears to be a factor that amplifies the bacteria's pathogenicity. Several virulence factors, including a nonclassical endotoxin, have been suggested as the amplifying agent.
  b. Lyme disease occurs in **three stages**, although the duration of each stage is not well defined. The disease rarely is fatal.
    (1) The first stage of Lyme disease is characterized by the development of a distinctive expanding skin lesion. The lesion usually occurs about 7 days after the tick bite, but the incubation period can range from 3–32 days.
      (a) A papule appears on the skin, followed by the characteristic skin lesion, known as **erythema chronicum migrans**, in 80% of patients. The lesion may be circular or oblong, and it may expand to many inches in diameter. Some people do not notice the lesion, or it sometimes may be absent.
      (b) Joint stiffness, headache, stiff neck, and other systemic symptomatology often accompany the lesion. Lymphadenopathy occasionally occurs.
    (2) The second stage of Lyme disease is characterized by the development of **multiple skin patches**.
      (a) Arthritis often occurs during the second stage and more commonly involves one rather than both knees.
      (b) Central and peripheral nervous system disorders begin to occur.

    (3) **The third stage** of Lyme disease typically occurs 3–4 weeks after the initial erythema chronicum migrans. Arthritis and neurologic symptoms become chronic.

        (a) Most patients with **neurologic involvement** display symptoms of meningitis or meningoencephalitis. About half of the patients develop mononeuritis multiplex characterized by weakness and sensory changes in peripheral or cranial nerves. Examination of the **cerebrospinal fluid** reveals a few hundred leukocytes with a predominance of mononuclear cells. The protein is 2–5 times normal, but glucose is only slightly elevated.

        (b) **Cardiac involvement** may include either bradycardia or tachycardia. Congestive heart failure may occur, but most cardiac abnormalities disappear in 3–4 weeks.

    c. **Incidence of symptomology.** Overall, about 60% of patients with Lyme disease develop joint involvement, at least 15% develop neurologic symptoms, and 8% have cardiac involvement. Blockage of arteries to the heart has led to a rare fatality.

3. **Diagnosis.** Erythema chronicum migrans is thought to be diagnostic, especially when followed by arthritis or neurologic symptoms. A number of serologic tests for diagnostic use are commercially available, but their total effectiveness in diagnosis is open to question.

4. **Therapy.** Lyme disease can be treated early with tetracycline. Penicillin V, vancomycin, and the cephalosporins also appear to be effective. Later in the disease, large doses of antibiotics are necessary for prolonged periods. *B. burgdorferi* is resistant to rifampin.

STUDY QUESTIONS

Directions: Each question below contains five suggested answers. Choose the **one best** response to each question.

1. The high mortality rate associated with Weil's disease typically is the result of

(A) gummas
(B) jaundice
(C) pyrexia
(D) cardiac failure
(E) meningitis

2. By what means do humans usually acquire leptospirosis?

(A) Tick and flea bites
(B) Ingestion of contaminated food
(C) Contact with animal urine
(D) Shared eating utensils
(E) Insect vectors such as flies

3. Transovarian passage of spirochetes occurs in the vector of which of the following diseases?

(A) Yaws
(B) Pretibial fever
(C) Bejel
(D) Relapsing fever
(E) Leptospirosis

4. Recurrent bacteremia in relapsing fever apparently is due to

(A) immunologic selection of different subpopulations of bacteria
(B) a lack of bactericidal antibiotics for therapeutic use
(C) the ability of the bacteria to become sequestered in macrophages
(D) a failure of the bacteria to induce antibody formation
(E) instability of bacterial antigens resulting in antigenic shift

5. A primary characteristic of early stage Lyme disease is

(A) acute necrotizing ulcerative gingivitis
(B) erythema chronicum migrans
(C) a petechial pretibial rash
(D) regional lymphadenopathy
(E) meningoencephalitis

Directions: Each question below contains four suggested answers of which **one or more** is correct. Choose the answer

    A if **1, 2, and 3** are correct
    B if **1 and 3** are correct
    C if **2 and 4** are correct
    D if **4** is correct
    E if **1, 2, 3, and 4** are correct

6. Antibiotic therapy can cure which of the following forms of syphilis?

(1) Congenital syphilis
(2) Primary syphilis
(3) Secondary syphilis
(4) Tertiary syphilis

7. Treponemes are visualized by

(1) fluorescent microscopy
(2) silver stain
(3) dark-field microscopy
(4) Gram stain

8. Specific tests for the diagnosis of syphilis include the

(1) fluorescent treponema antibody absorbed serum test
(2) rapid plasma reagin card test
(3) *Treponema pallidum* immobilization test
(4) culture and biochemical identification procedure

Directions: The group of questions below consists of lettered choices followed by several numbered items. For each numbered item, select the **one** lettered choice with which it is **most** closely associated. Each lettered choice may be used once, more than once, or not at all. Choose the answer

    A if the item is associated with **(A) only**

    B if the item is associated with **(B) only**

    C if the item is associated with **both (A) and (B)**

    D if the item is associated with **neither (A) nor (B)**

Questions 9–11

Match each characteristic feature with the appropriate form or forms of relapsing fever.

(A) Sporadic relapsing fever

(B) Epidemic relapsing fever

(C) Both

(D) Neither

  9. Louse-borne

10. Low mortality rate

11. Prevented by an attenuated vaccine

ANSWERS AND EXPLANATIONS

1. The answer is B. [*II E 3*] Weil's disease is a severe form of leptospirosis caused by *Leptospira interrogans* serovar *icterohaemorrhagia*. The disease is characterized by severe jaundice, and it has a mortality rate approaching 25%. Fever and meningitis can occur in leptospirosis caused by any serovariety, but jaundice is responsible for the high mortality rate in Weil's disease. Gummas are the noninfectious lesions of tertiary syphilis; gumma formation in the heart can result in cardiac failure and death.

2. The answer is C. [*II E*] Human leptospirosis is caused by several serovarieties of *Leptospira interrogans*. Infection normally is acquired by direct contact with contaminated urine from a variety of animals. Rats, mice, dogs, cats, pigs, and numerous other animals are asymptomatic carriers of *L. interrogans* in their kidneys. Contact with soil or water contaminated with the urine of colonized animals also can result in human infection. No arthropod or insect vectors are involved in the transmission of leptospirosis; in fact, the bacteria may be able to penetrate intact skin. Upon human ingestion, *Leptospira* species probably are killed by the acidic environment of the stomach.

3. The answer is D. [*III A 4 a*] Ticks of the genus *Ornithodros* are the vectors of sporadic cases of relapsing fever. *Borrelia recurrentis* is passed transovarily in this tick from generation to generation and has been known to survive for extended periods of time. Humans acquire leptospirosis, including pretibial fever, through contact with the urine of animals serving as asymptomatic carriers of *Leptospira interrogans*. Flies are potential vectors in yaws, and bejel appears to be transmitted from human to human by the sharing of eating utensils and drinking vessels.

4. The answer is E. [*III A 5 b*] Relapsing fever caused by *Borrelia recurrentis* is characterized by three to ten recurrences of bacteremia and fever. The cause of these recurrences is not completely understood. It has been suggested that, as antibodies appear in the circulation following infection, different subpopulations of *B. recurrentis* might be immunologically selected. However, experimental inoculation of a single antigenic type of *B. recurrentis* still results in the relapsing fever syndrome. In addition, antigenic instability is a recognized characteristic of the bacteria. Effective treatment is difficult to identify because of the relapsing nature of the disease. Some evidence suggests, however, that if effective chemotherapy is administered early in the disease, no relapses occur. If early treatment results in complete eradication of the bacteria, antigenic shift may be the cause of recurrent bacteremia.

5. The answer is B. [*III B 2 b*] Lyme disease is a recently recognized disease caused by *Borrelia burgdorferi* and transmitted to humans by ticks. The early stage of Lyme disease is characterized by a distinctive expanding skin lesion known as erythema chronicum migrans. The lesion occurs in at least 80% of people with the disease. Acute necrotizing ulcerative gingivitis once was thought to be caused by *Treponema vincentii* in conjunction with *Bacteroides melaninogenicus*, but now it is believed that herpesvirus is the etiologic agent. A pretibial rash is the characteristic sign of Fort Bragg fever caused by *Leptospira interrogans*. Although some lymph node involvement and meningitis may occur occasionally in Lyme disease, these are not characteristic features.

6. The answer is A (1, 2, 3). [*I A 7 a, b*] Primary and secondary syphilis respond to penicillin therapy and other antimicrobial agents. Congenital syphilis in the neonate is treated with the same therapeutic regimen. Congenital syphilis can be prevented if the infected mother is treated with appropriate antibiotics before the fourth month of pregnancy. Tertiary syphilis cannot be cured, as it appears to be primarily an immune response to treponemal antigens. Treatment of tertiary syphilis is supportive.

7. The answer is A (1, 2, 3). [*I A 1 b*] *Treponema* species are very thin spirals with a diameter of about 0.2 μm. The resolution of the light microscope also is approximately 0.2 μm. In addition, because the treponemes do not take up the aniline dyes such as crystal violet, the Gram stain is ineffective. Dark-field illumination allows visualization of live treponemes and observation of their characteristic motility. Fluorescently labeled antibody also is used to visualize the treponemes, but this method is expensive and time consuming. The cells also may be observed with the light microscope, using a silver stain. The silver coats the cell, increasing its diameter and visibility with the light microscope.

8. The answer is B (1, 3). [*I A 6 b (2) (a), (b)*] A variety of specific tests are employed in the diagnosis of syphilis. *Treponema pallidum* has not been cultured in the laboratory with any consistency to date; therefore, the definitive diagnosis of syphilis depends upon serologic tests. The most widely used definitive test is the fluorescent treponemal antibody absorbed serum test. In this test, the patient's serum is absorbed with antigens extracted from the Reiter strain and then fluorescently labeled with specific antibodies. The *T. pallidum* immobilization test also is specific for the diagnosis of syphilis. Live *T. pallidum* grown in rabbits is exposed to the patient's serum. If specific antibodies are present in the serum, the *T. pallidum* are immobilized within 18 hours. Although specific, this test is cumbersome, time consuming, expensive, and difficult to control. The rapid plasma reagin card test is a nonspecific test used for screening people for syphilis.

9–11. The answers are: 9-B, 10-A, 11-D. [*III A 4 b, d, 7*] Louse-borne relapsing fever is transmitted from human to human when a body louse takes a blood meal from an infected human and then bites other humans. The disease occurs in epidemic proportions. Crowding, stress, and poor personal hygiene appear to be predisposing factors. Sporadic cases of relapsing fever are transmitted to humans either by a tick bite or by contamination of a small lesion by the infected blood of a crushed tick.

The mortality rate is low in sporadic cases of tick-borne relapsing fever. Because this form of disease usually occurs in endemic areas, there probably is a degree of herd immunity. Epidemic, louse-borne relapsing fever has a mortality rate approaching 30%.

No vaccines are available for the prevention of relapsing fever. The antigenic variability of *Borrelia recurrentis* makes treatment difficult and makes the development of vaccine a complex procedure.

14
Mycobacteria
Gerald E. Wagner

I. GENERAL CHARACTERISTICS OF MYCOBACTERIA. *Mycobacterium* species generally are slow-growing, aerobic bacteria that have an unusual cell wall composition. Some species are saprobes, whereas others are significant pathogens for humans and animals. The most important human pathogens are *Mycobacterium tuberculosis* and *Mycobacterium leprae*.

A. Morphology

1. **Cellular structure and composition**
   a. Mycobacteria are **slender rods** measuring 0.2–0.4 μm in diameter and 2–10 μm in length. Individual cells may be straight or slightly curved. In culture, mycobacteria typically form clumps because of their hydrophobic surfaces, and individual cells may appear as plump rods.
   b. The **cell walls** of mycobacteria are unique.
      (1) They contain *N*-glycolylmuramic acid rather than *N*-acetylmuramic acid.
      (2) The lipid content of the cell wall is very high (i.e., as much as 60% of the dry weight). These high molecular weight lipids account for the hydrophobicity of the mycobacteria.

2. **Staining characteristics.** The hydrophobic mycobacteria are impermeable to basic aniline bacteriologic stains.
   a. The **acid-fast staining** character of mycobacteria is a result of the cell wall constituents and, specifically, of mycolic acid.
   b. The Ziehl-Neelsen stain is the most commonly employed acid-fast staining technique. Carbol fuchsin, a red dye, in 5% phenol is driven into the cells with heat. The cells then are decolorized with 1% hydrochloric acid in ethanol and counterstained with methylene blue. Mycobacteria retain the red dye during the decolorization step.
   c. The finding of acid-fast slender rods is a definitive characteristic of mycobacteria.

B. Cultural characteristics

1. **Differentiation of species** of mycobacteria is made on the basis of nutritional requirements, optimal growth temperatures, generation times, pigmentation, and a few biochemical tests.

2. **Runyon classification.** The classic Runyon grouping separates species according to pigment production and growth rate (Table 14-1). It generally is not used clinically because it does not classify *M. tuberculosis* or *M. leprae*.

C. Pathogenicity. Several species of mycobacteria have been associated with disease in humans. Most of these species cause serious, life-threatening disease. Disease processes tend to be the result of a delayed-type hypersensitivity response to mycobacterial antigens. No toxins have been associated with mycobacteria.

1. *M. tuberculosis* and *M. leprae* are strict human pathogens in natural situations. Infections are transmitted from person to person, and fomites may be involved.

2. *Mycobacterium kansasii*, *Mycobacterium avium-intracellulare*, and *Mycobacterium scrofulaceum* cause tuberculosis-like disease. The infections by these etiologic agents typically occur in debilitated or compromised patients.

3. *Mycobacterium marinum* and *Mycobacterium ulcerans* are opportunistic pathogens of humans. They cause cutaneous infections that usually are self-limited.

Table 14-1. Runyon Classification of Mycobacteria

Group	Pigment Production	Growth Rate (days)	Representative Species
I—Photochromogens	Produce bright yellow pigment only in the presence of light	10–21	*Mycobacterium kansasii*
II—Scotochromogens	Produce yellow to orange pigment independent of light	10–21	*Mycobacterium scrofulaceum*
III—Nonchromogens	Usually do not produce pigment	10–21	*Mycobacterium avium-intracellulare*
IV—Rapid growers	Vary in pigment production	3–7	*Mycobacterium fortuitum*

II. TUBERCULOSIS is a worldwide health problem. It is primarily a pulmonary disease, but disseminated disease can occur by hematogenous spread.

A. **Etiology.** Human tuberculosis is caused primarily by *M. tuberculosis. Mycobacterium bovis* was an important cause of tuberculosis in the United States prior to pasteurization and occasionally still is isolated from clinical cases. *Mycobacterium africanum* is the cause of many cases of tuberculosis in Africa; the disease is analogous to that caused by *M. tuberculosis.*

B. **Morphology**

1. **Cellular morphology.** *M. tuberculosis* is a slender rod that is strongly acid fast. Irregular staining often gives it a beaded appearance of connected acid-fast granules.

2. **Colonial morphology**
   a. Colonies of *M. tuberculosis* are buff to tan in color, raised, and waxy in appearance. The colonies are difficult to emulsify because of their hydrophobicity.
   b. **Cord factor** is a glycolipid that causes the cells to grow in parallel chains that form entwined bundles. Cording is characteristic of virulent strains of *M. tuberculosis*; however, the relationship of cord factor to virulence, if any, is unknown.

C. **Growth requirements.** *M. tuberculosis* is a slow-growing, fastidious bacterium.

1. **Enriched complex media** are required for isolation from clinical specimens (i.e., primary isolation). **Löwenstein-Jensen medium** is the most commonly employed mycobacterial isolation medium. It contains 60% egg in a nutrient base, with malachite green added to inhibit the growth of contaminating bacteria.

2. The **optimal growth temperature** is 37° C, and the bacilli will not grow at room temperature (i.e., 20°–25° C). Growth occurs aerobically and can be enhanced by an atmosphere containing 5%–10% carbon dioxide.

3. **Generation times** for isolates of *M. tuberculosis* are 6 hours or longer. Primary isolation may require 4 weeks or more for the appearance of macroscopic colonies.

D. **Pathogenicity**

1. *M. tuberculosis* is highly virulent and infectious in certain circumstances. Typically, only 10 cells injected intradermally or intramuscularly are required to cause fatal disease in guinea pigs, the standard animal model for studying tuberculosis.

2. The virulence of *M. tuberculosis* is due to its ability to multiply inside macrophages and by the production of large amounts of lactic acid, carbon dioxide, and a low pH in the tissue. Dead bacterial cells have a very low toxicity; however, hypersensitivity to tuberculoproteins is an important factor in the pathogenesis of tuberculosis.

E. Pathogenesis and clinical manifestations. Human tuberculosis usually results from the inhalation of respiratory droplets containing the etiologic agent. Tuberculosis is seen in several clinical forms that depend, to a large extent, upon the status of the host's sensitivity to the bacilli.

1. **Primary tuberculosis** is the initial reaction to infection in the previously unexposed or unsensitized individual. The periphery of the midzone of the lung is the most common site of infection.
   a. Inhaled bacilli are phagocytized by macrophages and transported to the hilar lymph nodes draining the localized site of infection. The bacteria multiply intracellularly, producing only a minor inflammatory response.
   b. Dissemination of the phagocytized bacilli occurs through the lymphatic vessels and the bloodstream. Any organ system can be infected, and microscopic **tubercles** (i.e., granulomata) develop.
   c. Tubercles are formed by the cell-mediated hypersensitivity response 2–6 weeks after infection.
      (1) The tubercle has a central area of epithelioid cells (i.e., modified macrophages) and giant cells, with a surrounding area of lymphocytic infiltration.
      (2) The central area often becomes semiliquid through caseous necrosis. Bacterial multiplication within this area is slow or stops completely.
   d. Most primary infections are controlled by the cell-mediated immune system. The lesions heal by fibrosis, and calcification sometimes occurs. The bacilli in the lesions slowly die, although a few may remain viable for as long as 20 years or more.
   e. Clinical symptomatology is absent or appears as a mild influenza-like syndrome. The primary site of infection and enlarged hilar lymph nodes occasionally can be detected radiographically.

2. **Reactivation tuberculosis** occurs in sensitized individuals with lowered immunity to *M. tuberculosis*. It is the most commonly diagnosed form of the disease in people over 50 years old living in western countries. Men have the highest incidence of disease.
   a. Reactivation tuberculosis may occur 20 years or longer after primary infection. The appearance of clinical disease frequently is associated with stress, underlying disease, or debilitating illness.
   b. The apex of the lung is the most common site of infection; other body areas of high oxygen tension also are frequently the sites of reactivation. The kidneys, bone and bone marrow, lymph nodes, brain, meninges, and the intestines may be involved.
   c. Lesions appear as spreading, coalescing tubercles with large areas of caseous necrosis.
      (1) In the lung, bronchi and small blood vessels may be eroded to form cavities, and blood-tinged sputum containing caseous material is expectorated.
      (2) Lesions and sputum contain numerous *M. tuberculosis*. Contaminated respiratory droplets are a major source of infection in others.
   d. Reactivation tuberculosis is characterized by a productive cough often with hemoptysis, significant weight loss, night sweats, and chronic fever.

3. **Disseminated tuberculosis**, also known as **miliary tuberculosis** because of the numerous small tubercles, usually occurs in the very young or elderly or in other immunocompromised patients.
   a. Initiation of the disease typically occurs when a necrotic tubercle erodes into a blood vessel. Hematogenous spread of the bacteria leads to fulminating infection in every organ system.
   b. The systemic symptomatology resembles that of reactivation tuberculosis. Meningitis and brain involvement are common. This form of tuberculosis is invariably fatal.

F. Diagnosis. Strong presumptive evidence of mycobacterial infection correlates with the microscopic finding of acid-fast bacilli. A confirmed diagnosis is made on the basis of culture results.

1. **Microscopic evidence**
   a. The finding of acid-fast slender rods in carefully obtained sputum from a patient with appropriate symptomatology is sufficient evidence to begin antituberculosis therapy.
   b. The Ziehl-Neelsen stain, or a modification of this procedure, is employed routinely.
   c. The resistance of mycobacteria to chemicals allows for the addition of digestive agents, such as *N*-acetylcysteine and weak sodium hydroxide, to make staining and visualization easier.

2. **Culture**
   a. Löwenstein-Jensen medium is employed for primary isolation of *M. tuberculosis*.

b. Clinical specimens can be treated with germicides or selected antibiotics to enhance the isolation of *M. tuberculosis.*

c. Cultures must be incubated for 6–8 weeks before being discarded as negative.

3. Identification

   a. Positive cultures are identified as *M. tuberculosis* by cultural and microscopic morphology and by a series of biochemical tests.

   b. The **production of niacin** is unique to *M. tuberculosis.*

G. Screening for tuberculosis is accomplished best with the **tuberculin skin test**. This test is of major epidemiologic importance; although it is used in conjunction with other parameters to aid in the diagnosis of tuberculosis, it is not a diagnostic test.

1. The tuberculin skin test involves the intradermal injection of **purified protein derivative (PPD)**. This heterologous mixture of proteins is precipitated from broth cultures of *M. tuberculosis* and is standardized in reference to tuberculin units (TU).

   a. Administration of PPD is accomplished most efficiently by intradermal injection of 5 TU. This is known as the **Mantoux test**.

   b. The **tine test** also administers 5 TUs but, although easier to administer (especially to children), is a less effective testing procedure.

2. Induration at the injection site occurs within 48–72 hours in a sensitized person. It may or may not be accompanied by erythema.

   a. Induration of 10 mm or greater indicates a positive reaction. This means only that the person has been exposed to *M. tuberculosis*, or another species with cross-reacting proteins, and has become sensitized. It does not indicate active disease.

   b. Induration of 5–10 mm is a questionable reaction. Frequently in this situation, the Mantoux test is repeated using 10 TU.

   c. Induration of less than 5 mm is a negative test. Erythema alone is not indicative of sensitivity to tuberculoproteins. Negative reactions occasionally are due to anergy.

   (1) A negative test result may be helpful as a diagnostic aid in countries where primary tuberculosis has a low incidence. Negative tests in these regions rule out the diagnosis of tuberculosis in an immunocompetent person.

   (2) A negative result does not automatically rule out a diagnosis of tuberculosis. An infected, anergic person would test negative to PPD.

   (3) Repeated testing with PPD does not induce a negative-to-positive conversion in the skin test results.

H. Therapy. Tuberculosis responds to a variety of antibacterial agents, some of which are used only as antituberculosis compounds. Prophylaxis can involve either chemotherapy or immunotherapy.

1. Antituberculosis drugs

   a. First-line drugs for treating tuberculosis include isoniazid (INH), ethambutol, streptomycin, pyrazinamide, and rifampin. Two or more of these agents routinely are used in combination to prevent selection of drug-resistant mutants.

   b. Alternative drugs for treating tuberculosis include kanamycin, cycloserine, para-aminosalicylic acid, ethionamide, viomycin, capreomycin, and thioacetazone.

2. Prevention

   a. INH almost always is used as a single chemoprophylactic agent. The drug is administered orally for up to 1 year following the finding of skin test conversion in the absence of indications of active progressive disease.

   b. Immunoprophylaxis can be achieved, to a degree, by intradermal vaccination with an attenuated strain of *M. bovis* known as **bacille Calmette-Guérin (BCG)**.

   (1) BCG vaccination routinely is used only in populations at particular risk in the United States and other countries where the incidence of tuberculosis is relatively low. The vaccine is administered routinely in some countries where the incidence of tuberculosis is high.

   (2) Occasional complications occur from the vaccination, and skin testing should not be used after effective vaccination because of the potential for large necrotic reactions.

3. **Treatment of active tuberculosis** generally is initiated in the hospital and followed by treatment at home for 1 year or longer. Effective antituberculosis drugs eliminate the infectious phase within 1–2 weeks and has eliminated the need for isolation hospitals and sanitoriums.

I. **Epidemiology.** The incidence and rate of tuberculosis in western countries have declined dramatically during the last century. In the United States, the incidence has decreased to a current rate of about 20,000 cases of active tuberculosis reported annually.

1. Typical clinical cases of tuberculosis in western countries involve men over 50 who most likely experienced primary infection at least 20 years earlier.

2. Socioeconomically disadvantaged persons living in crowded urban areas have the highest incidence of tuberculosis. The inner city areas of Chicago and Washington, D.C. have the highest incidence of infection in the United States.

3. Patients with acquired immunodeficiency syndrome (AIDS) are highly susceptible to tuberculosis. The influx of immigrants from less developed countries into specific geographic areas frequently has led to an increased incidence of tuberculosis. Native Americans also have a high incidence of the disease.

4. Sporadic localized epidemics of tuberculosis in children are seen in the United States, with a few of these children developing clinical and radiographic evidence of active tuberculosis.

III. **TUBERCULOSIS-LIKE DISEASES** can be caused by several other species of mycobacteria, and the incidence of these mycobacterial infections has become more recognized with the decline in tuberculosis. The sources of human infection by these agents are environmental reservoirs; person-to-person transmission apparently does not occur. In the past, the agents of tuberculosis-like disease were referred to as the atypical mycobacteria, but this term no longer is appropriate.

A. *M. kansasii* produces a pulmonary infection that can closely resemble tuberculosis. The environmental reservoir has not been identified definitively.

1. **Microbiology.** *M. kansasii* is a photochromogen that produces yellow to orange pigment after about 2 weeks incubation in light. Biochemical tests, in addition to the photochromicity, are used to identify *M. kansasii*.

2. **Pathogenesis and clinical manifestations**
   a. *M. kansasii* causes a slowly progressing, cavitary pulmonary disease.
   b. Cervical lymphadenitis (i.e., scrofula) and skin lesions are common clinical findings. Occasional disseminated disease occurs in debilitated patients.

3. **Epidemiology.** In the United States, infections by *M. kansasii* are most common in urban areas of Illinois, Oklahoma, and Texas. Disease does not occur in the Southeast. Men with chronic obstructive lung disease most frequently develop clinical disease.

4. **Diagnosis**
   a. A confirmed diagnosis is made by repeated isolation and identification of *M. kansasii* from a patient with appropriate symptomatology.
   b. *M. kansasii* antibodies cross-react with *M. tuberculosis* antigens, resulting in positive PPD tests in infected people.

5. **Therapy.** Clinical disease is treated with the usual antimycobacterial agents. The response usually is good with long-term treatment involving rifampin as part of the therapeutic regimen.

B. *M. avium-intracellulare* **complex** is a group of closely related mycobacteria that primarily cause disease in birds and mammals. Occasional human infection occurs, and symptomatic disease most commonly involves the respiratory tract.

1. **Microbiology.** Strains of bacteria in the *M. avium-intracellulare* complex generally do not produce pigment, although pigment may develop in some older cultures. The species within the complex are slow growing.

2. **Pathogenesis and clinical manifestations**
   a. Cavitary pulmonary disease that resembles tuberculosis is most common in humans.

b. Renal and skin lesions, cervical lymphadenitis, and chronic osteomyelitis also occur.

c. Infection frequently is superimposed on other pulmonary diseases, such as emphysema. Disseminated disease caused by the *M. avium-intracellulare* complex is a serious problem in AIDS patients.

3. Epidemiology. Although *M. avium-intracellulare* is found throughout the United States, surveys indicate that the highest incidence of infection is in the Southeast, where the bacteria exist in soil and water. In general, infections occur in rural areas and affect older men.

4. Diagnosis

a. A definitive diagnosis is made by culture and identification of the mycobacteria from sputum or biopsy specimens.

b. There is some cross-reaction with PPD. **PPD-B** is an antigen preparation that has been used experimentally; it is derived from a culture of *M. intracellulare.*

5. Therapy. The species in the *M. avium-intracellulare* complex are relatively resistant to antituberculosis agents; drug therapy frequently requires supplementary surgical procedures. About 20% of patients experience relapse within 5 years following treatment.

C. *M. scrofulaceum* is distributed worldwide in environmental sources. It typically causes disease in children.

1. Microbiology. *M. scrofulaceum* is a scotochromogen that produces yellow colonies after about 2 weeks incubation.

2. Pathogenesis and clinical manifestations

a. *M. scrofulaceum* is the most common cause of granulomatous cervical lymphadenitis in children. The disease is characterized by enlarged lymph nodes, which may ulcerate or form draining sinus tracts.

b. The enlarged nodes are painless, and there is little, if any, systemic symptomatology.

c. *M. scrofulaceum* may cause lung disease, producing a disease syndrome that resembles pulmonary tuberculosis.

3. Epidemiology. *M. scrofulaceum* has been isolated from moist environmental sources throughout the world. Human disease is most common in rural, agrarian environments.

4. Diagnosis. A confirmed diagnosis is made by culture and identification of *M. scrofulaceum* from biopsy specimens from the enlarged lymph nodes. Cross-reactivity with PPD is not observed.

5. Therapy. Effective treatment usually involves surgical excision of the infected lymph node. *M. scrofulaceum* generally is resistant to antituberculosis drugs.

IV. MYCOBACTERIAL DISEASE THAT DOES NOT RESEMBLE TUBERCULOSIS is caused by several species of free-living mycobacteria. Humans usually are incidental hosts, and the disease process generally involves cutaneous or subcutaneous tissue.

A. *M. marinum* causes tuberculosis in fish. It is widely distributed in fresh and salt water; it is easily isolated from the slime on rocks and other submerged objects and on rough walls of swimming pools.

1. Microbiology. *M. marinum* is a photochromogen that grows at 30° C but not at 37° C.

2. Pathogenesis and clinical manifestations

a. Infection typically follows an abrasion that is contaminated by the bacteria. The classic infection occurs on the hand or forearm traumatized on a swimming pool wall.

b. Infection is characterized by a localized granulomatous lesion. Ulceration eventually occurs, but the lesion usually heals spontaneously within a few weeks. Chronic infection occasionally occurs.

3. Therapy. Chronic infection caused by *M. marinum* usually responds to tetracycline and some of the antituberculosis drugs.

B. ***M. ulcerans*** causes chronic subcutaneous ulcers in humans. Most cases are seen in the tropics, particularly in parts of Africa, New Guinea, and northern Australia. Sporadic cases are reported worldwide. The source and mode of transmission are unknown.

1. **Microbiology.** *M. ulcerans* is a slow-growing mycobacterium that grows at 30° C but not at 37° C.

2. **Pathogenesis and clinical manifestations.** Infection results in severe ulceration of the skin and subcutaneous tissue (a condition known as **Buruli ulcer**). The ulceration progresses and can become extensive unless treated effectively.

3. **Therapy.** Surgical excision of lesions is the most successful means of treatment. Antibacterial chemotherapy usually is ineffective.

C. ***Mycobacterium fortuitum* complex** is a group of free-living, rapid-growing mycobacteria that only rarely cause human disease. **Injection-site abscesses** among drug abusers are the most common form of disease. Infection also has been associated with **implanted devices**, such as heart valves and breast prostheses. Pulmonary infection occurs occasionally.

V. HANSEN'S DISEASE (i.e., **leprosy**) is a disease of apparent mycobacterial origin. The disease occurs in several clinical forms.

A. Etiology. Acid-fast rods that morphologically resemble *Mycobacterium* species are seen in every case of Hansen's disease, which led to the etiologic agent being designated *M. leprae*. The source or reservoir of the organism is unknown.

1. *M. leprae* never has been cultured in the laboratory, either in artificial media or in tissue culture. Numerous reports of successful culture have not been confirmed.

2. *M. leprae* obtained from human biopsy material will grow in the nine-banded armadillo. However, the disease produced in this experimental animal does not closely resemble human infections. Infection apparently occurs naturally in armadillos, and leprosy can be induced experimentally in some species of monkeys.

B. Pathogenesis and clinical manifestations. The disease occurs in two major clinical forms: tuberculoid leprosy and lepromatous leprosy. There are numerous nonpolar forms, and the characteristics of the two major types merge in an intermediate form. However, progression from one polar form to the other does not occur.

1. **Tuberculoid leprosy** is characterized by macules or extensive plaques on the trunk, face, and limbs. The lesions have a raised erythematous edge and a dry, hairless center that may be hypopigmented.
   a. *M. leprae* invades sensory nerves, causing a patchy anesthesia in the lesions. Few mycobacteria are found in the granulomatous lesions. Epithelioid cells and giant cells are abundant, and lymphocytic infiltration is extensive in the lesions.
   b. Cell-mediated immunity is intact in tuberculoid leprosy. Patients show a delayed-type hypersensitivity reaction to **lepromin**, a *M. leprae* antigen derived from leprous tissue and analogous to tuberculin.
   c. The disease frequently is characterized by evidence of slow progression and healing occurring simultaneously. Localized anesthesia often leads to injury and severe secondary bacterial infection.

2. **Lepromatous leprosy** is characterized by the formation of large, diffuse granulomatous lesions. Facial disfigurement is common.
   a. Massive numbers of *M. leprae* are seen in lepromatous lesions. Sensory nerve involvement is diffuse and patchy anesthesia is common. The bacteria probably disseminate hematogenously, although the disease does not appear to be manifest in the deeper organs.
   b. Mycobacteria accumulate in subcutaneous tissue, particularly in loose folds of the skin and in the ear lobes. Thickening of the loose skin of the forehead, lips, and ears results in the classic **leonine facies**.
   c. Cell-mediated immunity is deficient in lepromatous patients, who are anergic to the lepromin skin-test antigen.

C. Diagnosis. In endemic areas, Hansen's disease is diagnosed primarily on the basis of clinical signs and symptoms. Scrapings or biopsies from lesions reveal large numbers of acid-fast bacilli in lepromatous disease and a typical cell-mediated response in tuberculoid disease.

D. Therapy for Hansen's disease must extend over many years, if not throughout life, to prevent relapse.

1. Dapsone (i.e., diamino-diphenylsulfone, or DDS) is the major drug employed in endemic areas. It is relatively inexpensive and usually effective; mutational resistance now has reached significant levels in some areas.

2. Clofazimine is a newer compound that has been successfully employed.

3. Rifampin is highly effective, but is prohibitively expensive in endemic regions.

4. Combination treatment with three or more antileprotic drugs is showing promise in preventing recurrence due to drug-resistant strains of *M. leprae.*

E. Epidemiology. It is estimated that there are between 15 and 20 million cases of leprosy worldwide. The disease occurs primarily in the tropics and subtropics, but a few cases occur annually in the United States and in western Europe. In the United States, the largest population of people with Hansen's disease is in New York City, but there is no evidence of transmission of the disease.

1. The mode of transmission of Hansen's disease is unknown; however, in endemic areas, close contact with infected persons increases the risk of developing the disease. The incubation period apparently is 5–20 years.

2. Climate, sanitation, racial differences, nasal droplets, and biting insects all have been suggested as factors in the epidemiology of the disease.

3. A recent hypothesis suggests that there is a high incidence of subclinical infection in endemic areas, with clinical disease occurring in only a small number of infections. Culture of the etiologic agent will be a critical breakthrough in the study of Hansen's disease.

STUDY QUESTIONS

Directions: Each question below contains five suggested answers. Choose the **one best** response to each question.

1. A 23-year-old Olympic high diver received an abrasion of the right forearm by striking the side of a swimming pool. A granulomatous lesion developed at the site of the injury and eventually ulcerated. Spontaneous healing occurred in about 5 weeks. The most likely etiologic agent of the lesion is

(A) *Mycobacterium intracellulare*
(B) *Mycobacterium ulcerans*
(C) *Mycobacterium kansasii*
(D) *Mycobacterium scrofulaceum*
(E) *Mycobacterium marinum*

2. The most common etiologic agent of granulomatous cervical lymphadenitis in young children is

(A) *Mycobacterium kansasii*
(B) *Mycobacterium scrofulaceum*
(C) *Mycobacterium fortuitum*
(D) *Mycobacterium avium*
(E) *Mycobacterium ulcerans*

3. A biochemical test that is useful in distinguishing *Mycobacterium tuberculosis* from other species of mycobacteria is the

(A) fermentation of erythritol
(B) hydrolysis of mycolic acid
(C) photochromatogenic reaction
(D) production of niacin
(E) Ziehl-Neelsen reaction

4. All of the following statements regarding the tuberculin skin test are true EXCEPT

(A) induration of 10 mm or greater indicates a positive reaction
(B) the Mantoux technique is most reliable
(C) repeated testing can cause a negative-to-positive conversion
(D) a negative reaction does not rule out a diagnosis of tuberculosis
(E) the test has greater epidemiologic than diagnostic value

Directions: Each question below contains four suggested answers of which **one or more** is correct. Choose the answer

A if **1, 2, and 3** are correct
B if **1 and 3** are correct
C if **2 and 4** are correct
D if **4** is correct
E if **1, 2, 3, and 4** are correct

5. First-line drugs for the treatment of tuberculosis include

(1) ethambutol
(2) streptomycin
(3) isoniazid
(4) rifampin

6. Etiologic agents of tuberculosis-like syndromes in humans include

(1) *Mycobacterium kansasii*
(2) *Mycobacterium ulcerans*
(3) *Mycobacterium avium*
(4) *Mycobacterium marinum*

7. True statements about Hansen's disease include

(1) Hansen's disease in the United States usually is transmitted to humans through contact with animals
(2) clinical disease exists in two distinct forms
(3) the disease is diagnosed by the culture and biochemical identification of *Mycobacterium leprae*
(4) the clinical presentation of Hansen's disease appears to be controlled largely by cell-mediated immunity

Directions: The group of questions below consists of lettered choices followed by several numbered items. For each numbered item, select the **one** lettered choice with which it is **most** closely associated. Each lettered choice may be used once, more than once, or not at all. Choose the answer

    A if the item is associated with **(A) only**
    B if the item is associated with **(B) only**
    C if the item is associated with **both (A) and (B)**
    D if the item is associated with **neither (A) nor (B)**

Questions 8–11

Match each clinical characteristic of Hansen's disease with the correct clinical form of disease.

(A) Tuberculoid leprosy
(B) Lepromatous leprosy
(C) Both
(D) Neither

8. Localized anesthesia

9. Positive lepromin skin test

10. Lesions containing numerous acid-fast bacteria

11. Clinical duplication of human disease in the armadillo

ANSWERS AND EXPLANATIONS

1. The answer is E. [*IV A 2 a, b*] Of the *Mycobacterium* species listed, *Mycobacterium marinum* is the only one likely to cause a self-limited cutaneous infection. *M. marinum* is easily isolated from the slime that covers the walls of swimming pools and submerged rocks in both freshwater and marine environments. The classic scenario of infection involves a hand or forearm that is abraded on the side of a swimming pool and becomes contaminated by *M. marinum*. A granulomatous lesion develops and frequently ulcerates. The lesion usually heals spontaneously within a few weeks, although chronic infection may occur. *Mycobacterium intracellulare* and *Mycobacterium kansasii* typically produce tuberculosis-like pulmonary infections, and *Mycobacterium scrofulaceum* occasionally causes pulmonary disease. *Mycobacterium ulcerans* produces a severe, progressive ulceration of cutaneous and subcutaneous tissue.

2. The answer is B. [*III C 2 a*] *Mycobacterium scrofulaceum* is the most common cause of granulomatous cervical lymphadenitis (also known as scrofula) in young children. The infection probably occurs by the inhalation of mycobacteria that eventually localize in the cervical lymph nodes. *Mycobacterium kansasii* and *Mycobacterium avium* also cause cervical lymphadenitis in adults and children, but not as frequently as does *M. scrofulaceum*. Cervical lymphadenitis is not a characteristic of disease caused by *Mycobacterium fortuitum* or *Mycobacterium ulcerans*.

3. The answer is D. [*I A 2; II F 3 b*] *Mycobacterium tuberculosis* is differentiated from other mycobacteria on the basis of growth characteristics, colony morphology, and a variety of biochemical tests. The production of large amounts of niacin is a characteristic that is unique to *M. tuberculosis*. All mycobacteria contain mycolic acid in their cell walls and are found to be acid fast by the Ziehl-Neelsen stain. Photochromogenicity is an old classification aid for mycobacteria, but it does not differentiate between individual species. Fermentation reactions are not characteristic of mycobacteria.

4. The answer is C. [*II G*] The tuberculin skin test uses a heterologous mixture of tuberculoproteins known as purified protein derivative, which is derived from the culture filtrate of *Mycobacterium tuberculosis*. The most efficient means of administering the antigen is by the intradermal injection of 5 tuberculin units (TU) of PPD—a technique referred to as the Mantoux test. A delayed-type hypersensitivity reaction characterized by induration of 10 mm or more within 48–72 hours indicates a positive test. A negative test does not necessarily rule out the diagnosis of tuberculosis, because the patient may have disease and be anergic. The tuberculin skin test primarily is an epidemiologic screening method to identify persons who have been exposed to *M. tuberculosis*; the test has only marginal diagnostic value. Repeated testing with PPD does not cause a negative-to-positive conversion in the person being tested.

5. The answer is E (all). [*II H 1 a*] A variety of antimicrobial agents inhibit the growth of *Mycobacterium tuberculosis*. Several of these compounds, such as isoniazid, are considered to be strictly antituberculosis drugs. First-line antimycobacterial agents for the treatment of human tuberculosis include isoniazid, ethambutol, streptomycin, pyrazinamide, and rifampin. Two or more of these agents routinely are used in combination to decrease the risk of selecting for drug-resistant strains of *M. tuberculosis*.

6. The answer is B (1, 3). [*III A, B*] Several species of *Mycobacterium* are capable of causing tuberculosis-like syndromes in humans; some of these etiologic agents also are associated with disease in other animals. *Mycobacterium kansasii* can produce a pulmonary disease that closely resembles tuberculosis. *Mycobacterium avium*, which causes disease in birds and mammals, produces cavitary pulmonary disease resembling tuberculosis in humans. *Mycobacterium ulcerans* and *Mycobacterium marinum* produce cutaneous infections in humans. The infections typically do not disseminate, as *M. ulcerans* and *M. marinum* are incapable of growing at 37° C.

7. The answer is C (2, 4). [*V B 1 a, b, 2 b, c*] Hansen's disease apparently is caused by the acid-fast bacterium, *Mycobacterium leprae*. The bacteria have not been successfully cultured in the laboratory, and humans appear to be the only natural reservoir of infection. The disease exists in two major clinical forms: tuberculoid leprosy and lepromatous leprosy. Cell-mediated immunity, or the lack of it, plays a significant role in the development of symptomatology in both forms of the disease. In tuberculoid leprosy, cell-mediated immunity is intact and there are few mycobacteria in the macular lesions. In lepromatous leprosy, cell-mediated immunity is inhibited and there are numerous mycobacteria in the large granulomatous lesions. A few cases of Hansen's disease are diagnosed each year in the United States, but the mode of transmission of the disease is unknown.

8–11. The answers are: 8-C, 9-A, 10-B, 11-D. [*V A 2, B 1 a, b, 2 a*] Both of the major clinical forms of Hansen's disease—tuberculoid leprosy and lepromatous leprosy—cause localized anesthesia due to infiltration of the nerves by the mycobacteria. This anesthesia frequently leads to serious injury without the knowledge of the patient. The presence of anesthetic lesions also is useful in diagnosing the disease.

Cell-mediated immunity is intact in tuberculoid leprosy, and the patient responds with a positive lepromin skin test. Lepromin is a heterologous mixture of mycobacterial proteins derived from biopsy material; it is analogous to purified protein derivative (PPD)—the antigen used in tuberculin skin tests.

Massive numbers of acid-fast *Mycobacterium leprae* are present in the lesions of lepromatous leprosy. The massive bacterial infiltration is responsible for the thickening of loose skin on the face and neck, which results in the classic leonine appearance of patients. Tuberculoid leprosy, in contrast, is characterized by few mycobacteria in the lesions.

M. leprae obtained from biopsy material from patients with lepromatous leprosy will grow in the nine-banded armadillo. However, the disease produced in armadillos does not clinically resemble human disease.

15
Obligate Intracellular Parasites: Chlamydiae and Rickettsiae

Gerald E. Wagner

I. *CHLAMYDIA* SPECIES. The genus *Chlamydia* consists of obligate intracellular parasites. Two species, *Chlamydia trachomatis* and *Chlamydia psittaci*, are pathogenic for humans. Once believed to be large viruses, these bacteria have a cell wall resembling that of gram-negative bacteria. They have both DNA and RNA, they replicate by binary fission, and they are susceptible to antibacterial agents.

A. Morphology. The chlamydiae are small, round-to-ovoid cells that vary morphologically during their replication cycles.

    1. The **cell wall** consists of peptidoglycan and muramic acid with an outer lipid layer resembling that of gram-negative bacteria.

    2. The **genome** of the chlamydiae is much smaller than that of most other bacteria.

B. Growth characteristics. The chlamydiae cannot be grown on artificial or complex bacteriologic medium.

    1. Yolk sac inoculation. Chlamydiae can be grown in the laboratory by inoculation of the bacteria into the yolk sac of embryonated chicken eggs; some strains of *C. trachomatis*, however, do not grow in yolk sacs.

    2. Tissue culture. Growth also can be achieved in some lines of tissue culture cells.
        a. Radiation of or the addition of cycloheximide to the tissue cells inhibits growth and allows the chlamydiae to compete better for nutrients.
        b. Visualization. The chlamydiae can be visualized in tissue culture cells with the light microscope when the cells are stained with Giemsa stain. *C. trachomatis* is more easily detected if stained with iodine (see section I C 3). Individual bacteria are not observed, but characteristic inclusion bodies are detected within the host cells.

    3. Energy source. The chlamydiae cannot synthesize adenosine triphosphate (ATP) nor oxidize nicotinamide-adenine dinucleotide phosphate (NADP). They are dependent upon the host cell for energy production.

C. Replication cycle. The chlamydiae replicate by binary fission, but they undergo morphologic variation during the replication cycle.

    1. Elementary bodies measuring about 0.3 μm in diameter infect the host cell, which they enter by **endocytosis**, a process similar to phagocytosis. A vacuole derived from the host cell membrane surrounds the elementary body, which is recognized by its electron-dense center.

    2. Initial bodies form in about 1 hour through a process of metabolic change and reorganization of the elementary bodies. The molecular basis of this process is not understood.
        a. The initial bodies, which measure about 1 μm in diameter and are less dense than elementary bodies, begin to replicate within the endocytic vacuole.
        b. Energy for replication is derived from the ATP generated by the host cell.
        c. Initial bodies reorganize and condense after 24–72 hours to form multiple new elementary bodies.
        d. The host cell ruptures and frees the elementary bodies, which are capable of infecting other host cells.

3. *C. trachomatis* synthesizes large amounts of glycogen, which surrounds the bacteria within the endocytic vacuole. This structure is referred to as the **intracellular inclusion body**. The inclusion body can become large enough to displace the nucleus of the host cell. It is visible when stained with iodine in histologic preparations.

4. *C. psittaci* does not synthesize glycogen. The chlamydiae rupture the vacuole, and the bacterial cells surround the host nucleus. Visualization of this species is more difficult.

D. **Antigenic composition.** Lipopolysaccharide is a group antigen that is common to all chlamydiae.

1. **Serotyping** is based on specific cell wall proteins; each serotype is associated with a particular disease. Fluorescently labeled antibodies are used in serotyping, but the procedure is not performed routinely in the clinical laboratory.

2. **Type-specific antigens** of the chlamydiae induce protective antibodies in experimental animals.

E. **Epidemiology.** The chlamydiae are susceptible to environmental conditions, and they survive for only a short time outside the host. Transmission of *C. trachomatis* is by direct personal contact among humans. *C. psittaci* is pathogenic for birds and domestic fowl and is transmitted to humans by inhalation of bacteria in droplets or dust.

F. **Pathogenicity.** The chlamydiae are highly infectious. The pathogenesis of diseases caused by these bacteria is not understood. Different strains of both *C. trachomatis* and *C. psittaci* show different degrees of virulence.

1. **Phagocytized chlamydiae** prevent fusion of lysosomes to the phagosome and, thus, escape intracellular destruction by lysosomal enzymes.

2. **Heat-labile toxin** is produced by both species of chlamydiae. The protein toxin is lethal to mice when it is injected intravenously. Each serotype of chlamydiae produces a specific toxin.

3. **Competition for nutrients** from the chlamydiae may be another cause of damage and death of the host cell. *C. trachomatis* may exist in a latent state and be reactivated if the host becomes immunosuppressed.

G. **Diseases caused by *C. trachomatis***

1. **Eye infections.** Two distinct syndromes involving the eye are caused by different antigenic types of *C. trachomatis.* **Trachoma** is a chronic infection usually occurring in underdeveloped countries; it often leads to blindness. **Inclusion conjunctivitis** is an acute infection with worldwide distribution.
   a. **Trachoma** is caused by *C. trachomatis* serotypes A, B, Ba, and C. The infection generally is acquired in infancy from the mother or other close contact as an acute conjunctivitis. Trachoma is a leading cause of blindness.
      (1) **Pathogenesis.** The initial acute conjunctivitis usually heals spontaneously. Reinfection or persistence of the original infection produces corneal scarring and conjunctival deformity as a result of the severe inflammatory response. Corneal vascularization occurs, and blindness frequently is the result of 15–20 years of chronic inflammation.
      (2) **Therapy** for trachoma is difficult. Public health control measures appear to be the most appropriate action. Topically and parenterally administered antimicrobial agents are used to prevent infection and to eradicate persistent disease. Improved household and public hygiene reduces the incidence of infection.
      (3) **Epidemiology.** Trachoma is a major problem in most regions of Africa. Native American populations in the southwestern United States also are affected. It is estimated that over 400 million people worldwide have trachoma and that 20 million of them are blind.
   b. **Inclusion conjunctivitis** is an acute disease that occurs throughout the world in both infants and adults. It is the most common form of neonatal conjunctivitis in the United States, affecting up to 6% of infants.
      (1) **The disease is acquired by direct contact during the birth process.** Inclusion conjunctivitis is caused by *C. trachomatis* serotypes D–K. These are the same serotypes that are responsible for most genital infections.

(2) The **onset** of inclusion conjunctivitis usually is within 25 days of birth. It is characterized by a mucopurulent discharge. The infection may disseminate and cause pneumonia in infants.

(3) The disease is less common in adults than in children and usually is associated with concurrent genital infection. The nasopharynx, rectum, and genital tract may be colonized.

(4) Therapy. Topical antimicrobial agents can be used to treat the conjunctival infection, but systemic therapy is advisable to prevent colonization of other tissues. The disease may heal spontaneously in neonates and escape medical attention.

2. Genital infections. *C. trachomatis* causes a sexually transmitted infection of the urogenital tract similar to that caused by *Neisseria gonorrhoeae*. Three serotypes are capable of causing a disease known as **lymphogranuloma venereum**.

   a. Disease in men. *C. trachomatis* serotypes D–K cause about 40% of cases of **nongonococcal urethritis** in men in westernized countries. It also is the most common cause of **epididymitis** in sexually active men under 35 years of age.

      (1) Nongonococcal urethritis presents the **same clinical picture as gonorrhea**. The male patient has dysuria and a purulent urethral discharge. Microscopic examination of the discharge reveals neutrophils but no bacteria. *C. trachomatis* can be isolated by swabbing the urethra.

      (2) Urethritis caused by *C. trachomatis* **may occur simultaneously with gonorrhea**. Persistence of symptoms with the disappearance of gonococci from a gram-stained smear after antigonococcal therapy is indicative of concomitant urethritis caused by *C. trachomatis*.

   b. Disease in women. *C. trachomatis* in the genital tract of women can cause urethritis, cervicitis, and salpingitis. Uterine and cervical infections may be asymptomatic.

      (1) Symptomatic infections are characterized by a mucopurulent urethritis or cervicitis.

      (2) *C. trachomatis* is a frequent cause of **salpingitis**, a painful inflammation of the fallopian tubes usually referred to as **pelvic inflammatory disease (PID)**. PID caused by *C. trachomatis* generally is the result of ascending infection from the cervix.

      (3) Epidemiology. *C. trachomatis* infection in women is most common in young adults, especially those in the lower socioeconomic population and those with multiple sexual partners.

      (4) Incidence. Between 5% and 10% of women seen by private gynecologists for prenatal examinations are colonized by *C. trachomatis*. The infection accounts for about 25% of cases of sexually transmitted diseases in women, and it may account for up to 25% of cases of PID.

        (a) Approximately 30%–50% of men who have sexual contact with women who have active cervicitis develop urethritis within 2–6 weeks.

        (b) Chlamydial disease develops in about 50% of infants born to mothers with cervicitis. Although inclusion conjunctivitis is the most common manifestation, 5%–10% of the infants develop pneumonia.

   c. Lymphogranuloma venereum (LGV) is caused by *C. trachomatis* serotypes L1, L2, and L3. These serotypes are not associated with any other form of chlamydial disease.

      (1) Primary lesion. LGV begins as a small ulcer on the genitalia; the lesion often is not noticed. The inguinal lymph nodes become swollen, and in 2–6 weeks they may suppurate and form draining sinuses. The initial lesion usually has disappeared by this stage.

      (2) Disseminated disease. LGV is one of the few chlamydial infections that can disseminate. The peritoneum may become involved, especially in women. Infection of the lower bowel in homosexual men causes a syndrome of ulcerative colitis.

      (3) Epidemiology. LGV is sexually transmitted. It occurs principally in Africa and South America and is rare in North America. Like other venereal diseases, the incidence is highest in sexually active young adults, and it is prevalent in homosexual men.

3. Laboratory diagnosis of *C. trachomatis* infections can be made microbiologically or serologically.

   a. The chlamydiae pose a significant **threat to laboratory personnel** because of their high degree of infectivity. Isolation, therefore, rarely is performed in the routine clinical laboratory; it is done instead in a specialized laboratory.

    b. Inoculation of McCoy cells pretreated with 5'-iodo-2'-deoxyuridine (IDUR) is the most widely employed method of culture. Cells are observed for characteristic intracellular inclusion bodies.

    c. Specimens for culture include corneal scrapings and exudate from conjunctivitis, urethral exudate from genital infections, and discharge from draining sinuses in LGV. Direct staining of corneal scrapings and other specimens may be helpful, but cytoplasmic inclusion bodies are not seen in every case.

    d. Serologic detection of immunoglobulin M (IgM) antibodies is useful in the diagnosis of the pneumonia syndrome in infants. Serology is of little diagnostic benefit for sexually active adults, in whom high IgG antibody titers may be the result of past infections with *C. trachomatis.*

H. Infections caused by *C. psittaci*. Psittacosis occurs naturally in a variety of both wild and domestic birds. Frequently, the disease is latent but is activated by the stress of captivity and shipping. **Human psittacosis is primarily an occupational disease of poultry workers.** Turkeys are a major source of human infection. Keepers of pet psitticines (i.e., parrots and parakeets) also may contact psittacosis.

    1. The route of infection is by the inhalation of respiratory discharge or dust from contaminated fecal material.

    2. The symptomatology is that of an acute pulmonary infection and is characterized by fever, headache, malaise, myalgia, and a nonproductive, hacking cough. Bilateral interstitial pneumonia is seen on chest roentgenograms.

    3. Systemic involvement occasionally develops. Encephalitis, myocarditis, and hepatitis have been described as complications. Hepatomegaly and splenomegaly are characteristic of systemic involvement.

    4. Diagnosis of human psittacosis is based upon the clinical symptoms and a history of association with birds. A fourfold increase in complement-fixing (CF) antibody to chlamydial group antigen during the illness is considered confirmatory.

    5. Therapy and control. Tetracycline and erythromycin are effective therapeutic agents if they are given early in the disease. The incidence of psittacosis in the United States has been reduced dramatically by the addition of antimicrobial agents to poultry feed and by strict quarantine regulations for imported birds.

II. RICKETTSIAE. The rickettsiae are obligate intracellular bacteria. They generally have animal reservoirs and are transmitted to humans by vectors that play important roles in their life cycles. Pathogenic rickettsiae are divided into the spotted fever group and the typhus group. *Coxiella burnetii* is a rickettsia-like bacterium that causes a systemic disease.

A. Morphology. The rickettsiae are small (i.e., 0.3–0.5 μm), gram-negative bacteria. Most frequently they are seen as pairs of rod-shaped cells with tapered ends. They also may be seen as single coccobacilli. The rickettsiae stain poorly with bacteriologic stains, but they can be visualized readily in tissue with the Giemsa stain.

B. Growth characteristics. Rickettsiae grow only in eukaryotic cells, and laboratory cultivation requires the use of tissue culture cell lines or embryonated eggs. These bacteria exhibit several unique growth characteristics.

    1. Intracellular growth. Rickettsiae enter the host cell by endocytosis, a process similar to phagocytosis except that the bacteria expend energy (in the form of ATP). By producing a phospholipase, the bacteria destroy the phagosome and begin to grow in the cytoplasm of the host cell. The host cell eventually lyses and releases the rickettsiae, which are capable of infecting other cells.

    2. Basis of intracellular parasitism. Rickettsiae apparently require coenzyme A, nicotinamide-adenine dinucleotide (NAD), and energy from the host cell. Energy is obtained by a mitochondria-like exchange transport system in which host ATP is exchanged for rickettsial adenosine diphosphate (ADP).

    3. Free rickettsiae cease metabolic activity and begin to leak intracellular constituents. The result is a loss of infectivity within a short period of time.

C. Pathogenicity. Most of the rickettsiae are transmitted to humans by the bite of an arthropod vector. The bacteria invade the vascular epithelial cells and become widely disseminated.

1. **Clinical characteristics** of rickettsial disease are fever, headache, and a rash. The rash is due to focal areas of infection that cause hyperplasia and inflammation of the vascular epithelium. The result is thrombosis and blockage of the small blood vessels.

2. **Toxic properties.** The experimental injection of whole rickettsia cells into animals produces a syndrome that is similar to the endotoxin shock produced by other gram-negative bacteria. The importance of this toxic effect in the pathogenesis of human disease is not known.

3. *C. burnetii* produces an atypical rickettsial disease. The bacteria are capable of surviving in the environment free of host cells. Human disease is caused by inhalation of aerosolized bacteria from infected animal tissue. The disease is characterized by fever and pneumonia.

D. Diseases caused by rickettsiae

1. **The spotted fever group of diseases** are caused by bacteria that are antigenically related to *Rickettsia rickettsii*. Some of these diseases are endemic to different areas of the world, and their names frequently reflect their endemic areas.
   a. **Rocky Mountain spotted fever** is the most important rickettsial disease in the United States. More than 1000 cases occur each year. It is seen throughout the country; the highest incidence of Rocky Mountain spotted fever is in the Middle Atlantic states.
      (1) An increased incidence of the disease in recent years is attributed to increased outdoor activity and greater exposure to ticks. About 60% of cases occur in children under the age of 15 years. More than 70% of patients have a known history of tick bite.
      (2) Several species of ticks are the natural vectors of Rocky Mountain spotted fever; each prevails in a particular region. *R. rickettsii* is transferred transovarially in ticks. The rickettsiae can be found in every stage of the tick's development and are transmitted to mammals during a blood meal. Infected adult ticks have been shown to survive 4 years without feeding.
      (3) **Pathology**
         (a) **Symptomatology.** Rocky Mountain spotted fever usually becomes clinically evident 2–6 days after the tick bite. A **skin rash** is the most characteristic sign, although fever, headache, myalgia, toxicity, and mental confusion also occur.
            (i) The **rash** usually develops 2–3 days after exposure and first appears on the wrists and ankles before spreading to the trunk within a few hours. The appearance of the rash on the palms and soles of the feet is considered diagnostic.
            (ii) **Vascular lesions** throughout the body are the primary pathologic sign of Rocky Mountain spotted fever. They are obvious in the skin, where the small vessels are blocked and blood leaks into the surrounding tissue, forming the rash. Vasculitis with more serious consequences occurs in the deeper organs, particularly the adrenal glands.
         (b) **Complications.** Extreme muscle tenderness occasionally occurs. Other serious complications include encephalitis, renal failure, heart failure, thrombocytopenia, disseminated intravascular coagulation, and vascular collapse.
      (4) **Diagnosis** is made on the basis of the clinical syndrome and history. Tests for early diagnosis of Rocky Mountain spotted fever are not available in the routine clinical laboratory.
         (a) **Serologic tests** indicating a rise in CF antibody titers between acute and convalescent sera are diagnostic. A rising or a single high antibody titer to *Proteus* organisms OX-19 and OX-2 (i.e., **the Weil-Felix reaction**) is presumptive evidence of infection.
         (b) **Immunofluorescent identification** of *R. rickettsii* in skin biopsy specimens is used in special circumstances.
      (5) **Antibiotic therapy** usually includes tetracycline or chloramphenicol, both of which are highly effective if administered during the first week of the disease. Untreated Rocky Mountain spotted fever has a mortality rate of up to 7%.
   b. **Rickettsialpox** is a benign rickettsial disease caused by *Rickettsia akari*. The disease is transmitted from animal to animal and to humans by a **rodent mite**.
      (1) The **major reservoir** of rickettsialpox is the **house mouse**, and the disease frequently occurs in urban areas. Control of mice is the best method of disease prevention in humans.

      (2) A biphasic syndrome characterizes the human disease.

          (a) Initial phase. At the site of inoculation (i.e., the mite bite), a papulovesicular lesion develops into a black eschar within 5 days. Fever, malaise, and other systemic symptoms develop as the bacteria disseminate.

          (b) Second phase. A generalized papulovesicular rash appears throughout the body. Each lesion develops into a black eschar.

      (3) Therapy. Rickettsialpox is a self-limited disease that begins to wane spontaneously in about 1 week. Tetracycline therapy shortens the disease to 1 or 2 days. No deaths due to rickettsialpox have been reported.

2. The typhus group of diseases are febrile illnesses. They generally are transmitted to humans from animal reservoirs by lice, mites, and fleas.

  a. Louse-borne typhus is caused by *Rickettsia prowazekii* and is transmitted to humans, the primary reservoir, by the body louse. Crowded conditions and poor personal hygiene predispose to epidemics of louse-borne typhus.

      (1) Transmission. The disease spreads from human to human when the body louse acquires the etiologic agent during a blood meal from a bacteremic person.

          (a) Large numbers of bacteria infect the louse, and the incubation period in the louse is 5–10 days.

          (b) The louse defecates when it takes a blood meal from a new host. The contaminated feces are rubbed into the bite wound when the host scratches the area. *R. prowazekii* survives in dried feces and can infect mucous membranes or can be inhaled.

      (2) Symptomatology. Louse-borne typhus is characterized by fever, headache, and rash that begin 1–2 weeks after inoculation. The rash begins on the trunk and spreads to the extremities. Headache, malaise, and myalgia are prominent features of the illness. Complications include myocarditis and central nervous system dysfunction.

      (3) Diagnosis. The **Weil-Felix reaction** is positive with *Proteus* OX-19 but not with OX-2. The disease usually is diagnosed by clinical symptoms and history.

      (4) Therapy and control. Tetracycline and chloramphenicol provide effective chemotherapy, but louse control is more effective in preventing the disease. The mortality rate in untreated cases increases with the age of the patient and ranges from 10%–60%.

  b. Brill's disease is a relapse of louse-borne typhus that commonly occurs many years after the primary infection.

      (1) Brill's disease usually is seen in immigrants from eastern Europe who had typhus during World War II. The syndrome is milder and of shorter duration than the primary disease.

      (2) Precipitating factors are unknown, but *R. prowazekii* probably remains sequestered in cells of the reticuloendothelial system. Antibody titers to *Proteus* OX-19 are absent.

  c. Murine typhus is caused by *Rickettsia typhi.* The vector for human infection is the rat flea. Humans are accidental hosts, and only about 50 cases per year are seen in the United States, principally in Texas and other Gulf states. The disease resembles louse-borne typhus in pathogenesis, symptomatology, and serology.

  d. Scrub typhus is caused by *Rickettsia tsutsugamushi,* and humans become infected from larvae of rodent mites called **chiggers**.

      (1) Epidemiology. The larvae feed on plant debris and are picked up when a person comes in contact with low trees and brush. The disease is endemic in the southwestern Pacific regions, Southeast Asia, and Japan.

      (2) Symptomatology. Scrub typhus is characterized by a slow onset. An initial necrotic eschar at the site of the bite occurs in only 50%–80% of cases.

          (a) Fever gradually increases over a period of 1 week, but it may reach the extreme temperature of 40.5° C.

          (b) Headache, generalized lymphadenopathy, and a rash, which is maculopapular and generally erupts suddenly, are features of the disease.

          (c) Hepatomegaly and conjunctivitis may develop as complications.

      (3) Diagnosis of scrub typhus is made on the basis of the clinical syndrome, which may be confused with dengue, malaria, leptospirosis, and other diseases. Antibody titers to *Proteus* OX-K can be detected in about 50% of cases, but these titers may be low.

      (4) Therapy. Tetracycline and chloramphenicol are effective therapeutic agents.

  e. Q fever is caused by *C. burnetii,* which differs from other rickettsial organisms in that the bacteria are not killed by drying and can survive for long periods of time outside the host cell.

(1) Epidemiology. Q fever is a zoonosis of livestock, rodents, and marsupials. The disease may be asymptomatic in animals. In some endemic areas, 50%–75% of herds are infected.

   (a) *C. burnetii* grows well in the placenta of pregnant animals. At parturition, the bacteria contaminate the soil and apparently enter a sporogenic cycle.

   (b) The soil can remain contaminated for months under normal conditions. At a constant temperature of 40° C, dried fomites have remained contaminated for more than 1 year.

   (c) Human disease is acquired by the respiratory route rather than by the bite of an arthropod. It is primarily an occupational disease of people who handle animals or animal products. Epidemics of Q fever have been reported in slaughterhouses and in the textile industry.

(2) Symptomatology. Q fever is characterized by the sudden onset of fever, chills, and headache 9–20 days after inhalation of *C. burnetii*.

   (a) A dry cough with patchy interstitial pneumonia develops in some patients. No typical rash develops.

   (b) The disease is systemic, and hepatomegaly frequently occurs. Liver function tests usually are abnormal.

(3) Complications of Q fever include myocarditis, pericarditis, and endocarditis. The disease rarely is fatal.

(4) Diagnosis of Q fever usually is made by detecting CF antibodies to *C. burnetii* antigens. Immunofluorescence techniques also have been employed in the diagnosis. The Weil-Felix reactions are negative.

(5) Therapy. Q fever can be treated successfully with tetracycline.

STUDY QUESTIONS

Directions: Each question below contains five suggested answers. Choose the **one best** response to each question.

1. What characteristic of *Chlamydia trachomatis* makes it visible within host cells?

(A) Its unique Gram stain characteristics
(B) The synthesis of large amounts of glycogen
(C) Its requirement for host-cell ATP
(D) The aggregation of elementary bodies
(E) A thick cell wall that is visible without stain

2. The infectious particle of chlamydiae is the

(A) inclusion body
(B) glycogen body
(C) elementary body
(D) aggregate body
(E) initial body

3. The chlamydial infection that is most likely to result in disseminated disease is

(A) lymphogranuloma venereum
(B) psittacosis
(C) trachoma
(D) inclusion conjunctivitis
(E) urogenital infection in women

4. Intracellular survival of *Rickettsia* species is achieved by which of the following mechanisms?

(A) Prevention of phagosome formation
(B) A protective mucopeptide capsule
(C) Shutting down host-cell protein synthesis
(D) Release of cytotoxic lipopolysaccharide
(E) Production of a phospholipase

5. What vector is responsible for the transmission of Rocky Mountain spotted fever to humans?

(A) Body louse
(B) Mite
(C) Chigger
(D) Rat flea
(E) Tick

Directions: Each question below contains four suggested answers of which **one or more** is correct. Choose the answer

A if **1, 2, and 3** are correct
B if **1 and 3** are correct
C if **2 and 4** are correct
D if **4** is correct
E if **1, 2, 3, and 4** are correct

6. Infectious diseases caused by *Chlamydia trachomatis* serotypes D–K include

(1) inclusion conjunctivitis
(2) infantile pneumonia
(3) cervicitis
(4) lymphogranuloma venereum

7. Factors that help in the diagnosis of Rocky Mountain spotted fever include

(1) a rash on the palms
(2) the Weil-Felix reaction
(3) history of a tick bite
(4) the handling of animals

8. Bacteria that can survive for prolonged periods outside a host cell include

(1) *Rickettsia rickettsii*
(2) *Chlamydia psittaci*
(3) *Chlamydia trachomatis*
(4) *Coxiella burnetii*

Directions: The group of questions below consists of lettered choices followed by several numbered items. For each numbered item select the **one** lettered choice with which it is **most** closely associated. Each lettered choice may be use once, more than once, or not at all.

Questions 9–14

Match each disease with the bacteria that cause it.

(A) *Rickettsia rickettsii*
(B) *Rickettsia prowazekii*
(C) *Rickettsia tsutsugamushi*
(D) *Rickettsia akari*
(E) *Coxiella burnetii*

9. Rickettsialpox

10. Rocky Mountain spotted fever

11. Brill's disease

12. Scrub typhus

13. Louse-borne typhus

14. Q fever

ANSWERS AND EXPLANATIONS

1. The answer is B. [*I B 2 b (2)*] The easiest way to visualize *Chlamydia trachomatis* within tissue cells is by staining the glycogen with iodine. The glycogen of the inclusion body is seen as a brown ovoid or sphere. The chlamydiae also can be seen with the light microscope when they are stained with the Giemsa stain. The cell wall of chlamydiae resembles that of gram-negative bacteria, but the Gram stain does not distinguish the bacteria adequately within host cells. The obligate intracellular nature of chlamydiae, indicated by their requirement for host-cell ATP, does not affect the staining characteristics.

2. The answer is C. [*I C 1*] Elementary bodies are the infectious particles of chlamydiae. These bodies, measuring about 0.3 μm in diameter, enter the host cell by a process known as endocytosis. They are surrounded by a vesicle derived from the host cell membrane. *Chlamydia trachomatis* produces large amounts of glycogen, and the accumulation of glycogen is referred to as an inclusion body. Initial bodies form about 1 hour after the host cell is infected. These initial bodies replicate by binary fission and then condense to form elementary bodies.

3. The answer is A. [*I G 2 c (2) (3)*] Lymphogranuloma venereum (LGV) is caused by *Chlamydia trachomatis* serotypes L1, L2, and L3. LGV is one of the few chlamydial infections that can disseminate, primarily via the lymphatic vessels. The disease is venereally transmitted and typically occurs in the sexually active young adult of Africa and South America. LGV is prevalent among homosexual men. Psittacosis is a pulmonary infection characterized by flulike symptomatology. Trachoma and inclusion conjunctivitis are caused by different serotypes of *C. trachomatis*, and they are localized to the eyes. Cervicitis and pelvic inflammatory disease, as well as urethritis in men, are the most common chlamydial infections in the United States and Europe.

4. The answer is E. [*II B 1*] *Rickettsia* species enter the host cell by a process similar to phagocytosis except that the bacteria expend energy in the form of ATP. Normally, an endocytic vesicle formed from the host cell membrane fuses with lysosomes containing lytic enzymes and other cell-destroying chemicals. This fused vesicle is known as the phagolysosome, and it is responsible for the intracellular death of bacteria. However, the rickettsiae produce a phospholipase that destroys the membrane-bound phagosome. Chlamydiae survive intracellulary by preventing the fusion of the endocytic vesicle and the lysosome. The rickettsiae do not have a capsule, and their lipopolysaccharide endotoxin probably contributes to the clinical symptomatology of infection. Although the rickettsiae require host cell ATP for energy production, the effect on host cell protein synthesis appears to be minimal.

5. The answer is E. [*II D 1 a (2)*] Several species of ticks serve as vectors for the transmission of Rocky Mountain spotted fever to humans. *Rickettsia rickettsii* is passed transovarially in ticks, and infected adult ticks have been shown to survive for up to 4 years without a blood meal. Humans acquire disease when an infected tick takes a blood meal. The body louse is the vector of *Rickettsia prowazekii*, the etiologic agent of louse-borne typhus. The larvae of rodent mites, known as chiggers, are the vectors for *Rickettsia tsutsugamushi*, the etiologic agent of scrub typhus. The rat flea is the vector for *Rickettsia typhii*, the etiologic agent of murine typhus.

6. The answer is A (1, 2, 3). [*I G 1 b (1), 2 b (4) (b)*] *Chlamydia trachomatis* serotypes D–K are the cause of urogenital infections in men and women. Uterine and cervical infections frequently are asymptomatic, although cervicitis may be clinically overt and may lead to salpingitis. These serotypes also are the cause of inclusion conjunctivitis; approximately 50% of infants born to mothers who carry *C. trachomatis* develop the disease. In addition, approximately 10% of these infants develop pneumonia. Lymphogranuloma venereum is caused by *C. trachomatis* serotypes L1, L2, and L3.

7. The answer is A (1, 2, 3). [*II D 1 a (1), (3) (a) (i), (4) (a)*] The appearance of a rash on the palms and soles of the feet, in conjunction with other appropriate findings, is essentially diagnostic for Rocky Mountain spotted fever. More than 70% of patients who develop the disease have a history of tick bite. The Weil-Felix reaction measures serum antibody titers to antigens of a *Proteus* species. Rising titers or a single high titer of antibodies to OX-19 and OX-2 is suggestive of Rocky Mountain spotted fever. The disease is not associated with the handling of animals. An increased incidence of disease in recent years is thought to be due to increased participation in outdoor activities.

8. The answer is D (4). [*II D 2 e*] *Coxiella burnetii* is classified as a rickettsia-like bacterium, although it differs from the rickettsiae in several respects. *C. burnetii* is the etiologic agent of Q fever, and it is acquired through the inhalation of aerosols rather than from transmission by arthropod vectors, as are the rickettsiae. The bacteria may remain infective in contaminated soil under normal conditions for several months. Fomites have remained contaminated for more than 1 year. *Rickettsia* and *Chlamydia* species depend upon the host cell for ATP; they can survive for 1 hour or less outside an appropriate host cell.

9–14. The answers are: 9-D, 10-A, 11-B, 12-C, 13-B, 14-E. [*II D 1 a, b, 2 a, b, d, e*] Rickettsial disease in humans classically is divided into two types: the spotted fever group and the typhus group.

The spotted fever group of diseases are caused by a variety of rickettsiae, all of which are antigenically related to *Rickettsia rickettsii*; the disease names frequently reflect the geographic region where the disease is endemic. Rickettsialpox, a benign disease caused by *Rickettsia akari*, and Rocky Mountain spotted fever, a clinically important disease caused by *R. rickettsii*, are characteristic human diseases of the spotted fever group.

Brill's disease is a relapse of louse-borne typhus caused by *Rickettsia prowazekii*. The disease typically occurs many years after the primary infection, and it tends to be milder and shorter in duration.

Rickettsia tsutsugamushi is the etiologic agent of scrub typhus. The disease is endemic to the southwest Pacific region, Japan, and Asia.

Louse-borne typhus is caused by *R. prowazekii*. It is transmitted from human to human by the body louse, and crowded living conditons predispose people to infection in endemic areas.

Q fever is caused by the inhalation of the rickettsia-like bacteria *Coxiella burnetii*. The bacteria is endemic in animals, infecting the placenta and at parturition contaminating the soil and fomites.

16
Miscellaneous Pathogenic Bacteria: *Mycoplasma, Legionella, Corynebacterium,* and Others

Gerald E. Wagner

I. *MYCOPLASMA* AND *UREAPLASMA* SPECIES. The smallest free-living bacteria are found in the genera *Mycoplasma* and *Ureaplasma*, which belong to the family Mycoplasmataceae. These microorganisms resemble other bacteria with one exception: they lack cell walls. They primarily are plant and animal pathogens and are ubiquitous in nature. Although a variety of species have been associated with humans, the three species widely recognized as human pathogens are *Mycoplasma pneumoniae, Mycoplasma hominis,* and *Ureaplasma urealyticum.*

A. Cellular morphology. The mycoplasmataceae are about 0.2 μm in diameter. They are very pleomorphic and may assume coccoid, filamentous, and large multinucleated forms. They do not stain with routine bacteriologic stains.

1. **Cell membrane.** Individual cells lack a cell wall but are surrounded by a unique, three-layered membrane that contains sterols.
   a. The **sterols** are not synthesized by the mycoplasmas but are derived from the medium or tissue in which the bacteria are growing.
   b. **Neuraminic acid** in the cell membrane of *M. pneumoniae* serves as receptor sites for host cells. These receptor sites also are responsible for a phenomenon known as **hemadsorption**—the binding of added erythrocytes to cultivated bacterial cells.

2. **Genetic structure.** DNA is replicated by the mechanism used by all bacteria. The genome is smaller than that of other bacteria, but other cellular components are analogous.

B. Growth characteristics

1. **Colonies.** *Mycoplasma* and *Ureaplasma* species require several days to produce minute colonies on specialized complex media. Colonies of *M. pneumoniae* on agar have a characteristic inverted fried-egg appearance.

2. **Species differentiation.** Differences in metabolic characteristics among the species allow easy differentiation.

3. **Oxygen requirements.** *M. pneumoniae* grows aerobically; other species are facultative anaerobes.

C. Pulmonary disease. *M. pneumoniae* is a human respiratory pathogen. It accounts for about 20% of all cases of pneumonia.

1. **Epidemiology.** *M. pneumoniae* is distributed throughout the world. It is responsible for pneumonia epidemics at about 6-year intervals in both civilian and military populations.
   a. The **prime age** for *M. pneumoniae* infection is between 5 and 15 years. The incidence of infection among teenagers is higher than that in the general population, and about 35% of teenage pneumonia is due to *M. pneumoniae*. Disease in children less than 6 months old is rare.
   b. Owing to a variable incubation period (2–15 days), the disease usually occurs in sporadic outbreaks in families and communities. The attack rate is about 60% for exposed susceptible hosts, most of whom develop symptoms of respiratory infection.

2. **Infectivity and pathogenicity.** The dose needed to produce infection in 50% of human volunteers (i.e., the ID_{50}) is one colony-forming unit when it is inhaled. About 100 colony-forming units are required to produce disease by intranasal instillation.

> **a. Primary infection.** The bacteria attach to protein receptors on the respiratory epithelium and interfere with ciliary action. The result is desquamation and an inflammatory response.
> **b. Inflammation** begins in the bronchial and peribronchial tissue but spreads to the alveoli. The infiltration of lymphocytes, plasma cells, and macrophages thickens the alveoli. An exudate is common.
> **c. Contagiousness.** *M. pneumoniae* generally is shed into the upper respiratory tract for 2–8 days before overt symptoms of disease are seen. Shedding may continue for as long as 14 weeks.

3. **Clinical disease.** *M. pneumoniae* causes pharyngitis, bronchitis, bullous tympanic membrane infection, and pneumonitis. Arthritis, meningoencephalitis, hemolytic anemia, and cutaneous rashes occasionally have been associated with *M. pneumoniae* infection.
   a. **General characteristics.** Clinical disease is less severe than other forms of bacterial pneumonia; it frequently is referred to as an **atypical or walking pneumonia**. Hospitalization rarely is necessary.
   b. **Pneumonia**
      (1) Disease **onset** is slow and often inapparent. Fever, headache, and malaise are experienced for 2–4 days before the occurrence of respiratory symptoms.
      (2) A **nonproductive cough** is the characteristic pulmonary symptom.
      (3) **Sputum** that is expectorated is mucoid and contains neutrophils. Bacterial cells are not visible on Gram stain.
      (4) **Chest roentgenography** reveals a lower lobe pneumonia, although multiple lobes occasionally are involved. Small pleural effusions are seen in about 25% of patients.
   c. **Pharyngitis** caused by *M. pneumoniae* is characterized by sore throat and fever. It is indistinguishable from pharyngitis caused by viruses or by *Streptococcus pyogenes*.
   d. **Tracheobronchitis** is a mild respiratory infection caused by *M. pneumoniae*. Headache, fever, malaise, and cough are commonly seen. There usually is roentgenographic evidence of lung involvement.
   e. **Stevens-Johnson syndrome and erythema multiforme** are allergic responses that may develop as a result of *M. pneumoniae* infection.

4. **Immunity.** Both local and systemic immune responses are seen in *M. pneumoniae* infections.
   a. **Incomplete immunity.** Immunity against infection following recovery is incomplete, and disease tends to be more severe in older people than in young children.
   b. **Immune response.** The incompleteness of immunity suggests that the clinical manifestations of the disease may be due to a cell-mediated response to bacterial antigens rather than to the actual invasion of *M. pneumoniae*.

5. **Laboratory diagnosis.** Normal staining techniques and culture of clinical specimens are of little help in diagnosis.
   a. **Specific diagnosis** normally is made by the serologic detection of complement-fixing (CF) antibodies. A fourfold increase in titer or a single high titer is indicative of disease.
   b. **Detection of nonspecific cold agglutinins** may be helpful; most patients develop high titers of these during *M. pneumoniae* infection.

D. **Urogenital infections.** Infections of the human urogenital tract are caused by *M. hominis* and *U. urealyticum*. These bacteria are considered indigenous flora of the urogenital tract under normal circumstances.

1. *M. hominis* generally is associated with **postabortion or postpartum fever**.
   a. The bacteria can be isolated from the blood of about 10% of women who develop this infection. The disease probably is self-limited, but antibiotic therapy is recommended.
   b. A significant number of cases of **pelvic inflammatory disease (PID)** are probably caused by the seven antigenic varieties of *M. hominis*.

2. *U. urealyticum* is the only species in the genus *Ureaplasma*. It is distinguished from other mycoplasmataceae by its ability to produce **urease**. It was formerly known as T- (i.e., tiny) mycoplasma.
   a. **Epidemiology.** *U. urealyticum* is found in the urogenital tract of sexually active adults. It rarely can be isolated before puberty. It is transmitted sexually and colonizes 80% of men and women who have had three or more sexual partners.

> **b. Clinical disease.** It has been difficult to associate *U. urealyticum* with specific human disease because of high colonization rates.
>> **(1) Urethritis.** It is now thought that *U. urealyticum* accounts for more than 50% of cases of nongonococcal urethritis seen in men.
>> **(2) Postpartum fever.** *U. urealyticum* has been associated with postpartum fever.
> **c. Therapy.** Tetracycline therapy is recommended because of its additional activity against the chlamydiae. Tetracycline resistance recently has been reported among isolates of *U. urealyticum*, and spectinomycin is used as a secondary agent.

II. *LEGIONELLA PNEUMOPHILA* AND OTHER LEGIONELLAE.

Legionnaires' disease (legionellosis) is the name given to an outbreak of disease that first was documented among American Legion conventioneers at Philadelphia in 1976. The etiologic agent was identified as a new genus and species, *Legionella pneumophila*. Retrospective serologic analyses of clinical material from mysterious outbreaks of pneumonia in the 1940s and 1960s indicate that *L. pneumophila* was the etiologic agent. The same organism also is responsible for Pontiac fever, a different respiratory syndrome first documented in Pontiac, Michigan, in 1968. Other species in the genus have been associated with pneumonia and with Pontiac fever, and still others may be nonpathogenic.

A. Cellular morphology. *L. pneumophila* is a thin, gram-negative rod that measures 0.5 µm in diameter and 2–20 µm in length. It tends to display pleomorphism.

> **1. Ultrastructurally**, *L. pneumophila* resembles gram-negative bacteria and has two triple-layered membranes (i.e., envelopes). Flagella are observed in the polar, subpolar, and lateral positions. Inclusion granules are present, but the bacteria do not form spores.

> **2. Staining properties.** *L. pneumophila* does not stain well in tissue with the Gram stain or with common histologic stains. A simple Gram stain that omits the decoloration and counterstain steps allows visualization of the bacterial cells. Silver impregnation stains are the most useful.

B. Growth characteristics. *L. pneumophila* does not grow on routine bacteriologic media without the addition of cysteine and ferric ions. Soluble iron is provided in the form of hemoglobin or ferric pyrophosphate.

> **1.** Colonies on an appropriate **agar medium** have a ground-glass appearance. The bacteria grow best under aerobic conditions.
> **a.** *L. pneumophila* require 3–5 days for growth at an optimal pH of 6.9 and at 35° C.
> **b.** The bacterial growth is highly sensitive to alterations in the pH of the medium.

> **2.** A satisfactory **liquid medium** has not been developed for clinical laboratory use. *L. pneumophila* grows well in **broth medium** containing alveolar macrophages and in **water** containing photosynthetic bacteria, algae, and amoebae.

C. Biochemical characteristics. The metabolic and biochemical activity of legionellae does not classify the organisms with any other family of bacteria. These microorganisms are weakly oxidase positive, and they produce catalase, gelatinase, and β-lactamase.

D. Antigenic characteristics. Identification of legionellae depends primarily upon antigenic properties. Chromatographic analysis also is employed to identify cellular fatty acid profiles.

> **1. Eight serogroups** of *L. pneumophila* have been described. Serogroup 1 is responsible for more than 80% of human disease. Specific antisera are used in immunofluoresent techniques for serogrouping.

> **2. Taxonomy.** Several bacteria that were originally designated legionella-like organisms have been assigned species names. These include *Legionella bozemanii*, *Legionella micdadei*, *Legionella dumoffii*, and at least six others. Suspected new species generally are assigned temporary names or alphabetic designations until they are recognized officially.

E. Legionnaires' disease is a severe pneumonia caused by *L. pneumophila*. The disease has an incubation period of 2–10 days and a mortality rate as high as 60%. The attack rate is estimated to be fewer than 1% of exposed people.

1. **Epidemiology.** In the United States, about 25,000 *L. pneumophila* infections occur every year. Most do not cause Legionnaires' disease, and many are undetected.
   a. **Environmental sources.** *L. pneumophila* has been isolated from both air-conditioning systems and natural freshwater sources.
      (1) Disease outbreaks are associated with **air-conditioning systems** of large buildings. Cooling towers, evaporative condensers, and other components of hotels, hospitals, and factories have been implicated. Lawn-sprinkling systems, showers, and disrupted soil at building sites also have been identified as sources of infection.
      (2) **Freshwater sources.** The bacteria grow in association with blue-green bacteria or, perhaps, inside algae and free-living amoebae. They can survive for more than 1 year in nonchlorinated tap water.
   b. **Transmission to humans.** *L. pneumophila* causes human disease when it is inhaled in aerosolized water or soil particles.

2. **Pathogenicity.** Legionnaires' disease is characterized by focal lung involvement of the alveoli and alveolar septae.
   a. A **cellular exudate** consisting of neutrophils and macrophages is present in the lungs. *L. pneumophila* usually is seen inside the macrophages.
   b. **Focal necrosis** occurs in the alveoli and pleura, but the bronchi are not involved.

3. **Clinical disease.** Legionnaires' disease rapidly becomes a severe toxic pneumonia.
   a. **Initial symptoms** are myalgia and headache. A rapidly rising fever ensues.
   b. **Variable signs and symptoms** include chills, vomiting, diarrhea, pleuritic chest pain, mental confusion, and delirium. An initially dry cough usually develops into a productive one, although sputum production is not prominent. Liver dysfunction frequently occurs.
   c. **Pulmonary symptomatology.** Bilateral lung involvement may be seen on radiologic examination. Patchy or interstitial infiltration is present and may progress to a nodular consolidation.
   d. **Prognosis.** Severe cases of disease progress to shock and respiratory failure within 1 week of worsening symptoms. Although the overall mortality rate is 15%, mortality reaches 60% in some outbreaks.

4. **Diagnosis.** Legionnaires' disease is suspected when a severe progressive pneumonia cannot be attributed to other causes. Diagnostic tests for legionellosis are not available in the routine clinical laboratory. Specimens and smears are sent to a reference laboratory when legionellosis is suspected.
   a. **Confirmation.** Direct examination and culture of affected tissue is the best means of confirming the diagnosis.
      (1) **Specimens.** A transtracheal aspirate, lung aspirate, and lung biopsy provide the best specimens.
      (2) **Stains.** Direct immunofluorescent stains of smears are rapid but are positive in only about 50% of cases that are proven by silver stains and culture.
   b. **Serology.** A significant rise in antibody titer during the disease course also is diagnostic. An indirect immunofluorescent technique usually is employed. A single high serum titer is presumptive evidence because, in some communities, 25% of the population have serum antibodies.

5. **Therapy.** A variety of antibiotics show in vitro activity against *L. pneumophila*. These include rifampin, aminoglycosides, quinolones, tetracycline, erythromycin, chloramphenicol, and β-lactamase-resistant penicillins. In vivo efficacy is difficult to determine because of the frequency of underlying disease in affected individuals. Supportive therapy appears to be beneficial.

F. **Pontiac fever** is clinically distinct from Legionnaires' disease. It is a nonpneumonic febrile illness caused by *L. pneumophila*.

1. **Epidemiology.** Pontiac fever is a self-limited, nonfatal infection. The attack rate appears to be greater than 90% of exposed individuals. Some form of this disease accounts for most of the 25,000 yearly cases of legionellosis in the United States.

2. **Disease syndrome.** Pontiac fever initially resembles Legionnaires' disease with headache and myalgia.
   a. The **clinical picture** is the same in all outbreaks of Pontiac fever, whereas Legionnaires' disease presents a highly variable picture.

      b. Approximately half of the people with Pontiac fever develop a **dry, nonproductive cough**.

      c. Prognosis. The disease does not progress to pneumonia, and there is full recovery within 2–5 days. The Pontiac, Michigan, outbreak in 1968 produced 144 documented cases with no deaths.

G. Other *Legionella* species

1. *L. bozemanii, L. micdadei, L. dumoffii*, and several other identified, species are the etiologic agents of legionella pneumonia.

2. *Legionella feeleii* has been associated with Pontiac fever.

3. *Legionella oakridgensis* appears to be an environmental isolate only. It has not been associated with any human disease.

III. *CORYNEBACTERIUM* SPECIES.
Several *Corynebacterium* species are members of the indigenous flora of the skin, nasopharynx, oropharynx, urogenital tract, and intestinal tract. These species generally are referred to as **diphtheroids**. The primary human pathogen in the genus is *Corynebacterium diphtheriae*.

A. Cellular morphology.
Corynebacterium species are small, gram-positive rods that display pleomorphism.

1. Some strains have **club-shaped ends**, and some have intracellular polyphosphate granules known as **metachromatic granules** because they stain redish purple with methylene blue.

2. *Corynebacterium* species frequently remain attached after division, which gives them a Chinese character-like appearance when viewed microscopically.

B. Growth characteristics.
Colonies on blood agar are 1–3 mm in size and grow best under aerobic conditions, although many strains grow microaerophilically and anaerobically in enriched media.

C. Biochemical characteristics.
Most of the *Corynebacterium* species do not hemolyze blood. They produce catalase and ferment carbohydrates to form primarily lactic acid. *C. diphtheriae* produces a potent exotoxin; species that are pathogenic for animals also produce toxins.

D. Epidemiology

1. **Transmission** of human disease is by droplet dissemination, by direct contact with cutaneous carriers, and occasionally by contact with fomites.

2. Nasopharyngeal and cutaneous **carriage** of toxigenic and nontoxigenic *C. diphtheriae* by healthy people can persist for life.

3. **Incidence.** As a result of mandatory immunization, only about 200 cases of clinical diphtheria occur in the United States each year. Most cases occur in Native Americans, migratory workers, and the transient poor.

E. Clinical disease.
Diphtheria is caused by toxigenic *C. diphtheriae* that carries the *tox*$^+$ gene. Pharyngeal and cutaneous forms of the disease are seen.

1. **Pharyngeal diphtheria** is acquired via the respiratory route. Pharyngitis and tonsillitis are the most common manifestations, but nasal diphtheria, which resembles a common cold, also occurs. Deep tissues and organs are not infected.

    a. The typical **incubation period** is 2–4 days, after which a sore throat, fever, and malaise develop. An exudate forms on the tonsils, pharyngeal walls, uvula, or soft palate.

    b. A **pseudomembrane** forms as the infection progresses. The grayish white membrane is composed of lymphocytes, plasma cells, cellular debris, fibrin, and bacteria. The membrane tenaciously adheres to the tissue and can extend from the oropharynx to the larynx and into the trachea.

      (1) Cervical adenitis frequently occurs, adding to the obstruction of the trachea.

      (2) A **tracheostomy** may be required in severe cases. Uncomplicated cases gradually resolve, and the membrane is expectorated in 5–10 days.

(3) Mechanical obstruction of the trachea by the pseudomembrane can cause **suffocation**.

2. **Cutaneous diphtheria.** Toxigenic and nontoxigenic *C. diphtheriae* can colonize small breaks in the skin and produce a characteristic pathologic process.
   a. **Epidemiology.** Cutaneous diphtheria frequently has been associated with lesions caused by insect bites, especially in the southeastern and southern United States and in the tropics.
   b. **Clinical presentation.** The lesion may appear as a simple pustule or a chronic, nonhealing necrotic ulcer. *Staphylococcus aureus* and *Streptococcus pyogenes* frequently can be isolated from the lesion along with *C. diphtheriae*.
      (1) *C. diphtheriae* **remains localized** in the cutaneous lesion and cannot be isolated from the blood.
      (2) **Systemic effects** produced during cutaneous diphtheria are due to the absorption of **exotoxin** into the tissues.

3. **Toxigenic properties.** The complications and mortality rate associated with diphtheria are due to the systemic effects of the **exotoxin**.
   a. **Toxin production.** Production of diphtheria exotoxin is the result of the *tox*$^+$ gene. The gene is carried by a bacteriophage known as the β-phage. *C. diphtheriae* strains that are lysogenic for this phage produce the exotoxin.
   b. **Physical structure of the toxin.** The toxin molecule is a protein consisting of fraction A and fraction B.
      (1) **Fraction B** is required for attachment of the toxin to the host cell membrane and for transport of fraction A into the cell.
      (2) **Fraction A** inhibits protein synthesis by binding to and inactivating elongation factor 2.
   c. **Primary effects of toxin on tissue**
      (1) The toxin **inhibits protein synthesis in cardiac muscle**, resulting in both structural and functional damage. Cardiac insufficiency can cause death.
      (2) **Demyelination** caused by the toxin can affect both peripheral and cranial nerves. The resulting paralysis of the nerves usually is reversible as the myelin sheath reforms.

4. **Diagnosis.** The initial diagnosis of diphtheria is made solely on the basis of clinical symptoms.
   a. Confirmation of the diagnosis is made by isolation and identification of the etiologic agent as toxigenic *C. diphtheriae*.
   b. Toxin production can be determined by in vitro precipitin tests with antitoxin and by in vivo animal protection tests.

5. **Therapy**
   a. **Antitoxin** is administered immediately upon development of signs and symptoms of diphtheria.
   b. **Antimicrobial therapy** may help reduce the number of *C. diphtheriae*, but because they are noninvasive it can be difficult to inhibit their growth.

6. **Immunization.** Treatment of diphtheria toxin with formalin produces a nontoxic, immunogenic substance (i.e., **a toxoid**) that usually is administered throughout early childhood as part of the diphtheria and tetanus toxoids and pertussis vaccine (i.e., the **DPT vaccine**). Booster immunizations at 10-year intervals maintain immunity.

IV. ***LISTERIA MONOCYTOGENES.*** Human listeriosis usually is a disease of neonates and of immunocompromised people. The etiologic agent is *L. monocytogenes*, a facultative intracellular pathogen.

A. **Morphologic and biochemical characteristics**

1. **Cellular morphology.** *L. monocytogenes* is a gram-positive rod that resembles *Corynebacterium* species in some respects. It has a characteristic tumbling motility caused by peritrichous flagella.

2. **Colonial morphology.** Colonies are small, smooth, and surrounded by a narrow zone of complete hemolysis on sheep blood agar.

3. **Metabolic activity.** *L. monocytogenes* is catalase positive but otherwise relatively inactive biochemically.

   B. Epidemiology. *L. monocytogenes* is distributed widely in nature. It is primarily an animal pathogen usually associated with humans as transient nonpathogenic flora. Other features of the epidemiology are not understood completely.

   C. Clinical disease. *L. monocytogenes* usually produces **meningitis** or **sepsis**. The most likely candidates for infection are neonates and immunocompromised people.

   1. **Neonatal and fetal infections.** Colonization of the maternal vagina by *L. monocytogenes* can cause neonatal meningitis if the infant is colonized as it passes through the birth canal. Intrauterine colonization of the fetus causes **infantiseptica granulomatosis**; the fetus typically is **aborted**, or a **stillbirth** occurs. The infected fetus is characterized by the presence of disseminated abscesses and granulomas.

   2. **Immunocompromised people** develop septicemia caused by *L. monocytogenes.* Alcoholics appear to be especially susceptible to infection. An animal source of the etiologic agent usually cannot be found.

   D. Immunity. The major immune response to infection by *L. monocytogenes* involves T lymphocytes and cellular immunity. A humoral immune response also can be detected. The bacteria appear to be capable of growing inside mononuclear phagocytes.

   E. Diagnosis and therapy. Listeriosis usually is diagnosed by culturing the bacteria from blood, lesions, or cerebrospinal fluid. *L. monocytogenes* is susceptible to natural and semisynthetic penicillins, erythromycin, and chloramphenicol.

V. *ERYSIPELOTHRIX RHUSIOPATHIAE* is a gram-positive rod that morphologically and antigenically resembles the *Listeria* species.

   A. Epidemiology. *E. rhusiopathiae* is distributed widely in animals and decaying organic matter. Infection is an occupational hazard of fishermen, butchers, veterinarians, and others who handle animals.

   B. Clinical disease. Human **erysipeloid** is acquired by traumatic inoculation of *E. rhusiopathiae* into the skin.

   1. **Clinical manifestations.** An erythematous swelling of the skin spreads from the site of inoculation. The disease usually remains localized in the skin, progresses slowly, and is very painful.

   2. **Therapy.** Penicillins are the treatment of choice for erysipeloid. *E. rhusiopathiae* also is susceptible to quinolones, erythromycin, and the tetracyclines.

VI. THE ACTINOMYCETACEAE FAMILY includes human pathogens and saprobic species in the genera *Actinomyces* and *Nocardia*. It also includes saprobic soil bacteria in the genus *Streptomyces*, which are medically important because of their production of many antibiotic substances.

   A. *Actinomyces* species. The actinomycetes are common bacteria in soil. For many years they were classified as fungi. The most common etiologic agent of human disease in this genus, *Actinomyces israelii*, is a member of the indigenous flora.

   1. **Cellular morphology.** *Actinomyces* species are gram-positive rods that branch to form filaments. Individual cells may demonstrate acute branching that gives them X and Y configurations. The cells stain irregularly.

   2. **Growth characteristics.** *Actinomyces* species are strict anaerobes. They grow best on blood agar, but 4–10 days' incubation may be required to produce visible colonies. *A. israelii* produces a yellow to orange pigment.

   3. **Biochemical characteristics.** Differentiation of the many species of actinomycetes is based on carbohydrate fermentation patterns and other biochemical tests for metabolic activity.

   4. **Epidemiology.** The actinomycetes are common inhabitants of anaerobic microenvironments of the oropharynx and gastrointestinal tract. These bacteria, particularly *A. israelii*, have a

distinctive ability to adhere to mucosal surfaces. Human actinomycosis usually occurs after trauma and inoculation of the bacteria deep into devitalized tissue.

5. **Clinical disease.** Actinomycosis occurs when the indigenous bacteria transgress the epithelial barrier and invade tissues with low oxygen tension. The different forms of actinomycosis correlate with the original site and conditions of tissue invasion.
   a. **Cervicofacial actinomycosis** is the most common form of disease. It usually is associated with poor dental hygiene or trauma to the mouth and jaw, including tooth extraction.
   b. **Thoracic actinomycosis** is very rare and is initiated either by aspiration of infected material or by invasive extension from cervicofacial or abdominal disease. Lung abscesses can form, and the pleura, mediastinal structures, and chest wall may become involved.
   c. **Abdominal actinomycosis** also is rare and generally occurs following surgery or other trauma to the intestine. Perforation of ulcers or diverticuli can initiate the disease process. Erosion or blockage of vital organs may occur in chronic disease.
   d. **Pelvic involvement** occasionally occurs by extension of infection from the abdomen. Cases of primary pelvic involvement have been associated with the use of contraceptive intrauterine devices (IUDs).

6. **Clinical manifestations.** The feature common to all types of actinomycosis is formation of draining sinus tracts. These tracts form by erosion of tissue, and they frequently extend to the external surfaces of the body, where they spontaneously discharge through the skin.
   a. **Sulfur granules** are seen in the exudate from the sinus tracts. They are small (0.3 mm) amorphous colonies of *A. israelii* held together by tissue exudate.
      (1) The tips of bacterial filaments, which are visible on the outer edges of the granule, characteristically are swollen. This phenomenon is believed to result from an immunologic response to bacterial antigens.
      (2) "Sulfur" refers to the yellowish orange color of the granules.
   b. **Purulent exudate** is abundant, but sulfur granules are infrequent. Individual cells of *A. israelii* very rarely are found in the exudate, although contaminating gram-negative rods are common.
   c. Actinomycosis typically is a **chronic disease** characterized by low-grade fever and initially by nondescript symptomatology. Abdominal and pelvic actinomycosis are the most insidious forms, and some cases have persisted for more than 1 year.

7. **Diagnosis.** A strong presumptive diagnosis is made on the basis of sulfur granules found in purulent discharge from draining sinus tracts. Additional physical findings and a history of trauma assist in making the diagnosis. Culture and identification of the etiologic agent confirms the diagnosis.

8. **Therapy.** Actinomycosis responds well to penicillin G, although 4–6 weeks of high-dose therapy usually is required. Tetracycline, erythromycin, and clindamycin also have proven to be effective therapeutic agents.

B. *Nocardia* **species.** The Actinomycetaceae family also includes the medically important species, *Nocardia asteroides* and *Nocardia brasiliensis*. These two pathogenic species and the saprobic *Nocardia* species are common in the environment.

1. **Cellular morphology.** *Nocardia* species are gram-positive rods that demonstrate true branching. The pathogenic species stain poorly and irregularly with the Gram stain, and they are weakly acid fast.

2. **Growth characteristics.** *Nocardia* species are not fastidious, but they may require several days' incubation to yield macroscopic colonies.
   a. *Nocardia* species are **strict aerobes**.
   b. **Colonies** on agar medium initially are chalky in appearance and form a crater-like depression. Continued incubation produces wrinkled colonies with white to orange pigmentation.

3. **Biochemical characteristics.** Identification of *Nocardia* species is a complex process that uses tests for the degradation of casein, tyrosine, xanthine, and other substrates. A battery of additional tests that are not employed routinely for identification of bacteria also must be performed. Identification of species typically requires several weeks.

4. **Epidemiology.** *Nocardia* species are prevalent in soil. They are not considered indigenous flora in humans, although they occasionally have been isolated in small numbers from healthy people. Nocardiosis is rare; more than half the cases occur in immunocompromised patients.

5. **Clinical disease.** *Nocardia* species produce human disease by two distinct routes—pulmonary and subcutaneous. Nocardiosis typically occurs in patients with underlying disease and in those who are naturally or chemically immunosuppressed.

   a. **Pulmonary nocardiosis** is the most common form of disease in humans. *N. asteroides* usually is the etiologic agent of this form of nocardiosis, and the disease is acquired by inhalation of the bacteria.

      (1) **Pulmonary lesions** develop into multiple confluent abscesses by the process of acute inflammation with suppuration and destruction of the parenchyma.

      (2) There is little fibrosis, and **dissemination to other organs**, particularly the brain, is common.

   b. **Subcutaneous disease.** Direct inoculation of nocardiae into the skin or subcutaneous tissue typically produces a superficial pustule at the site of the trauma.

      (1) Progression of the infection results in a disease process resembling actinomycosis with the formation of draining sinus tracts and granules.

      (2) *N. brasiliensis* is the most frequent etiologic agent of this localized form of nocardiosis.

6. **Diagnosis** is relatively easy because the bacteria usually can be found throughout the lesion.

   a. A **presumptive diagnosis** can be made on the basis of clinical findings in conjunction with microscopic morphology and staining characteristics of the bacteria.

   b. A **confirmed diagnosis** of nocardiosis is made through culture and identification of the etiologic agent.

7. **Therapy.** The administration of sulfonamides and drainage of abscesses usually is an effective treatment. Pulmonary, cutaneous, or brain abscesses caused by nocardiae are the few remaining indications for systemic sulfonamide therapy. A significant proportion of patients do not respond to chemotherapy, and nocardiosis rarely subsides spontaneously. Antibiotic susceptibility testing is difficult and is not performed routinely.

STUDY QUESTIONS

Directions: Each question below contains five suggested answers. Choose the **one best** response to each question.

1. Which organism causes a highly contagious pneumonia that is transmitted from human to human?

(A) *Ureaplasma urealyticum*
(B) *Listeria monocytogenes*
(C) *Mycoplasma pneumoniae*
(D) *Nocardia asteroides*
(E) *Legionella pneumophila*

2. Which bacterium has the staining characteristic referred to as weakly acid fast?

(A) *Nocardia asteroides*
(B) *Erysipelothrix rhusiopathiae*
(C) *Legionella pneumophila*
(D) *Ureaplasma urealyticum*
(E) *Actinomyces israelii*

Directions: Each question below contains four suggested answers of which **one or more** is correct. Choose the answer

A   if **1, 2, and 3** are correct
B   if **1 and 3** are correct
C   if **2 and 4** are correct
D   if **4** is correct
E   if **1, 2, 3, and 4** are correct

3. Characteristics of *Legionella pneumophila* include which of the following?

(1) It is acquired by inhalation of contaminated droplets from air conditioners
(2) It occurs in nature associated with blue-green bacteria, algae, and amoebae
(3) It requires complex media containing cysteine for growth
(4) It is a thin, poorly staining, gram-positive rod with flagella

4. The pseudomembrane formed during clinical diphtheria characteristically

(1) adheres tenaciously to the pharyngeal wall
(2) requires toxin for its formation
(3) may extend from the oropharynx to the larynx
(4) occurs during the paroxysmal stage of disease

5. Characteristics of the Mycoplasmataceae family include which of the following?

(1) They form minute colonies on laboratory media
(2) Their cell membrane contains sterols
(3) They are the smallest free-living bacteria
(4) They synthesize a unique cell wall

6. True statements regarding diphtheria toxin include which of the following?

(1) It is coded for by the *tox*$^+$ gene
(2) The toxin is a complex molecule of two fractions
(3) It is an exotoxin mediated by the β-phage
(4) Diphtheria toxin causes demyelination of nerves

7. Antitoxin administration is the principal form of therapy for

(1) listeriosis
(2) Legionnaires' disease
(3) erysipeloid
(4) diphtheria

Directions: The group of questions below consists of lettered choices followed by several numbered items. For each numbered item select the **one** lettered choice with which it is **most** closely associated. Each lettered choice may be use once, more than once, or not at all.

Questions 8–13

Match each disease description with the most appropriate etiologic agent.

(A) *Mycoplasma pneumoniae*
(B) *Mycoplasma hominis*
(C) *Listeria monocytogenes*
(D) *Actinomyces israelii*
(E) *Legionella pneumophila*

8. Draining sinus tracts
9. Postpartum fever
10. Walking pneumonia
11. Septic abortion
12. Fatal pneumonia
13. Neonatal meningitis

ANSWERS AND EXPLANATIONS

1. The answer is C. [*I C 1 b, 2*] *Mycoplasma pneumoniae* causes a relatively mild pneumonia. It is responsible for approximately 20% of all pneumonia and for about 35% of pneumonia in teenagers. The disease is transmitted from human to human and occurs worldwide. It has been estimated that the ID_{50} of inhaled *M. pneumoniae* in humans is one colony-forming unit. This makes mycoplasma pneumonia one of the most contagious pneumonias known. *Legionella pneumophila* causes Pontiac fever and Legionnaires' disease, which may have respiratory symptoms but are not known to be transmitted from human to human. Likewise, *Nocardia asteroides* is the cause of a pneumonitis resulting from the inhalation of contaminated material from the environment. *Ureaplasma urealyticum* has been associated with pelvic inflammatory disease and urethritis. *Listeria monocytogenes* is the etiologic agent of septic abortion and disseminated disease.

2. The answer is A. [*VI B 1*] *Nocardia* species are gram-negative rods demonstrating true branching. They stain poorly and irregularly with the Gram stain and are observed in clinical specimens most easily with a modified acid-fast stain using dilute acid as a decolorizing agent. They are referred to as weakly acid fast. *Erysipelothrix rhusiopathiae* and *Actinomyces israelii* are gram positive. *Legionella pneumophila* is a poorly staining gram-negative bacterium with no acid-fast characteristics. *Ureaplasma urealyticum* has no cell wall.

3. The answer is A (1, 2, 3). [*II A, B, E 1 a*] *Legionella pneumophila* was classified taxonomically in the late 1970s, when little was known of its growth and epidemiologic characteristics. Human disease is acquired by the respiratory route from the inhalation of contaminated droplets from air-conditioner systems, showers, sprinkler systems, and disturbed soil. In the environment, the bacteria have been found growing with, and perhaps inside of, blue-green bacteria, algae, and free-living amoebae. *L. pneumophila* can be grown in the laboratory in a liquid medium in association with macrophages. Solid agar medium routinely used for the culture of *L. pneumophila* must contain cysteine and ferric ions; some reports suggest that selenium also is a growth requirement. Ultrastructurally, the cell wall of *Legionella* species is that of gram-negative bacteria, but they stain poorly. The cells measure about 0.5 μm in diameter and 2–20 μm in length, and they are flagellated.

4. The answer is B (1, 3). [*III E 1 b*] Pharyngeal diphtheria is characterized by the formation of a pseudomembrane that may extend from the oropharynx to the larynx and into the trachea. The grayish white membrane adheres tenaciously to the pharyngeal wall, and attempts to remove it can cause serious bleeding. The pseudomembrane is formed from lymphocytes, cellular debris, plasma cells, and bacteria; toxin is not required for formation of the membrane, as evidenced by the fact that small pseudomembranes have been observed in cases of pharyngitis caused by nontoxigenic *Corynebacterium diphtheriae*. Pharyngeal diphtheria is not characterized by distinctive stages of disease as is pertussis, which has a catarrhal, paroxysmal, and convalescent stage.

5. The answer is A (1, 2, 3). [*I A 1, B 1*] *Mycoplasma* species and *Ureaplasma* species belong to the family Mycoplasmataceae. These are the smallest free-living bacteria known, measuring about 0.2 μm in diameter. They do not have cell walls, and the cell membrane contains sterols. These sterols are not synthesized by the bacteria, but rather must be obtained preformed from the growth environment. The mycoplasmataceae can be grown on a complex bacteriologic medium. It typically takes several days to form minute colonies when grown in the laboratory.

6. The answer is E (all). [*III E 3 a, b, c (2)*] The potent exotoxin produced by toxigenic strains of *Corynebacterium diphtheriae* is responsible for the systemic symptomatology associated with clinical diphtheria. The toxin is produced by lysogenic strains containing the tox^+ gene of the β-phage. The exotoxin is a complex molecule consisting of an A fraction, responsible for inhibition of protein synthesis, and a B fraction, responsible for attachment of the toxin to tissue cells and entry of the A fraction into the cell. The primary effects of diphtheria toxin are the result of the inhibition of protein synthesis. Cardiac structure and function are affected, and both peripheral and cranial nerves are demyelinated.

7. The answer is D (4). [*III E 5 a*] People who develop the signs and symptoms of diphtheria must be treated immediately with antitoxin. The systemic symptomatology of diphtheria, including myocardial insufficiency and paralysis, is the result of the potent exotoxin produced by toxigenic *Corynebacterium* *diphtheriae.* Antibiotics may be of some benefit in eliminating the toxin-producing bacteria. Antibiotics are the primary means of treating listeriosis, legionellosis, and erysipeloid. Identifiable exotoxins are not associated with these disease syndromes.

8–13. The answers are: 8-D, 9-B, 10-A, 11-C, 12-E, 13-C. [*I C 3 a, D 1; II E; IV E; VI A 6*] Draining sinus tracts discharging purulent exudate are characteristic of all forms of actinomycosis; the etiologic agent is *Actinomyces israelii.*

Postpartum fever and postabortion fever have been caused by *Mycoplasma hominis.* Both *M. homonis* and *A. israelii* are considered indigenous flora of humans.

Walking pneumonia is a common name given to pneumonia caused by *Mycoplasma pneumoniae.* The pneumonia is relatively mild and does not typically debilitate the patient.

Fatal pneumonia is caused by *Legionella pneumophila*; the mortality rate is reported as high as 60% in some outbreaks.

Listeria monocytogenes is transient flora in the vagina. Infants can acquire infection as they pass through the birth canal, and neonatal meningitis is the most common manifestation. Intrauterine infection of the fetus with *L. monocytogenes* typically causes abortion or stillbirth.

I. MYCOLOGY

A. Introduction. Fungi are eukaryotic microorganisms. Only about 15 genera and 100 species of the thousands of species of fungi generally are involved in human disease. Most of the medically important fungi are ubiquitous in nature.

B. Morphology. Structurally, fungi exist as either yeasts or molds. Some fungi are capable of existing as both forms at different times, depending upon the environment, nutrients, or other conditions. **Chitin** is a structural component of fungal cell walls.

  1. Yeasts are unicellular forms that are spherical or ovoid in shape. All yeasts are similar morphologically when viewed with the light microscope.

  2. Molds are complex, multicellular microorganisms with a variety of specialized structures with specific functions. The names of the structures vary with different genera (Figure 17-1 is a diagramatic representation of some of these structures).

    a. Hyphae (singular, hypha) are the structural units of molds. They are thread-like tubes containing the cytoplasm and organelles of the organism. The **mycelium** is a mass of hyphae forming the mold colony.

    b. Septa (singular, septum) are cross-walls in the hyphae, forming individual cells. The septa have pores that allow the movement of cytoplasm and even organelles between cells. Some lower fungi have **coenocytic hyphae** (i.e., they lack septa).

    c. The **conidiophore** or **sporangiophore**, depending upon the species, is a specialized hypha that bears the reproductive structures of some molds.

    d. The **vesicle** is the bulbous tip of the conidiophore or sporangiophore; it supports the reproductive structures of the mold.

    e. Sterigmata (singular, sterigma) are the flask-shaped structures on the vesicles, which bear the spores.

    f. Sporangia (singular, sporangium) are saclike structures that contain the spores.

    g. Conidia and spores vary in shape and size and are given different names according to the means by which they are borne or produced by the mold.

      (1) Conidia are asexual fungal spores. Some common types of conidia are microconidia, macroconidia, arthroconidia, and aleurioconidia.

      (2) Spores are sexual reproductive structures formed when positive and negative mating strains of fungi exchange DNA.

  3. Dimorphism is a characteristic of fungi that grow as molds in the natural environment and in laboratory culture and as yeasts or yeast-like structures in tissue. True dimorphic fungi of medical importance include the etiologic agents of systemic mycoses (i.e., *Histoplasma*, *Blastomyces*, *Coccidioides*, and *Paracoccidioides* species) and *Sporothrix schenckii*.

C. Reproduction among fungi occurs both asexually and sexually. It generally is accepted that the infectious agents of molds are the conidia and that sexual forms of fungi are not found in clinical material.

D. Classification and taxonomy of the medically important fungi is a progressive area of mycologic research.

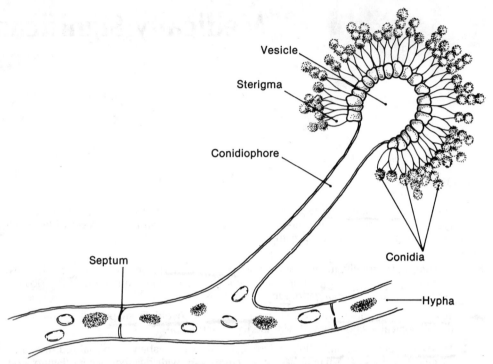

Figure 17-1. Diagramatic representation of some of the major structures of a filamentous fungus, or mold. The conidiophore is a specialized segment of hypha, which bears the asexual reproductive structures. This diagram is patterned after the structures seen in *Aspergillus* species.

1. Yeasts are identified by biochemical reactions based on the utilization of carbohydrates as well as other metabolic activities.

2. Molds are identified by their colonial and microscopic morphology. Microscopic details of specialized asexual reproductive structures are used to differentiate species of molds.

3. Most of the medically important fungi are placed in the form-class **Fungi Imperfecti** because their sexual forms have not been identified.

E. Cultivation. Cultural conditions and growth requirements for the medically important fungi differ from those required by most bacterial pathogens.

1. Most medically important fungi are not fastidious, and all grow aerobically. The standard laboratory culture medium is **Sabouraud's agar (SAB)** and variations of it, such as SAB plus antibacterial agents (SAB +). **Potato dextrose agar (PDA)** is another commonly used medium.

2. The **carbohydrate content** of SAB is much higher than that of bacterial culture media. SAB usually contains 3% dextrose or sucrose. This high carbohydrate concentration is inhibitory to most bacteria.

3. The **pH** of mycologic media usually ranges between 4.0 and 6.5. Bacterial pathogens usually are not capable of growing under these acid conditions.

F. Pathogenicity. The virulence of the medically important fungi is low. There is little evidence that disease-producing fungi release extracellular toxins or other virulence factors. Human-to-human transmission occurs rarely and generally only among the superficial mycoses and with *Candida* species.

II. SUPERFICIAL MYCOSES involve the keratin-containing structures of the body. The infections generally are considered cosmetic problems and are not life threatening. The disease processes very rarely spread to other tissues and, then, only in extremely immunocompromised individuals.

A. Etiology

1. The major cause of superficial mycoses are *Trichophyton* species, *Microsporum* species, and a single species of *Epidermophyton*, which collectively are referred to as **dermatophytes**. The various species of the dermatophytes are indistinguishable in tissue, appearing as strands of hyphae or conidia.

2. Other fungi involved in superficial mycoses include *Pityrosporum orbiculare*, *Pityrosporum ovale*, and *Candida* species.

B. Pathogenesis.
Infection with the agents of the superficial mycoses occurs with conidia and hyphal fragments and is restricted to keratinous structures (i.e., the stratum corneum, hair, and nails). The virulence of these fungi is low; injection into subcutaneous or deeper tissues in a healthy person does not result in disease. The pathogenesis of the superficial mycoses is not completely understood.

1. **The status of cell-mediated immunity appears to be important.** Other factors affecting susceptibility are not known.

2. **The ability to break down and metabolize keratin is characteristic** of all the etiologic agents. The enzyme involved in this process sometimes is considered as a virulence factor.

3. **The disease processes usually are described by their location on the body.** Dermatophytoses commonly are known as **ringworm** and diagnostically are referred to as **tineas**.
   a. **Tinea capitis** is infection of the hair and scalp.
      (1) Ectothrix infections occur when the fungus grows and produces conidia on the outside of the hair shaft.
      (2) Endothrix infections involve invasion of the hair shaft.
   b. **Tinea barbae** is infection of the bearded area of the face.
   c. **Tinea corporis** involves the trunk of the body.
   d. **Tinea cruris** is infection of the inguinal areas.
   e. **Tinea pedis** involves the feet and is commonly known as athlete's foot. **Tinea mannum** is infection of the hand.
   f. **Tinea unguium** is involvement of the nails.

C. Clinical manifestations
may include scaling of skin, hair loss, pruritis, and erythema or other discoloration. This symptomatology is characteristic, but it is not diagnostic.

1. **Dermatophytic infections** of the skin are characterized by pruritic, red, and raised skin in infected areas.

2. **Endothrix infections** typically cause breakage of the hair due to invasion of fungal hyphae into the shaft.

3. **Tinea unguium** is characterized by discoloration, thickening, and disfigurement of the nail.

D. Diagnosis.
Clinical signs and symptoms, microscopic examination of tissue samples, and the results of culture aid in the diagnosis of superficial mycotic diseases.

1. **Specimens for microscopic examination** include skin scales and hair.
   a. Skin scales are removed and placed on a microscope slide with a drop of 10% potassium hydroxide (KOH). After 10–15 minutes, the slide is examined for the presence of hyphae or conidia. The presence of hyphae during microscopic examination distinguishes the lesion from an allergic reaction or other etiology.
   b. Microscopic examination of hair easily distinguishes endothrix and ectothrix infections. An endothrix infection produces a hair with holes and pits in the shaft. Hyphae are seen surrounding the hair shaft in ectothrix infection.

2. **Cultures** on SAB, SAB+, and PDA utilize individual hairs or skin scales obtained by gently scraping the lesion with a sterile scalpel blade or glass slide. Colonies are identified by colonial and microscopic morphology.

E. Therapy.
The superficial mycoses typically are treated with topical compounds. Griseofulvin is an orally administered compound traditionally used for the treatment of refractive superficial mycoses.

1. Topical compounds
   a. Topical preparations containing **sulfur, selenium, coal tar derivatives**, and other chemical substances traditionally have been effective in treating routine superficial infections.
   b. Tolnaftate is a highly effective compound for the treatment of infections caused by the dermatophytes.
   c. Amphotericin B occasionally has been used as a topical agent for the treatment of superficial mycoses. Results generally have been unsatisfactory. The high nephrotoxicity of amphotericin eliminates its parenteral use in the treatment of superficial infections.

2. Oral compounds. Griseofulvin is an orally administered antibiotic employed in severe cases of superficial infection.
   a. A high fat diet is needed during treatment to ensure adequate gastrointestinal absorption. Griseofulvin apparently inhibits DNA synthesis.
   b. Griseofulvin has been used successfully to treat tinea unguium, which is difficult to treat with topical agents as they do not penetrate the nails. Removal of the nail and cleaning of the nail bed frequently is performed in conjunction with griseofulvin administration.

F. Epidemiology. Many of the etiologic agents of the superficial mycoses are ubiquitous, and human infections are common. Children are more susceptible than adults, and infections are more common in warm, moist climates.

   1. Geophilic dermatophytes are found in soil. Infections result from casual contact by a susceptible host with contaminated soil.

   2. Zoophilic dermatophytes are contracted from animal sources. Dogs, cats, and domesticated livestock are common sources of infection in humans.

   3. Anthrophilic dermatophytes infect only humans. Transmission of disease is by person-to-person contact.

III. SUBCUTANEOUS MYCOSES involve the subcutaneous tissue and rarely disseminate to cause fungemia or infections in muscular tissue. The diseases generally are chronic and eventually may involve the joints and bones.

A. Sporotrichosis is a chronic, generally self-limited disease of the lymphatic vessels and lymph nodes. The infection most often occurs in an arm or a leg.

   1. Etiology. The dimorphic fungus *Sporothrix schenckii* is the etiologic agent of sporotrichosis. *S. schenckii* is a small (3–5 μm) cigar-shaped yeast in tissue and a slow-growing, mouse-gray colony in laboratory culture.

   2. Pathogenesis
      a. *S. schenckii* disseminates from the original site of implantation (e.g., the hand, wrist, foot, or ankle) via the lymphatics. The replicating fungus stimulates an acute pyogenic response as well as granulomatous inflammation.
      b. Draining sinus tract formation is characteristic, but the fungus is scanty in the purulent drainage.
      c. The disease is chronic and rarely spreads beyond the axillary or inguinal lymph nodes.

   3. Diagnosis
      a. Microscopic examination
         (1) *S. schenckii* is difficult to find during microscopic examination of drainage and biopsy specimens, even when specially stained with silver stains.
         (2) The observation of **asteroid bodies**—a histopathologic structure composed of the yeast surrounded by rays of amorphous material—is considered diagnostic.
      b. Culture. A confirmed diagnosis is made by culture on routine mycologic media. Identification is based on microscopic and colonial morphology of the isolate.

   4. Therapy. Localized subcutaneous infection responds to orally administered potassium iodide; the mechanism of action is unknown. Systemic manifestations of sporotrichosis require amphotericin B or imidazole therapy.

   5. Epidemiology. *S. schenckii* has been isolated from soil, decaying organic matter, and plant surfaces. Infection occurs by traumatic implantation of contaminated material. The disease usually is related to an occupation or hobby of agriculture or gardening; the common name is **drunken rose gardener's disease**.

B. **Chromoblastomycosis** is a chronic, granulomatous infection typically confined to the skin and subcutaneous tissue of the feet and legs.

1. **Etiology.** Chromoblastomycosis is caused by several species in the genera *Phialophora* and *Fonsecaea*. The etiologic agents have darkly pigmented (i.e., dematiaceous) hyphae due to a melanin-like compound.

2. **Pathogenesis**
   a. Papular lesions appear at the site of inoculation, usually on the foot or leg. The lesion develops into a large wart-like structure, and **satellite lesions** generally form; lymphatic vessels usually are not involved.
   b. The lesions form slowly, often resembling heads of cauliflower. The lesions are painless, although secondary bacterial infection may cause intense pain.

3. **Diagnosis**
   a. **Microscopic examination.** Brown pigmented spherical fungal elements (i.e., sclerotic, or, "copper," bodies) with transverse and horizontal septa are found in scrapings treated with KOH and in histopathologic preparations.
   b. **Culture.** A confirmed diagnosis is made by culture and identification of the etiologic agent, but the formation of characteristic conidial structures may require weeks to months of incubation.

4. **Therapy.** Surgery and antifungal chemotherapy, usually 5-fluorocytosine (5-FC), are employed. Results generally are unsatisfactory, particularly in advanced chronic disease, and amputation may be the only alternative.

5. **Epidemiology**
   a. The etiologic agents of chromoblastomycosis are ubiquitous, occurring in soil, decaying vegetation, and on the surfaces of plants.
   b. The disease occurs worldwide but is most common in tropical areas, especially in adults who are nutritionally deficient.
   c. The disease process appears to be initiated by the traumatic inoculation of contaminated plant material into subcutaneous tissue.

C. **Mycetoma**, sometimes referred to as **Madura foot**, resembles chromoblastomycosis but usually lacks the large wart-like lesions.

1. **Etiology.** Mycetoma can be caused by a variety of fungi. *Pseudoallescheria boydii* (formerly *Allescheria* and *Petriellidium boydii*) is the most common etiologic agent in temperate climates. *Nocardia* species are common bacterial agents of mycetoma.

2. **Pathogenesis and clinical manifestations**
   a. Mycetoma typically begins in a wound or abrasion contaminated with soil or some other environmental source of the etiologic agent.
   b. The disease is characterized by chronic granulomatous inflammation, and the formation of draining sinus tracts is common.
   c. Feet and legs are the most common sites of infection, although other parts of the body may be involved. Occasional bone involvement may ensue, and secondary bacterial infections may be fatal.

3. **Diagnosis**
   a. **Microscopic examination.** Histopathologic preparations of biopsied tissue can reveal hyphae, but direct examination frequently is not successful as the fungus tends to form microcolonies in tissue known as **black grains** and **white grains**.
   b. **Culture.** Colonies for identification may appear in days to weeks in culture, depending upon the etiologic agent.

4. **Therapy.** Antifungal chemotherapy usually is unsuccessful. Amputation of the affected foot or leg frequently is the only cure.

5. **Epidemiology.** Mycetoma occurs worldwide, but it primarily is a disease of the tropics and subtropics. The etiologic agents vary from region to region, and usually are associated with decaying plant matter and soil.

D. **Rhinosporidiosis** is a subcutaneous disease that has been described in humans and a variety of wild and domesticated animals.

1. **Etiology.** The etiologic agent of rhinosporidiosis is *Rhinosporidium seeberi*. This presumed fungus has not yet been isolated or classified.

2. **Pathogenesis and clinical manifestations**
   a. Rhinosporidiosis is a chronic granulomatous disease characterized by the production of large polyps, tumors, papillomas, or wart-like lesions. The lesions are hyperplastic, highly vascularized, and friable.
   b. The nose and the conjunctiva are the most frequently affected areas. Infection rarely occurs in the anus, genitalia, ears, pharynx, or larynx.

3. **Diagnosis** is made on the basis of clinical symptomatology and the finding of spherules in affected tissue.

4. **Therapy.** Surgical excision of affected areas appears to be the only effective treatment. Localized injection of amphotericin B into the lesions has been used as an adjunct to surgery, but its efficacy is unproven.

5. **Epidemiology.** Rhinosporidiosis has been associated with people who bathe or work in stagnant water. Approximately 90% of the cases occur in India and Sri Lanka; cases also have been reported in South America, particularly from Brazil.

IV. SYSTEMIC MYCOSES, or **deep mycoses**, are noncommunicable infections that involve the internal organ systems. The respiratory tract characteristically is the route of infection. As a group, the systemic mycoses differ from opportunistic mycoses in that they occur in otherwise healthy individuals.

A. **Histoplasmosis** typically is a self-limited pulmonary infection. Primary histoplasmosis occasionally develops into chronic pulmonary disease and disseminated infection.

1. **Etiology.** Histoplasmosis is caused by the dimorphic fungus *Histoplasma capsulatum*. The fungus grows as a saprobe in the soil and has been isolated from the gastrointestinal tracts of birds and bats. African histoplasmosis is caused by the closely related *Histoplasma capsulatum* var. *duboisii*.
   a. **Cultural characteristics.** *H. capsulatum* is a slow-growing, fluffy white mold on laboratory medium at room temperature. The colonies are characterized by the formation of tuberculate macroconidia.
   b. **Tissue phase.** *H. capsulatum* is a small (3–4 μm) intracellular yeast in tissue. The yeast form of growth sometimes can be induced in the laboratory by incubating the culture at 37° C.

2. **Pathogenesis**
   a. Most cases of histoplasmosis occur from inhalation of conidia from soil contaminated with bird or bat excreta.
   b. The fungus establishes a focal infection in the lung, which is characterized by a granulomatous response.
   c. Dissemination of *H. capsulatum* occurs lymphohematogenously through the reticuloendothelial system.

3. **Clinical manifestations**
   a. **Primary histoplasmosis** is characterized by an influenza-like syndrome that may last several weeks. The focus of infection usually heals, forming a calcified "coin lesion."
   b. **Disseminated histoplasmosis** may involve any organ system due to the lymphohematogenous spread; the lungs, liver, spleen, and bone marrow are the most frequently affected.
      (1) Disseminated disease is characterized by fever, night sweats, weight loss, a productive cough, and enlargement of the reticuloendothelial organs. The course usually is chronic.
      (2) The appearance of mucocutaneous lesions is a sign of disseminated disease. Primary cutaneous histoplasmosis is very rare.

4. **Diagnosis.** Histoplasmosis frequently poses a diagnostic challenge. A confirmed diagnosis is made by culture and identification of *H. capsulatum*.
   a. **Differential diagnosis.** Histoplasmosis must be differentiated from other systemic mycoses, tuberculosis, and neoplasia. The differentiation between neoplasia and a nodule that enlarges over a period of several years in chronic pulmonary histoplasmosis may be particularly difficult.
   b. **Microscopic examination**

    (1) The yeast form of *H. capsulatum* is difficult to find on direct examination of sputum or other specimens. Smears of mucocutaneous lesions and biopsy of reticuloendothelial organs are the most productive samples.

    (2) The small size and intracellular location in phagocytic macrophages makes it difficult to observe *H. capsulatum* in clinical material. Oil immersion ($\times$ 1000) microscopy of specially stained specimens yields the best results.

  c. Culture

    (1) Sputum and biopsy material are cultured on routine mycologic media supplemented with cysteine and incubated at 30° and 37° C.

    (2) Repeated cultures often are necessary; 4 weeks or longer may be required for the development of macroscopic colonies.

  d. Skin testing is performed with **histoplasmin** derived from *H. capsulatum* hyphae.

    (1) The skin test yields a positive delayed hypersensitivity reaction approximately 3 weeks after infection.

    (2) A positive skin test indicates exposure only, and the test is most useful in epidemiologic studies.

    (3) Cross-reactions to histoplasmin may occur in other mycoses and tuberculosis.

  e. Serologic tests employ histoplasmin or yeast-cell antigens. Immunodiffusion and complement-fixation (CF) tests usually are used to aid in the diagnosis of all forms of histoplasmosis.

    (1) CF titers typically rise to 1:332 or greater in disease processes.

    (2) Skin testing with histoplasmin interferes with the serologic results.

5. Therapy

  a. Primary pulmonary histoplasmosis and localized lung lesions typically do not require treatment.

  b. Amphotericin B is the drug of choice for progressive pulmonary disease and disseminated histoplasmosis. Ketoconazole is being used increasingly in immunocompetent patients.

6. Epidemiology. Sporadic cases of histoplasmosis occur worldwide. Disseminated histoplasmosis occurs in less than 0.1% of infected persons.

  a. In the United States, histoplasmosis is endemic to the Midwest, particularly to the Mississippi and Ohio river valleys. More than 50% of the residents of the states drained by these rivers are skin-test positive.

  b. Aerosolization of conidia in soil contaminated with bird or bat feces is the typical mechanism of infection; bird roosts and bat caves are common sources.

B. Blastomycosis is a localized or systemic mycosis most commonly involving the lungs, skin, and other viscera. Human blastomycosis may occur as primary cutaneous lesions or as primary pulmonary, progressive pulmonary, or disseminated disease.

1. Etiology. Blastomycosis is caused by the dimorphic fungus *Blastomyces dermatitidis*.

  a. Cultural characteristics. Laboratory cultures consist of fine septate hyphae with round-to-oval conidia. Mold colonies grown at 30° C are white with a faint yellow reverse, and young cultures often closely resemble those of *H. capsulatum*. Growth at 37° C occasionally produces typical yeast-like colonies.

  b. Tissue phase. *B. dermatitidis* is a large (8–15 μm), thick-walled yeast in tissue. Budding cells have a characteristic broad-based neck.

2. Pathogenesis. Growth of *B. dermatitidis* in tissue stimulates a mixed inflammatory response. Neutrophils and macrophages infiltrate the focus of infection, and giant cells may be observed. The response becomes granulomatous with progressive disease.

3. Clinical manifestations

  a. Primary disease

    (1) Primary pulmonary blastomycosis typically is mild and difficult to diagnose. A progressively reactive response to skin test antigens may be the only indication of disease.

    (2) Advanced pulmonary disease occasionally develops and frequently resembles neoplasia, tuberculosis, or other mycoses. Pulmonary infection is characterized by fever, a productive cough, and chest pain. Hilar lymphadenopathy also may be present.

    (3) Primary cutaneous lesions occur, although lesions on the mucous membranes, such as those seen in histoplasmosis, are uncommon. Extensive necrosis and fibrosis may result in disfigurement.

b. Disseminated blastomycosis may produce osteomyelitis and involvement of cutaneous tissue and viscera. The urinary and genital tracts are most frequently infected, and prostate gland involvement is particularly common.

4. Diagnosis
   a. Microscopic examination. The presence of typical thick-walled, broad-necked, budding yeast in sputum, pus, or histopathologic preparations is strong presumptive evidence of blastomycosis.
   b. Culture and identification are needed for confirmation of the diagnosis. Macroscopic colonies may take 4 or more weeks to develop. The hyphae and conidia have no distinctive characteristics, and identification can be difficult.
   c. Serologic and skin tests
      (1) Antigens are available for CF tests, but CF titers are absent in up to 50% of cases of blastomycosis.
      (2) Reactions to skin test antigens are of poor quality. Skin testing is of no diagnostic value.

5. Therapy
   a. *B. dermatitidis* is highly sensitive to amphotericin B. However, treatment is required only in progressive pulmonary and disseminated blastomycosis.
   b. Ketoconazole also has been employed successfully.

6. Epidemiology. Less is known about the epidemiology of blastomycosis than about other mycoses, in part due to the lack of sensitive and specific skin test antigens.
   a. In the United States, geographically endemic areas appear to be in the north central and southeastern regions, although this has been the subject of some debate.
   b. *B. dermatitidis* probably occurs naturally in soil or decaying vegetation. The highest incidence of blastomycosis is in people who work or engage in other activities outdoors.

C. Coccidioidomycosis is a noncommunicable systemic mycosis that can occur as a primary infection in the lungs or as a progressive, granulomatous infection involving the skin, bones, joints, viscera, and meninges.

1. Etiology. Coccidioidomycosis is caused by the dimorphic fungus *Coccidioides immitis*.
   a. Cultural characteristics. *C. immitis* grows on standard mycologic media to form grayish colonies in about 1 week. The hyphae give rise to thick-walled arthroconidia with pointed corners.
   b. Tissue phase. Rather than distinct yeast forms, *C. immitis* produces spherules in vivo.
      (1) The spherules measure 12–100 μm in diameter and contain **endospores**, which serve as reproductive units that are released when the spherule ruptures.
      (2) Spherules may be produced in the laboratory by special tissue culture techniques.

2. Pathogenesis. *C. immitis* is one of the most infectious and virulent agents of mycosis. It is estimated that a single arthroconidium inhaled by a mouse is sufficient to produce fatal disease.
   a. *C. immitis* arthroconidia appear inhibited by natural defenses of the tracheobronchial tree, and they lodge in the alveoli after inhalation. Germination of arthroconidia results in spherule formation.
   b. A mixed inflammatory response of neutrophils and macrophages occurs, and giant cells form. The wall of the arthroconidia and the initial small spherules possess antiphagocytic properties.
      (1) Most cases of primary coccidioidomycosis are controlled by cell-mediated immunity.
      (2) In a few cases, the production of spherules continues, and an apparent defect in cell-mediated immunity allows dissemination of the disease.

3. Clinical manifestations
   a. Pulmonary coccidioidomycosis
      (1) More than 50% of cases of coccidioidomycosis are primary pulmonary infections that are asymptomatic or too mild to concern the affected person.
      (2) A few cases progress to "valley fever" in 1–3 weeks, with fever, cough, chest pain, arthralgias, malaise, and erythema nodosum or erythema multiforme.
      (3) Chest radiographs usually are negative but may reveal hilar lymphadenopathy.
   b. Disseminated coccidioidomycosis is an uncommon development; evidence of dissemination generally appears within the first year after primary infection.

> **(1) Coccidioidal meningitis** is a slowly progressive disease characterized by worsening headache, nuchal rigidity, fever, and other symptomatology of meningeal infection. Meningitis always is fatal if untreated.
>
> **(2) Osteomyelitis, acute arthritis**, and warty or ulcerative **skin lesions** also are common.

4. Diagnosis. Chronic pulmonary and disseminated coccidioidomycosis generally are easy to diagnose, although repeated direct examination of specimens may be necessary.

  a. Microscopic examination

  (1) Biopsy specimens of skin and visceral lesions usually reveal the thick-walled spherules on examination of KOH preparations.

  (2) Sputum typically contains small undeveloped spherules that are difficult to differentiate from cellular artifacts.

  (3) Cerebrospinal fluid (CSF) only rarely yields spherules on direct examination. The cellular response primarily is mononuclear, although large numbers of neutrophils sometimes are present.

  b. Culture. Diagnosis is confirmed by identification of arthroconidia and their conversion to spherules in tissue culture or in animals.

  (1) *C. immitis* is cultured routinely from sputum, skin lesions, and visceral lesions in cases of disseminated disease; standard mycologic media are used.

  (2) Special biohazard laboratory conditions are required for handling clinical specimens.

  c. Skin testing uses either **coccidioidin** (from the mycelial phase) or **spherulin** (from the tissue phase); spherulin appears to be the more specific skin test antigen.

  (1) Positive skin tests are seen after 1–4 weeks of pulmonary coccidioidomycosis.

  (2) Anergy often is associated with disseminated disease.

  d. Serologic testing is useful in the diagnosis and management of coccidioidomycosis. Skin testing does not interfere with serologic results, as it does in histoplasmosis.

  (1) Immunoglobulin M (IgM) antibodies develop in 50%–75% of cases of primary disease within 3 weeks. These antibodies persist in the serum for 2–4 months.

  (2) IgG antibodies, detected by the CF test, appear later than IgM antibodies and persist for the duration of infection. The titer of IgG antibodies varies directly with the severity and magnitude of the infection.

  (3) The presence of CF antibodies in CSF is important in the diagnosis and management of coccidioidal meningitis, as cultures of the CSF usually are negative.

  e. The exoantigen immunofluorescence test is a rapid means of differentiating *C. immitis* from other arthroconidia-forming fungi. The test is based on the specificity of certain mycelial antigens.

5. Therapy

  a. Primary coccidioidomycosis resolves without treatment.

  b. Chronic pulmonary and disseminated disease require chemotherapy, traditionally with amphotericin B. The new imidazole compounds (e.g., ketoconazole) are less toxic than amphotericin B and have shown some therapeutic promise, but the relapse rate appears to be significant.

6. Epidemiology. Coccidioidomycosis is endemic to the semiarid Lower Sonoran life zone of the southwestern United States, Mexico, Central America, and South America. Up to 90% of people living in these areas are skin-test positive. Infection frequently is associated with earth-moving activities.

  a. The incidence of coccidioidomycosis varies with race. Filipinos appear to be the most susceptible, blacks and other dark-skinned people less susceptible, and light-skinned races the least susceptible to infection.

  b. Pregnancy is a definite predisposing factor to infection and dissemination of coccidioidomycosis, as is corticosteroid therapy. Steroids, particularly estrogen, have been shown to stimulate the growth of *C. immitis* in vitro.

  c. Although coccidioidomycosis does not naturally occur outside endemic areas, sporadic infection is reported as a result of the transport of conidia to nonendemic areas. A visit of only a few hours' duration is sufficient to contract infection.

D. Paracoccidioidomycosis, also known as **South American blastomycosis**, is a systemic mycosis. Although primarily a progressive disseminated disease, benign, self-limited pulmonary disease can occur.

1. **Etiology.** Paracoccidioidomycosis is caused by the dimorphic fungus *Paracoccidioides brasiliensis*. The tissue phase of the fungus is a large (5–40 μm), yeast-like cell with multiple buds; the yeast phase also can be induced in the laboratory at 37° C.

2. **Clinical manifestations.** The site of primary infection appears to be the lungs.
   a. Mucocutaneous and cutaneous ulcers are the most prominent features of the disease. The chronic ulcers spread slowly and develop a granulomatous base.
   b. Regional lymph nodes, the lungs, and the reticuloendothelial organs may become involved.

3. **Diagnosis.** A strong presumptive diagnosis is made on the basis of the history, physical, and microscopic findings in patients living in endemic areas. Confirmed diagnosis is made by culture and identification of the etiologic agent as *P. brasiliensis*.

4. **Therapy.** Amphotericin B and imidazoles have been successful therapeutically. Sulfonamides traditionally have been used, but the relapse rate is high.

5. **Epidemiology.** Paracoccidioidomycosis is endemic to tropical and subtropical regions of Central and South America, and little is known about the epidemiologic features of the disease. There is a distinct predilection for males.

V. OPPORTUNISTIC MYCOSES are caused by fungi that ordinarily are not considered pathogenic for humans. The increased number of immunocompromised people in the population (e.g., AIDS patients) has led to a dramatic increase in the incidence of opportunistic mycotic infections. Only a few of the most important opportunistic mycoses are discussed in this section.

A. Candidiasis is a relatively common human infection that can take the form of superficial, mucocutaneous, or systemic disease. Candidiasis usually is the result of autoinfection during a metabolic or an immunologic disturbance.

1. **Etiology.** Several species in the genus *Candida* have been associated with human disease; *Candida albicans* is the most common cause of infections.
   a. **Cultural characteristics**
      (1) *C. albicans* typically grows as round-to-oval yeast cells that are 4–6 μm in diameter. Pseudohyphae and hyphae also are seen, especially at lower incubation temperatures (i.e., 22°–25° C) and on nutritionally poor media.
      (2) Asexual reproduction is by budding of the yeasts or by chlamydospore formation on hyphae.
   b. **Tissue phase.** *C. albicans* is not a dimorphic fungus, as both yeast and hyphae are seen in tissue. The hyphae are thought to represent the tissue-invasive form of the fungus.
   c. **Differential characteristics.** *C. albicans* should be differentiated from other *Candida* species because of its greater medical importance.
      (1) *C. albicans* forms **germ tubes** (i.e., incipient hyphae) when yeast cells are incubated at 37° C in serum. The molecular basis for this phenomenon is not known.
      (2) *C. albicans* forms thick-walled, terminal chlamydospores when incubated at 22°–25° C with decreased oxygen on a nutritionally poor medium, such as cornmeal agar. Formation of chlamydospores seems to be due to environmental stress.
      (3) Final differentiation of the species is made on the basis of carbohydrate assimilation and fermentation reactions as well as other biochemical tests.

2. **Pathogenesis.** The events leading to candidiasis are not clear, but a variety of anatomic, metabolic, chemical, and immunologic factors are known to predispose to infection. A breakdown in natural anatomic or immunologic defenses precedes *C. albicans* infection, and invasion occurs by the penetration of pseudohyphae into compromised tissue.
   a. Skin trauma, persistent wetness, and maceration increase the risk of superficial infection. Dry, intact skin is highly resistant to *C. albicans*.
   b. Disturbances in the indigenous flora induced by broad-spectrum antibiotics or other alterations of the microenvironment predispose to overgrowth of *C. albicans*.
   c. Diabetes mellitus and other metabolic diseases as well as hormonal changes during pregnancy or with the use of oral contraceptives predispose to infection.
   d. Immunosuppression with corticosteroids or other agents and T cell deficiency may result in fulminant candidiasis or chronic mucocutaneous infection.

3. **Clinical manifestations**
   a. **Superficial candidiasis** typically involves skin surfaces that are in close proximity—providing a warm, moist environment—or skin that frequently is immersed in water.
      (1) **Candida intertrigo.** The initial lesions are vesicular pustules that enlarge, rupture, and cause fissures. Involved areas are characterized by white, scalloped borders of necrotic epidermis.
      (2) **Diaper rash** frequently is caused by *C. albicans* originating from the lower gastrointestinal tract. The rash is characterized by scaly macules or vesicles in association with intense burning and pruritis.
      (3) **Paronychia** is a localized inflammation around and under the nails; it frequently is caused by *C. albicans*, especially when the hands are immersed in water frequently (e.g., in dishwashers and laundry workers). Thickening and discoloration of the nails are characteristic, and the nail may be lost.
   b. **Mucocutaneous candidiasis** typically involves the vagina or the oral cavity. Disease usually results from metabolic changes in the host or alteration of the indigenous flora.
      (1) **Oral candidiasis**, or **thrush**, may occur during the use of broad-spectrum antibiotics and in immunocompromised individuals. White, cheesy plaques form on the mucosal surfaces.
      (2) Vulvovaginitis caused by *C. albicans* is common in women using oral contraceptives and during the last trimester of pregnancy. Infection is accompanied by a thick yeasty-smelling discharge, and vaginal itching and discomfort.
      (3) **Chronic mucocutaneous candidiasis** is a rare condition associated with T cell deficiency. The skin, mucous membranes, scalp, and nails usually are involved to varying degrees; the most serious form is characterized by the formation of thick, horny plaques on the skin and nails (i.e., **candida granuloma**).
   c. **Disseminated candidiasis** is caused by parenchymal invasion by pseudohyphae with resultant microabscesses, and occasionally by a chronic granulomatous response. Untreated disseminated disease is fatal.
      (1) Deep organ infection usually occurs in the setting of organ transplantation, heart surgery, intravenous catheterization, prosthetic implantation, hyperalimentation, long-term steroid therapy, or immunosuppressive therapy. Rarely, the cause of disseminated disease is a superficial infection.
      (2) The most common sites involved are the kidneys, brain, heart, and eyes. Multiple foci of infection frequently occur, especially when fungemia is present due to a continuous source (e.g., a contaminated catheter).
      (3) The symptomatology associated with disseminated candidiasis is not characteristic enough to distinguish it from systemic bacterial infections.

4. **Diagnosis**
   a. Superficial or mucocutaneous candidiasis is diagnosed by finding the fungus in tissue scrapings and by culture.
   b. Disseminated candidiasis is difficult to diagnose because of the inaccessibility of the infected organs. Definitive diagnosis is made by the histopathologic demonstration of the invasion of tissue by the yeast, establishing a causal relationship.

5. **Therapy.** *C. albicans* is susceptible to several antifungal agents. The therapy used depends upon the type and severity of infection.
   a. Superficial infections generally are treated with topical nystatin. Tolnaftate, used in dermatophytic infections, is not active against *C. albicans*.
   b. Mucocutaneous candidiasis usually is treated with nystatin or miconazole rinses or douches. Oral ketoconazole has been effective in controlling chronic mucocutaneous candidiasis. Amphotericin B may be required in severe cases at risk for dissemination.
   c. Amphotericin B is the drug of choice in disseminated candidiasis, although 5-FC and the imidazoles also have been used successfully. Combination therapy with amphotericin B and 5-FC is beneficial in some cases.

6. **Epidemiology**
   a. *C. albicans* is indigenous to the oral cavity, gastrointestinal tract, female genital tract, and, occasionally, the skin.
   b. Disease processes usually are caused by the indigenous flora where there is a breakdown in physical or immunologic defenses, particularly in the cell-mediated immune system.

    c. Person-to-person transmission is uncommon, but thrush can be transferred from mother to child during birth and by breast feeding; vulvovaginitis occasionally is acquired venereally.

B. Cryptococcosis usually is seen as a disseminated progressive disease in immunocompromised patients. Meningeal infection is common, and it produces the most recognizable form of the disease.

  1. Etiology. Cryptococcosis is caused by *Cryptococcus neoformans*, the only encapsulated yeast of medical importance. Other species, such as *Cryptococcus laurentii*, occasionally have been the apparent etiologic agents of human disease.

    a. Cultural characteristics. *C. neoformans* is an encapsulated, round-to-oval yeast measuring 4–6 μm in diameter. It grows as a yeast on routine mycologic media at both 25° C and 37° C. The large polysaccharide capsule gives the colonies a shiny, mucoid appearance.

    b. Tissue phase. *C. neoformans* grows as a yeast in tissue. The capsule typically is very large during tissue growth, giving the entire fungus a diameter of 25 μm or more.

  2. Pathogenesis. Cryptococcosis most frequently occurs in immunocompromised or metabolically disturbed patients. Inhalation of yeast cells, or perhaps spores from the filamentous sexual stage, initiates the disease process.

    a. Inhaled organisms establish a primary focus of infection in the lungs, with eventual extension to the hilar lymph nodes.

    b. Most infections are arrested, and there is spontaneous healing of the primary lesions. Occasionally, primary pulmonary infection does not resolve and disseminated disease occurs.

    c. The inflammatory response to cryptococci varies and is dependent upon the cell-mediated immune status of the patient. Patients with T cell deficiency or dysfunction particularly are at risk for disseminated disease.

    d. The capsule of *C. neoformans* appears to be antiphagocytic. It elicits little, if any, antibody response. No toxins have been identified with *C. neoformans*; the extracellular enzyme **phenoloxidase** is a possible virulence factor.

  3. Clinical manifestations

    a. Primary cryptococcosis

      (1) Primary pulmonary cryptococcosis apparently is asymptomatic or so mild that medical attention is not sought. Primary pulmonary cryptococcosis rarely is diagnosed.

      (2) Primary cutaneous lesions are rare.

    b. Cryptococcal meningitis is the most recognizable form of disease. It is characterized by a slow onset and nonspecific initial symptomatology.

      (1) Intermittent frontal headaches of increasing intensity, dizziness, disturbances in vision, irritability, and mental confusion typically occur over a period of weeks to months in late stages of the disease.

      (2) The cardinal signs of meningitis eventually develop. Fever usually is present, and nuchal rigidity is common.

      (3) Seizures, papilledema, and cranial nerve signs may appear. The skin, bones, kidneys, and spleen also may be involved.

      (4) About 50% of patients with meningitis have residual neurologic damage.

  4. Diagnosis

    a. Microscopic examination

      (1) *C. neoformans* is easily recognized with **India ink** (i.e., negative) staining of CSF or other body fluids. The finding of encapsulated yeast in conjunction with appropriate symptomatology is diagnostic.

      (2) The **mucicarmine stain** also allows strong presumptive identification of the yeast in histopathologic preparations.

    b. Culture and biochemical identification of *C. neoformans* are routine clinical laboratory procedures.

    c. Serology. A latex bead agglutination test for identifying capsular polysaccharide is useful in diagnosing culture-negative cases of cryptococcosis.

  5. Therapy

    a. Amphotericin B, 5-FC, or a combination of the two has been used successfully in the treatment of disseminated cryptococcosis.

    b. The relapse rate is significant, and some cases become chronic with the need for frequently repeated therapy.

  6. Epidemiology
    a. *C. neoformans* can be isolated worldwide from the feces and nesting materials of pigeons and other birds and from contaminated soil. The birds, themselves, are not infected.
    b. AIDS, leukemia, and renal transplantation are important predisposing factors to infection.
    c. Some studies suggest that the source of infection sometimes is endogenous pulmonary or renal colonization.

C. Aspergillosis is the name applied to a variety of syndromes caused by any of about 150 species of *Aspergillus.*

  1. Etiology. *Aspergillus fumigatus, Aspergillus flavus, Aspergillus niger,* and *Aspergillus terreus* are the species most frequently involved in human disease.
    a. Cultural characteristics
      (1) *Aspergillus* species grow rapidly on all types of microbiologic media.
      (2) The vegetative hyphae are septate and branching. The arrangement of sterigmata and conidia aid in differentiating the species.
      (3) Conidial colors include white, beige, green, black, and many other shades.
    b. Tissue phase. Aspergillus species appear as dichotomously branching, septate hyphae in tissue. Conidial heads rarely are seen in tissue.

  2. Pathogenesis and clinical manifestations. The spectrum of human disease caused by aspergilli ranges from allergy to disseminated infection. Most disease processes are caused by the inhalation of conidia.
    a. Bronchopulmonary aspergillosis includes several syndromes.
      (1) Allergic bronchopulmonary aspergillosis. Allergy to the conidia of aspergilli is relatively common.
        (a) The presence of ungerminated conidia in the bronchi stimulates an IgE response.
        (b) It appears that 10%–15% of patients diagnosed as having asthma actually have allergy to aspergilli.
      (2) Noninvasive bronchopulmonary aspergillosis is characterized by germination of inhaled conidia and the growth of hyphae in the bronchi or lungs.
        (a) A typical bronchitis syndrome develops.
        (b) Eosinophilia and transient pulmonary infiltrates are common findings.
      (3) Invasive bronchopulmonary aspergillosis occurs when hyphae invade the bronchial walls or lung tissue.
        (a) A chronic inflammatory response is established, accompanied by eosinophilia and infiltrates.
        (b) Hematogenous dissemination of the fungus can cause diffuse infection, particularly in the spleen, liver, and kidneys.
    b. Aspergilloma, also known as **fungus ball,** is a noninvasive disease characterized by a granulomatous mass of entwined hyphae that may measure up to 2 cm in diameter.
      (1) Aspergilloma usually occurs in the lungs, colonizing a preformed cavity created by tuberculosis or other disease. Occasionally, aspergillomata occur in other organs, such as the brain.
      (2) Erosion of a blood vessel by the expanding aspergilloma may cause severe hemoptysis that can be fatal.
    c. Disseminated aspergillosis is most commonly seen in immunocompromised patients.
      (1) The fungus disseminates hematogenously and by direct extension, and any organ system may be involved.
      (2) Symptomatology frequently is chronic, with cough, fever, night sweats, weight loss, and malaise.
    d. Toxicosis in animals and humans has been attributed to several toxins produced by aspergilli. The **aflatoxins** produced by some strains of *A. flavus* growing on grains cause **hepatocarcinoma.**

  3. Diagnosis. Isolation of aspergilli from clinical specimens poses the dilemma of determining whether the fungus is the etiologic agent or a contaminant. Repeated isolation of the same species in conjunction with appropriate symptomatology is more definitive than the occasional isolation of aspergilli.

  a. **Bronchopulmonary aspergillosis**
    (1) Allergic bronchopulmonary aspergillosis often is diagnosed by skin testing and precip-
        itating antibody titers.
    (2) Bronchial washings are useful in the diagnosis of bronchopulmonary aspergillosis. Care
        must be taken during this and other procedures to avoid exogenous contamination.
  b. **Aspergillomata** are visible on radiographic examination.
  c. **Disseminated aspergillosis**
    (1) Disseminated aspergillosis is difficult to diagnose. Diagnosis frequently is not confirmed
        until the time of autopsy.
    (2) Blood and other body fluids generally are negative for culture. Hyphal fragments are too
        large to remain free in the circulation.
    (3) No skin test antigens are available to aid in the diagnosis.

4. **Therapy**
  a. Amphotericin B is the drug of choice for invasive and disseminated aspergillosis. There are
     a few reports of 5-FC being effective therapeutically.
  b. Corticosteroids have been used in noninvasive bronchopulmonary aspergillosis and as-
     pergilloma, but the benefits are unclear.
  c. Surgery currently is the most effective treatment for aspergilloma.

5. **Epidemiology.** The aspergilli are ubiquitous in the environment. Their conidia have been
   isolated from a variety of sources, including extreme natural environments, soil, air, sulfuric
   acid baths, and distilled water. Immunosuppressed patients are at the highest risk for infection.

D. **Zygomycosis**, also known as **phycomycosis** and **mucormycosis**, is caused by several genera
   belonging to the class Zygomycetes.

1. **Etiology.** Species of *Mucor*, *Rhizopus*, and *Absidia* are capable of causing zygomycosis. *Mucor*
   species are the most common cause of human disease.
  a. **Cultural characteristics.** Zygomycetes are rapidly growing fungi that produce **aerial hyphae**.
     The hyphae are coenocytic (i.e., nonseptate) and large. *Mucor* species have tan to brown
     conidia.
  b. **Tissue phase.** In tissue, all zygomycetes are large, nonseptate hyphae with irregularly spaced
     branching at right angles.

2. **Pathogenesis and clinical manifestations.** Mucor species and the other zygomycetes usually
   invade the upper respiratory tract following germination of inhaled conidia.
  a. Growth is rapid, and hyphae penetrate the cribriform plate, gaining access to the brain.
  b. Frontal headache of increasing intensity is a consistent complaint. Blindness is common
     because of optic nerve involvement, and the disease is rapidly fatal.
  c. Rapidly progressive pulmonary infections also may occur; the clinical picture resembles
     invasive aspergillosis.

3. **Diagnosis**
  a. Zygomycetes are easily isolated and identified, but their ubiquity poses the same dilemma
     as in aspergillosis. Repeated positive cultures are suggestive of infection.
  b. In tissue, the large coenocytic hyphae of the fungi are readily recognized.

4. **Therapy.** There is no effective treatment for fulminant zygomycosis.
  a. Surgery and antifungal therapy, usually with amphotericin B, is the prescribed treatment.
  b. Treatment has not been successful to date, and death generally occurs within 72 hours after
     definitive symptomatology (i.e., headache, visual disturbances) appear.

5. **Epidemiology.** The zygomycetes are ubiquitous in the environment. The immunosuppressed
   organ transplant patient and the uncontrolled ketoacidotic diabetic are at the highest risk for
   disease.

STUDY QUESTIONS

Directions: Each question below contains five suggested answers. Choose the **one best** response to each question.

1. A negative stain revealing the presence of an encapsulated yeast in cerebrospinal fluid (CSF) is diagnostic of

(A) histoplasmosis
(B) coccidioidomycosis
(C) candidiasis
(D) cryptococcosis
(E) blastomycosis

2. Combination chemotherapy with amphotericin B and 5-fluorocytosine has the highest success rate when used to treat

(A) cryptococcosis
(B) sporotrichosis
(C) coccidioidomycosis
(D) aspergillosis
(E) mycetoma

Directions: Each question below contains four suggested answers of which **one or more** is correct. Choose the answer

A   if **1, 2, and 3** are correct
B   if **1 and 3** are correct
C   if **2 and 4** are correct
D   if **4** is correct
E   if **1, 2, 3, and 4** are correct

3. Etiologic agents of superficial mycoses include

(1) *Trichophyton* species
(2) *Microsporum* species
(3) *Epidermophyton* species
(4) *Candida* species

4. Chemotherapeutic agents routinely used to treat dermatophytic fungal infections include

(1) selenium sulfide
(2) griseofulvin
(3) tolnaftate
(4) amphotericin B

5. Fungi that exhibit true dimorphism include

(1) *Histoplasma capsulatum*
(2) *Candida albicans*
(3) *Coccidioides immitis*
(4) *Cryptococcus neoformans*

6. *Aspergillus* species are etiologic agents of

(1) fungus ball
(2) endocarditis
(3) noninvasive bronchitis
(4) asthma

Directions: The group of questions below consists of lettered choices followed by several numbered items. For each numbered item select the **one** lettered choice with which it is **most** closely associated. Each lettered choice may be use once, more than once, or not at all.

Questions 7–11

Match each description of a fungus as it typically appears in tissue with the appropriate species.

(A) *Histoplasma capsulatum*
(B) *Blastomyces dermatitidis*
(C) *Coccidioides immitis*
(D) *Sporothrix schenckii*
(E) *Cryptococcus neoformans*

7. Yeast with a large polysaccharide capsule

8. Large, thick-walled, broad-necked budding yeast

9. Small, intracellular yeast

10. Small, cigar-shaped yeast

11. Spherules

ANSWERS AND EXPLANATIONS

1. The answer is D. [*V B 4 a (1)*] *Cryptococcus neoformans* is the only encapsulated yeast of medical importance, and its presence in cerebrospinal fluid (CSF) is diagnostic of cryptococcal meningitis. The yeast, with its large polysaccharide capsule, is easily recognized with a negative stain such as India ink. The colloidal suspension of carbon does not rapidly penetrate the capsule, resulting in the microscopic appearance of the yeast with a lighted halo around it on a black background.

2. The answer is A. [*V B 5 a*] Combination chemotherapy with amphotericin B and 5-fluorocytosine (5-FC) has been used successfully in the treatment of disseminated cryptococcosis and is the recommended course of therapy in many cases. Low doses of amphotericin B are administered to produce pores in the yeast cell membrane, which allows the diffusion of higher concentrations of 5-FC than could be transported inside the cell by the permease system. The use of lower doses of amphotericin B lessens the degree of nephrotoxicity and other side effects of the drug. This combination therapy also has been somewhat successful in the treatment of disseminated candidiasis; it has been less effective in the treatment of other fungal infections.

3. The answer is E (all). [*II A 1; V A*] Superficial mycoses, also called dermatophytoses, primarily affect the keratin-containing structures of the human body (i.e., the skin, hair, and nails). Three genera—*Trichophyton*, *Microsporum*, and *Epidermophyton*—comprise the dermatophytic fungi and are the etiologic agents of many superficial infections known as tineas. *Candida albicans*, although it may cause mucocutaneous and systemic disease, also may cause superficial infections of the skin. Most of the etiologic agents of superficial mycoses thrive on areas of the skin that are chronically warm and moist.

4. The answer is A (1, 2, 3). [*II E*] A variety of topical agents have been employed successfully in the treatment of dermatophytic fungal infections. These include preparations that contain selenium sulfide, coal tar derivatives, and sulfur compounds. Tolnaftate is a recently developed topical agent that shows excellent activity against the dermatophytes. Griseofulvin is an orally administered antibiotic that may prove to be effective for difficult to treat cases of tinea corporis and tinea unguium, but, as with all systemically administered agents, it carries a risk for undesirable side effects. Amphotericin B never should be administered intravenously for the treatment of dermatophytoses because of its association with severe nephrotoxicity and other side effects. Topical use of amphotericin B has not proven to be satisfactory in most cases of dermatophytosis.

5. The answer is B (1, 3). [*IV A 1, C 1*] Dimorphism is the term used to describe a fungus that exists as a mold in the environment and on routine laboratory culture and as a yeast or yeast-like organism when grown in tissue. Dimorphism is a characteristic of the etiologic agents of systemic mycoses, including *Histoplasma capsulatum* and *Coccidioides immitis*. *Candida albicans* is not a dimorphic fungus; both yeast and hyphal forms are seen in tissue and in culture. *Cryptococcus neoformans* typically is seen only as a yeast in medically significant situations and in routine laboratory culture, although the filamentous sexual stage has been identified recently.

6. The answer is E (all). [*V C 2 a (1), (2), b, c (1)*] All of the approximately 150 species of *Aspergillus* may be opportunistic pathogens for humans. Pulmonary infections are the most common type of human disease, and several forms are seen. Fungus ball, or aspergilloma, is a noninvasive disease characterized by a mass of hyphae growing in a pulmonary cavity that was created by another disease. Another form of noninvasive bronchopulmonary aspergillosis occurs when the conidia germinate and grow on the surface of the bronchi, producing a bronchitis. Studies have shown that about 10%–15% of persons with a diagnosis of asthma actually have an allergy to conidia of the aspergilli. Hematogenous dissemination of the fungus may occur and cause infection of any organ system, including the heart. Endocarditis may result from the hematogenous spread of the aspergilli from other foci of infection; it also may occur by direct inoculation of the conidia by the intravenous route or during heart surgery.

7–11. The answers are: 7-E; 8-B, 9-A, 10-D, 11-C. [*III A 1; IV A 1 b, B 1 b, C 1 b; V B 1 b*] *Cryptococcus neoformans* is the only encapsulated yeast of medical importance. It exists as a yeast in both tissue and in laboratory culture, and it typically retains its polysaccharide capsule during laboratory passage. *C. neoformans* is an opportunistic pathogen that causes systemic disease usually manifested by meningitis.

  Blastomyces dermatitidis is a dimorphic fungus that exists as a large, thick-walled yeast with a characteristic broad-based neck. This fungus may cause primary cutaneous infection, but it also is the etiologic agent of a serious systemic mycosis typically acquired by the respiratory route.

Histoplasma capsulatum is a dimorphic fungus that exists in tissue as a small yeast; it is an intracellular pathogen of lymphocytes. *H. capsulatum* is a common cause of primary histoplasmosis in certain geographic regions of the United States, but only a very small percentage of infected persons develop disseminated disease.

Sporothrix schenkii is a dimorphic fungus that appears as a small, cigar-shaped yeast in tissue. It typically is difficult to see in tissue samples, and silver stains are used to visualize the fungus microscopically. *S. schenkii* is the etiologic agent of a generally self-limited subcutaneous mycosis referred to as sporotrichosis.

Coccidioides immitis is a dimorphic fungus that exists as spherules in tissue. The spherule is a large, saclike sphere that contains endospores, each of which can give rise to a new spherule when released into the tissue. *C. immitis* is endemic to the semiarid regions of the southwestern United States. It is highly infectious and causes a serious systemic mycosis.

18
Viral Propagation, Structure, and Classification

David T. Kingsbury

I. THE NATURE AND STUDY OF VIRUSES

A. Basic properties of all viruses. Although extremely heterogeneous in size, shape, and behavior, viruses share several common features.

1. All viruses are **obligate intracellular parasites**. They contain no enzymes associated with energy metabolism and are entirely dependent on the host cell for biosynthesis of macromolecules.

2. Viruses contain only **one type of nucleic acid**—either DNA or RNA. The genome is enclosed in a protective shell formed of protein or a mix of protein and lipid.

3. Naturally occurring viruses **infect virtually all organisms** in nature.
   a. Bacteriophages (i.e., bacterial viruses) are very common in nearly all bacterial groups.
   b. Plant viruses include complete viruses as well as **viroids** (i.e., small, nonencapsulated circular RNA molecules).
   c. Animal viruses represent a widely varied group of agents that infect insects or vertebrates; some animal viruses infect both.

B. Propagation, detection, and quantitation of viruses

1. **Propagation.** Animal cell culture is the basic tool of animal virology. Animal cells of many types can be propagated in vitro under the proper cultural conditions. The physiologic requirements for growth include necessary amino acids and vitamins, a balanced salt solution, glucose, a buffer system, and supplementation with serum. Because many viruses show tissue and species specificity (e.g., poliovirus infects only primate cells), the cell culture system employed often is dictated by the virus under study.
   a. Cell culture. Dispersed cell suspensions frequently grow as attached monolayer cultures when placed on a suitable flat surface.
      (1) Primary and secondary cultures are the first and second cultures obtained after dispersed tissues are plated. These cultures often are mixed cell types, have a finite lifespan, and frequently cannot be subcultured further. They maintain the same number of chromosomes as the parent cells.
      (2) Cell strains can be redispersed and regrown a limited number of times (usually no more than 30–50 subcultures) without dying out or undergoing spontaneous transformation.
      (3) Continuous cell lines are cells that have survived a number of cultivations and have spontaneously transformed or cells that were derived from tumor tissue. They can be redispersed and regrown infinitely without further transformation.
   b. Organ culture. Not all cell types grow as monolayers; in some cases, differentiated cells can only be maintained in organ culture. An organ culture is a **suspension** of a tissue with a specialized function.

2. **Detection**
   a. The presence and biologic activity of viruses generally is recognized through the effects they have on the host itself (e.g., fever) or on host cells (as seen in culture). Cultured cells may undergo changes in morphology, growth, or reproduction or may be destroyed outright by the cytopathic effect of certain viruses.
   b. Viruses also may be detected directly through the measurement of viral antigens by immunofluorescence, enzyme immunoassay, or complement fixation (CF).

3. **Quantitation.** Viruses are measured in two ways: by direct measurement of infectivity and by quantitation of virus antigen. The determination of the infectious titer for viruses depends greatly on the quantitation method employed; whereas bacterial viruses have a particle-to-infectivity ratio (i.e., each virus particle will cause an infection) of approximately 1, the ratio for animal viruses rarely is less than 10 and usually is much higher.
   a. **Measurement of viral infectivity**
      (1) **Measurement of viral plaques** is the most direct form of viral quantitation. Viral plaques form on monolayers of susceptible cells at the site of inoculation of virus particles.
      (2) **Direct infectivity assays** may be used to estimate the **infectious dosage (ID)** or the **lethal dosage (LD)** of the virus being tested.
         (a) The ID_{50} designates a dilution that will infect 50% of cells inoculated.
         (b) The LD_{50} designates a dilution that will kill 50% of cells or animals inoculated.
   b. **Measurement of antigen**
      (1) **Quantitative hemagglutination** is a convenient method of measuring the amount of viral antigen present. This technique relies on the common tendency for animal viruses to adsorb to the surface of red blood cells of various animals (see also section II E 1).
      (2) **Quantitative electron microscopy** is used to count the total number of virus particles —not just infectious particles—in a suspension.

II. VIRAL MORPHOLOGY.

II. VIRAL MORPHOLOGY. All viruses consist of a **nucleic acid** genome enclosed in a protective shell, or **capsid**; the complex of nucleic acid and capsid is called the **nucleocapsid**. The complete virus particle is called the **virion**; the viral nucleocapsid may constitute the virion (a **naked capsid virus**), or it may be surrounded by a membrane-containing envelope (an **enveloped virus**).

A. **Nucleic acid.** Viruses contain only one type of nucleic acid—DNA or RNA—which may be single- or double-stranded.

   1. **Structure.** The nucleic acids extracted from purified viruses vary widely in structure and size.
      a. **Circularity and segmentation.** Most viral nucleic acids are single, linear molecules, but exceptions do exist.
         (1) The DNA of the papovaviruses is in the form of a double-stranded, covalently closed circle, or **supercoil**.
         (2) Several RNA viruses have segmented genomes varying from two segments (in the arenaviruses) to eleven segments (in the rotaviruses).
      b. **Polarity.** Single-stranded RNA of viruses exist in either messenger RNA (mRNA) or anti-mRNA polarities.
         (1) **Plus-strand RNA** is single-stranded RNA that can serve directly as mRNA; it is said to have **plus polarity**.
         (2) In order for **minus-strand RNA** to serve as mRNA, a complementary strand must be synthesized by viral RNA polymerase (transcriptase).
      c. **Base composition**
         (1) Viral DNAs vary in base composition from 36% guanine plus cytosine (G + C) in the poxviruses to 70% G + C in the herpesviruses.
         (2) Many viral DNAs and some RNAs have terminal redundancy of nucleotide sequences.
      d. **Linked molecules.** Some viral DNAs (e.g., those of the poxviruses) have cross-linking between strands, and several viral nucleic acids (e.g., those of the picornaviruses) are covalently linked to protein.

   2. **Infectivity.** Many viral nucleic acids are themselves infectious; if introduced into a suitable host cell, they contain all the genetic information needed to produce new viruses.

      a. The naked nucleic acids of most plus-strand RNA viruses and most DNA viruses (with the major exception of the poxviruses) are infectious.
      b. The RNAs from the double-stranded and from the minus-strand RNA viruses are not infectious.
      c. The host range usually is much wider for infectious nucleic acids than for virions.

B. **Capsid.** The protein shell, or capsid, is composed of numerous repeating subunits arranged in a highly ordered pattern.

1. The simplest structural component is a single protein molecule, or **protomer**.

2. Groups of protomers form a **capsomer**, which is the basic structural unit of a virus. A highly specific number of capsomers (depending on the size and morphology of the virion) assemble to form the capsid.

C. **Nucleocapsid.** In most viruses, the nucleocapsid has either helical or icosahedral symmetry.

1. **Helical symmetry.** In helical nucleocapsids, the relationship of the nucleic acid with the protein molecules yields a **single rotational axis**.
   a. Each helical virus type has a characteristic length, width, and periodicity of its nucleocapsid.
   b. **Tobacco mosaic** and **influenza viruses** are examples of viruses whose nucleocapsids have helical symmetry.

2. **Icosahedral symmetry.** In icosahedral viruses, the nucleic acid is condensed at the core of the structure and is surrounded by the protein coat.
   a. Icosahedral viral structures are characterized by 20 triangular faces, 12 vertices, 30 edges, and a precise five- to three- to twofold rotational symmetry.
   b. **Adenoviruses** are examples of complex icosahedral structures.

3. **Viruses without regular symmetry. Large viruses** (e.g., the brick-shaped poxviruses) and **bacteriophages** have complex structures that demonstrate no regular symmetry.

D. **Viral envelope.** Like cell membranes, viral envelopes contain lipid bilayers and virus-specific proteins.

1. **Lipids in viral envelopes** are mixtures of neutral lipids, phospholipids, and glycolipids. Except in the poxviruses, envelope lipids are derived from host cell membranes. The exact lipid composition of a virus varies depending on the host cell and the composition of the growth medium.

2. **Virus-specific envelope proteins** fall into two categories.
   a. **Glycoproteins** generally are found as surface structures such as spikes or hemagglutinin molecules. They are essential for infectivity.
   b. **Matrix proteins** are nonglycosylated proteins that form a structural layer at the inner surface of the viral envelope.

E. **Viral proteins.** Protein is the principal component, by weight, of all viruses. In addition to being the sole component of the capsid and the major component of the viral envelope, viral proteins also may have structural or enzymatic functions.

1. Many animal viruses (naked or enveloped) agglutinate red blood cells through the interaction of capsid or envelope proteins with receptors on the blood cell surface. This action, called **hemagglutination**, allows for easy quantitation of viruses (see section I B 3 b).

2. **Enzymes** often are present in virions.
   a. **Neuraminidase** in orthomyxoviruses and paramyxoviruses is located on a structural spike protein of the viral envelope.
   b. An RNA-directed RNA polymerase (**transcriptase**) is found in both minus-strand and double-stranded RNA viruses.
   c. An RNA-directed DNA polymerase (**reverse transcriptase**) is found in retroviruses.
   d. Picornavirus capsid proteins have a specific protease activity.

III. VIRAL CLASSIFICATION. The classification of viruses is based on fundamental properties of virus particles, such as size, nucleic acid type, symmetry, and number of capsomers.

A. **Major groups of DNA viruses**

1. **Parvoviridae**
   a. The parvoviruses are the smallest of the DNA viruses. They have a single-stranded genome with a molecular weight of $1.5–2.2 \times 10^6$.
   b. They have a naked capsid virion with icosahedral symmetry approximately 18–26 nm in diameter.

2. Papovaviridae

a. The papovaviruses have a circular, double-stranded genome with a molecular weight of $2.4–5 \times 10^6$.

b. They have a naked capsid virion with icosahedral symmetry approximately 45–55 nm in diameter.

3. Adenoviridae

a. The adenoviruses have a linear, double-stranded genome with a molecular weight of $20–30 \times 10^6$.

b. They have a naked capsid virion with icosahedral symmetry approximately 70–90 nm in diameter.

4. Herpesviridae

a. The herpesviruses have a linear, double-stranded genome with a molecular weight of $54–92 \times 10^6$.

b. They have an enveloped capsid virion with icosahedral symmetry approximately 100 nm in diameter.

5. Iridoviridae

a. The iridoviruses have a linear, double-stranded genome with a molecular weight of approximately 130×10^6.

b. They have an enveloped capsid virion with icosahedral symmetry approximately 130 nm in diameter.

6. Poxviridae

a. The poxviruses are the largest of the DNA viruses. They have a linear, double-stranded genome with a molecular weight of approximately 160×10^6.

b. The poxvirus particles have a complex coat that resembles the envelope of other viruses, they have no regular symmetry, and they are approximately 230×300 nm in size.

B. Major RNA viruses

1. Picornaviridae

a. The picornaviruses are the smallest of the RNA viruses. They have a single-stranded genome with a molecular weight of approximately $2.3–2.8 \times 10^6$.

b. They have a naked capsid structure with icosahedral symmetry approximately 24–30 nm in diameter.

2. Reoviridae

a. The reoviruses have a segmented, double-stranded genome with a molecular weight of approximately $12–15 \times 10^6$.

b. They have a naked capsid structure with icosahedral symmetry approximately 60–80 nm in diameter.

3. Togaviridae

a. The togaviruses have a nonsegmented, single-stranded genome with a molecular weight of approximately 4×10^6.

b. They have an enveloped capsid structure with icosahedral symmetry approximately 40–60 nm in diameter.

4. Orthomyxoviridae

a. The orthomyxoviruses contain a segmented, single-stranded genome with a molecular weight of approximately 4×10^6.

b. They have an enveloped structure containing a nucleocapsid with helical symmetry approximately 80–120 nm in diameter.

5. Paramyxoviridae

a. The paramyxoviruses have a nonsegmented, single-stranded genome with a molecular weight of approximately $6–8 \times 10^6$.

b. They have an enveloped structure containing a nucleocapsid with helical symmetry, and they are 150–300 nm in diameter.

6. Rhabdoviridae

a. The rhabdoviruses have a nonsegmented, single-stranded genome with a molecular weight of approximately $3–4 \times 10^6$.

    b. They have an enveloped virion with a helical nucleocapsid, forming a bullet-shaped structure that is 60×180 nm in size.

7. Retroviridae

    a. The retroviruses contain two copies of a single-stranded genome complexed with several **transfer RNA (tRNA)** molecules. The genome has a molecular weight of approximately $10\text{–}12 \times 10^6$.

    b. The retroviruses are enveloped viruses approximately 100 nm in diameter, with no known symmetry.

8. Arenaviridae

    a. The arenaviruses have a segmented genome with a molecular weight of approximately $3\text{–}5 \times 10^6$.

    b. They have an enveloped structure that contains a randomly coiled nucleocapsid and are 80–150 nm in diameter.

9. Coronaviridae

    a. The coronaviruses contain a nonsegmented, single-stranded genome with a molecular weight of approximately 9×10^6.

    b. They have an enveloped structure that contains a randomly coiled helical nucleocapsid and are 80–130 nm in diameter. The coronaviruses have unusual, club-shaped spike proteins that give them a distinctive electron microscopic appearance.

10. Bunyaviridae

    a. The bunyaviruses contain a segmented, single-stranded genome with a molecular weight of approximately $6\text{–}7 \times 10^6$.

    b. They have an enveloped structure that contains a randomly coiled helical nucleocapsid and are 100 nm in diameter.

STUDY QUESTIONS

Directions: Each question below contains five suggested answers. Choose the **one best** response to each question.

1. Which of the following groups of viruses contains an RNA-dependent DNA polymerase as a structural component of the virion?

(A) Adenoviruses
(B) Orthomyxoviruses
(C) Rhabdoviruses
(D) Reoviruses
(E) Retroviruses

2. Diploid cell strains have all of the following characteristics EXCEPT

(A) they continue to divide in culture but for only a limited number of generations
(B) they are obtainable from both fetal and adult tissue
(C) they require serum in the growth medium
(D) they are not capable of undergoing spontaneous transformation
(E) they usually grow as cell monolayers

3. The total number of viral particles in a suspension of animal viruses can best be determined through

(A) direct measurement of viral plaques
(B) direct infectivity assays
(C) quantitative electron microscopy
(D) quantitative hemagglutination
(E) none of the above

4. All of the following types of viruses have an infectious nucleic acid genome EXCEPT

(A) poliovirus
(B) herpes simplex virus
(C) influenza virus
(D) papovavirus
(E) togavirus

Directions: Each question below contains four suggested answers of which **one or more** is correct. Choose the answer

A   if **1, 2, and 3** are correct
B   if **1 and 3** are correct
C   if **2 and 4** are correct
D   if **4** is correct
E   if **1, 2, 3, and 4** are correct

5. The various nucleic acid configurations found as viral genomes include

(1) unsegmented, double-stranded DNA
(2) unsegmented, double-stranded RNA-DNA hybrid
(3) segmented, double-stranded RNA
(4) segmented, double-stranded DNA

6. The presence of viruses in an infected cell culture may be detected through

(1) the cytopathic changes in the cultured cells
(2) the ability of the cytoplasmic membrane of the infected cells to adsorb red blood cells
(3) the presence of viral proteins in the infected cell monolayers.
(4) the altered nutritional requirements noted in the infected cells

Directions: The group of questions below consists of lettered choices followed by several numbered items. For each numbered item select the **one** lettered choice with which it is **most** closely associated. Each lettered choice may be use once, more than once, or not at all.

Questions 7–11

For each statement below, select the tissue culture type most likely to demonstrate the property described.

(A) Organ cultures
(B) Primary cell cultures
(C) Continuous cell lines
(D) Cell strains
(E) None of the above

7. These frequently contain multiple cell types but form monolayers that continue to divide in culture for a few generations

8. These can be subcultured about 30–50 times before dying out or undergoing spontaneous transformation

9. These will not grow as monolayer cultures but maintain virus sensitivity

10. These maintain some forms of differentiated tissue markers but propagate indefinitely

11. These maintain a very high degree of differentiated cell markers

ANSWERS AND EXPLANATIONS

1. The answer is E. [*II E 2 c*] The most striking characteristic of the retrovirus group is the presence of reverse transcriptase, which is an RNA-dependent DNA polymerase. This enzyme is essential to retrovirus replication and is responsible for the unique biologic properties of this virus group. Orthomyxoviruses and rhabdoviruses are single-stranded, minus-strand RNA viruses, and reoviruses are double-stranded RNA viruses; the virion of each of these virus groups contains an RNA transcriptase. Adenoviruses contain neither transcriptase nor reverse transcriptase.

2. The answer is D. [*I B 1 a (2)*] Diploid cell strains are dispersed cell suspensions that grow as monolayer cultures, which can be redispersed and regrown for a limited number of generations (30–50 subcultures) without dying out or undergoing spontaneous transformation. Diploid cell strains may be started from tissue obtained from individuals of any age, although fetal tissue generally is more effective. Like all tissue and organ cultures, diploid cell strains require nutrients found in serum for long-term maintenance and growth.

3. The answer is C. [*I B 3 b (2)*] Animal viruses are characterized by a particle-to-infectivity ratio of greater than 1. Therefore, any procedure that measures infectivity (e.g., viral plaque assay; direct infectivity assays) underestimates the actual number of viral particles by anywhere from 10 to 10,000. The only way to measure the number of viral particles is to count them in the electron microscope. By combining detection and quantitation techniques it is possible to determine the particle-to-infectivity ratio. Quantitative hemagglutination measures viral antigens regardless of whether they are in assembled virus particles; however, the technique is still somewhat insensitive due to the requirement for numerous particles to bridge red blood cells.

4. The answer is C. [*II A 2 a; III A 2, 4, B 1, 3*] Influenza virus is a minus-strand RNA virus that contains a transcriptase that is essential for viral replication. Herpes simplex virus and the papovaviruses are double-stranded DNA viruses that do not require their own polymerase and are fully infectious. Poliovirus (a picornavirus) and the togaviruses are plus-strand RNA viruses with infectious RNA genomes.

5. The answer is B (1, 3). [*II A 1*] Although a virus may contain any one of several nucleic acid configurations, the genome never contains more than one type of nucleic acid (i.e., DNA or RNA, not a hybrid). The RNA or DNA may be single or double stranded. RNA genomes also may be segmented; no virus has been identified with a segmented DNA genome.

6. The answer is A (1, 2, 3). [*I B 2, 3 b (1)*] The presence of viruses in an infected cell culture can be recognized through structural changes that occur during viral infection, such as the cytopathic effect that viruses have on some host cells, through modification of the surface of the cultured cells so that they adsorb red blood cells, and by the presence of viral proteins in the infected cell monolayers. Although many measurable changes occur during virus infection, none has been easily identified as nutritional.

7–11. The answers are: 7-B, 8-D, 9-A, 10-C, 11-A. [*I B 1*] Since viruses are obligate intracellular parasites, living cells are needed for virus propagation. Animal cells propagated in culture, regardless of their origin, have complex nutritional requirements, many of which are not defined. The need for these trace-level undefined nutrients is met best through the inclusion of serum in the tissue culture medium.

Primary cell cultures are derived directly from tissues. Primary cell lines often contain multiple cell types but are capable of only limited growth in culture (i.e., a few subcultures). Primary and secondary cell cultures are referred to as diploid.

Some secondary cell cultures are capable of surviving many (about 30–50) subcultures before a natural aging process ensues, leading to senescence and eventual death of the cell lines. In some cases, the cell strains undergo spontaneous transformation to become continuous cell lines.

Organ culture permits the maintenance of differentiated cells of several types in a single culture and is required for the propagation of some viruses. Although the cells in organ cultures usually do not divide, they may be maintained for long periods of time.

Tumor cells, when established in culture, frequently maintain a few of the differentiated characteristics of the stem cell of the tumor. In addition, tumor cells are transformed and become immortal continuous cell lines.

The most effective way to maintain maximal differentiation of cells is to leave them in their surrounding tissue matrix and culture them in their original organ configuration.

19

Viral Replication and Genetics

David T. Kingsbury

I. GENERAL FEATURES OF VIRAL REPLICATION

A. Virus-host interaction. All viruses interact with host cells in one of two ways.

1. In **lytic virus-host interaction**, the most frequent and best understood relationship, the invading virus multiplies and eventually kills and lyses the host cell.

2. In **stable virus-host interaction**, the invading virus does not kill the host cell. Instead, the viral genetic information becomes associated with the host cell's genetic information. Examples are lysogeny in bacteria and viral transformation in animal cells.

B. Viral replication cycle. The viral replication cycle described here occurs in lytic virus-host interaction.

1. **Adsorption** is the first stage in the interaction of virus and host.
   a. The adsorption stage is **temperature independent**—that is, it does not require energy—and it has **two phases**.
      (1) **Ionic attraction** between virus and host is the first phase of adsorption.
      (2) **Physical alignment** of the virus particle with the appropriate cell surface receptor is the second phase of adsorption.
   b. **Host and tissue specificities**, where they exist, are expressed at the adsorption stage of replication. Poliovirus, for example, adsorbs only to cells of human or primate origin, because only these cells have the appropriate cell surface receptors.
      (1) Only cells of the primate central nervous system (CNS) and intestinal tract have poliovirus receptors.
      (2) Infectious poliovirus RNA may infect nonprimate cells (e.g., chicken embryo fibroblasts), but it replicates only once because these cells lack receptors for viral adsorption, which is necessary for additional cycles of replication.
   c. The **multiplicity of infection (MOI)** is the number of infectious virus particles adsorbed to a cell. Animal cells may contain as many as 10^5 cell surface receptors for some viruses.

2. **Penetration and uncoating.** The penetration process begins almost immediately after stable adsorption of virus to the cell surface and is intimately associated with the uncoating process (i.e., removal of the virion capsid or viral envelope from the nucleic acid or nucleoprotein core).
   a. The penetration and uncoating process is **temperature dependent** and follows **two pathways**.
      (1) Virus particles appear to fuse with or become engulfed by the cytoplasmic membrane and are transported directly into the cytoplasm as naked nucleocapsids (in the case of enveloped viruses) or as partially uncoated particles (in the case of most naked capsid viruses).
      (2) Virus particles may be phagocytized and appear internally in the cell in vacuoles, which eventually break down, releasing the virus into the cytoplasm.
   b. The poxviruses have a unique and highly specialized uncoating mechanism involving engulfment into phagocytic vacuoles in which the viruses are broken down into cores that require a special virus-encoded enzyme for further activation.

3. **Synthesis.** The synthetic process, which begins once uncoating is complete, is a highly structured series of events that are different for each viral group.
   a. The process is initiated by **synthesis of** the class of **macromolecule** necessary to sustain further replication.

 (1) Plus-strand RNA viruses initiate **protein** synthesis.
 (2) Minus-strand RNA viruses, double-stranded RNA viruses, and DNA viruses initiate **nucleic acid** synthesis.
 b. For some viruses, the synthetic process is **temporally regulated**.
   (1) The **early period** is devoted to synthesis of materials needed for the initiation of nucleic acid synthesis.
   (2) In the **late period**, viral genomes and proteins needed for formation of new viral particles are produced.
 c. During the synthetic period, viral components and, in some cases, empty capsids or nonenveloped nucleocapsids begin to accumulate.

4. Eclipse period. From the final stage of adsorption through penetration, uncoating, and synthesis marks the eclipse period, during which it is not possible to recover infectious virions from the cell. The eclipse period ends with the assembly process.

5. Assembly of the complete intracellular virus is the next step in the replication cycle.
 a. Capsids become associated with viral nucleic acid.
 b. For enveloped viruses, complete nucleocapsids become associated with specific viral membrane sites on the cytoplasmic membrane.

6. Release of the virion progeny is the final stage of the infection cycle.
 a. **Naked capsid viruses and poxviruses** that have accumulated within the cell are released in a fairly rapid burst, and the host cell disintegrates.
 b. **Enveloped viruses** are released more gradually; completed nucleocapsids migrate to the cytoplasmic membrane and bud through virally modified membrane patches.

II. THE REPLICATION CYCLE IN PARTICULAR VIRAL GROUPS

 A. Plus-strand RNA viruses. The major groups here are picornaviruses, flaviviruses, and togaviruses.

 1. Poliovirus, the most thoroughly studied picornavirus, is a small, naked capsid virus containing a single molecule of plus-strand RNA. Adsorption of poliovirus is limited to human or primate cells containing a specific receptor. Replication occurs entirely in the cell cytoplasm.
  a. After adsorption, penetration, and uncoating, the **first biosynthetic step** in replication **is protein synthesis**.
   (1) Poliovirus RNA is both **multigenic** and **monocistronic**; therefore, there is a single site for the binding of ribosomes and the initiation of protein synthesis.
   (2) Poliovirus RNA is translated into a **single giant polypeptide** with a molecular weight between 200,000 and 300,000.
    (a) The precursor polypeptide is post-translationally processed into at least **12 smaller proteins**.
    (b) Although the enzymes involved in the specific cleavage of the polyprotein are not known, it appears that one or more of the virion proteins possesses **protease** activity.
   (3) The **pactamycin mapping technique** has been used to determine the genetic map and the order of all the virion polypeptides.
    (a) Pactamycin inhibits the initiation of translation but has no effect on elongation.
    (b) Radioactive labeling can be used to determine the order of genes in the viral genome. Radioactive amino acids are incorporated into the protein, and the time of synthesis of each portion of the polyprotein molecule is measured relative to the time of initiation of synthesis of that molecule.
  b. **RNA synthesis** is the **second biosynthetic stage** of poliovirus replication.
   (1) RNA replication begins as soon as an adequate amount of **polymerase** has been formed.
   (2) RNA replication involves **two stages**.
    (a) The parental strand is transcribed into a **minus-strand copy**. The number of minus strands produced is strictly controlled.
    (b) The **minus strand serves as template for** multiple rounds of **transcription into plus-strand progeny**. Plus-strand synthesis appears to proceed uncontrolled.
  c. During the **assembly process**, empty capsids mature and associate with viral RNA molecules.
   (1) Virion RNA is covalently bonded at the 5′ end to a viral protein of 5000 molecular weight.

(2) Empty capsids of poliovirus contain a large precursor polypeptide (VP0), which is not cleaved into its final parts (VP2 and VP4) until the final stage of virion maturation.

d. The infected cell is **lysed**, and the **progeny** of viral particles **are released**.

2. **Togaviruses** replicate in a cycle somewhat different from the one for picornaviruses.

   a. After uncoating, approximately **two-thirds of the viral RNA is translated into a single protein**, which is subsequently **cleaved** into functional units.

   b. A newly synthesized virion **polymerase transcribes the parental plus strands into minus strands**, which then serve as templates for the synthesis of **two types of plus strand**: a full-sized strand, which can serve as a template for protein synthesis or as a virion RNA, and a shorter strand, about one-third the length of the full-sized strand. The shorter strand encodes the virion capsid proteins that are formed on that RNA via the synthesis of a large polyprotein.

   c. **Maturation** of the togaviruses involves the formation of complete enveloped virions by the **budding** of naked nucleocapsids through viral-specific regions of the cell membrane.

B. **Minus-strand RNA viruses and double-stranded RNA viruses.** Viruses in these two groupings follow the same fundamental scheme of replication with some minor differences.

1. In the **minus-strand RNA viruses**, the **rhabdoviruses, paramyxoviruses, and orthomyxoviruses** are the most widely studied groups. The rhabdoviruses and paramyxoviruses have an unsegmented genome, whereas the orthomyxoviruses contain eight separate RNA molecules. All minus-strand RNA viruses are enveloped and have helical symmetry.

   a. **Adsorption** appears to involve interaction with **specific receptors** on the cell surface.

   b. **Penetration**

      (1) **Rhabdoviruses and orthomyxoviruses penetrate** the cell **through phagocytosis** and subsequently **become uncoated in the phagocytic vesicle**.

      (2) The **paramyxoviruses penetrate through membrane fusion** and subsequently release their nucleocapsids into the cytoplasm.

   c. **Uncoating** activates an RNA directed **RNA polymerase (transcriptase)**, from which **messenger RNA (mRNA) is synthesized**—the first biosynthetic step. The template for the virion-associated transcriptase is the viral nucleoprotein, not naked viral RNA. The requirement for virion-associated transcriptase makes it **impossible to obtain infectious RNA** from minus-strand viruses.

      (1) **Two different transcripts** are produced.

         (a) **Full-sized plus-strand copies** are transcribed from the minus-strand template and are subsequently used as templates for further minus-strand synthesis.

         (b) **Incomplete plus-strand copies** are also produced. They contain polyadenylic acid on the 3′ termini and are capped at the 5′ ends and function as mRNA for protein synthesis.

      (2) For **rhabdoviruses and paramyxoviruses**, transcription is strictly a **cytoplasmic** event.

      (3) For **orthomyxoviruses**, transcription occurs in the **nucleus**.

         (a) Orthomyxovirus transcription requires active transcription of the host cell and can be inhibited by actinomycin D and α-amanitin, both of which inhibit DNA-dependent RNA synthesis.

         (b) The orthomyxoviruses require newly capped cellular RNA molecules as primers for transcription of viral mRNA.

   d. **Viral protein synthesis** is the second biosynthetic step. It follows association of mRNA molecules with ribosomes.

      (1) Virion proteins are synthesized in approximately the same molar proportions as those found in intact virions.

      (2) Control of the ratios of virion polypeptides is held at the transcriptional stage of replication.

      (3) Enveloped viruses encode nucleoproteins, which associate directly with viral RNA to form nucleocapsids, and envelope proteins, which are synthesized in association with membrane-bound ribosomes and are quickly inserted into the membrane following biosynthesis.

      (4) Envelope proteins are glycosylated during synthesis and also during transmembrane transport.

      (5) Most of the spike glycoproteins on the viral surface are synthesized with a hydrophobic leader sequence, which aids in transport across the cellular membrane and is removed in the final configuration of the spike protein.

    e. The viral **envelope proteins replace normal cellular proteins** in the membrane, eventually forming a patch of viral-specific envelope on the cell surface.

    f. **Nucleocapsids migrate** to the viral-specific areas of the cell membrane.

      (1) Interactions between nucleoproteins and membrane proteins are the basis for recognition of viral-specific regions.

      (2) The specific recognition signal leads to an accurate alignment of the nucleocapsid with the membrane.

    g. At **maturation** the virus **buds** from the cell surface.

      (1) Budding is more efficient in some cell types than in others.

      (2) The budding process does not appear to have any significant effect on the integrity of host cells.

2. **Double-stranded RNA viruses** are members of the **Reoviridae** family, which includes reoviruses and rotaviruses. All double-stranded RNA viruses are naked icosahedral viruses. Their replication cycle differs from that for minus-strand RNA viruses in the following particulars:

    a. The mRNA produced by double-stranded RNA viruses does not contain a polyA 3′ tail.

    b. The parental RNA remains associated with virion proteins in the form of a subviral particle.

    c. Synthesis of mRNA, mediated by a virion-associated polymerase, proceeds for several hours before new double-stranded molecules appear.

    d. Double-stranded RNA molecules are synthesized in RNA-protein complexes through the production of a complementary strand formed to complexed mRNA, not through a DNA-like mechanism of direct double-stranded replication.

    e. The viruses are assembled in the cytoplasm and do not bud from the cell surface.

C. **Double-stranded DNA viruses.** Papovaviruses, adenoviruses, and herpesviruses share basic principles of replication; the poxviruses have a very different strategy. In general, the genetic complexity varies more than 50-fold among the double-stranded DNA viruses.

    1. **Papovaviruses and adenoviruses**

      a. The replication process of papovaviruses and adenoviruses can be divided into well-defined **early and late phases**.

        (1) During each phase, different portions of the viral genome are transcribed into mRNA.

        (2) The onset of viral DNA replication marks the usual distinction between early and late synthesis.

        (3) Adsorption and penetration and uncoating proceed as for RNA viruses, except that **final uncoating always occurs in the nucleus and not in the cytoplasm**.

      b. **Early phase.** Synthesis of mRNA is the first biosynthetic process in DNA virus replication. Host RNA polymerase II is used for this process, and the transcriptional programs are highly regulated.

        (1) **Transcription** of DNA viruses **is identical to that of host DNA**. It involves large transcripts that are eventually processed through RNA splicing into functional mRNA molecules.

          (a) Early transcription involves a complex set of genes that act as regulatory molecules controlling the expression of the late genes or as essential enzymes in the replication of viral DNA.

          (b) The use of bacterial restriction endonucleases to map many DNA viral genomes has provided a physical basis for understanding the regulation of DNA virus gene expression.

        (2) A common feature of the **papovaviruses** is that the early genes are all found in the same region of the DNA and share many sequences; however, they vary according to the intron regions that are removed during RNA splicing.

        (3) Early mRNA, following its synthesis in the nucleus, is spliced, capped, and polyadenylated, after which it is transported to the cytoplasm.

        (4) Early mRNA is translated in the cytoplasm to form early gene products.

        (5) Early proteins are transported back into the nucleus, where they act as regulatory proteins during the early period. They are also involved in initiating viral DNA synthesis.

        (6) The amount of the viral genome necessary for early functions ranges from one-third to one-half of the DNA.

      c. **DNA synthesis** begins once the early genes have been synthesized and the enzymes necessary for initiation of DNA synthesis are present.

(1) In the nonlytic replication (transformation) cycle for both papovaviruses and adenoviruses, only the early gene products appear; they change the regulation of gene expression in the host cell.

(2) The mechanism of synthesis of viral DNA progeny varies among the different DNA viruses on the basis of size and structure of viral genomes.

(3) Newly synthesized viral DNAs appear to have a slightly different physical structure than the parental molecules and therefore serve as suitable templates for late synthesis.

d. Late phase. During **late transcription**, that portion of the viral DNA turned off during the early period is actively transcribed.

    (1) Late transcription continues to resemble cellular transcription with active splicing of the transcripts.

    (2) Late viral gene products are those, such as capsid proteins, that are involved in the formation of new viral particles.

    (3) As occurs with early gene products, during late transcription some viruses use the same DNA sequences for the production of more than one gene product.

    (4) Late mRNA transcripts are transported to the cytoplasm, where they are used for the synthesis of late gene products.

    (5) Late gene products are transported back to the nucleus or nuclear membrane for eventual formation into virus particles.

e. Capsid formation and the **maturation** of complete particles **occurs in the nucleus**. Papovaviruses and adenoviruses are naked capsid viruses that are released without budding.

2. Herpesviruses differ in replication slightly from papovaviruses and adenoviruses because their **genetic complexity is much greater**.

  a. Transcription is not as strictly divided into early and late periods but instead into **abundance classes**, which vary during early and late replication.

  b. There are **three classes of herpesvirus transcripts**—alpha, beta, and gamma.

  c. The herpesviruses mature by **budding** through a virally modified inner nuclear membrane, which serves as the source of the viral envelope.

3. Poxviruses replicate in a substantially different manner from other DNA viruses. Poxvirus replication can be separated into **three distinct stages**—pre-early, early, and late.

  a. The **pre-early stage** begins immediately after the loss of the viral envelope. Pre-early synthesis is required for the production of a viral-encoded enzyme that is necessary for the further uncoating of the viral core.

    (1) Unlike other DNA viruses, **the poxvirus itself contains an RNA polymerase**, which is responsible for transcription of as much as one-half of the genome during the pre-early and early stages of infection.

    (2) Uncoated poxvirus DNA initiates the formation of intracytoplasmic inclusions, which are centers for virus replication and assembly.

  b. In the **early stage** of replication, approximately one-half of the viral DNA is transcribed.

    (1) Early gene products are principally **enzymes** involved in viral DNA replication.

    (2) A limited number of structural **proteins** are produced during early replication.

  c. In the **late stage**, which coincides with the beginning of DNA replication, a new transcriptional pattern uses the second half of the genome.

    (1) Translational controls block the translation of early mRNA.

    (2) Viral **assembly proceeds entirely in the cytoplasm**. There is de novo membrane synthesis in the cytoplasm for the formation of intact virions.

    (3) Upon **maturity**, the viruses are released by a single **burst** mechanism.

D. Retroviruses are plus-strand RNA viruses in which each virion contains two copies of the viral genome. The unusual feature of retrovirus replication is that it uses a double-stranded DNA intermediate.

1. A few **transfer RNA (tRNA)** molecules of specific acceptor types are associated with the viral RNA, apparently in a hydrogen-bonded complex.

2. The virion nucleocapsids contain an **RNA-dependent DNA polymerase (RDDP or reverse transcriptase)**.

3. The tRNA molecules within the virion appear to act as primers for the initiation of complementary DNA synthesis, which proceeds from the viral RNA template immediately after uncoating.

4. A **double-stranded DNA copy** is produced and becomes associated with the host DNA through a covalent attachment.

5. Viral mRNA is produced from the integrated DNA provirus, and at least two different messenger populations are produced, the exact number depending on the site of the RNA splicing event.

6. Unspliced full-sized RNA copies become associated with the core nucleoproteins [**group antigen proteins (gag)**] and eventually **bud** from the cytoplasmic membrane at the site of virus-specific membrane protein replacements.

    a. Retroviruses are **rarely cytopathic** for their host cells.

    b. Budding and release of the virus into the surrounding medium can go on for years.

III. VIRAL GENETICS AND VIRAL GENOME INTERACTIONS. The tools of genetics have been extremely important in understanding the biology and biochemistry of viruses, especially the bacterial viruses.

A. Mutants, both naturally occurring and induced, are the essential elements for the study of viral genetics.

    1. Conditional lethal mutants are the most widely studied class of viral mutants.

        a. Temperature-sensitive mutants fail to grow at high (nonpermissive) temperatures. These mutants are applicable to the study of all virus groups.

        b. Suppressible chain termination mutants have been most useful in the study of bacterial viruses in which suppressor tRNA-carrying bacterial strains have been characterized.

    2. Mutant classes with other recognizable phenotypes have also been used.

        a. Mutants with **altered host ranges** in the form of tissue cell lines of susceptible animals have been useful.

        b. Plaque size and plaque morphology mutants are often useful markers for genes for capsid proteins.

        c. Drug-resistant mutants have been recognized in both RNA and DNA viruses.

        d. Enzyme-deficient mutants, such as the thymidine kinaseless mutants of herpesvirus, have been useful for locating these and related genes physically.

B. Genetic interactions among viruses take several different forms, regardless of the virus group. The presence of one virus may cooperate or interfere with the replication of another one.

    1. Cooperative interactions

        a. Recombination and **reassortment** of genes among fragmented genomes occur easily.

            (1) Double-stranded DNA viruses recombine readily at rates appropriate for the size of their genomes.

            (2) Poliovirus and the virus causing foot-and-mouth disease appear to have limited recombination, even though both are small, single-stranded RNA viruses.

            (3) Genetic reassortment of fragmented viral genomes, both single- and double-stranded, occurs at high rates during mixed infections.

        b. In **complementation**, viral genomes interact indirectly.

            (1) In a cell infected at a nonpermissive temperature by two different temperature-sensitive mutants of a virus, complementation occurs when a functional gene product of one mutant serves as a replacement for the temperature-sensitive gene product of the other.

            (2) Helper-dependent replication is a form of complementation in which a virus lacking a gene product necessary for growth is able to use the gene product supplied by a coinfecting virus.

        c. Phenotypic mixing, phenotypic masking, and pseudotyping constitute a third form of genetic interaction.

            (1) In phenotypic mixing, closely related viruses (e.g., different serotypes of poliovirus or polio- and coxsackieviruses) coinfect a cell and produce progeny that are a mixture of viral capsid proteins forming a mosaic.

            (2) In phenotypic masking, closely related viruses coinfect a cell and the genome of one virus is encapsulated in the coat of the other.

    (3) In pseudotyping, a viral core of one type of enveloped virus becomes engulfed in the envelope of another type. This situation is sometimes called phenotypic masking.

2. Interfering interactions

  a. In **heterologous interference**, infection by one virus entirely precludes the replication of a second virus within the same cell.

    (1) Inhibition of adsorption between viruses by blocking or destroying receptors appear to be one mechanism for heterologous interference.

    (2) Prevention of mRNA translation of any heterologous mRNA in the infected cells appears to be a second mechanism.

  b. **Homologous interference**, mediated by **defective interfering (DI) particles**, is a common occurrence in many viruses, especially in those that have been repeatedly passaged in vitro at high multiplicities of infection.

    (1) DI particles contain the normal viral proteins but only a subgenomic fragment of the viral nucleic acid.

    (2) DI particles are helper dependent for replication and need the homologous virus as a helper.

    (3) DI particles contain a limited amount of genetic information and frequently express some viral genes, but their main activity is specific interference with the homologous virus.

    (4) The mechanism of interference is unknown, but because DI nucleic acids are always internal deletions of the parental strand but retain the polymerase binding sites on the ends of the molecules, it seems likely that the smaller molecules replicate more rapidly and compete with the full-sized genomes for resources such as polymerase molecules.

STUDY QUESTIONS

Directions: Each question below contains five suggested answers. Choose the **one best** response to each question.

1. All of the following stages of viral replication are temperature-dependent processes EXCEPT

(A) adsorption
(B) penetration
(C) uncoating
(D) assembly
(E) release

2. Which of the following factors can overcome the adsorption barrier of poliovirus for avian cells?

(A) Trypsination of the cells
(B) Use of infectious RNA
(C) Treatment of the cells with neuraminidase
(D) Treatment of the virus with neuraminidase
(E) None of the above

3. What is the proper sequence of events in the replication of a plus-strand RNA virus?

(A) Uncoating, transcription, protein synthesis, viral RNA synthesis
(B) Uncoating, protein synthesis, copy RNA synthesis, viral RNA synthesis
(C) Transcription, protein synthesis, uncoating, viral RNA synthesis
(D) Protein synthesis, uncoating, copy RNA synthesis, viral RNA synthesis
(E) None of the above

4. What is the proper sequence of events in the replication of herpesviruses?

(A) Uncoating, transcription, protein synthesis, viral DNA synthesis
(B) Uncoating, protein synthesis, transcription, viral DNA synthesis
(C) Transcription, protein synthesis, uncoating, viral DNA synthesis
(D) Protein synthesis, uncoating, transcription, viral DNA synthesis
(E) None of the above

5. What is the major function of the pre-early gene products in poxvirus replication?

(A) A new polymerase activity
(B) Uncoating the viral core
(C) Antilysosomal activity
(D) Breakdown of host mRNA
(E) None of the above

6. Which of the following intermediates of the retroviral replication cycle serves as the template for viral mRNA?

(A) The input viral RNA
(B) The initial DNA copy
(C) The RNA-DNA hybrid molecule
(D) The double-stranded DNA copy
(E) The integrated DNA provirus

7. A HeLa cell is infected with two different temperature-sensitive mutants of poliovirus, and the cells are incubated at the nonpermissive temperature. The result is a normal virus yield of each type. This is an example of

(A) phenotypic mixing
(B) complementation
(C) recombination
(D) pseudotyping
(E) site-specific mutagenesis

8. Phenotypic mixing occurs between

(A) closely related naked capsid viruses
(B) naked capsid and enveloped viruses
(C) closely related DNA viruses
(D) all viruses that infect the same cell
(E) only DNA and RNA viruses

9. The defective interfering (DI) particles of vesicular stomatitis virus have all of the following characteristics EXCEPT

(A) the presence of all the normal proteins found in a nondefective virus
(B) the ability to replicate by themselves in monkey kidney cells
(C) a shortened RNA genome
(D) the ability to alter the course of infection by normal virus particles
(E) a sedimentation constant different from that of the wild-type virus

Directions: The group of questions below consists of lettered choices followed by several numbered items. For each numbered item select the **one** lettered choice with which it is **most** closely associated. Each lettered choice may be use once, more than once, or not at all.

Questions 10–13

A major feature of the replication strategy of RNA viruses is the mechanism that deals with the monocistronic nature of mammalian mRNA. For each of the viruses listed below select the appropriate principal replication mechanism.

(A) Post-translational cleavage of polyproteins
(B) Fragmented message produced from a non-segmented genome
(C) Segmented message produced from a segmented genome
(D) Spliced message produced from a provirus copy
(E) None of the above

10. Vesicular stomatitis virus (a rhabdovirus)

11. Influenza virus (an orthomyxovirus)

12. Reovirus

13. Coxsackievirus (a picornavirus)

ANSWERS AND EXPLANATIONS

1. The answer is A. [*I B 1 a*] Adsorption of viruses to the surface of host cells involves the intramolecular recognition of a surface structure with a receptor, which leads to a lower energy state; the process does not require energy. All subsequent stages of viral replication, beginning with the penetration of the cell membrane, require some type of enzymatic or other energy-consuming process and therefore are highly temperature dependent.

2. The answer is B. [*I B 1 b*] Infectious RNA extracted from poliovirus can infect a wide variety of cells; however, only a single round of replication occurs because the result of the infection is intact poliovirus, which regains its original host range limitation. Treatment of the cells or virus with proteases or neuraminidase has no effect on susceptibility to poliovirus. In some cases, such treatment removes the receptor for some viruses, thereby reducing cellular susceptibility. For example, neuraminidase treatment of cells removes the receptor for influenza virus.

3. The answer is B. [*II A 1*] The replication cycles of different viruses vary according to the nature of the viral nucleic acid. Plus-strand RNA is the same as the viral messenger RNA (mRNA), and following adsorption and penetration the viral RNA is used as a messenger as soon as it is uncoated. No other biochemical processes are necessary. Once the new viral proteins are produced, viral RNA synthesis is initiated, resulting in additional rounds of protein synthesis, viral capsid formation, and maturation.

4. The answer is A. [*II C 2*] The replication cycles of different viruses vary according to the nature of the viral nucleic acid. Following adsorption and initial penetration, the herpesvirus nucleocapsid migrates to the nucleus, where final uncoating occurs. The next step in replication is transcription by the host RNA polymerase followed by protein synthesis and viral DNA synthesis. Following these events, there is an additional round of (late) transcription and protein synthesis prior to maturation.

5. The answer is B. [*II C 3 a*] Pre-early synthesis of poxvirus begins immediately following the loss of the outer viral envelope, and it is responsible for the production of enzymes required for the breakdown of the viral core. Once the pre-early gene products take effect, a wide variety of specific and nonspecific metabolic changes occur within infected cells, including the appearance of new polymerases and mRNA breakdown.

6. The answer is E. [*II D 6*] During retroviral replication, the plus-strand viral RNA goes through a series of replication and copying steps, resulting in the production of a DNA copy that is inserted into the host cell genome. Retroviral RNA is a plus-strand RNA and, therefore, cannot be used as the template for further synthesis. Each of the intermediates of the retroviral replication cycle is present in very low concentrations and for only a short period of time. The integrated DNA provirus acts as a host gene for transcription and serves as a stable source of different transcripts throughout the life of the cell.

7. The answer is B. [*III B 1 b*] Complementation is the sharing of viral gene products in such a way that the defects of one viral temperature-sensitive mutant are offset by the functional gene product from the homologous gene encoded by a coinfecting virus. Recombination, phenotypic mixing, pseudotyping, and site-specific mutagenesis all are processes that occur during replication; their effects are only visible on the following generation of virus particles. Pseudotypic and phenotypic mixing are not heritable changes and appear for only one generation, whereas recombination and mutagenesis lead to stable heritable changes.

8. The answer is A. [*III B 1 c (1)*] Phenotypic mixing is the result of the mixing of the capsid proteins of very closely related viruses (e.g., polio- and coxsackieviruses). The interchange of the proteins has only minimal structural effects. Phenotypic mixing requires the exchange of essentially identical structural subunits (capsomers) and is restricted to naked capsid viruses, always viruses that are very closely related.

9. The answer is B. [*III B 2 b*] The defective interfering (DI) particles of the vesicular stomatitis virus have all the proteins found in wild-type virus but have a truncated RNA, thereby leading to the formation of shortened virus particles. The DI particles are unable to replicate without the help of the wild-type particle, with which they interfere during growth.

10–13. The answers are: 10-B, 11-C, 12-C, 13-A. [*II A 1 a (2), B 1 b, c (1), 2*] The rhabdoviruses have an unsegmented single-stranded genome with a negative polarity. As a result, the mRNA is synthesized in fragments from the full-sized template.

Influenza virus is the prototypic orthomyxovirus and contains a segmented minus-strand RNA. Following infection, the virion-associated transcriptase makes plus-strand copies from each of the segments, and these copies are used for protein synthesis as well as for templates for RNA synthesis.

The reoviruses resemble the orthomyxoviruses in the regulation of their early transcription. These viruses have a segmented double-stranded genome, which is transcribed by a virion-associated transcriptase.

Coxsackievirus is a picornavirus and shares its replication mechanism with poliovirus, which produces a single large polyprotein from the plus-strand viral RNA. This polyprotein is cleaved into functional viral proteins through a combination of cellular and viral proteases.

20
Interferon and Antiviral Agents
David T. Kingsbury

I. INTERFERONS are naturally occurring **small proteins with antiviral properties**. They can be produced by almost any type of cell in response to infection by a virus or to exposure to several other substances (inducers).

 A. Three types of interferon—α, β, and γ—are produced by each animal species. What follows and the data in Table 20-1 are true for human interferons.

   1. Interferon α is produced by leukocytes.

   2. Interferon β is produced by fibroblasts.

   3. Immune interferon, or **interferon** γ, appears to be the product of stimulated immune lymphocytes.

 B. Interferons are host specific; each species produces unique interferons that, while they may closely resemble those of another species, are not effective antiviral agents.

 C. Mechanism of action. The principal role of interferons in resistance is to induce an antiviral state that is usually the result of inhibition of translation of viral messenger RNA (mRNA).

   1. Interferons induce an antiviral state within the cell several hours after they are administered or induced. They do not interact directly with the virus and are not themselves antiviral.

   2. Interferons attach to specific receptors on the cell surface; they do not enter the cell. Cells from patients with Down syndrome have more than the usual number of interferon receptors on their surface and show a proportionately greater responsiveness to interferons.

   3. New cellular proteins mediate the antiviral state.
      a. A protein kinase phosphorylates elongation factor 2 in the presence of double-stranded RNA.

Table 20-1. Properties of Human Interferons

| Property | Type of Interferon | | |
	Alpha	Beta	Gamma
Cellular origin	Pheripheral leukocytes	Fibroblasts	Lymphocytes
Inducing agent	Virus infection; dsRNA	Virus infection; dsRNA	Specific immune stimulation; mitogens
Number of genes	14–20	1	1
Glycoprotein	No	Yes	Yes
Molecular weight	17,000	17,000	17,000
Number of amino acids in active protein*	143	145	146
Stable at pH2	Yes	Yes	No

*After removal of the signal peptide.

b. An unusual enzyme that is activated by double-stranded RNA synthesizes an adenine trinucleotide with a 2'-5'-phosphodiester linkage called two-five A (pppA2'p5'A2'p5'A), which activates an endonuclease that cleaves mRNA.

4. In mixtures of interferon-sensitive and interferon-resistant cells, the antiviral state of the sensitive cells spreads to the resistant cells.

5. Interferons also may affect several other viral replication steps.
   a. Uncoating of infectious virus particles is blocked in some cases, and the assembly of mature viral particles in others.
   b. The stability and methylation of viral RNA are altered in some infections.

D. Method of production. Cells in cultures and in living animals and humans produce interferons in response to virus infection or to treatment with chemically defined compounds.

1. Either infectious or inactivated virions may stimulate cells to produce interferons.

2. Interferons are produced very late in the infection cycle, usually about the time that viral maturation begins.

3. The best interferon-inducing viruses are those that replicate slowly and do not block host protein synthesis. Because interferons are produced from cellular genes, viruses that block cellular mRNA or protein synthesis are poor interferon producers.

4. The best synthetic inducers are purified, double-stranded RNA and synthetic polyribonucleotides with a high molecular weight, which are resistant to enzymatic degradation. Particularly effective is **poly (I:C)**, which is a polymer consisting of one chain of polyriboinosinic acid and one chain of polyribocytidylic acid.
   a. Poly (I:C) has shown some effectiveness in cases of localized virus infection; systemic administration is highly limited because of the compound's toxicity and rapid destruction by blood and tissue enzymes.
   b. Poly (I:C)-treated cells begin to produce interferon after 2 hours of exposure.

5. Double-stranded replicative intermediates of RNA viruses and complementary overlapping transcripts of DNA viruses are probably responsible for interferon induction in the course of active infection.

E. Effectiveness. Although interferon is beneficial in the treatment of many viruses, its value is still largely untested. Several **barriers to its usefulness** are known.

1. For maximum effect, it must be administered before virus infection occurs so that an antiviral state exists when infection is initiated.

2. At high dosages interferon leads to several side effects such as fever and myalgia.

3. As a prophylactic for influenza and the common cold, interferon has shown limited clinical success.

II. CHEMICAL INHIBITORS OF VIRUS REPLICATION

A. Amantadine (1-adamantanamine hydrochloride) is a symmetrical amine of peculiar structure that has antiviral activity against **influenza A** viruses and, to a limited extent, **rubella** virus.

1. The drug has only prophylactic value and to be effective must be given prior to or immediately following influenza exposure.

2. This drug has no serious side effects unless it is administered in very high dosages. (A methyl derivative, **rimantadine**, is equally effective and is even less toxic.)

B. Methisazone (N-methylisatin-β-thiosemicarbazone) is representative of a group of compounds called **thiosemicarbazones**.

1. Methisazone and related derivatives are active against the **poxviruses** and, to a lesser extent, against **other DNA viruses**.

2. These drugs are most effective as prophylactic agents and are not effective in developed viral disease.

C. **Purine and pyrimidine analogs** make up a series of drugs with potent antiviral activity because they interfere with viral DNA synthesis.

1. **Adenine arabinoside** (ara-A; vidarabine) is an analog used to treat human infections caused by herpes simplex, herpes zoster, and cytomegaloviruses. Ara-A inhibits viral DNA polymerase.

2. **Cytosine arabinoside** (ara-C; cytarabine) is a compound similar to ara-A, but it shows a much higher toxicity and less enzyme selectivity.

3. **Acycloguanosine** (acyclovir) is a guanosine analog with high selectivity for virus-infected cells. It is effective in the treatment of primary herpes simplex virus (HSV) infections. This drug's selectivity is based on its preferential phosphorylation by herpes thymidine kinase rather than by a cellular enzyme.

4. **Ribavirin** (virazole) is a purine analog that shows antiviral activity against both RNA and DNA viruses by interfering with viral nucleic acid polymerases. It has been approved by the FDA for the treatment of severe respiratory syncytial virus in children even though it shows adverse side effects, including immunosuppression.

5. **Dideoxynucleosides** such as 3'-azidothymidine (AZT), dideoxycytosine, and dideoxyinosine are highly selective for retroviral reverse transcriptases. AZT is currently the only FDA-approved treatment for human immunodeficiency virus infection [acquired immune deficiency syndrome (AIDS) and AIDS-related disease].

6. **Halogenated derivatives of deoxyuridine** are more toxic and less selective than subsequently developed drugs (e.g., acycloguanosine).
   a. **Method of action.** These halogenated compounds are readily phosphorylated by viral thymidine kinase and incorporated into viral DNA, leading to the synthesis of altered viral proteins.
   b. The **toxicity** of these drugs prohibits their systemic administration; topical application is safe and effective.
   c. **5-Iodo-2'-deoxyuridine (IUDR)** was one of the first and most effective antiviral compounds synthesized. It is used to treat corneal lesions due to **HSV**.

7. **Trifluorothymidine** (viroptic), a thymidine analog related to IUDR, has been more effective than IUDR as a topical treatment for corneal lesions in **HSV** infection.

D. **Derivatives of the antibacterial drug rifamycin**, primarily **rifampin**, have been shown to have antiviral activity. They inhibit reverse transcriptase in the **retroviruses** and interfere with **poxvirus** maturation. These drugs have limited usefulness in systemic disease; their high **toxicity** prevents administration of antiviral dosages.

STUDY QUESTIONS

Directions: Each question below contains five suggested answers. Choose the **one best** response to each question.

1. Cells may produce interferon in response to either virus infection or chemical inducers. Which of the following factors is true of both classes of inducer?

(A) Induction occurs approximately 2 hours following exposure

(B) The best inducers block host RNA synthesis

(C) The best inducers are or resemble double-stranded RNA replication intermediates

(D) The presence of specific receptors on the cell surface is essential

(E) None of the above

2. All of the following statements regarding interferon treatment are true EXCEPT

(A) to be maximally effective, interferon must be given prior to virus infection

(B) because it is a natural product, interferon has no side effects, even when administered in high doses

(C) interferon has been successful in limited clinical trials against influenza

(D) the chemical inducer, poly (I:C), is highly toxic following systemic administration

(E) poly (I:C) is one of the best chemical inducers of interferon, but it is rapidly destroyed by blood and tissue enzymes

3. Several types of interferons have been identified and characterized. All of the following statements regarding interferons are correct EXCEPT

(A) interferon α is produced by leukocytes

(B) interferon γ is produced by stimulated lymphocytes

(C) interferon β is produced by fibroblasts

(D) different interferons are highly virus specific

(E) interferons act on infected cells, not directly on susceptible viruses

Directions: The group of questions below consists of lettered choices followed by several numbered items. For each numbered item select the **one** lettered choice with which it is **most** closely associated. Each lettered choice may be use once, more than once, or not at all.

Questions 4–7

For each mechanism of action listed below, select the antiviral drug with which it is most closely associated.

(A) Methisazone

(B) Adenine arabinoside (ara-A)

(C) Acycloguanosine

(D) Rifampin

(E) Amantadine

4. Blocks the replication of poxviruses

5. Inhibits influenza virus infection when given prior to or immediately following infection

6. Inhibits the reverse transcriptase of the retroviruses

7. Inhibits viral DNA replication through incorporation into the viral DNA

ANSWERS AND EXPLANATIONS

1. The answer is C. [*I D*] The most effective interferon inducers are double-stranded RNA replication intermediates and their apparent synthetic analogs, high molecular weight polyribonucleotides. These compounds also are somewhat resistant to host enzymes, which break down single-stranded RNA molecules; therefore, they persist longer under physiologic conditions. In DNA viruses, the active molecules appear to be partially double-stranded RNA transcripts, probably produced prior to splicing or other processing. Partially double-stranded transcripts are especially frequent in late virus replication. Synthetic inducers frequently act in about 2 hours; however, many viruses take considerably longer to induce interferon production. Viruses and inducers that block host RNA synthesis are poor inducers because they require new messenger RNA (mRNA) synthesis. Interferons bind to specific receptors on the cell surface; however, many inducers do not have such a requirement.

2. The answer is B. [*I D 4 a, E*] Extensive clinical studies have clearly demonstrated that interferon, even in its purest form, has a variety of physiologic activities on normal, virus-infected, and tumor cells. These activities have led to the development of unpleasant side effects in patients treated with high interferon doses, regardless of the route of inoculation. Side effects can be minimized through the more gradual administration of interferon (i.e., by administering the same dose over a 24-hour period).

3. The answer is D. [*I C 1*] Interferons are not directly antiviral, but they induce an antiviral state in treated cells. Therefore, interferons are not virus specific but are specific for the species of origin of the treated cells. Three interferons have been identified—α, β, and γ—which are produced by leukocytes, fibroblasts, and stimulated immune cells, respectively.

4–7. The answers are: 4-A, 5-E, 6-D, 7-C. [*II A, B, C, D*] Methisazone is a member of the thiosemicarbazone family; together with other family members, it shows significant activity against the poxviruses. These compounds are active against other DNA viruses, but to a far lesser extent. Unfortunately, they are only effective as prophylactic agents and do not act on established infections.

Amantadine is a drug of proven effectiveness against influenza virus infection if the drug is administered prior to or immediately following virus infection. The drug appears to block the adsorption of the virus to sensitive cells and, therefore, blocks infection at the initial stages. If infection develops, the drug is rendered useless because its action is limited to the adsorption stage.

Rifampin is a derivative of the antibacterial drug, rifamycin. This class of antibiotic drugs exerts its activities through the inhibition of RNA polymerase. In the case of rifampin, the activity is selectively directed against an unusual RNA-dependent DNA polymerase activity found only in retroviruses. Rifampin is of limited usefulness in human disease because of its toxicity; it lacks selectivity for virally infected cells and is cross-reactive with cellular RNA polymerase.

Acycloguanosine is a member of a class of purine and pyrimidine analogs that is incorporated into viral DNA and interferes with DNA replication. Acycloguanosine is especially selective because it is an unusually good substrate for the herpesvirus-induced thymidine kinase and a relatively poor substrate for normal cellular enzymes. Its differential activity leads to a favorable therapeutic index and lowered toxicity. Adenosine arabinoside (ara-A) also inhibits viral DNA synthesis but does so through an interaction with the viral DNA polymerase.

21
DNA Viruses
David T. Kingsbury

I. PAPOVAVIRUSES. The papovaviruses are DNA viruses that infect both humans and animals; however, human disease only infrequently is associated with this virus group.

A. Structure

1. Papovaviruses are small (about 45 nm in diameter), **naked capsid viruses with icosahedral symmetry**.

2. Papovavirus DNA is approximately 5 kilobases long and consists of a covalently **closed circular, double-stranded molecule** that generally is associated with cellular histones.

B. Human papovaviruses

1. **Human papilloma (wart) virus (HPV)** causes **skin warts (verrucae)**—the most common human infection associated with papovaviruses.
   a. **Clinical disease**
      (1) Warts typically are **benign epithelial tumors**. However, specific serologic types of HPV (i.e., HPV-6, HPV-11) are associated with **anogenital warts (condylomata acuminata)** and are found—along with other rare types (e.g., HPV-16, HPV-18)—in all cervical biopsies that show precancerous change. HPV-16 and HPV-18 likely cause human invasive **cervical carcinoma**.
      (2) The **incubation period** for the development of a wart is thought to be 1–6 months, with peak virus concentration at 6 months.
   b. **Epidemiology.** Warts are more common in children than in adults. HPV transmission is thought to be by direct contact or autoinoculation.
   c. **Diagnosis.** The human wart virus has not been propagated in culture. Wart tissue is the only source of mature virus particles.

2. **BK virus** originally was isolated from the urine of a patient receiving immunosuppressive drugs.
   a. **Clinical disease.** Infection appears to be limited to renal allograft patients on long-term immunosuppressive therapy; the condition is not serious.
   b. **Epidemiology.** BK virus appears limited to humans. Serologic surveys suggest that the virus is widely distributed in the population; 75% of children 6 years of age and older have antibodies to BK virus.
   c. **Diagnosis.** BK virus can be cultivated in African green monkey kidney cells and in primary human fetal kidney cells.

3. **JC virus** originally was isolated from the brain of a patient with progressive multifocal leukoencephalopathy. JC virus and BK virus share many biologic properties, including agglutination of human, guinea pig, and chicken erythrocytes as well as several intracellular antigens.
   a. **Clinical disease.** JC virus is highly oncogenic when inoculated intracerebrally into Syrian hamsters; however, in humans the virus is not etiologically associated with malignancies but with progressive multifocal leukoencephalopathy, a rare disease characterized by progressive neurologic defects terminating in death.
   b. **Diagnosis.** JC virus is difficult to isolate because it requires fetal glial cells for growth.

C. Animal papovaviruses. **Rodent polyomavirus** and **simian virus 40 (SV40)**, a latent virus found in the rhesus monkey kidney, are the most widely studied papovaviruses.

1. Both polyomavirus and SV40 are oncogenic in laboratory animals and, as such, have served as models for the study of DNA tumor viruses.

2. Viruses closely related to SV40 have been isolated from humans but are considered not to be associated with disease.

II. ADENOVIRUSES are DNA-containing viruses that induce latent infections of the tonsils, adenoids, and other lymphoid tissues of humans.

A. Structure

1. Adenoviruses are simple, **naked capsid viruses with icosahedral virions** that are 60–90 nm in diameter. There are 252 capsomers, including 240 **hexons** (which comprise the faces) and 12 **pentons** (which comprise the vertices). The pentons are complex, consisting of a **polygonal base** with an **attached fiber**.

2. Each virion contains a single, **linear, double-stranded DNA molecule**, which varies in size from 30 to 38 kilobase pairs, depending on the serotype. The linear DNA, in association with proteins, forms the dense **viral core**, which is a unique structure seen by means of electron microscopy.

B. Antigenicity. The many serotypes of adenoviruses are differentiated on the basis of their **immunologic reactivities**, which reside in the hexon and penton proteins.

1. The **hexons contain family-reactive determinants**, which cross-react with all but a few (chicken) adenoviruses. Hexons also contain a **type-specific antigenic determinant**, which is exposed when the hexons are assembled in the virions.

2. The **pentons contain** minor virion antigens and a **family-reactive soluble antigen**, which is found in infected cells. The purified **fibers contain** a major **type-specific antigen**.

C. Clinical disease

1. **Infections.** Most infections caused by adenoviruses are **acute and self-limited**; most children are infected with serotypes 1, 2, and 5 early in life without apparent symptoms.
   a. **Acute respiratory disease**, the most common adenovirus infection, is an influenza-like illness seen in military training camps. Serotypes 4, 7, and 21 are most commonly associated with this infection.
      (1) The highest incidence occurs in the late fall and winter, but outbreaks can occur year-round.
      (2) Although typically self-limited, acute respiratory disease may be complicated by pneumonia in some cases.
   b. **Pharyngoconjunctival fever** is primarily a disease of infants and children. Serotypes 3, 5, 7, and 21 usually are involved.
      (1) The disease occurs most commonly in the summer months, and the source of infection often is a contaminated swimming pool.
      (2) The conjunctivitis does not lead to corneal damage.
   c. **Epidemic keratoconjunctivitis.** Adenovirus types 8 and 19 formerly were the primary causes of this infection, but type 37 recently has emerged as the leading cause.
      (1) Epidemic keratoconjunctivitis may occur as a result of dust-induced conjunctival trauma or the use of improperly cleaned optical instruments.
      (2) Corneal erosions may occur and be severe enough to affect vision.
   d. **Pneumonia** in infants has been associated with infection by adenovirus types 3, 7, and 21.
   e. **Gastroenteritis** in children is associated with a noncultivatable adenovirus (candidate serotype 38) in 5%–15% of cases.

2. **Undifferentiated tumors** develop with inoculation of several serotypes of human adenoviruses into newborn hamsters, rats, and mice. However, adenoviruses of humans generally do not cause disease in laboratory animals.

D. Epidemiology

1. Adenoviruses are ubiquitous. Humans serve as the only known reservoir for the strains that infect humans.

2. Adenovirus infections are spread person-to-person through respiratory and ocular secretions.

E. Therapy and prevention. There is no specific treatment for adenovirus infection.

1. Effective prevention of acute respiratory disease has been achieved with a live-virus vaccine containing the predominant serotypes associated with the disease, but the vaccine has been discontinued due to the oncogenic potential of some adenoviruses.

2. Proper chlorination of pools may help to decrease the spread of pharyngoconjunctival fever.

III. HERPESVIRUSES.
The herpesviruses are a group of structurally similar, ubiquitous viruses that infect both humans and animals. They are the most common causes of human viral infections. Important human pathogens include herpes simplex virus types 1 and 2 (HSV 1 and HSV 2), varicella-zoster virus (VZV), cytomegalovirus (CMV), and Epstein-Barr virus (EBV).

A. Structure and replication

1. The herpesviruses are **enveloped, icosahedral viruses** that are 180–200 nm in diameter and contain a **linear, double-stranded DNA molecule**.
   a. The genome ranges in size from 154 kilobase pairs, for HSV, to 231 kilobase pairs, for CMV.
   b. In addition, the base composition of the viral DNAs varies from 46% guanine plus cytosine (G + C), for VZV, to 70% G + C, for HSV 2.

2. Characteristic of enveloped viruses, the herpesviruses are relatively **unstable at room temperature** and are rapidly inactivated by lipid solvents.

3. All human herpesviruses **replicate within the nucleus of infected cells**. The capsid gains its protein envelope as it passes through the inner nuclear membrane.

B. Herpes simplex virus

1. **Antigenicity.** The two major immunologic variants, **HSV 1** and **HSV 2**, can be differentiated by serologic typing, by DNA homology, and to some extent by clinical disease pattern.
   a. Specific antibodies can be assayed either by neutralization or complement fixation (CF).
   b. HSV 1 and HSV 2 share only 40%–46% of their DNA sequences.

2. **Epidemiology.** Humans are the only known reservoir for HSV 1 and HSV 2, although experimentally the viruses can infect a variety of animals and cell cultures.
   a. Because of the lability of the viruses, most primary infections require direct contact with lesion material or infectious secretions.
   b. Antibody studies have confirmed the clinical observation that HSV 1 (generally associated with oral lesions) is acquired early in life, and HSV 2 (generally associated with genital lesions) is acquired after the onset of sexual activity.

3. **Clinical disease**
   a. **Primary infection**
      (1) The **incubation period** for a primary herpes infection is from 2 to 20 days, depending upon the infected site and the infecting strain of virus.
      (2) The most common lesions occur on the lips and mouth (**primary gingivostomatitis** leading to **recurrent herpes labialis**) and on the genitals.
      (3) Less commonly, HSV infections involve the eyes, skin, or brain. Keratoconjunctivitis may be primary or recurrent; blindness may occur. HSV encephalitis may be the result of a recurrence but generally is associated with primary infection and may be severely damaging or fatal.
   b. **Recurrent lesions.** The most striking feature of HSV infection is its propensity for recurrence.
      (1) Recurrence appears to involve activation of a noninfectious form of the virus from neurons of cervical ganglia (in HSV 1 infection) or sacral ganglia (in HSV 2 infection).
      (2) Herpes lesions tend to recur at the site of the primary lesion in response to a variety of inducing agents (e.g., local trauma, sunlight, stress, menstruation).

4. **Immunity.** Circulating antibodies appear to contribute little to the control of HSV infection, since recurrences proceed in the presence of high antibody titers.
   a. Primarily, the spread of HSV is from cell to cell, which is very efficient in the presence of antibody.

    b. Some evidence suggests that cell-mediated immunity may play a significant role in the course of HSV infection.

  5. Diagnosis of HSV infection usually is done clinically. However, with the increased infection of immunosuppressed patients, laboratory diagnosis more often is attempted.

    a. The quickest and simplest test is the demonstration of characteristic multinuclear giant cells in scrapings from lesions.

    b. The propagation of virus in tissue culture or the direct demonstration of viral antigens in vesicular fluid or scrapings also is used.

  6. Prevention and therapy

    a. Control is not feasible due to the high frequency of inapparent infection and the absence of an effective vaccine. However, it is important to avoid contact with active herpetic lesions.

    b. The course of primary infection and encephalitis can be altered significantly with drugs that interfere with viral DNA synthesis, such as acyclovir and vidarabine. Therapy for recurrent HSV infection has been ineffective.

C. Cytomegalovirus

  1. Epidemiology. Although CMV infections are widespread, illness rarely occurs with this infection. Human CMV appears to be confined to humans and cultured human cells; no laboratory animals seem to be susceptible to infection.

    a. The incidence of CMV appears to be increased during the perinatal period and during early adulthood. In addition, patients with neoplastic disease or AIDS, transplant recipients, and steroid immunosuppressed patients often have local and disseminated CMV disease.

    b. Transmission is not well understood. The virus is believed to be spread by transplacental passage or infection at birth; during nursing, blood transfusion, or sexual intercourse; and, perhaps, even by person-to-person spread through respiratory secretions and urine.

  2. Clinical disease

    a. Perinatal infection

      (1) The most common CMV infections appear to be congenital and neonatal subclinical diseases acquired from the mother before or during birth.

      (2) On rare occasions, newborns develop a severe, often fatal illness associated with infection of the salivary glands, brain, kidneys, liver, and lungs. The disease is characterized by huge intranuclear inclusion bodies in infected cells, hence the name cytomegalovirus.

    b. Long-term latent infections caused by CMV may be activated later in life by pregnancy, multiple blood transfusions, or immunosuppression. CMV infection in immunocompromised patients can be severe and involve many organs, most commonly the lungs, liver, gastrointestinal tract, and eyes.

  3. Diagnosis of CMV infection is made most readily by viral isolation in human embryonic fibroblast tissue cultures. Using modern diagnostic tools such as immunofluorescence and DNA hybridization, viral replication can be detected in 24–36 hours. Direct immunologic tests are less satisfactory because of their lower sensitivity. Direct examination of urine sediment, however, can be useful.

  4. Prevention and therapy. At present, there are no well-developed regimens for treating and preventing CMV infections and disease. However, some drugs that inhibit viral DNA synthesis (e.g., ganciclovir) may be promising.

D. Epstein-Barr virus is the etiologic agent of infectious mononucleosis; it also is associated with the African form of Burkitt's lymphoma (see Chapter 23, section II C 1) and with nasopharyngeal carcinoma in men of certain ethnic groups in southern China.

  1. Epidemiology. EBV is ubiquitous and is limited to humans.

    a. Spread of EBV is from human to human via respiratory secretions, primarily through oral contact. Less commonly, EBV is transmitted by blood or venereally.

    b. Persons from developing countries or from lower socioeconomic classes are exposed to EBV at an early age and typically develop asymptomatic infections that go unnoticed. Among persons from developed countries or from middle to upper socioeconomic classes, transmission and primary infection usually are delayed to adolescence or young adulthood.

2. **Clinical disease.** The virus replicates in the oropharynx and in lymphatic tissues. The genome resides in a latent form in B cells; latent EBV infection is common in a large portion of the population.
   a. **Infectious mononucleosis** is an acute infectious disease affecting lymphoid tissue throughout the body. It is a disease of middle- and upper-class teenagers, with a peak incidence at 15–20 years of age.
      (1) The **incubation period** for infectious mononucleosis is 30–50 days in adults and as little as 10–40 days in young children.
      (2) Initial **symptoms** of the disease include enlarged and tender lymph nodes and abnormal lymphocytes in the blood. Fever, sore throat, and fatigue are common, and many patients have splenomegaly. Less frequently, infections such as hepatitis and meningitis occur.
   b. **Chronic, persistent, or reactivated EBV infection** may take many clinical forms and is less common than acute mononucleosis.
      (1) A "chronic mononucleosis" syndrome has been described, which is characterized by persistent fatigue with or without physical or laboratory findings.
      (2) Chronic active EBV infection in immunocompromised patients (e.g., AIDS and renal transplant patients) typically may present as a progressive, multiorgan lymphoproliferative disorder or a CNS lymphoma.

3. **Immunity**
   a. **Specific antibodies** to EBV are formed during acute infection and persist for years.
   b. **Nonspecific heterophile antibody** (which agglutinates sheep and horse red blood cells) also appears and is useful as an early diagnostic test.

4. **Diagnosis.** Since EBV cannot be conveniently propagated in culture, diagnosis relies on the identification of specific antibodies directed against the virus proteins.
   a. An **indirect immunofluorescence assay** is used to detect antibodies against both a capsid antigen and a noncapsid early antigen.
   b. A **CF antigen**, which is present in homogenates of continuous cell lines, also is useful for detecting a specific immune response.

5. **Therapy.** Since infectious mononucleosis generally is mild and self-limited, therapy usually is symptomatic; no treatment modalities directed specifically at EBV have been tried.

E. **Varicella-zoster virus** is a typical herpesvirus; it is the cause of both varicella (chickenpox) and herpes zoster (shingles).

1. **Epidemiology**
   a. **Varicella** is a highly contagious primary infection in a host without VZV immunity. The disease usually occurs in an epidemic form in children less than 10 years old; adults usually have antibody to VZV.
      (1) Transmission is by respiratory droplets or contact with lesions.
      (2) The disease is most common during cold weather seasons.
   b. **Zoster** occurs due to reactivation of VZV infection and is primarily a disease of older adults and immunocompromised persons; it is rare in children.
      (1) The incidence of zoster rises considerably with advancing age (to almost 20% in the ninth decade) and with increasing degree of immunocompromise.
      (2) The disease occurs with equal frequency year-round.

2. **Clinical disease**
   a. **Varicella** usually is a mild, self-limited illness in children. In uncommon cases of adult infection, the disease generally is more severe and may have a mortality rate as high as 20%.
      (1) The virus enters the respiratory tract, where it begins to replicate. Eventually, the virus invades local lymph nodes and causes a primary viremia.
      (2) After a 14- to 16-day incubation period, fever develops followed by a papular rash of the skin and mucous membranes.
      (3) The papules rapidly become vesicular and begin to itch but remain painless (in contrast to the rash in zoster).
   b. **Zoster** occurs primarily as a reactivation of VZV infection in adults with circulating antibodies.

       (1) The disease develops from an inflammatory stimulation of a sensory ganglia of spinal or cranial nerves. The virus appears to remain latent in ganglionic nerve cells and, following activation, travels back along the nerve fiber to the skin.

       (2) Zoster presents as a unilateral, painful vesicular rash along the affected sensory nerve, which may be accompanied by fever and malaise. The rash may last for 2–4 weeks, with pain persisting for weeks or months.

       (3) Clinically, episodes of zoster may be triggered by trauma, drugs, neoplastic disease, or immunosuppression. Immunosuppressed persons are at risk for disseminated disease, which may be fatal.

  3. Diagnosis of VZV infection rarely requires laboratory intervention. However, when needed, the principal tests involve detecting a change in antibody titer between paired sera.

  4. Therapy
    a. Varicella typically is self-limited and requires no specific treatment. Symptomatic measures (for itching) often are used.
    b. Antiviral drugs (e.g., acyclovir, vidarabine), which interfere with herpesvirus DNA replication, have shown promise in the treatment of disseminated zoster in immunocompromised patients.

  5. Prevention. Passive immunization with varicella-zoster immune globulin (VZIG) may be achieved and is indicated for persons at high risk for severe infection. There is some fear that immunization of children may lead to a diminished immunity in adulthood and an increased risk for serious infections.

IV. POXVIRUSES are the largest and most complex human viruses, which cause a variety of human diseases with characteristic vesicular lesions. Variola (smallpox) virus and vaccinia virus are most representative of the group.

  A. Structure

    1. Poxviruses are enveloped DNA viruses that lack regular morphology but appear brick shaped; they range from 250 to 390 nm in length and from 200 to 260 nm in width.

    2. Poxviruses contain a linear, double-stranded DNA molecule that is 230–307 kilobase pairs in length; unlike other DNA viruses, they replicate entirely in the cytoplasm of infected cells.

  B. Antigenicity. All poxviruses are immunologically related by an **internal antigen**; however, they can be divided into discrete genera on the basis of more specific antigens, DNA homology, morphology, and natural host range.

    1. The genus *Orthopoxvirus* includes variola and vaccinia viruses as well as the viruses responsible for cowpox and monkeypox. Although variola virus has a host range limited to humans and, in some cases, to monkeys, vaccinia virus can replicate in chicken embryo cells and in the skin of rabbits, calves, and sheep.

    2. The genus *Parapoxvirus* includes the viruses responsible for paravaccinia (milker's nodes) and orf, which normally affect cattle and sheep, respectively, but occasionally are transmitted to humans.

  C. Variola (smallpox) virus

    1. Antigenic structure. There are two major strains of smallpox virus, which are almost indistinguishable and closely resemble vaccinia virus.
      a. The antigens of the poxviruses may be measured easily by hemagglutination inhibition and by neutralization as evidenced by reduction of the cytopathic effects in tissue culture or by reduction of pock formation on the chorioallantoic membrane of chicken embryos.
      b. Following immunization or infection, antibodies develop to each of the viral antigens; the time of appearance and duration of each group of antibodies depend upon the nature of the specific viral antigen.

2. **Epidemiology.** In 1977, the World Health Organization declared global eradication of smallpox. Historically, the disease can be traced to ancient times. It affected all levels of society, typically in epidemic waves. Since humans were the only hosts and natural reservoir, and spread was by direct person-to-person contact, immunization of the world's population directly led to eradication.

3. **Clinical disease.** Although smallpox occurred in two forms—**variola major** (the more virulent form, with a mortality rate of about 25%) and **variola minor**, or **alastrium** (the less virulent form, with a mortality rate below 1%)—aside from severity, the two are alike. The viral particles, which replicate and mature entirely in the cytoplasm, appear to have a toxin-like property, which apparently plays a role in cell necrosis.

   a. The development of smallpox begins with replication of the virus in the upper respiratory tract and regional lymph nodes. A transient viremia follows, during which the virus spreads to internal organs, where extensive propagation occurs.

   b. A second viremia occurs, marking the end of the incubation period and onset of the toxemic phase, which includes prodromal rashes, fever, and a flu-like syndrome. The incubation period is about 12 days, and the prodrome lasts 2–4 days.

   c. The virus spreads to the skin and multiplies, leading to skin eruptions that appear following the prodromal period. The skin lesions eventually become vesicular and pustular.

4. **Vaccination** against smallpox is highly effective, as evidenced by the worldwide eradication of the disease. The procedure no longer is used routinely.

   a. Vaccination involves the intradermal administration of a live virus preparation with an antigenic composition identical to that of variola. Protective immunity develops 7–10 days after vaccination.

   b. Although relatively safe, vaccination occasionally causes generalized infection that can be fatal. The infection can take one of many forms (e.g., eczema vaccinatum), depending on the immune status of the person being vaccinated.

D. Vaccinia virus once was the virus used to vaccinate against smallpox, which it closely resembles. Vaccinia in humans almost always is a purposefully acquired infection to provide immunity to smallpox and, in normal persons, is mild and localized. However, complications can be severe, or even fatal, in immunocompromised persons. Its use no longer is recommended.

E. Other poxviruses that cause human infection

1. **Molluscum contagiosum virus** causes an unusual skin disease that primarily affects children and young adults. The causative agent is an **unclassified poxvirus**.

   a. The disease usually is spread among persons living in close contact and may be transmitted sexually.

   b. Molluscum contagiosum is characterized by formation of many small, shiny, painless papules on the skin and mucous membranes. The disease is benign, self-limited, and preventable with good hygienic practices.

2. **Cowpox** is a self-limited occupational disease of humans, which is acquired from the udders of infected cows. It usually appears on the hands and fingers.

3. Several very rare diseases of humans—monkeypox, milker's nodes, contagious pustular dermatitis, and tanapox—are caused by members of the poxvirus group.

STUDY QUESTIONS

Directions: Each question below contains five suggested answers. Choose the **one best** response to each question.

1. Which of the infections listed below is most likely to be associated with adenovirus?

(A) Gastroenteritis
(B) Pharyngoconjunctivitis
(C) Encephalitis
(D) Meningitis
(E) Myocarditis

2. Which of the following statements is true of most primary infections with herpes simplex virus type 1 (HSV 1)?

(A) They occur in preschool- and elementary school-aged children
(B) They have dermatome distribution
(C) They are associated with keratoconjunctivitis
(D) They can be prevented with a live attenuated virus vaccine
(E) Recurrence involves activation of the virus from neurons of sacral ganglia

3. Which of the infections listed below is a complication arising from vaccination against smallpox?

(A) Eczema vaccinatum
(B) Subacute sclerosing panencephalitis
(C) Giant cell pneumonia
(D) Orchitis
(E) None of the above

4. Varicella-zoster virus (VZV) is best described as being

(A) a member of the poxvirus group
(B) a cause of parotitis
(C) latent for years following primary infection
(D) associated with Koplik's spots
(E) treatable with methisazone

5. All of the following viral infections likely occur due to activation of latent infection in an immunosuppressed patient EXCEPT

(A) shingles (zoster)
(B) CMV pneumonia
(C) adenovirus keratoconjunctivitis
(D) BK virus infection of the kidney
(E) disseminated CMV infection

6. All of the following statements regarding acute respiratory disease caused by adenovirus are true EXCEPT

(A) it is an acute and self limited disease
(B) it is rare in healthy young adults; it is common in geriatric nursing homes
(C) serotypes 4, 7, and 21 commonly are associated with this infection
(D) this disease can be complicated by pneumonia
(E) there is no clinically acceptable vaccine for prevention of this disease

Directions: Each question below contains four suggested answers of which **one or more** is correct. Choose the answer

    A if **1, 2, and 3** are correct
    B if **1 and 3** are correct
    C if **2 and 4** are correct
    D if **4** is correct
    E if **1, 2, 3, and 4** are correct

7. Routine laboratory procedures used to diagnose viral infection include

(1) measurement of antiviral titers in paired sera taken at early and late stages of infection
(2) propagation of the suspected virus in cell culture or by animal inoculation
(3) examination of viral lesions for a direct demonstration of viral cytopathic changes, viral particles, viral antigen, or nucleic acid
(4) assessment of immune lymphocyte function, especially CD 4$^+$ T cells

8. True statements about clinical cytomegalovirus (CMV) disease include

(1) the most common CMV infections are mild, subclinical diseases
(2) latency is activated by immunosuppression, as in AIDS
(3) human CMV disease cannot be studied in laboratory animals
(4) treatment with RNA polymerase inhibitors is highly successful

9. Factors that eventually enabled public health workers to eradicate smallpox include

(1) its extremely limited geographic distribution in the Indian subcontinent
(2) the antigenic identity of the two major forms as well as their cross-reactivity with the vaccine strain
(3) the relative instability of the virus
(4) the absence of an animal reservoir, which limited all infections to humans

Directions: The group of questions below consists of lettered choices followed by several numbered items. For each numbered item select the **one** lettered choice with which it is **most** closely associated. Each lettered choice may be use once, more than once, or not at all.

Questions 10–13

For each disease or syndrome described below, select the most likely viral agent.

(A) JC virus
(B) Epstein-Barr virus
(C) Adenovirus
(D) Varicella-zoster virus
(E) Herpes simplex virus type 2

10. The primary infection almost always occurs in children, whereas the recurrent form is rare in children

11. The agent associated with a chronic brain infection but not with any malignant form of disease in humans, although it is highly oncogenic in hamsters

12. The most common clinical form is a disease of middle- and upper-class teenagers

13. Most infections of childhood are not clinically apparent but result in latent infection or development of antibodies

ANSWERS AND EXPLANATIONS

1. The answer is B. [*II C 1 a, b*] The most common adenovirus infection is acute respiratory disease, which usually is seen in military training camps. The second most common infection due to adenovirus—and the most common one in children—is pharyngoconjunctival fever. It is a disease that usually is seen in the summer months. Although adenovirus is swallowed in respiratory secretions and can infect the cells of the gastrointestinal tract, gastroenteritis only rarely is associated with such infections.

2. The answer is A. [*III B 2 b*] Herpes simplex virus type 1 (HSV 1) infection usually is acquired prior to the age of 5 years; HSV 2 generally is acquired later in life. Despite years of work on potential herpesvirus vaccines, currently there is no effective vaccination for either serotype. HSV infections usually are limited to the site of primary infection and do not show the dermatome distribution of varicella-zoster infection. Recurrence appears to involve activation of a noninfectious form of the virus from neurons of cervical ganglia (in HSV 1 infection) or sacral ganglia (in HSV 2 infection). Keratoconjunctivitis may be attributed to certain strains of adenovirus; it is not associated with herpesviruses.

3. The answer is A. [*IV C 4 b*] The vaccination for smallpox is a highly effective and very *safe* procedure. In the years just prior to smallpox eradication, however, the rate of vaccination reactions exceeded the incidence of smallpox. The most common side effect is an eczema resulting from vaccinia virus infection (eczema vaccinatum).

4. The answer is C. [*III E 2 b (1)*] Varicella-zoster virus (VZV) is the etiologic agent of chickenpox, although it is a member of the herpesvirus group. As with other members of this group, primary infection usually leads to lifelong latency with a high frequency of recurrence of some type of infection from the endogenous virus. The recurrent disease of VZV is shingles. Parotitis is associated with the childhood disease mumps, not chickenpox, and Koplik's spots are inflamed submucosal glands observed in measles. Methisazone is most useful for treating poxviruses but has shown no usefulness in VZV infection.

5. The answer is C. [*II C 1 c*] Almost all DNA viruses establish a long-term latent infection, and most people carry many such viruses. However, the acute epidemic adenoviral syndromes (e.g., keratoconjunctivitis) are the result of primary infection. Cytomegalovirus (CMV) pneumonia and the more fully disseminated disease, as well as BK virus infection, occur almost exclusively in immunosuppressed patients (e.g., transplant recipients) and are the result of activation of latent infection. Shingles (zoster) occurs in the absence of known immunosuppression but not commonly. The highest incidence of shingles is in older persons, cancer patients, and other immunosuppressed persons.

6. The answer is B. [*II C 1 a, E 1*] Acute respiratory disease principally is a disease of young adult males and is often seen in military recruit training centers. The disease generally is mild and self limited, although in some cases pneumonia may develop as a complication. Serologic types 4, 7, and 21 are most commonly associated with this disease. An effective live-virus vaccine was developed several years ago; however, its use was discontinued when it was recognized that some types of adenovirus are able to cause malignancies in hamsters. No acceptable replacement vaccine has been developed, and immunization is no longer practiced.

7. The answer is A (1, 2, 3). [*III B 5, C 3, D 4, E 3*] Definitive viral diagnostic testing is becoming increasingly important with the influx of new and more effective antiviral agents. To take maximum advantage of these agents, rapid diagnosis is essential. Therefore, many new diagnostic tests have appeared, which rely on sensitive detection of viral antigens or nucleic acids in materials taken directly from the patient. In many cases these new techniques are applied to early detection of viral multiplication in inoculated tissue cultures. Often the measurement of antibody rise in acute and convalescent sera is the only procedure available and this approach, while time-consuming, is still widely used. The assessment of immune cell function is not broadly applicable in viral diagnostics and has no place in routine laboratory diagnosis. This procedure is valuable in the monitoring of the clinical status of patients infected with human immunodeficiency virus (HIV).

8. The answer is A (1, 2, 3). [*III C 1, 2 a (1), b, 4*] Cytomegalovirus (CMV) is acquired predominantly through congenital (in utero) exposure or exposure at birth due to an active infection in the mother. In most cases, congenital and neonatal CMV infection is subclinical. A large proportion of the population has long-term latent CMV carriage as a result of inapparent infection. Latent infections can develop into serious clinical disease following immunosuppression due to such factors as AIDS or steroid immunosuppressive treatment. Human CMV cannot be readily grown in any laboratory animal, although it can be cultured on human fibroblasts. RNA polymerase inhibitors have proven to be useless in treating CMV infection; however, new compounds specifically directed toward viral DNA synthetic enzymes are showing some promise.

9. The answer is C (2, 4). [*IV C*] The eventual eradication of smallpox, a disease that at one time or another has affected all parts of the world, was a result of a massive immunization effort in all of the endemic areas. This massive undertaking was possible because a highly effective and easily administered vaccine was available and because the only reservoir for the virus was humans, all of whom developed symptomatic disease when infected. Several factors were important in the global eradication campaign. The vaccine strain and all forms of the disease strain of virus shared antigens and, therefore, induced cross-reactivity. Another significant factor was the unusual stability of the virus, which allowed room temperature transport of the virus into remote areas if needed.

10–13. The answers are: 10-D, 11-A, 12-B, 13-C. [*I B 3 a; II C 1; III D 2 a, E 1*] Varicella-zoster virus (VZV) is the etiologic agent of varicella (chickenpox), which almost always is a childhood disease. The recurrent form of VZV infection is zoster (shingles), which almost always is a disease of adults rather than children. Infection presents particularly difficult problems in older persons due to a decrease in immune function with age.

JC virus is one of a group of papovaviruses isolated from humans, most of whom do not have overt clinical disease. JC virus, however, is usually associated with the neurologic disease known as progressive multifocal leukoencephalopathy. JC virus never is associated with malignant disease in humans despite its high oncogenic potential in hamsters.

In lower socioeconomic classes Epstein-Barr virus (EBV) infection occurs at an early age and generally results in asymptomatic infection with antibody production and probably lifelong latent infection. Children of higher socioeconomic classes (and better sanitary conditions) are exposed to the virus much later in life, usually in the form of a mild epidemic. Infection in the teen years can lead to the development of mononucleosis.

Most adenovirus infections of childhood are asymptomatic or so benign that they go entirely unnoticed. The result of these infections, however, is latent infection of adenoid and tonsillar tissues. Adenoviruses originally were isolated from "normal" tonsils.

22
RNA Viruses
David T. Kingsbury

I. ORTHOMYXOVIRIDAE. The orthomyxoviruses are the influenza viruses that cause occasional worldwide epidemics of acute respiratory disease. They are subclassified as immunologic types A, B, and C on the basis of the antigenic properties of their major **nucleocapsid protein (NP)** and viral envelope **matrix protein (M protein)**. The antigens of each type are unique and do not cross-react with those of the other types. Influenza A causes far more serious illness than the other types; influenza C has been associated only with mild, nonepidemic illness.

A. Structure. The orthomyxoviruses are **enveloped, single-stranded, segmented RNA viruses with negative polarity genomes.**

   1. The **virions** usually are 80–120 nm in diameter.

   2. The **genome** of influenza A and B viruses is 13,588 nucleotides in length and consists of eight individual segments of RNA. Each segment encodes a different viral protein.
      a. One segment is the precursor of two **nonstructural proteins (NS$_1$ and NS$_2$)**, which exist only in infected cells.
      b. The other segments encode **virion proteins**.
         (1) The **M protein**, which is located on the inner surface of the viral membrane, is the most abundant protein in the viral particle.
         (2) The **NP** is a single phosphoprotein species associated with the RNA.
         (3) **Hemagglutinin** and **neuraminidase** are surface spike glycoproteins.
         (4) Three large **internal peptides (P$_1$, P$_2$, and P$_3$)** are associated with virus transcription and replication and exist in very small numbers in the virion.

B. Replication occurs primarily in the **cytoplasm** of infected cells. The nucleus is obligatorily involved, however, and appears to be the site of viral RNA synthesis. The host apparently must constantly supply fresh RNA transcripts, the 5' ends of which are used as a supply of capped 5' termini for viral messenger RNA.

C. Antigenic variation. In addition to the NP and M protein variations that produce influenza A, B, and C, the influenza viruses are characterized by a high frequency of immmunologic variations within the subtypes. These **variations are caused by the surface spike glycoproteins, hemagglutinin, and neuraminidase**, and they lead to the appearance of new serologic types that are subject to epidemic spread. Two types of antigenic variation have been demonstrated.

   1. **Antigenic drift** results from mutations that are selected because they are less susceptible to the antibodies most common in the population at the time.

   2. **Antigenic shift** constitutes the appearance of a **new antigenic type** unrelated or only distantly related to earlier types. It occurs infrequently and has been identified only in **influenza A.**
      a. Antigenic shift probably results from genetic recombination between a human and an animal strain of influenza.
      b. Influenza A viruses have undergone **four major antigenic shifts since 1933.**
         (1) The antigenic types are based on serologic reactivity as measured by neuraminidase inhibition (NI) or hemagglutinin inhibition (HI).
         (2) The first strain isolated from a human (in 1933) has been designated H$_0$N$_1$. In 1947, a new hemagglutinin type, H$_1$N$_1$, emerged. Variants designated H$_2$N$_2$ and H$_3$N$_2$ have emerged in the past 30 years.

D. Clinical infections. Symptoms typically appear suddenly 1–2 days after exposure.

1. **Major symptoms** include a rapid rise in body temperature to 102° F with accompanying myalgias and, on occasion, sore throat, headache, cough, and nasal congestion.

2. **Other symptoms** include marked lassitude, moderate anorexia, and a variety of other minor complaints.

3. **Pneumonia is the most common serious complication** of influenza infection. In most cases it is caused by secondary bacterial infection of an already weakened person.

4. In its **most serious form**, influenza may cause a disease with explosive onset accompanied by uncontrollable coughing and shortness of breath with subsequent hypoxia. There is a marked lung infiltrate involving one or several lobes, and the patient eventually dies from extreme hypoxia.

E. Epidemiology and immunity. Both influenza A and influenza B may cause epidemic illness when they are established in a nonimmune or only partially immune population.

1. **Transmission occurs primarily by aerosolization.**
   a. Virus-laden droplets are spread from person to person by sneezing and coughing.
   b. Illness spreads rapidly in confined populations in, for example, nursing homes, classrooms, and ships where crowded conditions prevail.

2. **Immunity** to a new influenza variant depends on immunity to the previous variant circulating in the population and on the relatedness of the two variants.
   a. **Most major epidemics and pandemics are due to antigenic shifts** that produce new influenza subtypes antigenically and distantly related to types prevailing in the past.
   b. Antigenic drifts may cause any degree of illness from subclinical to severe. Individual response varies immensely.
      (1) Levels of immunoglobulin A (IgA) and serum IgG and cellular immunity are among the factors determining individual resistance.
      (2) Nonimmune factors such as genetics and nutrition probably also play a role.

F. Laboratory diagnosis. Influenza is difficult to diagnose because it produces viral respiratory syndromes similar to those produced by many other viruses. The virus must be isolated or the virus antigen demonstrated in nasal secretions or epithelial cells. To determine the antigenic identity of the influenza subtype, virus isolation is necessary.

G. Therapy and prevention. Immunization and chemical prevention are available and partially successful. Treatment offers only relief of symptoms and generally does not change the course of the illness.

1. **Amantadine** is useful both for relief of symptoms and for prevention. It enhances the effectiveness of immunization.

2. **Immunization** is about 70% effective. Difficulties stem from the mechanics of vaccine production and selection of the appropriate immunizing strain.
   a. Inactivated vaccine produced from purified, egg-grown virus is the preparation in common use. This vaccine is expensive to produce and causes allergic reactions in people with hypersensitivity to hens' eggs.
   b. Because influenza virus exhibits frequent antigenic variation, the proper antigenic mixture for immunizing a population in any given flu season cannot be determined until the onset of that particular disease cycle.

II. PARAMYXOVIRIDAE. The paramyxoviruses include four significant causes of human disease in infancy and childhood: parainfluenza, mumps, measles, and respiratory syncytial viruses. Parainfluenza and mumps viruses are antigenically related.

A. Structure. The paramyxoviruses are **enveloped, minus-strand RNA viruses with an unsegmented genome** of approximately 5–6 × 10^6 daltons.

1. The **virion** ranges in diameter from 100 to 800 nm and averages 125–250 nm.

2. The **genome** is RNA associated with a single polypeptide species, the NP.

3. The **envelope** may contain three proteins: the **HN protein**, which has both hemagglutinin and neuraminidase activity; the **F protein**, which has both hemolytic and cell-fusion activity; and the **M protein**, which forms the inner layer of the viral envelope. The HN and F proteins are glycoproteins; the M protein is nonglycosylated.

B. Replication occurs entirely in the host cell **cytoplasm**.

1. It is insensitive to both actinomycin D and amanitin, which affect host RNA synthesis.

2. Paramyxoviruses contain their own RNA-dependent RNA polymerase (transcriptase).

3. The native F protein requires a proteolytic cleavage to be completely mature; viruses that incorporate the immature form are noninfectious until treated with protease.

C. Parainfluenza viruses are important causes of human respiratory disease throughout the year but especially in fall and winter.

1. **Antigenicity. Four major serotypes** of parainfluenza are based on the HN, F, and NP protein antigens.
   a. **Types 1, 2, and 3** show substantial serologic relatedness as well as cross-reactivity with mumps virus. **Type 4** is not related to these types and has two subtypes.
   b. Most adults have circulating neutralizing antibody to all four serotypes; however, few have the critical IgA antibodies necessary to prevent primary infection. **Reinfection by the same type is common** despite the presence of circulating antibody to it.

2. **Clinical infections.** Parainfluenza viruses cause a variety of diseases, primarily in young children and infants.
   a. **Laryngotracheobronchitis (croup)** is commonly associated with serotypes 1 and 2, which tend to cause biennial epidemics, primarily in the fall.
   b. **Bronchiolitis and pneumonia** are associated with serotype 3, which primarily infects infants less than 1 year old.
   c. **The common cold in subclinical form** is the usual result of adult infection with any parainfluenza type.

3. **Epidemiology.** Parainfluenza viruses are spread through respiratory secretions. Closed populations including young children are particularly at risk. Type 3 is the most prevalent serotype.

4. **Prevention.** Neutralizing IgA is the only effective protection; induction of IgG antibodies does not offer significant protection.

D. Mumps virus causes a clinically unique disease that typically is a **parotitis** of acute onset involving one or both glands.

1. **Antigenicity.** Mumps virus is very much like the parainfluenza viruses. It contains a surface HN glycoprotein, a surface F glycoprotein, and an internal NP, which is also called the cytoplasmic-soluble, or S, antigen. Mumps virus constitutes a single antigenic type.

2. **Clinical infections**
   a. After the usual incubation period of 16–18 days, **bilateral or unilateral parotitis** usually appears. With or without parotitis, other organs may be involved. The salivary glands become infected via blood-borne virus (viremia), which appears 3–5 days before the onset of symptoms.
   b. The **viremic phase** leads to widespread dissemination of virus throughout the body.
      (1) Virus is shed in the urine for approximately 10 days after the onset of disease.
      (2) Occasional complications of mumps viremia include **orchitis**, which occurs in 20%–35% of postpubertal male patients.

3. **Epidemiology**
   a. Mumps virus is **transmitted in saliva and respiratory secretions**, and the route of infection is the respiratory tract.
   b. Humans are the only known natural hosts for mumps; however, the virus has been transmitted experimentally to monkeys and chicken embryos.

   4. **Laboratory diagnosis.** Diagnosis usually is made solely from clinical observation. If required, viral isolation may be done in chick embryos or cell cultures.

   5. **Prevention.** Mumps is prevented by immunization. The infectious attenuated virus leads to the development of antibody in 95% of antibody-free recipients.

E. **Measles virus** causes one of the most highly infectious diseases known. It is almost universally a disease of childhood, and early infection results in permanent immunity.

   1. **Antigenicity.** All measles strains examined to date belong to a single type.
      a. Antibodies are produced to the three major virion proteins—the hemagglutinin protein, the F protein, and the NP protein.
      b. Antibodies directed to the virion surface glycoproteins (i.e., the hemagglutinin and F proteins) are cytotoxic to measles virus-infected cells, which have viral antigen on their surface.

   2. **Clinical infections.** Measles is an acute disease with a characteristic fever and exanthematous rash.
      a. The virus multiplies in the respiratory epithelium and regional lymphoid tissue for 10–12 days.
      b. In a stage of **viremia**, the virus spreads hematogenously to lymphoid tissue and skin.
         (1) The viremia is accompanied by **prodromal symptoms**: conjunctivitis, coryza, headache, low-grade fever, sore throat, and Koplik spots.
         (2) Large **syncytial giant cells** are the most characteristic cytopathic change; these cells appear throughout the body.
      c. The measles **rash** follows the viremia. It appears to result from the interaction between virus-infected cells and either sensitized lymphocytes or antibody–complement complexes.
      d. **Circulating antibodies** to the measles virus appear approximately 10–14 days after infection, at about the same time that the measles rash appears. **Cell-mediated immunity** also appears approximately 10–14 days postinfection.
      e. **Serious complications** of measles virus infection can occur.
         (1) **Bronchopneumonia and otitis media** can occur with or without accompanying bacterial infections.
         (2) **Encephalomyelitis** occurs in approximately 1 in 2000 cases. It begins 5–7 days after the onset of the rash. The mortality rate associated with measles encephalomyelitis is approximately 10%.
         (3) **Giant cell pneumonia** is a rare consequence of measles virus infection; it occurs only in immunodeficient individuals.
         (4) **Subacute sclerosing panencephalitis** is a progressive, degenerative neurologic disease of children and adolescents with a characteristic pattern of findings, including mental and motor deterioration, myoclonic jerks, and electroencephalographic dysrhythmias. This rare disease usually results in death within a year.

   3. **Epidemiology**
      a. Measles virus is a readily transmissible disease in both humans and monkeys. It appears not to occur in other species, although it is serologically related to canine distemper and rinderpest, which do.
      b. It is **transmitted via respiratory secretions and urine** during the prodromal phase and when the rash appears.

   4. **Prevention**
      a. Measles has been impossible to control by simply isolating infected individuals.
      b. **Immunization** of children with a live attenuated virus vaccine **induces a form of protective immunity that may last as long as 10 years** and has significantly reduced the occurrence of measles in immunized populations.

F. **Respiratory syncytial virus (RSV)** is a major cause of lower respiratory tract disease in infants and young children.

   1. **Structure.** RSV appears to be **closely related to parainfluenza viruses and measles virus**; however, there are several important differences.
      a. RSV does not contain detectable hemagglutinin, and it lacks hemadsorptive, hemolytic, and neuraminidase activity, even though the virus has surface projections resembling those of the parainfluenza viruses.

b. RSV particles are very fragile and, in purified preparations, take several forms, including a thin, filamentous form.

2. Antigenicity
   a. RSV appears to have **three minor types**; however, a high degree of cross-reactivity among them suggests that no significant amount of antigenic drift has occurred. The antigenic differences are associated with a specific surface antigen.
   b. **Immunity** against reinfection is short-lived and depends entirely on the level of secretory IgA in nasal secretions, not on circulating IgG concentrations.

3. Clinical infections
   a. RSV causes **yearly worldwide epidemics** of respiratory tract infection in infants and young children. Adults are also infected, but their symptoms are mild or inapparent.
   b. **The major impact is during the first 6 months of life.** Approximately one-third of infants develop antibodies in the first year of life.
   c. The initial **site of virus multiplication** is the epithelium of the upper respiratory tract.
      (1) In older children and adults the virus does not spread beyond this site.
      (2) In infants less than 8 months old, the virus spreads into the lower tract (i.e., the bronchi, bronchioli, and lung parenchyma), partially owing to the absence of IgA antibodies in the respiratory tract.

4. Epidemiology. RSV spreads rapidly, especially in closed communities that include young children. It causes disease only in humans and chimpanzees.

5. Laboratory diagnosis
   a. RSV is difficult to isolate in tissue culture; however, it grows in some human and simian cell lines.
   b. Its predominant cytopathic property in culture is the formation of syncytia and giant cells.

6. Prevention. No effective method of controlling RSV infection has been identified, and all attempts to invoke either public health measures or preventive immunization have failed.

III. PICORNAVIRIDAE. Four members of the picornavirus family cause significant human disease: polioviruses, coxsackieviruses, echoviruses, and rhinoviruses. The first three of these are enteroviruses.

A. Structure. The picornaviruses are small, **naked capsid, plus-strand RNA viruses with an unsegmented genome** of approximately 2.5×10^6 daltons (7500 bases). They have icosahedral symmetry.

   1. The **virions** are approximately 22–30 nm in diameter and are resistant to extraction with lipid solvents.

   2. The **genome** is a polyadenylated, single RNA molecule with a single protein molecule, termed **VPg**, covalently attached to the 5' end. Carefully extracted naked picornavirus RNA remains infectious even after the VPg molecule is removed by protease digestion.

B. Replication of picornaviruses occurs entirely in the host cell **cytoplasm** and is resistant to inhibitors of cellular RNA synthesis.

   1. The initial biochemical event after uncoating is the initiation of viral protein synthesis from the plus-strand RNA.

   2. The viral proteins are translated as a single polyprotein, which is subsequently cleaved into individual virion proteins.

   3. The capsids are assembled and filled in the cytoplasm, and the virions are released upon lysis of the infected cell.

C. Polioviruses. Three distinct types of poliovirus (**1, 2, and 3**) can be differentiated by such techniques as neutralization. The three types have identical physical and biologic properties, and they have in common 36%–52% of their nucleotide sequences.

1. **Structure.** The individual viral particles consist of aggregates of four capsid proteins plus a single molecule of VPg attached to the viral RNA.

2. **Antigenicity.** Polioviruses are markedly stable. Although minor serologic variants have been observed, they are rare.

3. **Clinical infections.** The major sequence of events in poliovirus infection has been well understood for several years.
   a. Infection is initiated by ingestion of infectious virus, and primary replication occurs in oropharyngeal and intestinal mucosae.
   b. The virus drains into the cervical and mesenteric lymph nodes and then into the blood, leading to a transient viremia and finally to systemic spread.
   c. Replication continues in a variety of nonneural sites, leading to a persistent viremia, during which the virus spreads to the central nervous system (CNS).
   d. Serious paralytic disease is an unusual development resulting from infection of either the anterior horn cells of the spinal cord (spinal poliomyelitis) or the cells of the medulla or brain stem (bulbar poliomyelitis).
   e. The full course of the disease partly depends on host factors that may predispose to or discourage nervous system infection.

4. **Epidemiology**
   a. Polioviruses are **found worldwide and spread rapidly**, especially in densely populated areas with poor sanitation standards.
      (1) In countries with low hygienic standards, virtually everyone has antibody to poliovirus by the age of 5 years.
      (2) In countries with improved sanitation, children do not contact the virus at an early age and thus lack protective antibody.
   b. The **incidence and severity of paralytic disease increase dramatically with age**. Children older than 10–15 years are more likely than younger children to develop crippling disease.
   c. **Infection occurs primarily in the summer** and is spread in the feces. Infected individuals excrete large amounts of the virus in the feces for up to 5 weeks.
   d. **Transmission is primarily by person-to-person contact** through pharyngeal secretions, but the disease sometimes is spread by infected water sources.
   e. Humans are the only known natural hosts, but the viruses are adsorbed by and propagate in selected primate cells.

5. **Laboratory diagnosis.** Simian tissue cultures are widely used for growth of polioviruses. Human embryonic kidney cell cultures are also used.

6. **Prevention.** Control of poliovirus disease has been achieved through widespread **immunization** with either killed or live attenuated virus vaccines.
   a. In the United States, the live attenuated virus vaccine is widely used and is effective in reducing both the paralytic disease and the rate of alimentary disease. It also has several public health advantages, including herd immunity.
   b. Outside the United States, the killed vaccine is most widely used.

D. **Coxsackieviruses** are differentiated from other enteroviruses in the picornavirus group by their much **greater pathogenicity for suckling mice** than for adult mice.

1. **Classification**
   a. **Main groups.** Coxsackieviruses are divided into two groups on the basis of lesions induced in suckling mice.
      (1) **Coxsackievirus A** produces a diffuse myositis with inflammation and necrosis of voluntary muscle fibers.
      (2) **Coxsackievirus B** produces focal areas of brain degeneration, focal necrosis in skeletal muscle, and inflammatory damage in the pancreas, dorsal fat pads, and, on occasion, the myocardium.
   b. **Subgroups.** Each group has several serologically distinct subgroups. Group A has 23 subgroups, and group B has 6. Subgroups are identified by means of a type-specific antigen.
      (1) The serotypes have some cross-reactivity, but there is **no group-specific antigen**.
      (2) Type-specific antibodies are induced during infection. In humans they appear in the serum approximately 1 week after the onset of the disease and peak at 3 weeks postinfection.

2. **Clinical infections.** Most coxsackievirus infections of humans are mild and frequently asymptomatic. Rarely, serious infection results in severe disease.
   a. **Myocarditis of the newborn** has a high mortality rate; it is caused by group B coxsackieviruses, which also may cause a mild interstitial focal myocarditis and, infrequently, a valvulitis in infants and children.
   b. **Herpangina** is an acute disease associated with certain group A coxsackieviruses. It is characterized by fever of sudden onset, headache, sore throat, dysphagia, anorexia, and, occasionally, a stiff neck.

3. **Epidemiology**
   a. All serotypes of coxsackievirus are apparently **found worldwide**. They are highly infectious within families and closed communities.
   b. The greatest epidemic spread occurs in the **summer and fall**.
   c. **Transmission appears to be by the fecal-oral route and from** nasal and pharyngeal **secretions**. They enter through the mouth and nose, multiply locally, and spread viremically somewhat as polioviruses do.

4. **Laboratory diagnosis.** Suckling mice must be inoculated.

5. **Prevention.** No effective control or immunization procedures against coxsackieviruses currently are available.

E. **Echoviruses** are identified by their ability to cause cytopathic changes in tissue culture without causing any detectable lesions in suckling mice.

1. **Classification.** At least **32 serotypes** of echoviruses are distinguished on the basis of a specific antigen in the viral capsid and neutralization by type-specific antibodies.
   a. Twelve serotypes are capable of hemagglutination, and several show a subsequent spontaneous elution.
   b. There is no group-specific antigen; however, several type-specific antigens show some cross-reactivity.

2. **Clinical infections.** Originally thought to be orphan viruses, echoviruses have subsequently been clearly associated with a variety of human illnesses.
   a. The most common human echoviral conditions include aseptic meningitis, rash, fever, and enteritis.
   b. Less common conditions include acute respiratory infection, myocarditis, pleurodyna, paralysis, and encephalitis.

3. **Pathogenesis.** Echoviruses usually are acquired by ingestion but on occasion by respiration. In both cases, the virus usually remains confined to the site of primary infection.
   a. Occasionally the virus spreads, probably via the blood.
   b. In serious echovirus disease, the virus generally can be isolated from the affected organ.
   c. The pathologic effects of many echoviruses are unknown.

4. **Epidemiology.** Echoviruses spread in patterns similar to those for the polioviruses and coxsackieviruses, though at a somewhat lower rate.

5. **Prevention.** There is no known means of controlling echoviruses.

F. **Rhinoviruses** are associated with a group of acute, afebrile upper respiratory diseases usually identified as the common cold.

1. **Stability.** The rhinoviruses differ in stability from the enteroviruses.
   a. They are very unstable at low pH and are readily inactivated at pH levels of 3–5.
   b. They are unusually stable at a temperature of 50° C and have a higher buoyant density in cesium chloride than the other picornaviruses.

2. **Classification.** The rhinoviruses fall into two broad categories based on the primate cell type in which they replicate.
   a. **H group viruses** multiply and produce cytopathic changes in a limited number of diploid cell lines, primary human embryo cells, and a special strain (strain R) of HeLa cells.
   b. **M group viruses** multiply and produce cytopathic changes in primary rhesus monkey kidney cells, human embryonic kidney cells, and a variety of continuous human cell lines.

3. Antigenicity
    a. There are at least 113 immunologically distinct groups of rhinoviruses based on a single type-specific antigen. There is no group-specific antigen.
    b. In humans, natural rhinovirus infections stimulate the production of type-specific neutralizing antibodies for at least 2 years; however, immunity is effective only against homologous challenge.
    c. Immunization with 113 different rhinoviruses is not practical.

4. Replication.
The rhinoviruses differ from the enteroviruses in requiring an incubation temperature of 33° C for maximal replication as well as for primary virus isolation. Temperatures greater than 37° C block a late step in rhinovirus multiplication.

5. Clinical infections.
In humans, infections appear to be confined to the respiratory tract. The disease syndrome is limited to that termed the common cold and, in very rare cases, bronchopneumonia.

6. Epidemiology.
Rhinovirus transmission in a family unit usually begins when a child introduces the virus, which spreads rapidly via nasal secretions. Spread is highest in the fall, winter, and early spring, when rhinoviruses are the major, though not the only, agent associated with the common cold.

IV. RHABDOVIRIDAE.
The rhabdoviruses include more than 25 variants that infect mammals, fish, insects, and plants. The only significant human pathogen in the rhabdovirus group is the rabies virus. The biochemical properties, mechanism of replication, and genetics of rhabdoviruses have been studied largely in vesicular stomatitis virus, an organism usually found in horses and cattle but occasionally also found in humans. Because several rhabdoviruses replicate in arthropods as well as in mammals, rhabdoviruses at one time were classified as arboviruses.

A. Structure.
Rhabdoviruses are bullet-shaped, **enveloped**, **minus-strand RNA viruses** with helical symmetry.

    1. They average 180 nm in length and 75 nm in width and contain a single stranded RNA molecule of $3.5-4.6 \times 10^6$ daltons.

    2. The viral particle is rounded at one end and flat at the other, and its surface is covered by regularly spaced projections with knob-like structures.

    3. The virion core is symmetrically wound within the viral envelope and runs along the long axis of the particle.

    4. The viral envelope is composed of a lipid bilayer covered by external surface glycoprotein projections.
        a. The membrane consists of a single **surface glycoprotein (G)** and two nonglycosylated **M proteins (M_1 and M_2).**
        b. The nucleocapsid RNA is complexed with many copies of the core protein (NP) and a few copies of the virion transcriptase, which is composed of a large (L) protein and a smaller (NS) protein.

B. Replication
is confined to the **cytoplasm** of the host cell.

    1. After adsorption and uncoating, the virus RNA is transcribed by the virion RNA-dependent RNA polymerase (transcriptase).

    2. The plus-strand mRNA then serves as a template for further RNA synthesis via a new polymerase activity (the replicase) and serves as the messenger for synthesis of viral protein.

C. Rabies
is an encephalitis causing neuronal degeneration of the brain and spinal cord. In humans, it is almost always fatal.

    1. Antigenicity. Rabies viruses from all sources appear to be a single immunologic type.
        a. Fixed virus (a virus form that is yielded by serial passage in laboratory animals) and **street virus** (a form that is freshly isolated from an infected animal) are immunologically indistinguishable.

    b. Antibodies directed against the surface glycoprotein projections are responsible for neutralization.

    c. Antibodies against the nucleocapsid are recognized by complement fixation but play little role in protection.

2. Pathogenesis

    a. The **virus is introduced into the body** through a break or abrasion of the skin, usually a bite inflicted by a rabid animal. It also may enter through the respiratory tract by means of an aerosol created by a dense population of rabid bats.

        (1) The virus replicates in muscle and connective tissue, where it may remain for days or months.

        (2) The infection progresses along the axoplasm of peripheral nerves to the basal ganglia and the CNS, where it multiplies further and causes a severe encephalitis.

    b. The **incubation period** usually is 3–8 weeks but can be as short as 6 days.

    c. **Symptoms** are first noted in the prodromal period and include irritability and abnormal sensations at the wound site.

    d. **Clinical disease** becomes apparent with a change in muscle tone leading to such problems as difficulty in swallowing.

    e. **Negri bodies** (i.e., cytoplasmic inclusions located within affected neurons) are the most characteristic microscopic finding and are diagnostic for the disease.

3. Epidemiology. All mammals are susceptible to rabies infection. Dogs, cats, cattle, skunks, bats, foxes, and squirrels are the most common hosts.

    a. **Dogs and cats** are the largest source of human infection. Cattle are the most commonly infected domestic animals.

    b. The major source of rabies is the wildlife reservoir, which carries what is called **sylvan rabies**; skunks, foxes, raccoons, and bats are the predominant hosts. (The virus may have a long latency period in bats.)

    c. In the United States, **animal immunization** programs and leash laws have reduced rabies somewhat. Worldwide, the incidence of rabies continues to increase.

4. Prevention of rabies **must entail prevention in domestic animals** as well as in humans.

    a. Control of rabies in the wildlife reservoir appears to be impossible.

    b. For humans, **two vaccines** are currently available: one produced from virus propagated in chicken and duck embryos and inactivated by beta-propiolactone, and one produced from virus grown in human diploid fibroblasts. Humans are usually immunized only after exposure to the virus.

5. Therapy for persons who have been bitten by an animal suspected to be rabid is designed to confine the virus to the site of entry.

    a. Vaccine produces frequent side effects and should not be given to a person who has had minimal contact with a questionable source.

    b. A patient who has had documented exposure should be treated with either the duck or, preferably, the diploid cell vaccine and with human immunoglobulin.

V. TOGAVIRIDAE AND FLAVIVIRIDAE. Both togaviruses and flaviviruses were previously classified as togaviruses. Two togavirus genera—*Alphavirus* (formerly called group A arboviruses) and *Rubivirus*, which has one member, the virus of rubella (German measles)—and all the flaviviruses contain human pathogens.

A. Structure. Both groups are spherical, **enveloped viruses** with an icosahedral core containing a **single, plus-strand RNA** molecule.

1. The RNA molecule of alphaviruses and flaviviruses measures 4×10^6 daltons (12 kilobases); in rubivirus, it measures $2.5–3.0 \times 10^6$ daltons.

2. The alphaviruses are approximately 45–75 nm in diameter, the flaviviruses 37–50 nm, and the rubella virus 60 nm.

3. The viral RNA is infectious after extraction from the virion and, like most eukaryotic messenger RNAs, contains a 3' polyadenylic acid tail.

4. The surface of these viruses is covered with glycoprotein spikes that contain hemagglutinins.

B. Replication. Togaviruses and flaviviruses both replicate in the **cytoplasm** of infected cells. The general sequence of events is common to group members, but the length of the cycle varies from relatively short for the alphaviruses to somewhat long for the flaviviruses.

  1. After adsorption and uncoating, the virion RNA serves as the messenger RNA (mRNA) for the formation of viral proteins, which are formed by proteolytic cleavage of a polyprotein.

  a. In **alphavirus** replication, two viral mRNA populations appear during the main stage. One is the **42S mRNA**, which encodes the nonstructural proteins; the other is the **26S mRNA**, which encodes the virion structural proteins.

  (1) The 26S mRNA is identical to the 3′ terminus of the 42S mRNA.

  (2) Amplification of the 26S mRNA species leads to increased production of the specific mRNA population that is required in the highest concentration during replication.

  b. In flavivirus replication, only one species of mRNA is observed. The entire flavivirus genome is translated into a single polyprotein, which is subsequently processed by proteolytic cleavage.

  2. The viruses **mature by budding** from the cell membrane, which appears to be the site where viral particles are assembled.

C. Antigenicity. The togaviruses are divided into genera on the basis of their immunologic reactivities. Many togaviruses, however, cannot be placed in an established immunologic group. Flaviviruses are separated on the basis of both serologic and biochemical properties.

  1. The togaviruses are **grouped on the basis of hemagglutination inhibition (HI) reactions**; members of one group (e.g., the alphaviruses) cross-react among themselves but not with members of the other groups.

  a. Within the alphaviruses and flaviviruses, organisms also fall into immunologic subgroups whose members show cross-reactivity within the hemagglutinins.

  b. Isolates of a given virus cultured at different times may be immunologically distinct.

  c. For rubivirus, only a single antigenic type has been observed.

  2. Neutralizing antibodies appear in infected individuals approximately 7 days after infection, after which solid **immunity** develops and **may last a lifetime**.

  a. Inactivation by β-propiolactone or formalin does not destroy the antigenicity of most of these viruses.

  b. Immunization with a single serotype often elicits the formation of cross-reactive antibodies, which may be markedly stimulated by exposure to heterologous antigen.

  c. Immunization against dengue and yellow fever (both flaviviruses) must employ the live attenuated virus; inactivation in these cases impairs the virus's ability to induce an antibody response.

D. Alphavirus and flavivirus pathogenesis. In humans, these viruses produce illness ranging from subclinical to rapidly fatal. It follows a general pattern with only minor variation between the two groups.

  1. An **infected mosquito** bites a warm-blooded host and injects virus from its salivary glands into the blood or lymph of the host.

  2. Virus is removed from the blood or lymph by reticuloendothelial cells, which become the site of viral replication. A **viremia initiates the systemic phase** of the disease.

  3. The virus subsequently **invades specific host tissues**—the CNS in the encephalitides, the skin and blood vessels in the hemorrhagic fevers, and the skin, muscles, and viscera in dengue and yellow fever.

  a. Some alphaviruses (e.g., the Chikungunya and Mayaro viruses) are confined to a systemic phase with strictly constitutional symptoms.

  b. **Encephalitic alphaviruses may vary in the severity** of the encephalitic phase. **Eastern equine encephalitis (EEE)** virus may cause a very severe disease, and **western equine encephalitis (WEE)** often is very mild.

  c. The flaviviruses produce **three types of clinical syndrome**.

  (1) An **encephalitis** similar to that caused by alphaviruses is produced by such flaviviruses as the St. Louis, Japanese B, and Russian spring-summer viruses.

(2) Severe systemic disease involving internal organs such as liver and kidneys is represented by yellow fever.

(3) A **milder systemic disease** with muscle pains and a rash that may be hemorrhagic is caused by the viruses of dengue and West Nile fever and by some tick-borne flaviviruses.

E. Alphavirus and flavivirus epidemiology. Alphaviruses and flaviviruses have in common a **mosquito or other arthropod vector**.

1. For the **alphaviruses**, humans and horses are merely accidental hosts.
   a. In the normal infection cycle for EEE and WEE, the virus is transmitted from arthropods to birds, with snakes and frogs as possible secondary sources.
   b. In the normal infection cycle for Venezuelan equine encephalitis, the virus is transmitted to small mammals and mosquitos rather than birds.

2. The **flaviviruses** have a somewhat more varied epidemiologic pattern.
   a. The **St. Louis encephalitis virus** is the most common flavivirus in the United States. Like many members of the group, it is transmitted from mosquitos to birds; humans are an accidental dead-end host.
   b. In the two forms of **yellow fever** (urban and jungle), humans or monkeys are major intermediates. **Dengue** resembles yellow fever in this respect.
   c. **Tick-borne viruses** have several unique epidemiologic features. Humans are a major part of the infection cycle, but the virus may be transmitted transovarially from tick to tick and, by means of goat milk, from goats to humans.

F. Rubella virus, the agent of rubella (German measles), is classified as a togavirus based on its biochemical and morphologic properties.

1. Replication. Rubella virus replicates in cell cultures of various origin. The steps in multiplication are similar to those of other togaviruses with the exception that nucleocapsids do not accumulate in the cytoplasm.
   a. Cytopathic changes are not detectable in most cell cultures infected with rubella virus, although some cellular changes may be seen in primary human cells.
   b. Continuous carrier cultures are not uncommon and frequently show many chromosomal breaks and abnormalities.

2. Pathogenesis. Rubella resembles measles except that it has a milder clinical course and shorter duration.
   a. Transplacental fetal infection induces serious damage to the fetus.
      (1) Tissues of all germ layers are damaged as a result of both cell death and persistent infection.
      (2) Fetal death is not uncommon and occurs most often when the fetus is infected during the first trimester.
      (3) Infants that survive have a variety of neurologic and other congenital abnormalities collectively referred to as the **rubella syndrome**.
   b. Rubella rash appears approximately 14–25 days after infection. During the incubation period, viremia occurs and virus is disseminated throughout the body, including the placenta during pregnancy. As a rule, few symptoms are experienced.

3. Epidemiology. Rubella is a highly **contagious** disease spread by nasal secretions. Because infection often is inapparent, viral dissemination may be widespread before it is recognized.
   a. Virus can be isolated from nasopharyngeal secretions as early as 1 week before and as late as 1 week after the appearance of the rash. These secretions probably are the main mechanism of transmission.
   b. Infants infected in utero may appear normal, but they excrete high levels of virus and may present a danger to expectant mothers and health-care personnel.
   c. Minor **epidemics** occur every 1–2 years, and major epidemics occur every 6–9 years.

4. Prevention. Widespread **vaccination** with a live attenuated virus and serologic **screening** of women planning to conceive have significantly reduced both postnatal and congenital rubella in the United States.

VI. BUNYAVIRIDAE. The bunyavirus family is a large collection of **arthropod-borne** viruses grouped on the basis of immunologic, chemical, and morphologic similarities. Bunyaviruses fall into **at least six subgroups** on the basis of immunologic reactivities, and **at least three genera (supergroups)** are recognized.

   A. Structure. Bunyaviruses are spherical, **enveloped virions** with a helical core containing three distinct, linear, **minus-strand RNA** segments.

     1. The envelopes are covered with glycoprotein projections that have hemagglutinating activity.

     2. The virions contain an RNA-dependent RNA polymerase.

   B. Replication. Available data suggest that bunyavirus replication resembles that of the orthomyxoviruses.

   C. Antigenicity. Like the togaviruses, bunyaviruses are immunologically differentiated by hemagglutination assays.

     1. Complement fixation (CF) and neutralization assays also reveal many cross-reactivities within the groups.

     2. **A common family antigenic component** has been identified by complement fixation assays.

   D. Epidemiology. Like the togaviruses and flaviviruses, bunyaviruses are accidentally transmitted to humans by a **mosquito vector**.

     1. **California encephalitis virus** is the most widely studied bunyavirus. It causes a prominent clinical disease characterized by fever, headache, and CNS involvement. This virus has been isolated from a variety of small animals, but the exact warm-blooded reservoir has not been identified.

     2. **Monkeys and other forest mammals** are a **reservoir** for a large number of bunyaviruses that are associated with mild disease in humans.

VII. ARENAVIRIDAE. The arenaviruses are a group of seemingly unrelated viruses grouped on the basis of morphologic, immunologic, and clinical characteristics.

   A. Structure. The name for the arenaviruses derives from their distinctive **pebbly appearance**, which is created by a number of electron-dense granules visible by means of electron microscopy.

     1. The arenavirus virions are round or pleomorphic and **enveloped**. The viral RNA is segmented, **single-stranded**, and of **negative polarity**.

     2. The virions contain several types of **both virus- and host-derived ribonucleoproteins**. Those derived from the host cell behave like ribosomes and have all of the types of host ribosomal RNA.

   B. Replication events, so far as they are known, resemble those of other minus-strand RNA viruses.

   C. Antigenicity. Immunologic relationships and subgroupings are established by CF assays. Species are defined by neutralization titrations.

   D. Pathogenicity. Arenaviruses commonly induce a **chronic carrier state** in their natural hosts.

     1. The viruses may be isolated from urine, blood, and organs.

     2. The **human diseases** associated with the arenaviruses range from a mild influenzalike illness caused by **lymphocytic choriomeningitis (LCM) virus** of mice to the serious diseases of Bolivian hemorrhagic fever, which is associated with **Junin and Machupo viruses**, and Lassa fever, which is associated with the highly contagious **Lassa virus**.

E. Epidemiology

1. LCM virus of mice is found in the Americas and in Europe. It is spread to humans in excretions of the naturally infected rodents.

2. Other arenaviruses are apparently confined to small regions of Africa and Central and South America.

3. The Tacaribe viruses, which have been isolated from bats and cricetid rodents, appear to resemble LCM virus of mice.

VIII. CORONAVIRIDAE. The introduction of ciliated, human embryonic tracheal cultures revealed a new group of human viruses, the coronaviruses, which are associated with the common cold and other respiratory diseases.

A. Structure. The coronaviruses are named from their appearance in the electron microscope, which reveals moderately pleomorphic, spheric, or elliptic virions covered with distinctive, club-shaped projections.

1. These **enveloped** viruses have an outer membrane that is 7–8 nm thick. Within the envelope the coronaviruses have a loosely wound, helical nucleocapsid.

2. The surface projections, which create the striking solar corona appearance in the electron microscope, are composed of two glycoproteins.

3. The single molecule of **plus-strand RNA** is complexed with many molecules of a single protein species.

B. Replication of coronaviruses is imperfectly understood but is known to be **slow** and confined to the **cytoplasm** of infected cells. The optimal temperature for growth is 32°–33° C, which is similar to the temperature requirement for rhinovirus replication.

C. Antigenicity. Coronaviruses of humans appear to be totally unrelated to coronaviruses of animals; they are limited not only to humans but to a very restricted group of ciliated epithelial cells.

1. Human coronaviruses form **at least three antigenic groups** based on neutralization assays.

2. The fact that the immunologic group correlates with the culture system used for the original isolation indicates that these viruses have a **strict host range**.

D. Pathogenicity. The coronaviruses are associated with the illness known as the **common cold**, and the syndrome they cause cannot be differentiated from that caused by the rhinoviruses. Unlike most other respiratory viruses, coronavirus infects older children and adults more commonly than younger children.

IX. REOVIRIDAE. The reoviruses have been subclassified into several genera, three of which contain members that infect humans: *Orthoreovirus*, *Rotavirus*, and *Orbivirus*.

A. Structure. The reoviruses contain an unusual **double-stranded RNA** genome consisting of 10 or 11 distinct, double-stranded fragments that are distributed in three distinct and highly reproducible size classes.

1. The viruses have **icosahedral symmetry** and are 78–80 nm in diameter.

2. The virions are surrounded by **two capsids** and have **no envelope**.

B. Replication of the reoviruses has been carefully studied.

1. After adsorption and penetration, the viruses become associated with lysosomes whose proteolytic enzymes digest the capsid proteins to create a subviral particle without releasing the viral RNA. The subviral particle is larger than the viral core.

2. The virion RNA does not act as a messenger but is transcribed to produce a single-stranded mRNA. This process occurs entirely within the subviral particle, which leaves the lysosome after the uncoating is complete. The production of mRNA is accomplished by a virion-associated transcriptase located within the core particle.

3. Approximately 75% of newly synthesized plus-strand RNA becomes associated with polyribosomes and is used in protein synthesis. The remainder of the plus-strand RNA is converted into double-stranded virion RNA. This unique conservative mechanism of synthesis is in marked contrast to double-stranded DNA synthesis.

4. Infectious virions begin to form inclusion bodies 6–7 hours after infection, and the assembly process proceeds for several more hours before the virion is released from the cell.

C. **Orthoreoviruses** are **widely distributed in nature**; antibodies have been found in virtually all mammals.

1. The genus consists of **three immunologic types** based on the M_1 outercapsid protein. The different immunotypes are related by a group of 3 or 4 cross-reacting antigens.

2. Orthoreoviruses are regularly isolated from the **feces** and **respiratory secretions** of apparently healthy humans as well as from patients with minor upper respiratory and gastrointestinal symptoms.

D. **Rotaviruses.** All of the human rotaviruses examined to date belong to **a single immunologic group**. This group has been identified as a major worldwide cause of sporadic **acute enteritis in infants and young children**. The human and animal isolates of rotavirus are closely related immunologically.

1. **Pathogenicity.** Rotavirus infection is responsible for almost half of the serious illnesses causing hospitalization of infants. Its incidence is particularly high in the fall and winter months.
   a. Common **symptoms** of rotavirus infection in children under 2 years of age include severe diarrhea, fever, and, occasionally, vomiting.
   b. The characteristic symptoms and high rates of virus excretion during and following acute illness are due to the fact that the **duodenal mucosa is the primary site of virus replication**.

2. **Prevention.** There are no known measures for controlling rotaviruses.

E. **Orbiviruses. Colorado tick fever virus** is the only orbivirus that is associated with human disease. This virus is immunologically unrelated to both the orthoreoviruses and the rotaviruses.

1. Colorado tick fever is the **only tick-borne viral disease of the United States**. It occurs specifically in the western United States and is usually associated with the tick, *Dermacentor andersoni*.

2. Colorado tick fever is an acute, febrile, nonexanthematous infection with muscle pains, especially of the back and legs. The disease course generally is short and recovery complete.

3. The major means of **prevention** is avoidance of infected ticks.

STUDY QUESTIONS

Directions: Each question below contains five suggested answers. Choose the **one best** response to each question.

1. Which of the following statements best describes antigenic drift in influenza viruses?

(A) It appears to be associated with minor antigenic changes in the viral hemagglutinin.
(B) It appears to be associated with minor antigenic changes in the viral capsid protein.
(C) Antigenic drift is responsible for pandemics of influenza virus.
(D) It usually results from recombination between influenza viruses of human and avian sources.
(E) Antigenic drift is responsible for the appearance of new antigenic types of influenza viruses.

2. Infectious virus can be isolated from the stool for 3 or more weeks following infection with which of the following viruses?

(A) Herpes simplex virus
(B) Poliovirus
(C) Smallpox virus
(D) Influenza virus
(E) Hepatitis B virus

3. Control of rabies in the United States relies on which of the following factors?

(A) Immunization of humans beginning in late childhood
(B) Mass immunization of wild animal populations that harbor the virus in nature
(C) Immunization of domestic animals and pets
(D) A bat colony eradication program
(E) None of the above

4. Which of the following diseases is characterized by the presence of Negri bodies in host cells?

(A) Rabies
(B) Infectious mononucleosis
(C) Congenital rubella
(D) Mumps
(E) Varicella

5. The alphaviruses and flaviviruses have almost worldwide distribution and often are quite virulent. All of the following statements regarding these viruses are correct EXCEPT

(A) many "species" appear to exist, several of which cannot be placed in established immunologic groups
(B) humans are the principal reservoir for the viruses
(C) the viruses can replicate in both vertebrates and arthropods
(D) infected mosquito saliva may contain high titers of virus
(E) some of these viruses may replicate in several different tissues in an infected arthropod

Directions: Each question below contains four suggested answers of which **one or more** is correct. Choose the answer

 A if **1, 2, and 3** are correct
 B if **1 and 3** are correct
 C if **2 and 4** are correct
 D if **4** is correct
 E if **1, 2, 3, and 4** are correct

6. Statements that accurately describe respiratory syncytial virus (RSV) include

(1) it is a strongly hemagglutinating virus for a variety of red blood cell types
(2) immunity to RSV is short lived and dependent on IgA levels
(3) it has a well-defined morphology that is easily identified in the electron microscope
(4) RSV is associated with lower respiratory tract infection

7. Statements that correctly describe the properties of bunyaviruses include

(1) the RNA core has helical symmetry and is found as distinct fragments rather than a single RNA molecule
(2) the transmission of the bunyaviruses is similar to that of the togaviruses
(3) monkeys and other forest mammals are a reservoir for many bunyaviruses
(4) the virions contain an RNA-dependent RNA polymerase

8. The epidemiology of rabies involves a variety of animals, including

(1) foxes
(2) skunks
(3) bats
(4) cattle

9. Alphaviruses are common in which of the following reservoirs?

(1) Horses
(2) Birds
(3) Dogs
(4) Small wild mammals

Directions: The group of questions below consists of lettered choices followed by several numbered items. For each numbered item select the **one** lettered choice with which it is **most** closely associated. Each lettered choice may be use once, more than once, or not at all.

Questions 10–13

Each of the following phrases describes a virus in terms of a specific property. Match each phrase to the virus it best describes.

(A) Rhinovirus
(B) Coxsackievirus A
(C) Rubella virus
(D) Rotavirus
(E) Poliovirus

10. Virus that causes myositis in suckling mice

11. Virus associated with severe infantile gastro-enteritis

12. Virus that requires a temperature of 33° C for maximal replication

13. Virus that induces carrier cultures of cells infected in vitro; the cells frequently show chromosomal aberrations

ANSWERS AND EXPLANATIONS

1. The answer is A. [*I C 1*] Antigenic variation occurs in two ways—antigenic drift and major antigenic shift—and it plays a significant role in the epidemic spread of influenza. Antigenic drift, which involves minor changes in either the hemagglutinin or neuraminidase, is not significantly associated with the pandemic spread of viruses. Pandemics are due to major antigenic shifts resulting from either recombination between strains or a major mutation leading to a new antigenic type. These recombinants frequently result from combinations between human and avian strains of influenza.

2. The answer is B. [*III C 4 c*] Poliovirus is a typical member of the enterovirus group. The virus spreads via the fecal-oral route rather than the respiratory route—the route used by herpes simplex, smallpox, influenza, and hepatitis B viruses. Virus replication occurs in the epithelial lining of the gastrointestinal tract, from which it is released into the gut and subsequently excreted in the feces.

3. The answer is C. [*IV C 4*] Two effective vaccines against rabies are available for use in humans; however, because of problems associated with immunization they generally are used only after exposure to the virus. To date, it has been impossible to devise measures that effectively provide immunity to the wild animal populations that harbor the virus, although research on suitable mechanisms is underway. Presently, prevention relies on mass immunization of domestic animals and pets. Although bats are a reservoir of the virus, they are not an important enough source of infection to be singled out for eradication.

4. The answer is A. [*IV C 2 e*] Intracellular inclusion bodies composed of high concentrations of viral particles or viral cores occur with several different viral infections. One type of intracellular inclusion, the Negri body, is characteristic of, and has become the common marker for, rabies infection. The presence of these inclusions in the brain tissue of an animal or human suspected of having rabies is diagnostic; however, the absence of Negri bodies does not rule out rabies infection. Infectious mononucleosis, congenital rubella, mumps, and varicella are characterized by a variety of cytopathic changes, but none of these viral diseases is associated with the neuronal aggregates known as Negri bodies.

5. The answer is B. [*V E*] Alphaviruses and flaviviruses are distinguished serologically, and a wide variety of serologic variants exists. Both groups of viruses are transmitted by arthropods, which are essential to their life cycle, and both replicate in the arthropod. Humans are accidental hosts for all alphaviruses and flaviviruses; the major reservoirs are wild animals or birds, depending on the particular virus. These viruses replicate in a variety of arthropod tissues leading to their prolonged presence in the population. Some flaviviruses survive many months in ticks as a result of transovarian transmission.

6. The answer is C (2, 4). [*II F*] Respiratory syncytial virus (RSV) is a significant cause of lower respiratory tract infection in children and infants. Generally, immunity to infection is short lived, since the only effective immunity is local immunoglobulin A (IgA) rather than circulating IgG. Structurally, RSV does not have a hemagglutinin on its surface, and several morphologies are observed.

7. The answer is E (all). [*VI A–D*] Epidemiologically, the bunyaviruses are identical to the togaviruses; with both viruses, a mosquito vector accidentally transmits the disease to humans. Biochemically, however, the bunyaviruses differ considerably from the togaviruses. The most striking biochemical property of the bunyaviruses is their fragmented (i.e., three-fragment) genome. The virions of bunyaviruses contain an RNA-dependent RNA polymerase.

8. The answer is E (all). [*IV C 3*] Rabies is established in virtually every wild mammal population in North America. All common domestic animals are susceptible, including cattle. The most common human exposure comes from household pets, but it is not unusual in some parts of the country for rabid racoons and skunks to enter urban areas.

9. The answer is C (2, 4). [*V E 1*] The alphaviruses are the most common togaviruses in the United States. Although these viruses often are found in association with infected horses, the horse, like man, is not a reservoir but a dead-end host. The common alphavirus reservoirs are birds and, to a lesser extent, small mammals. In all cases, the vehicle for alphavirus transmission is an infected mosquito.

10–13. The answers are: 10-B, 11-D, 12-A, 13-C. [*III D 1 a (1), F 3; V F 1; IX D*] Coxsackieviruses are typical picornaviruses, which are small, naked capsid RNA viruses. Coxsackieviruses can be differentiated from other picornaviruses only on the basis of immunologic testing and their ability to cause disease in suckling mice. The exact clinical pattern of disease in the suckling mice helps to differentiate the group of coxsackievirus responsible.

For many years, the double-stranded RNA viruses were not clearly associated with any form of viral disease but frequently were isolated from feces of normal persons. However, a subgroup of these viruses, the rotaviruses, has been identified as the major cause of sporadic acute enteritis in infants and young children.

Many viruses have biologic properties that appear to have resulted from adaptation to the environment. Viruses of the upper respiratory tract (e.g., the rhinoviruses) replicate in tissue that generally has a lower mean temperature than the rest of the body owing to the cooler, outside air that passes over the respiratory epithelial tissue on the way to the lungs. The rhinoviruses have adapted to the degree that maximal replication occurs at a temperature that is most similar to that of the respiratory epithelium (i.e., 33° C).

Both the paramyxoviruses and rubella virus are RNA viruses that have a very high rate of noncytopathic infection of cultured cells, which leads to chronically infected carrier cultures. Rubella-infected cells are unique in the appearance and the frequency of chromosomal abnormalities in these cultured carrier cells. This chromosomal breakage is thought to be analogous to the chromosomal damage seen in the congenital rubella syndrome.

23

Miscellaneous Viruses: Hepatitis, Oncogenic, AIDS, and Slow (Unconventional) Viruses

David T. Kingsbury

I. HEPATITIS VIRUSES. Hepatitis is an infectious disease that appears sporadically and can be transmitted either through the intestinal-oral route (commonly known as **infectious hepatitis**, or **hepatitis A**) or by parenteral inoculation with infected blood or blood products (known as **serum hepatitis**, or **hepatitis B**). The two forms of hepatitis are compared in Table 23-1. The viruses associated with these diseases are known as **hepatitis A virus (HAV), hepatitis B virus (HBV)**, and **hepatitis C virus (HCV**, formerly referred to as **non-A, non-B hepatitis)**. HCV presently is the major cause of serum hepatitis in the United States. A fourth virus, **delta-associated virus (DAV)**, causes chronic hepatitis—an occasional sequel to serum hepatitis; DAV requires HBV as a helper for replication.

A. HAV recently has been propagated in tissue culture in a reproducible manner, and the viral nucleic acid has been cloned and sequenced.

1. Structure. The physical and chemical properties of HAV are those of a **picornavirus** (see Chapter 22; section III); it is currently classified as **enterovirus 72**.
   a. HAV is a **naked capsid virus** approximately 27 nm in diameter.
   b. It has a **single-stranded RNA** genome.
   c. HAV particles have the density of picornavirus particles.
   d. It has three major virion polypeptides.

2. Antigenic composition. The major viral antigen is termed HAAg. Available data suggest that HAV has a single immunologic type and that lifelong immunity follows infection.

B. HBV has not been successfully grown in a cell culture system. It has many unique properties and has not been placed in an existing viral group, but rather is placed in a new group, the **hepadnaviruses**.

Table 23-1. Comparison of Hepatitis Types A, B, and C

	Hepatitis A	**Hepatitis B**	**Hepatitis C**
Incubation period (days)	15–40	60–160	60+
Transmission route	Fecal-oral	Parenteral inoculation	Parenteral inoculation
Onset	Acute	Insidious	Insidious
Most common age of incidence	Childhood; young adulthood	Adulthood	Adulthood
Virus size (nm)	27	42	80
Phase			
Incubation	Found in feces and blood	Found only in blood*	Found only in blood
Acute	Found in feces and blood	Found only in blood*	Found only in blood

*May persist for years.

1. Structure. Morphologically, three distinct structures have been observed in the serum of infected humans. All three structures contain the **HBV surface antigen (HB$_s$Ag)**.

   a. Spherical particles approximately 22 nm in diameter are the most common particles seen in infected serum.

   b. Filamentous forms approximately 22 nm in diameter and 50–230 nm in length also are quite common. Treatment of the filaments with a nonionic detergent converts them into structures resembling spherical particles.

   c. The Dane particle, the least common of the three structures, is the only one with traditional viral morphology. It is infectious.

   (1) The Dane particle is a complex, double-layered, spherical particle 42 nm in diameter with a dense, 22-nm core.

   (2) It contains a partially **double-stranded, circular DNA molecule** that is approximately 3600 nucleotides in length.

   (a) The two DNA strands each have gaps, although one strand is almost complete.

   (b) The DNA is associated with a DNA polymerase, the activity of which is dependent upon the addition of all four nucleoside triphosphates and magnesium ion but not primer DNA.

   (c) Endogenous DNA polymerase is capable of filling the open regions of each DNA strand.

2. Antigenic composition

   a. Dane particles, which are thought to be complete viruses, contain two major antigens, **HB$_s$Ag** and the **core antigen (HB$_c$Ag)**.

   (1) **HB$_s$Ag** is identical to the Australia antigen, which was the first antigen identified as associated with HBV. It appears in the serum during the incubation period.

   (2) **HB$_c$Ag** is of a single antigenic type and is found only in the core of the Dane particle.

   b. The **e antigen (HB$_e$Ag)** is a third antigen found in some HB$_s$Ag-positive sera. It is distinct from the particles but appears to be associated with the presence of Dane particles.

   (1) HB$_e$Ag appears in serum during the incubation period, just after the appearance of HB$_s$Ag.

   (2) The function of HB$_e$Ag is unknown, but it is probably the **most reliable diagnostic indicator** of active infection. Patients with HB$_e$Ag are most likely to be efficient transmitters of the disease.

   c. Antibodies directed against HB$_s$Ag and HB$_c$Ag appear during infection, and an increase in antibodies to the HB$_s$Ag is clearly correlated with immunity.

3. Replication. Because HBV has not been grown successfully on a cell culture system, only fragmentary information about the biochemistry of its replication is available.

   a. Synthesis of DNA and assembly of virus cores appear to occur in the nucleus of infected cells.

   b. HB$_s$Ag is abundant in the cytoplasm of infected cells and is associated with cell membranes and the endoplasmic reticulum.

   c. The data imply that the completed Dane particle is assembled by means of maturation of a core budding through cytoplasmic membranes.

C. HCV (formerly known as non-A, non-B hepatitis), like HBV, is transmitted by intravenous administration of infected blood or serum. There may be more than one non-A, non-B hepatitis virus; however, the isolate designated as HCV has been partially characterized as a member of the togaviruses and is associated with the vast majority of cases of non-A, non-B hepatitis.

1. The virus is a **single-stranded RNA** virus with a genome of approximately 10,000 nucleotides.

2. The viral RNA has been cloned and partially sequenced, and the major antigens expressed in vitro.

D. Clinical disease. Both infectious and serum hepatitis vary in severity from inapparent infections and nonicteric hepatitis to serious liver degeneration. Serum hepatitis is the more severe disease, with a more gradual onset, longer duration, and higher mortality rate than infectious hepatitis.

1. **Acute disease.** Although **jaundice** is not universally seen, it is the predominant symptom, and the **liver** is the organ primarily affected.

   a. **Preicteric stage.** Jaundice usually is preceded by anorexia and malaise, often with abdominal discomfort and nausea. Headache and myalgia are frequent. In HBV infection, rash and arthritis sometimes are seen.

   b. **Icteric stage.** The main symptom accompanying jaundice is fatigability. The jaundice lasts from a few days to 2–3 months.

2. **Sequelae**

   a. Between 3% and 5% of adults and 10%–15% of children infected with **HBV** become **chronic HB$_s$Ag carriers**, and most continue to exhibit some evidence of **chronic liver disease**. Neither the carrier state nor chronic hepatitis has been seen after HAV infection.

   b. There is a strong correlation between **hepatocarcinoma** and prior infection with either **HBV** or **HCV**.

E. **Pathogenesis**

   1. **HAV** undergoes primary **replication in the gastrointestinal tract**, leading to viremia that spreads to the liver, kidney, and spleen.

      a. Virus is shed in feces and is present in the blood throughout the preicteric period, peaking in titer just before the appearance of jaundice.

      b. Antibodies to HAV appear in the blood when the viral titer decreases and liver damage appears.

   2. The initial **site of HBV replication is unknown**. Replication in the hepatocytes has been observed as early as 2 weeks after infection.

      a. During the second half of the incubation period (a total of 40–180 days), infectious virus is found in the blood, urine, semen, feces, and nasopharyngeal secretions.

      b. It is not certain whether immune reactions play a role in HBV pathogenesis; however, there is a correlation between the onset of clinical illness and the inital appearance of anti-HBV antibodies.

F. **Epidemiologically**, infectious and serum hepatitis are distinct.

   1. Epidemics of infectious hepatitis occur when contaminated food or water is ingested by populations that lack immunity.

   2. Serum hepatitis is endemic but not epidemic. It is an occupational disease among health care personnel.

   3. The incidence of serum hepatitis increases with age.

G. **Laboratory diagnosis** of hepatitis is based upon direct demonstration of hepatitis virus antigens in the feces, blood, or liver biopsy tissue.

H. **Therapy and prevention.** No specific treatment is available for any of the hepatitis virus infections. Prevention is the only effective means of control.

   1. **Careful screening of blood and blood products** is essential for HBV control, as is careful handling of all medical supplies that have had contact with blood.

   2. Both infectious and serum hepatitis may be prevented or attenuated by **passive immunization** with pooled human gamma globulin.

   3. An **active vaccine against HAV** may soon be available since the virus has been propagated in tissue culture.

   4. An **active vaccine against HBV** has been developed using genetic engineering techniques. It has been approved by the United States Food and Drug Administration and is in use in many parts of the world.

II. ONCOGENIC VIRUSES. Viruses with oncogenic potential are found in most DNA virus groups but only in one RNA virus group, the retroviruses.

A. Papovaviruses (see also Chapter 21, section I). The papovaviruses **polyomavirus** and **simian virus 40 (SV40)** are the most widely studied of the DNA tumor viruses. These small viruses have a double-stranded, circular DNA genome of approximately 5 kilobase pairs; it is highly regulated during infection.

1. Replication, which takes place in the cell nucleus, is divided into early and late phases.
   a. Early transcription of viral DNA is carried out by cellular RNA polymerase II. Early transcription products are three nonstructural proteins known as **tumor (T) antigens**.
   b. Late transcription products are viral capsid proteins.

2. Infection
   a. A productive infection results when the virus invades a **permissive** cell. Productive infections lead to replication of viral DNA and production of infectious virions with an accompanying cell lysis and death.
   b. A nonproductive, or **abortive**, **infection** results when the virus invades a **nonpermissive** cell or when **defective virus particles** with deletions in the viral genome affecting some replication processes invade **permissive** cells. In tissue culture, nonproductive or abortive infections transform the host cell so that it resembles virus-induced tumor cells.

3. Transformed clones of papovavirus-infected cells are identified on the basis of their acquisition of properties characteristic of tumor cells, such as anchorage independence and the production of plasminogen activator.
   a. Transformation is accompanied by **integration of viral DNA into host cell DNA.**
   b. Transcription of integrated viral DNA (the **provirus**) is limited to **early transcription** with the **production of the three T antigens** (large, middle, and small).
      (1) The **large T antigen** appears to be essential for the establishment and maintenance of transformation.
      (2) The three T antigens share significant nucleotide sequences, and the final configuration of each antigen depends upon the RNA processing (splicing) events within the cell.
      (3) Some form of T antigen finds its way into cellular membranes, where it exerts an unknown effect.

4. Oncogenicity. The human papillomaviruses (HBV) induce both benign warts, which begin as a proliferation of dermal connective tissue, and malignant carcinomas, which result from rare transformed cells.
   a. Human warts often regress spontaneously by means of some form of immunologic intervention.
   b. Sexually transmitted anogenital warts are associated with HPV-6 and HPV-11. Both men and women have shown malignant transformation of infected cells.
   c. In **epidermodysplasia verruciformis**, about 25% of patients develop malignancies at the site of the red-brown macular plaques.
   d. In **precancerous cervical biopsies**, almost 100% contain HPV antigens and DNA from HPV-6, HPV-11, HPV-16, HPV-18, or on rare occasions other limited serotypes.

B. Adenoviruses (see also Chapter 21, section II). Despite their clear oncogenic potential in newborn hamsters and their wide distribution in humans, the adenoviruses do not appear to be oncogenic in humans.

C. Herpesviruses (see also Chapter 21, section III). The herpesvirus group is implicated in neoplastic disease in humans. Several members of the group induce neoplasia in animals such as chickens and monkeys.

1. Epstein-Barr virus (EBV) of humans is widespread among adults and almost always is isolated from tissues affected by Burkitt's lymphoma or nasopharyngeal carcinoma.
   a. Burkitt's lymphoma is a B-cell lymphoma usually seen in children. It is endemic in central Africa and New Guinea and sporadic in the rest of the world.
      (1) Unlike normal B cells, lymphoma B cells grow into lymphoblastoid cell lines in culture and maintain many of their original properties.
      (2) All cell lines contain multiple copies of EBV DNA as well as an EBV-induced nuclear antigen (EBNA).

b. The exact role of EBV in Burkitt's lymphoma and nasopharyngeal carcinoma (a common form of cancer among males of some Chinese ethnic groups) remains a mystery; however, there appears to be some close, probably etiologic, relationship.

2. Herpes simplex virus type 2 (HSV 2) was postulated in the etiology of cervical cancer on the basis of seroepidemiologic and virologic studies; however, a causal relationship here is less well established than with EBV and Burkitt's lymphoma, and an HSV 2 etiology is very unlikely.

3. Marek's disease virus of chickens causes a virulent lymphoma that can be prevented by immunization of the birds with a closely related nononcogenic virus.

4. Several herpesviruses are able to transform cells in vitro.

D. Retroviruses. The RNA-containing retroviruses are the most widespread and widely studied of the tumor viruses.

1. Genome structure and replication
   a. Retroviruses contain two copies of a positive polarity, single-stranded RNA molecule that is approximately 10 kilobases long. It may be shorter in defective forms of the viruses.
   b. A unique enzyme, **reverse transcriptase (RNA-dependent DNA polymerase)**, produces a double-stranded DNA copy of the viral RNA.
      (1) This DNA intermediate is integrated into host cell DNA to provide a proviral copy.
      (2) Replication proceeds through transcription of the proviral copy of the viral DNA.

2. Protein antigens. The retrovirus genome encodes three groups of proteins.
   a. Group antigen proteins (gag proteins) make up the viral core together with viral RNA. The core proteins form the group- and species- (i.e., feline or murine) specific antigens and may also include an interspecies determinant in some cases.
   b. The pol (polymerase) protein is the reverse transcriptase.
   c. The envelope glycoproteins (env proteins) are the membrane surface glycoproteins. They are type specific, and they define the individual virus species.

3. Assay of retroviruses is difficult because few of them are cytopathic. Biologic properties such as syncytia formation and the presence of viral antigens (immunofluorescent plaques) are commonly relied on in assays as is measurement of reverse transcriptase activity.

4. Classification. The retroviruses may be subdivided into groups based on morphology as determined with the electron microscope. Morphologic classification largely correlates with biologic properties.
   a. Type C oncoviruses are the principal group of oncogenic retroviruses.
      (1) The group has two main **categories**.
         (a) Nondefective leukosis viruses have low oncogenic potential.
         (b) Usually defective leukosis and sarcoma viruses have high oncogenic activity.
      (2) Type C retroviruses may be **transmitted either vertically or horizontally**.
         (a) Vertical transmission is genetic transmission of the virus as an endogenous proviral copy to all offspring.
         (b) Horizontal transmission is direct animal-to-animal spread by infection.
   b. Types B and D oncoviruses also exist.
      (1) The only widely studied type B oncovirus is the mouse mammary tumor virus, which causes a hormonally dependent mammary adenocarcinoma.
      (2) Type D oncoviruses have been isolated from several primates; their true oncogenic potential is not well established.

5. Oncogenesis. In the highly oncogenic defective viruses, a portion of the viral genome is frequently replaced with host DNA, which encodes some information that changes the growth control of the cell, the oncogene (**onc gene**).
   a. The first onc gene to be characterized fully was **arc**.
   b. About 35 onc genes have now been identified in various retroviral genomes.
   c. Each viral oncogene (**v-onc**) has a cellular homolog (**c-onc**) known as a **proto-oncogene**.

6. Isolation of a type C retrovirus from humans with T cell leukemia has suggested that oncogenic retroviruses infect all species.
   a. Human T-cell leukemia viruses (HTLV) have been observed in several foci of disease and in different geographic locations. They appear to have low oncogenic potential.

(1) **HTLV-I** is the etiologic agent of adult T cell leukemia and is endemic in southwestern Japan and the Caribbean Basin, where it behaves like an infectious disease.

(2) **HTLV-II** is an uncommon virus associated with human lymphomas.

b. Because HTLV-I is transmitted sexually and through blood products, the number of seropositive people in the population has gradually increased. The virus is found particularly in intravenous drug users, hemophiliacs, and homosexual or bisexual males.

III. HUMAN IMMUNODEFICIENCY VIRUS (HIV), a retrovirus of the lentivirus group, is the etiologic agent of acquired immune deficiency syndrome (AIDS).

A. History. AIDS was first recognized as a disease syndrome in 1981, and HIV was identified as its cause in 1984.

1. Viruses isolated by Montagnier in France (lymphadenopathy-associated virus, or **LAV**) and by Gallo (**HTLV-III**) and Levy (AIDS-related retrovirus, or **ARV**) in the United States were found to be identical and to be associated with AIDS. The virus is now universally known as **HIV**.

2. HIV antibodies were found in sera drawn in parts of Africa in the early 1970s, leading most investigators to believe the disease originated in Africa and spread to the Caribbean and then to the United States.

3. **Two strains** of HIV have been identified.
   a. **HIV-1** is the predominant isolate in clinical AIDS and is found in central Africa and other regions of the world.
   b. **HIV-2** generally is limited to West and Central Africa and has not demonstrated the virulence of HIV-1.

B. Clinical disease

1. **AIDS.** Clinically, AIDS is characterized by opportunistic infections and malignancies in the absence of any known cause for immune deficiency.
   a. The most frequent **opportunistic infections** are *Pneumocystis carinii* pneumonia, disseminated cytomegalovirus (CMV) pneumonia, and disseminated atypical mycobacteria. A large number of others are commonly seen.
   b. The most common **malignancy** is Kaposi's sarcoma.

2. **AIDS-related complex (ARC)** is seen in many people who test positive for HIV. ARC may eventually develop into clinical AIDS but in many cases does not.

C. Pathogenesis. Like other lentiviruses, HIV tends to maintain a long-term latent infection and is somewhat restricted in its in vivo replication.

1. The **latent (incubation) period** appears to range from 5 to 65 months, according to findings in transfusion-related AIDS.

2. **Modes of transmission** are well documented. They include sexual contact, exposure to blood or blood products, and from mother to child in utero or during nursing.
   a. The virus has been isolated from blood, semen, cerebrospinal fluid (CSF), saliva, and mother's milk of infected patients but has only been transmitted in blood, semen, and human milk.
   b. Transmission of HIV is thought to be **cell mediated** and not the result of free virus in acellular fluids.

3. **Replication.** HIV specifically replicates in T-lymphocytes carrying the **CD4 antigen** on their surface.
   a. Loss of immune function appears to result from depletion of the CD4$^+$ subset of T cells.
   b. The CD4 molecule also is the receptor for virus attachment.
   c. All CD4$^+$ cell types, including peripheral blood monocytes, some tissue macrophages, skin Langerhans' cells, and brain microglial cells, are susceptible to HIV infection.
   d. The pathogenicity of HIV is directly related to its specificity for CD4$^+$ cell types.

D. Epidemiology. AIDS has reached epidemic proportions within several high-risk groups—primarily homosexual men, intravenous drug users, hemophiliacs who receive frequent blood products, sexual partners (men and women) of AIDS patients, and infants born of women with AIDS in pregnancy.

    1. In **Africa**, however, AIDS involves primarily heterosexual men and women.

    2. HIV shows a **high attack rate**—25%–30%—in contrast to HTLV-I, which has an attack rate of 1 in 1000, or 0.1%.

E. Laboratory diagnosis of AIDS involves the demonstration of HIV antibody by immunoassay followed by a confirmatory Western blot assay.

F. Therapy. The only currently FDA-approved treatment for AIDS is the administration of **3′-azidothymidine (AZT)**; however, almost 100 other potential drugs are in or nearing clinical trials.

IV. SLOW (UNCONVENTIONAL) VIRUSES are a poorly characterized group of agents sometimes associated with **subacute, progressive disease**. These agents must be differentiated from conventional viruses responsible for such subacute, progressive illnesses as subacute sclerosing panencephalitis (measles virus), progressive congenital rubella (rubella virus), and subacute encephalitis (herpes simplex virus).

A. Clinical disease

    1. Scrapie, a disease of sheep and goats, is the most common disease associated with an unconventional virus.

    2. Human diseases associated with unconventional viruses are **kuru, Creutzfeldt-Jakob disease, and the Gerstmann-Straussler syndrome**.
        a. Spongiform encephalopathy is common to these chronic, subacute, neurologic diseases.
        b. The immune system appears not to recognize the infectious agent.
        c. Histologically, the brain tissue shows a widespread astrocytic hypertrophy and a marked status spongiosis. There is no inflammation of the brain.

B. In the absence of a tissue culture system, these agents have been studied only in laboratory animals.

C. The **prion protein (PrP)**, a unique hydrophobic protein, has been identified in both scrapie and Creutzfeldt-Jakob disease. The protein's role in the disease and its mode of replication are not understood.

    1. PrP is encoded in the DNA of the host and not by a nucleic acid carried in association with the protein, as in the case of conventional viruses.

    2. A cellular homolog of PrP is expressed on normal neurons, but it is slightly different in physical properties from the form associated with disease.

D. Transmission of the human diseases in this group is understood only in the case of kuru.

    1. Transmission of kuru is clearly a result of handling and consuming infected brain tissue. Cessation of ritualistic cannibalism in New Guinea has stopped the spread of kuru.

    2. The Gerstmann-Straussler syndrome is clearly a genetically transmitted disease, and there are familial clusters of Creutzfeldt-Jakob disease, which may be the result of genetic susceptibility or vertical transmission of the agent. Creutzfeldt-Jakob disease appears to be a sporadic, not an epidemic, illness.

STUDY QUESTIONS

Directions: Each question below contains five suggested answers. Choose the **one best** response to each question.

1. The incubation period for hepatitis A is

(A) less than 15 days
(B) 15–40 days
(C) 40–60 days
(D) 60–160 days
(E) more than 160 days

2. Three distinct forms of hepatitis B virus have been observed in serum from infected patients. All three forms share

(A) HB_sAg
(B) HB_eAg
(C) HB_cAg
(D) HB_sAg and HB_eAg
(E) HB_cAg and HB_eAg

3. Hepatitis C virus belongs to which of the following virus groups?

(A) Picornaviruses
(B) Herpesviruses
(C) Hepadnaviruses
(D) Togaviruses
(E) Retroviruses

4. Human papillomaviruses have been routinely demonstrated in

(A) malignant melanoma
(B) precancerous cervical biopsies
(C) epithelial carcinoma
(D) nasopharyngeal carcinoma
(E) none of the above

5. The retroviruses have many unique biologic and biochemical features. All of the following are properties of the retroviruses EXCEPT

(A) they are genetically diploid
(B) the particles contain an RNA-dependent DNA polymerase
(C) they require integration into the host genome for proper replication
(D) the RNA has minus polarity
(E) the retroviral group antigens are highly reactive between strains

6. The primary screening method in the detection of AIDS carriers is

(A) virus isolation
(B) Western blot followed by immunoassay
(C) immunoassay followed by Western blot
(D) immunoassay for viral antigen
(E) DNA hybridization for viral RNA

7. The unique properties of the unconventional viruses probably are a result of

(A) a hydrophobic protein component
(B) their unique nucleic acid
(C) their highly structured capsid
(D) a segmented DNA genome
(E) their carbohydrate capsule

8. Most of the obvious symptoms of AIDS are the result of

(A) infection by *Pneumocystis carinii*
(B) depletion of B cells
(C) depletion of $CD8^+$ T cells
(D) depletion of $CD4^+$ T cells
(E) HIV-infected macrophages

9. All of the following properties of transformed cells characterize transformation by SV40 virus EXCEPT

(A) activation by cellular genes
(B) integration of viral DNA into the cellular genome
(C) expression of a nuclear T antigen
(D) continuous budding of mature virus particles
(E) expression of a membrane T antigen

ANSWERS AND EXPLANATIONS

1. The answer is B. [*Table 23-1*] Hepatitis A has a short incubation period of between 15 and 40 days. The infection is transmitted by the fecal-oral route and takes hold very quickly. The virus replicates in the gastrointestinal tract and is shed in the feces during both the incubation and acute phases of the disease.

2. The answer is A. [*I B 1*] All three serologic forms of hepatitis B virus (HBV) contain hepatitis B surface antigen (HB$_s$Ag). The core antigen (HB$_c$Ag) is seen on only one of the three—the Dane particle, which is thought to be the complete virus. The early antigen (HB$_e$Ag) is seen only in association with infected cells and is the most diagnostic of current HBV infection.

3. The answer is D. [*I C*] Hepatitis C virus (HCV) is a recently isolated and characterized hepatitis virus that belongs to the togavirus group of RNA viruses. Hepatitis A virus (HAV) is a picornavirus, and HBV is a hepadnavirus. No major hepatitis viruses have been identified among the herpesviruses or retroviruses.

4. The answer is B. [*II A 4 d*] Evidence for the presence of several specific forms of papillomaviruses in precancerous cervical tissue has become quite convincing in the past few years. The only other site of routine observation of these viruses is in wart tissue. There is no established viral etiology for malignant melanomas or epithelial carcinoma. Nasopharyngeal carcinoma is associated with the presence of Epstein-Barr virus in many parts of the world.

5. The answer is D. [*II D 1*] The retroviruses are unique in their diploid genetic structure, with two identical RNA molecules per virion. In order to initiate an effective infection, the virion-associated reverse transcriptase must produce a double-stranded DNA copy of the viral RNA, and that copy must be integrated into the host genome. Infectious virus then is produced from the integrated copy. The various viral antigens are made from messenger RNA (mRNA) produced from the DNA copy, and these antigens include both group-specific and host-specific reactive antigens. The viral RNA has plus polarity.

6. The answer is C. [*III E*] The current method of screening patients for evidence of infection by human immune deficiency virus (HIV) is to perform an immunoassay on their serum to detect the presence of antibodies to HIV. Immunoassay-positive sera are retested using a Western blot to confirm that the reactive antibodies are HIV specific. Presently, the direct detection of the HIV virus itself either by cultivation or by immunoassay is not reliable. The detection of viral RNA in blood looks promising in some cases but is still not reliable enough to be anything more than a research tool.

7. The answer is A. [*IV C*] To date, no nucleic acid has been identified as part of the unconventional viruses; however, a unique protein—the prion protein—has been identified. This protein has the properties of a very hydrophobic molecule and probably accounts for most of the "unconventional" properties of this group of unusual pathogens.

8. The answer is D. [*III C 3*] The CD4$^+$ T cell is the primary target for HIV replication, and the depletion of this cell population accounts for most of the symptoms observed in AIDS patients. The CD4$^+$ cell is central to most functions of the immune system, and many symptoms appear to be more specific to other immune system functions, such as the loss of antibody-producing cells. However, these other cell populations generally are intact but cannot be properly activated or otherwise regulated. HIV-infected macrophages appear to be a significant reservoir of infectious virus in the body, but they alone do not account for the symptoms. *Pneumocystis carinii* infection is one of the symptoms frequently observed in fully developed AIDS, but it is not the cause of the other symptomatology.

9. The answer is D. [*II A 3 b*] Cell transformation by Simian virus #40 (SV40) is accompanied by the insertion of a copy of the viral genome into the cellular DNA. The expression of viral genes is limited to the early transcription products, which include the T antigens that serve as membrane proteins and those that serve as activators of cellular genes. Transcription is limited to the early regions; therefore, no viral capsid proteins are formed and no viral DNA is produced unless it is replicated as part of the cellular DNA. As a result, viral replication and viral production do not occur in these infected cells.

24
Medically Significant Protozoa

Gerald E. Wagner

I. SPOROZOA SPECIES. The sporozoa are unicellular protozoa that have a life cycle of alternating sexual and asexual reproduction. The major human diseases caused by sporozoa are malaria and toxoplasmosis. It is estimated that these two diseases affect about 35% of the world population.

A. Malaria. A number of ***Plasmodium* species** cause disease in vertebrates. Human malaria is caused by *Plasmodium malariae, Plasmodium falciparum, Plasmodium ovale*, and *Plasmodium vivax*.

    1. Life cycle. All of the plasmodia have essentially the same life cycle. The species differ, however, in their ability to invade specific subpopulations of erythrocytes.

        a. Sporogony is the sexual stage of the plasmodia life cycle. It occurs during a 1- to 3-week period in the gastrointestinal tract of mosquitoes.

            (1) Sporogony begins with the ingestion of male and female gametocytes by a female ***Anopheles* mosquito**, which acquires them in a blood meal from an infected human.

            (2) Fertilization results in a **zygote**, which invades the gut wall of the mosquito and becomes an **oocyst**.

            (3) **Sporozoites** develop, rupture the oocyst, and disseminate within the mosquito. Some of them penetrate the salivary glands, rendering the mosquito infectious for humans.

        b. Schizogony is the asexual stage of the plasmodia life cycle. It occurs during a 1- to 2-week period in human hepatocytes.

            (1) The hepatocytes usually are infected within 1 hour after the mosquito injects sporozoites into subcutaneous capillaries during a blood meal.

            (2) Each sporozoite multiplies to produce 2,000–40,000 **merozoites**. The merozoites rupture the hepatocyte and are released into the circulation.

        c. An **erythrocyte phase** is initiated by merozoites invading red blood cells.

            (1) The parasite appears as a ring-shaped **trophozoite** that enlarges and divides to become a multinucleated **schizont.**

            (2) The schizont develops into 6–24 merozoite daughter cells.

            (3) *P. vivax, P. ovale*, and *P. falciparum* cause lysis of the erythrocyte in 48 hours; *P. malariae* causes lysis in 72 hours. The released merozoites then infect other erythrocytes.

    2. Epidemiology

        a. Distribution. Human malaria occurs on a worldwide basis at latitudes between 45° N and 40° S and at altitudes below 1800 m.

            (1) *P. vivax* is the most widely distributed plasmodium in temperate climates. The relatively rare *P. malariae* also is found in temperate climates.

            (2) *P. falciparum* usually is seen in the tropics. The rare *P. ovale* is found only in Africa.

        b. Incidence

            (1) Malaria affects more than 125 million people in about 104 countries. Death generally occurs among infants and nonimmune adults.

            (2) The incidence of malaria is dependent upon the density of feeding mosquitoes and the number of infected people who serve as reservoirs of the parasite.

    3. Clinical manifestations. Fever, chills, anemia, and circulatory changes are seen in all forms of malaria. These manifestations are caused by the parasites invading the erythrocytes.

        a. Fever correlates with the lysis of erythrocytes and the release of merozoites.

            (1) Initially, the fever is erratic because lysis occurs at different times. As the cycle of the parasite becomes synchronized, the fever occurs at 48- to 72-hour intervals (e.g., *P. malariae* sporulates every 72 hours). The temperature may reach 40.0°–41.7° C.

(2) A **pyrogen** may be released at the time of lysis, although none has been identified.
- b. **Anemia** is the result of lysis and the phagocytosis of infected erythrocytes by cells in the reticuloendothelial system.
  - (1) **Blackwater fever** is massive hemolysis resulting in hemoglobinuria.
  - (2) **Intravascular hemolysis** is rare, but it can occur during falciparum malaria.
- c. **Circulatory changes** are caused by high temperature and other factors during the illness.
  - (1) Increased vasodilation results in decreased blood volume and hypotension.
  - (2) Vasospasm, increased blood viscosity, capillary obstruction by agglutinated erythrocytes, and intravascular coagulation can occur and can result in tissue hypoxia and infarction.
- d. **Acute glomerulonephritis** is seen in falciparum malaria, and **progressive renal failure** can occur during chronic *P. malariae* infection. These renal effects probably are due to the host's immune response.
- e. **Splenomegaly and thrombocytopenia** also are common in malaria.

4. **Immunity**
  - a. Invasion of the erythrocytes causes a rapid immunologic response that inhibits multiplication of the plasmodia to some degree and, thus, moderates the clinical manifestations.
  - b. Malaria has a long recovery period with increasing immunity. This period is characterized by recurrent multiplication of the parasite with subsequent symptomatology. The syndrome gradually subsides and eventually disappears.
  - c. Both T and B lymphocytes are required for immunity, but the exact mechanism is unknown.

5. **Laboratory diagnosis.** Malaria can be diagnosed by observing the parasites in erythrocytes of symptomatic patients. Table 24-1 lists differences in erythrocytes and parasites for the various *Plasmodium* species, as observed in peripheral blood.
  - a. Peripheral capillary or venous blood is used to make a smear that is stained with either Wright's or Giemsa stain.
  - b. Morphologic differences are used to differentiate *Plasmodium* species. Serologic methods generally are used in epidemiologic studies.

6. **Therapy.** Effective treatment of malaria requires elimination of the erythrogenic schizont (to terminate the acute attack), the hepatocytic schizont (to prevent relapse), and the circulating gametocyte (to prevent spread of the infection to other humans). No single drug accomplishes all three effects.
  - a. **Acute attacks** can be terminated with several antimalarial agents.
    - (1) **Chloroquine** is the most widely used agent. This 4-aminoquinoline compound is active against all four species of plasmodia, although some resistance has been reported.

Table 24-1. Erythrocyte and Parasite Characteristics for *Plasmodium* Species

Characteristic	*P. vivax*	*P. ovale*	*P. malariae*	*P. falciparum*
Erythrocyte				
Enlarged, pale	+	+	–	–
Ovoid, irregular	–	+	–	–
Schuffner's dots	+	+	–	–
Maurer's dots	–	–	–	+
Parasite				
All asexual stages seen	+	+	+	–
Band forms	–	–	+	–
Double chromatin dot	–	–	–	+
Rounded gametocytes	+	+	+	–
Banana-shaped gametocytes	–	–	–	+

+ = presence of the characteristic; − = absence of the characteristic.

    (2) Quinine, folate antagonists, and sulfonamides can be used in combination as effective therapy.

  b. The hepatic schizont of *P. vivax* and *P. ovale* is treated effectively with the 8-aminoquinoline, **primaquine**. This drug can induce massive hemolysis in patients with glucose-6-phosphate dehydrogenase (G6PD) deficiency.

  c. Gametocytes of *P. vivax*, *P. ovale*, and *P. malariae* are killed by the action of **chloroquine**. Primaquine is effective against this stage of *P. falciparum*.

7. Prevention. Malaria control is directed toward reducing the number of infected individuals and controlling the mosquito vector. A human vaccine currently is not available, but there is major effort and support for one.

B. Toxoplasmosis. *Toxoplasma gondii* is the etiologic agent of disease syndromes in both adults and neonates. The organism is an obligate intracellular sporozoan.

1. Life cycle

  a. Primary host. The definitive host of *T. gondii* is the domestic cat and other felines.

    (1) Ingested oocysts germinate, and the parasites enter the epithelial cells of the ileum. The intracellular trophozoite resides in a membrane-bound vesicle and undergoes schizogony.

    (2) Merozoites rupture the epithelial cells, invade adjacent cells, and undergo asexual reproduction until gametocytes eventually are formed. Fusion of male and female gametocytes results in oocyst formation.

    (3) The process in the primary host requires 1–3 weeks.

  b. Intermediate hosts. Oocysts are excreted in the feces of the feline host. Any warm-blooded animal that ingests the oocysts develops an intracellular infection of macrophages.

    (1) The macrophages carry the sporozoites to all organ systems via the lymphatics.

    (2) Continued schizogony causes rupture of the macrophages and the release of infective parasites that invade any nucleated host cell.

2. Epidemiology

  a. Incidence. The incidence of **exposure** to *T. gondii* increases with age. Serologic surveys indicate that 50% of American adults have been exposed.

  b. Distribution. Toxoplasmosis occurs worldwide and has been seen in almost all mammals and in many birds.

  c. Transmission. Human toxoplasmosis is transmitted in a variety of ways in different populations, geographic areas, and age groups. The most common means of transmission are ingestion of oocysts, ingestion of tissue cysts, and congenital acquisition.

3. Clinical manifestations. Most cases of toxoplasmosis are asymptomatic. Symptoms in overt disease generally are related to the status of the host.

  a. Normal, otherwise healthy, people most frequently develop a lymphadenopathy involving the cervical lymph nodes, although other nontender nodes may be detected.

    (1) Other symptomatology includes pharyngitis, fever, rash, hepatomegaly, splenomegaly, and an atypical lymphocytosis; the disease mimics infectious mononucleosis.

    (2) Meningoencephalitis, visceral involvement, pneumonitis, myocarditis, hepatitis, and chorioretinitis can develop in severe cases of toxoplasmosis.

  b. Immunocompromised people, including those with AIDS, acquire a serious, often fatal, form of toxoplasmosis.

    (1) Most commonly, administration of immunosuppressive agents reactivates latent disease.

    (2) Primary disease in these patients results in disseminated toxoplasmosis with necrotizing encephalitis, myocarditis, and pneumonitis.

    (3) Encephalitis occurs in about 50% of cases and is the apparent cause of death in 90% of fatal cases.

  c. Congenital toxoplasmosis has the characteristics of the most severe form of the disease because the fetal immune system is not fully developed. Involvement of the viscera and central nervous system (CNS) is common. Abortions and stillbirths are the usual results of transplacentally acquired disease.

4. Laboratory diagnosis. *T. gondii* can be viewed microscopically in, and can be cultured from, biopsy material or body fluids. Serologic tests also are used in the diagnosis of toxoplasmosis.

  a. Microscopy. When examining tissue samples, care must be taken to distinguish acute infection from the presence of the parasite in a latent form.

(1) Trophozoites can be viewed in tissue stained with Wright's or Giemsa stain.
(2) Electron microscopy and indirect immunofluorescent stains have been used for detecting *T. gondii* in brain tissue.
(3) Inoculation of animals with blood from suspected patients can assist in detecting acute toxoplasmosis. The animal tissues are examined histologically for the presence of the parasite.

b. **Serologic diagnosis**
(1) **Detection of immunoglobulin G (IgG).** Peak serum titers usually are obtained 4–8 weeks into the acute infection; values of 1:1000 or more are typical.
(a) Indirect immunofluorescence is the most widely used serodiagnostic procedure for detecting IgG antibody against *T. gondii.*
(b) The Sabin-Feldman dye test also is sensitive and specific.
(c) Paired sera must be used; a fourfold rise in titer indicates acute toxoplasmosis.
(2) **Detection of IgM.** IgM is seen early in the acute disease.
(a) IgM can be detected by indirect immunofluorescence; a single high titer (i.e., 1:80) is indicative of disease.
(b) The test for detecting IgM antibodies against *T. gondii* is not sensitive, is difficult to standardize, and is not applicable to neonates and immunosuppressed people.
(3) **Radioimmunoassay and ELISA tests** to detect antibody induced against *T. gondii* presently are being developed. The use of these methods to detect IgG especially will be useful in diagnosing toxoplasmosis in the neonate and immunocompromised person.

5. **Therapy and prevention**
a. Therapy is not required for uncomplicated toxoplasmosis in the immunocompetent host unless symptomatology is severe or persistent.
b. Patients with severe or ocular toxoplasmosis, pregnant women, and immunocompromised people should be treated with combinations of **sulfonamides and pyrimethamine**.
c. Washing of hands after handling raw meat, proper cooking of food, and avoidance of cat feces are important control measures during pregnancy and for immunosuppressed people.

II. RHIZOPODA SPECIES. Members of the class Rhizopoda (i.e., the **amoebae**) are the most primitive protozoans. Several genera are obligate parasites of the human alimentary tract, and some of these live under anaerobic conditions. *Entamoeba histolytica* is the only species that is seen regularly as a cause of human infections. Of the many free-living organisms, *Naegleria* and *Acanthamoeba* **species** are the only organisms associated with human disease.

A. **Amoebiasis** is a gastrointestinal infection caused by *E. histolytica*. The disease is seen worldwide and is spread by the fecal-oral route.

1. **Cellular morphology.** *E. histolytica* can be differentiated from other species living in the colon by its size, nuclear characteristics, and cytoplasmic inclusions. This amoeba exists in **trophozoite and cyst forms**.
a. *E. histolytica* is 12–20 μm in diameter. Invasive strains that cause disease usually are larger than commensal strains. The parasite has a sharply defined, clear ectoplasm with finger-like **pseudopods**.
b. The nucleus has a small central **karyosome**. Peripheral chromatin in the form of fine granules is regularly spaced around the nuclear membrane. Initially the cyst contains a single nucleus.
c. Invasive strains generally contain ingested erythrocytes in the cytoplasm. The cyst contains a glycogen vacuole and one or more cigar-shaped chromatid inclusion bodies.

2. **Life cycle.** Humans are the primary host of *E. histolytica*. The microaerophilic amoebae live in the colon and feed on bacteria and tissue cells. In commensal relationships, the cysts are passed in the stools, whereas trophozoites frequently are excreted during diarrhea syndromes.
a. The parasite is acquired by ingestion of fecally contaminated food or water containing cysts. The cysts pass through the stomach and reach the terminal ileum.
b. Within the ileum, the cysts break open and release the quadrinucleated protozoan. The parasite then divides to form eight trophozoites that are carried to the colon.

3. **Epidemiology**
a. **Incidence.** Surveys in the United States indicate an infection rate of 1%–5%. In tropical areas with poor sanitation, infection rates probably exceed 50% of the population.

    b. Distribution. Areas with the highest incidence include Central and South America, Southeast Asia, and western and southern Africa. Symptomatic disease is sporadic.

  4. Pathogenicity. Virulence is an unstable characteristic of *E. histolytica*. Cells maintained in culture lose their virulence, although it can be reactivated by animal passage.

    a. Virulent strains of *E. histolytica* have the unique ability to lyse intestinal epithelial cells after cell-to-cell contact; the exact mechanism of virulence is unknown.

      (1) These strains produce a cytotoxic enterotoxin and a variety of enzymes, but there is no clear relationship between these substances and virulence.

      (2) The invasive strains of *E. histolytica* belong exclusively to only 2 of 18 different zymodemes. The factors distinguishing these two groups have not been defined completely.

    b. Virulent strains are seen more frequently in the tropics than in temperate climates.

  5. Clinical manifestations

    a. General symptomatology. The most frequent signs and symptoms of infection are diarrhea, flatulence, and abdominal cramps.

      (1) The diarrhea is intermittent and usually contains blood and mucus.

      (2) Physical findings include abdominal tenderness. Sigmoidoscopy reveals ulcerative lesions.

    b. Amoebic dysentery usually occurs in debilitated people and pregnant women. It also may be induced by corticosteroid therapy.

      (1) Amoebic dysentery is characterized by severe abdominal pain, fever, and profuse bloody stools.

      (2) Complications include massive colonic hemorrhage and perforation of the bowel.

    c. Hepatic abscess is seen in about 5% of patients with clinical disease.

      (1) Hepatomegaly and point tenderness due to a single abscess in the upper outer quadrant of the right lobe are the most common signs.

      (2) Severe complications arise with extension of the abscess into the peritoneum or thoracic cavity.

    d. Asymptomatic disease. Most people with colonic involvement are asymptomatic. Spontaneous clearance of *E. histolytica* is common, and about 15% of the population in the United States are transient carriers during any year. Some asymptomatic people carry virulent strains of the parasite.

  6. Diagnosis

    a. Intestinal amoebiasis is diagnosed by determining the presence of *E. histolytica* in stools or sigmoidoscopic aspirates.

      (1) Smears for histologic stains should be fixed immediately for later staining and detection of the fragile trophozoites.

      (2) Wet mounts are prepared to detect characteristic motility of cysts.

      (3) *E. histolytica* must be differentiated from other commensal intestinal amoebae on the basis of morphologic criteria.

    b. Extraintestinal infections are more difficult to diagnose. Material from drained abscesses can be examined for trophozoites.

    c. Serologic tests may be difficult to interpret because the antibodies may persist for years.

      (1) Serologic tests are useful in diagnosing extraintestinal infection; more than 90% of patients with hepatic abscess demonstrate antibody titers.

      (2) Asymptomatic patients with intestinal infection generally do not have detectable serum antibodies.

  7. Therapy. Metronidazole is the drug of choice in the treatment of *E. histolytica* infections. Supportive therapy such as the replacement of fluid, electrolytes, and blood also is beneficial.

B. Primary amoebic meningoencephalitis is produced by the free-living **Naegleria and Acanthamoeba species**. Disease produced by the *Naegleria* species is almost always fatal, whereas the *Acanthamoeba* species produce a milder meningoencephalitis.

  1. Naegleria fowleri have been implicated as the etiologic agent in cases of primary amoebic meningoencephalitis. Serologic surveys suggest that asymptomatic infection is more common.

    a. Epidemiology. *N. fowleri* live in freshwater lakes and ponds. Cases of disease have been reported principally from the United States, Australia, Africa, Great Britain, and Czechoslovakia.

    b. Acquisition. Patients who have developed meningoencephalitis due to *N. fowleri* have a history of swimming or waterskiing in small, shallow lakes.
        (1) The infection occurs in the summer months.
        (2) Infection apparently from chlorinated pools and hot mineral springs also has been reported.
    c. Clinical manifestations
        (1) The amoebae cross the nasal mucosa and enter the CNS via the cribriform plate. A severe hemorrhagic inflammation is produced, and the infection spreads from the olfactory bulbs to other brain tissue.
        (2) Onset of the disease is sudden, and the course of the disease is rapid and fatal.
    d. Diagnosis. Primary amoebic meningoencephalitis cannot be distinguished from cases of purulent meningitis on the basis of clinical findings.
        (1) The cerebrospinal fluid (CSF) usually is bloody and contains many neutrophils. Careful microscopic examination of wet mounts usually reveals the amoebae.
        (2) Specific immunofluorescent stains or culture confirms the diagnosis.
    e. Therapy. The only drug that appears to be useful in treating meningoencephalitis caused by *N. fowleri* is **amphotericin B.** The disease is virtually always fatal.

  2. *Acanthamoeba* **species** cause a milder form of meningoencephalitis. The disease usually is seen in debilitated or immunocompromised people.
    a. Epidemiology. *Acanthamoeba* species are found in soil and in fresh and brackish water, and they have been isolated from the human oral cavity.
    b. Acquisition. The mechanism of human infection with *Acanthamoeba* species is unknown.
        (1) The parasite probably reaches the brain by hematogenous spread from a primary lesion in the respiratory tract or cutaneous tissue.
        (2) A history of swimming usually is absent.
    c. Clinical manifestations
        (1) A diffuse, sometimes metastatic, necrotizing, granulomatous encephalitis is characteristic. The course of the disease is more prolonged than in infection caused by *Naegleria* species.
        (2) The patient frequently recovers spontaneously, although death may occur in immunocompromised patients.
    d. Diagnosis. The CSF shows a mononuclear response. The amoebae may be seen in the CSF and in specimens from skin and corneal lesions.
    e. Therapy. The clinical efficacy of antimicrobial agents has not been determined. In vitro tests show susceptibility to sulfonamides, co-trimoxazole, polymyxin, and 5-fluorocytosine, but the clinical value of these agents is doubtful.

III. FLAGELLATE PROTOZOA. The flagellate protozoa are characterized by their specific types of motility, which is due to one or more flagella. Many genera of flagellate protozoa are human parasites, but only four regularly cause human disease. *Trichomonas* **and** *Giardia* **species** are luminal parasites; *Leishmania* **and** *Trypanosoma* **species** are tissue and blood parasites.

A. Trichomoniasis. *Trichomonas vaginalis* is the only species of the genus that produces human disease. Saprobic species—*Trichomonas hominis* and *Trichomonas tenax*—colonize the intestinal tract and the mouth. *T. vaginalis* is the etiologic agent of a sexually transmitted infection commonly seen in women.

  1. Morphology and growth characteristics
    a. *T. vaginalis* is an ovoid protozoan measuring 7 μm in diameter and 15 μm in length. It has a single elongated nucleus and a small cytosome.
    b. Four flagella exit the cell and a fifth bends back and runs posteriorly along an undulating membrane.
    c. The protozoan grows best on artificial media at a pH of 5.5–6.0 and under anaerobic conditions.

  2. Epidemiology
    a. Trichomoniasis is a cosmopolitan disease. It is estimated that 3 million women annually acquire the disease in the United States. Nonvenereal transmission is rare.

b. About 25% of sexually active women are infected, and the incidence is directly proportional to the number of sexual contacts.

c. Men are believed to be similarly infected, but confirmatory data are not available.

3. Clinical manifestations

a. *T. vaginalis* usually produces an irritating, **persistent vaginitis**. The symptomatology typically worsens during menses and pregnancy, probably because the vagina is more alkaline in those conditions.

(1) Vulvar itching and burning, dysuria, and dyspareunia are common symptoms of the vaginitis. The symptoms fluctuate in intensity over a period of weeks to months.

(2) About 75% of recently infected women develop an odiferous discharge.

(a) The copious discharge generally pools in the posterior vaginal fornix.

(b) The discharge is described classically as thin, yellow, and frothy, but in reality, it seldom has these characteristics.

(3) Physical examination reveals reddened vaginal and cervical mucosa.

(a) Petechial hemorrhages and extensive erosion are common in severe cases of vaginitis.

(b) A reddened, granular, friable endocervix is characteristic of trichomoniasis but is seen rarely.

b. **Infection in men** usually involves the urethra and prostate gland and, occasionally, the epididymis.

(1) Infections in men **usually** are **asymptomatic** because the parasite is effectively eliminated by urination.

(2) Recurrent dysuria and a slight, nonpurulent discharge are seen in symptomatic cases.

4. Laboratory diagnosis. Wet-mount preparations of specimens taken from the genital tract provide the best diagnostic test. The preparations are examined for protozoa with characteristic morphology and motility. Specimens also may be cultured in the laboratory.

5. Therapy. Metronidazole administered orally at recommended dosages is effective in curing trichomoniasis in 95% of patients. **Vinegar douches** frequently control the symptoms.

a. Metronidazole should not be used during the first trimester of pregnancy because of its potential teratogenic effects.

b. Treatment of sexual partners can prevent reinfection.

B. Giardiasis. *Giardia lamblia* is the etiologic agent of a diarrheal disease known as giardiasis. It is the most frequently identified intestinal parasite in the United States.

1. Morphology. *G. lamblia* is a large flagellate that measures 9–21 μm in length, 5–15 μm in width, and 2–4 μm in thickness. It has both trophozoite and cyst forms.

a. The trophozoites

(1) Two nuclei and a central parabasal body give the trophozoite the appearance of having eyes and a crooked mouth.

(2) Four pairs of flagella, located anteriorly, laterally, ventrally, and posteriorly, are responsible for the trophozoite's peculiar tumbling or falling-leaf motion.

(3) A large ventral sucker allows attachment to intestinal epithelium.

b. The cyst

(1) Cysts are oval and smaller than the trophozoites.

(2) The mature cyst contains four nuclei, a sucking disk, four parabasal bodies, and eight axonemes.

2. Habitat. *Giardia* species colonize the duodenum and jejunum, where they attach to epithelial cells. Free parasites are carried with fecal matter into the colon, where the infective cysts are formed.

3. Epidemiology

a. Giardiasis is most prevalent in areas with poor sanitation and in areas where personal hygiene is difficult.

b. The disease is transmitted by the fecal-oral route. The parasite has been isolated from 4% of stool samples submitted for pathologic examination in the United States.

c. All age-groups are affected, but the attack rate is more than 90% in **day-care nurseries.**

4. Clinical disease
a. Malabsorption syndrome
(1) Giardiasis primarily involves the malabsorption of lipids, carbohydrates, and vitamins from the gastrointestinal tract.
(2) Coating of the epithelium by large numbers of *G. lamblia* and damage to the microvilli probably are contributing factors, although the exact mechanism is not known.
b. Infectivity
(1) Susceptibility to giardiasis is related to strain virulence, inoculum size, production of hydrochloric acid in the stomach, and the immunologic status of the host.
(2) Fewer than 100 cysts produce infection.
c. Clinical manifestations
(1) Clinical giardiasis is characterized by diarrhea; many infections are asymptomatic.
(2) Onset is sudden and explosive 1–3 weeks after ingestion of the cysts.
(3) Abdominal cramps, distension due to gas, low-grade fever, nausea, and vomiting also are common.
(4) The disease resolves in 1–4 weeks.

5. Diagnosis
a. Giardiasis is diagnosed by observing either cysts in formed stools or trophozoites in diarrheal stools or by observing trophozoites in duodenal secretions or in jejunal biopsy specimens.
b. Repeated examination of specimens obtained on a weekly basis may be necessary in chronic cases.
c. Culture and serologic tests generally are not employed in the routine diagnosis.

6. Therapy. Quinacrine hydrochloride, metronidazole, and furazolidone are the three chemotherapeutic agents currently available in the United States for the treatment of giardiasis.
a. Quinacrine hydrochloride and metronidazole have a cure rate of 70%–95%.
b. Furazolidone has a lower cure rate but can be used in a liquid suspension for treatment of pediatric patients.

C. Leishmaniasis. *Leishmania* species are obligate intracellular parasites of mammalian tissue. Several species cause human disease. All species are morphologically identical, which causes some taxonomic confusion.

1. Etiology
a. *Leishmania tropica* produces a localized cutaneous ulcer known as **oriental sore** in Old World countries.
b. *Leishmania mexicana* produces a similar lesion known as **chiclero ulcer** in New World countries.
c. *Leishmania braziliensis* is the etiologic agent of **mucocutaneous (American) leishmaniasis**.
d. *Leishmania donovani* causes a disseminated disease known as **kala-azar**.

2. Common characteristics
a. Morphology
(1) The **promastigote** is the infectious form that resides in the gut of the intermediate insect vector. It measures 15–30 μm in length and 1.5–4.0 μm in width. The **kinetoplast** is located in the anterior end, where the flagellum leaves the cell.
(2) The **amastigote** inhabits the mammalian host. It is round to oval and measures 1.5–5.0 μm in diameter. The amastigote contains a clear nucleus and a central karyosome. There is no free flagellum, although the kinetoplast and axoneme are present.
b. Life Cycle. *Leishmania* species spend part of their lives in blood-sucking sandflies that inhabit animal burrows and feed at night. They are transmitted to humans and animals bitten by infected sandflies.
(1) Amastigotes that are ingested by sandflies from a blood meal on an infected mammal assume the promastigote form.
(2) The promastigotes multiply in the gut and eventually invade the buccal cavity of the sandfly.
(3) A blood meal on a human or animal injects the parasite into the skin.
(4) The injected promastigotes invade the cytoplasm of the mononuclear phagocytes of the new host, where they assume the amastigote form, multiply, and rupture the cell. Then they invade other monocytes or are ingested by a feeding sandfly.

3. **Localized cutaneous leishmaniasis** is caused by **L. tropica** and **L. mexicana**.
   a. **Epidemiology.** These protozoa cause a zoonotic infection of rodents in tropical and subtropical regions. Children are particularly susceptible, and the disease occurs in both rural and urban areas.
   b. **Clinical manifestations**
      (1) Lesions usually are seen on the extremities or on the face 1–2 months after the bite of the sandfly. The lesion first appears as a simple papule with regional lymphadenopathy.
      (2) Craters are produced when the papules ulcerate a few months after forming. The craters have raised erythematous edges and a granulating base; they are painless. Satellite lesions may form around the primary ulcer and fuse with it.
      (3) **Spontaneous healing** occurs in 3–6 months, and a hypopigmented, pitted scar is formed. Lesions occurring on the ears occasionally fail to heal and cause progressive destruction.
   c. **Diagnosis** of the disease is made on a clinical basis in endemic areas.
   d. **Immunity** is strain specific and permanent.

4. **Mucocutaneous leishmaniasis** is caused by **L. braziliensis**.
   a. **Epidemiology.** The organism naturally occurs in large forest rodents of Latin America. Human disease usually is seen in people who establish settlements in new areas of jungle.
   b. **Clinical manifestations**
      (1) Primary skin lesions resemble those produced by *L. tropica*. They appear 1–4 weeks after the sandfly bite and may heal spontaneously.
      (2) More typically, the primary lesion progressively enlarges over a period of weeks to years. Painful, destructive, metastatic lesions of the mouth and nose appear in 2%–50% of cases. The metastatic lesions occasionally appear many years after the primary lesion has healed.
      (3) Destruction of the nasal septum, hard palate, and larynx may occur. Deformity of the lips and cheeks also is seen.
      (4) Fever, weight loss, anemia, and secondary bacterial infections are common.
   c. **Diagnosis** is made by observing the etiologic agent in specimens taken from the lesions. Growth characteristics in culture can be used to differentiate *L. braziliensis* from other species.
   d. **Therapy** includes **pentavalent antimonials** and **amphotericin B.** Advanced disease is difficult to cure, and relapses are common.
   e. **Immunity.** Recovered patients are immune to reinfection.

5. **Disseminated visceral leishmaniasis** (i.e., **kala-azar**) is caused by **L. donovani**. The epidemiologic and clinical features of the disease vary with geographic location.
   a. **Epidemiology.** Kala-azar is seen in the tropical and subtropical areas of every continent except Australia. In general, rodents are the animal reservoir, and human disease is acute and often fatal.
      (1) In Eurasia and Latin America, the dog is the primary reservoir; human disease is subacute or chronic and principally involves children.
      (2) In India, humans are the only known reservoir, and the disease occurs in epidemic form at 20-year intervals.
   b. **Clinical disease**
      (1) The promastigotes of *L. donovani* enter the bloodstream of humans and are phagocytized by the histiocytes of the liver, spleen, bone marrow, lymph nodes, skin, and ileum. The parasites replicate in these organs and produce enlargement with atrophy.
      (2) **Symptomatology** generally appears 3–12 months after inoculation.
         (a) **Fever** usually is present.
            (i) Body temperature may rise abruptly or gradually.
            (ii) Fever persists for 2–8 weeks, subsides, and then reappears at irregular intervals.
         (b) **Diarrheal syndrome and malabsorption** also are common manifestations.
         (c) **Physical findings** include extensive splenomegaly, lymphadenopathy, hepatomegaly, and edema. The kala-azar (i.e., **black fever**) is a grayish pigmentation that develops on the face and head of light-skinned persons.
         (d) **Anemia** is common in advanced cases, and thrombocytopenia leads to petechial hemorrhages. Agranulocytosis occurs, and the leukocyte count typically is less than 4000/mm^3.

   c. Diagnosis is made by identifying the etiologic agent in aspirated material from bone marrow, spleen, liver, or lymph nodes. Serologic tests that are presently available lack sensitivity.

   d. Therapy

      (1) Pentavalent antimonials substantially decrease the mortality rate below the 75%–90% level seen in untreated cases. Initial therapy fails in about 30% of cases that occur in Africa, and another 15% have relapses.

      (2) Pentamidine and **amphotericin B** are used as second-line drugs.

D. *Trypanosoma* species resemble the leishmania in that their life cycle involves insect and animal hosts. Trypanosomes are the etiologic agents of **African trypanosomiasis (sleeping sickness)** and **American trypanosomiasis (Chagas' disease)**.

  1. Common characteristics

    a. Epimastigote. The trypanosomes exist as epimastigotes in the intestinal tract of insect vectors. In laboratory culture, the parasite produces epimastigotes on artificial media.

    b. Trypomastigote. The trypomastigote is the form in which the trypanosome circulates in the bloodstream of the animal host. An exception is seen in *Trypanosoma cruzi*, which forms an **amastigote**.

    c. Relative size. The trypanosomal forms are larger than those of the leishmania; the kinetoplast and origin of the flagellum are closer to the posterior end of the cell.

  2. African trypanosomiasis

    a. Etiology. African trypanosomiasis, or sleeping sickness, is caused by three subspecies of *Trypanosoma brucei*.

      (1) *T. brucei* subspecies *gambiense*, *T. brucei* subspecies *rhodesiense*, and *T. brucei* subspecies *brucei* are morphologically and serologically identical.

      (2) The subspecies can be differentiated on the basis of biologic and enzymatic characteristics.

    b. Life cycle

      (1) The infectious trypomastigote is ingested by the tsetse fly. It forms an epimastigote, undergoes replication in the gut and salivary glands, and becomes an infectious trypomastigote again within a period of a few weeks.

      (2) The parasite is transmitted to humans and other mammals by the bite of the tsetse fly. Humans and animals become reservoirs of the trypanosomes.

    c. Epidemiology. Sleeping sickness is confined to the central areas of Africa, where the tsetse fly is endemic. The range of this vector is bordered in the north by the Sahara Desert and in the south by the Kalahari Desert.

      (1) Gambian sleeping sickness, caused by *T. brucei* subspecies *gambiense*, is transmitted by the riverine tsetse fly of western and central Africa.

        (a) Humans appear to be the only reservoir of this trypanosome. The incidence of the disease corresponds to the proximity to water.

        (b) The infection rate is 2%–3% except in epidemics. The chronic nature of the infection maintains its presence in the population.

      (2) Rhodesian sleeping sickness, caused by *T. brucei* subspecies *rhodesiense*, is transmitted by the tsetse flies indigenous to the savannas of eastern Africa.

        (a) Small antelope serve as the major reservoir. Humans become infected when they graze livestock or hunt on the savanna.

        (b) Human-to-human and cattle-to-human transmission have been reported.

      (3) Disease attributed to *T. brucei* subspecies *brucei* is endemic in other geographic regions of central Africa. Sleeping sickness in each of these regions is transmitted by the variety of tsetse fly that is indigenous to the particular area.

    d. Clinical disease. Sleeping sickness is characterized by parasitemia and disseminated infection. Vasculitis, lymph node enlargement, and panencephalitis are major complications.

      (1) A **chancre** due to an inflammatory response appears at the site of inoculation. The chancre is seen 2–3 days after the fly bite.

        (a) Replicating parasites in the chancre disseminate through the lymphatic system and into the bloodstream.

        (b) A proliferative enlargement of the lymph nodes occurs.

      (2) Parasitemia characterized by recurrent symptomatology occurs 2–3 weeks after the bite.

(a) The patient experiences recurrent episodes of parasitemia accompanied by fever, skin rash, headache, tender lymphadenopathy, and mental disorder.

(b) The recurrent episodes are due to the emergence of new antigenic variants of the trypanosome after antibodies have eliminated the variant that caused the preceding parasitemia.

(3) Brain involvement

(a) Gambian sleeping sickness progresses slowly; brain involvement usually occurs several years after initial infection.

(b) Rhodesian sleeping sickness progresses more rapidly; brain and myocardial involvement are seen in 3–6 weeks. Coma, convulsions, and heart failure lead to death in 6–9 months.

(c) Brain damage generally is due to hemorrhage and a demyelinating panencephalitis.

(d) Attention span is lost, spontaneous activity decreases, tremors develop, speech becomes indistinct, sphincter control is lost, and seizures with transient paralysis occur.

e. Diagnosis. It is important to perform a lumbar puncture in patients with trypanosomiasis to determine whether brain involvement has developed.

(1) Microscopic examination. Lymph node aspirates, blood, and CSF are examined microscopically for trypomastigotes. Concentration of the parasites from blood or CSF by centrifugation can be of benefit in preparing wet mounts and stained smears.

(2) Culture. Inoculation of clinical specimens into mice or rats can enhance isolation of the trypanosomes in culture.

(3) Serologic tests. Specific trypanosomal antibodies and increased IgM levels also can be detected and are of diagnostic value.

f. Therapy

(1) Suramin and **pentadione** can be used to treat African trypanosomiasis if the CNS is not involved.

(2) A positive CSF warrants the use of **melarsoprol**, a highly toxic arsenical that crosses the blood–brain barrier.

3. American trypanosomiasis, or Chagas' disease, is caused by *T. cruzi* and is transmitted by species of true bugs in the family Reduviidae. The disease, which is acute in children and chronic in adults, is a febrile illness involving the heart or gastrointestinal tract.

a. Life cycle

(1) The life cycle of *T. cruzi* resembles that of *T. brucei*. The epimastigotes multiply in the midgut of different reduviids that inhabit distinct geographic areas of Central and South America.

(2) *T. cruzi* differs from *T. brucei* in that the trypomastigotes must invade mammalian tissue cells before they can multiply.

(3) The infectious trypomastigotes are inoculated into humans when the reduviid defecates while feeding. Scratching the bite area breaks the skin and allows entry of the parasite.

b. Epidemiology. Chagas' disease is estimated to afflict more than 7 million people in Central and South America.

(1) The disease is transmitted in rural areas, where the reduviid inhabits animal burrows and poorly constructed buildings. The reduviid usually bites near the eyes or lips of its sleeping host.

(2) Many wild and domestic animals, as well as humans, serve as reservoirs. Rats, cats, dogs, opossums, and armadillos can be infected and serve as reservoirs for human infection.

(3) Congenital disease and transfusion-transmitted disease are becoming more common in endemic areas.

(4) *T. cruzi* has been isolated from vertebrates and invertebrates in the southwestern United States. Serologic studies indicate that human exposure occurs, but no clinical cases have been reported in the United States.

c. Clinical disease. Chagas' disease is a febrile illness. Any nucleated host cell can be infected, and considerable tissue destruction usually occurs.

(1) An inflammatory chancre (i.e., a **chagoma**) occurs at the site of inoculation.

(a) Reddened eyes, swollen lids, and enlarged preauricular lymph nodes are seen 1–3 weeks after the bite and in conjunction with a chagoma near the eyes.

(b) Acute symptoms usually occur in children, although most infections in the younger age-group appear to be asymptomatic.

(2) **Parasitemia** eventually occurs and is characterized by fever, hepatomegaly, spleno-megaly, enlarged lymph nodes, peripheral edema, and a transient skin rash. The heart also may be involved.

(3) **Neonates** experience acute disease that frequently leads to meningoencephalitis. Myocardial and brain involvement leads to death in 5%–10% of untreated cases.

(4) **Chronic disease** is seen only in adults, who usually deny having acute disease. The disease is characterized by end-stage organ destruction.

    (a) The most serious form of the chronic disease involves **cardiac damage**. As many as 10% of some rural Latin American populations have cardiac manifestations of chronic disease.

    (b) Less commonly there is chronic involvement of the **gastrointestinal tract** in the form of megaesophagus and megacolon.

  d. Diagnosis

    (1) **Acute disease** is diagnosed by morphologically identifying the trypomastigotes of *T. cruzi* from peripheral blood. Procedures are the same as for sleeping sickness.

    (2) In **chronic disease**, isolation of the parasite is rare, and the diagnosis is made on the basis of clinical, epidemiologic, and serologic findings.

  e. Therapy. Nitrofuramox recently has been used in the treatment of acute disease, but its curative effect is debatable. There is no successful treatment for chronic Chagas' disease.

IV. PNEUMOCYSTIS CARINII is the most common cause of pneumonia in immunocompromised patients, particularly in those with AIDS. Little is known about this protozoan because the ability to culture it in the laboratory is limited.

A. Morphology and growth. Trophozoite and cyst forms of *P. carinii* have been identified.

  1. The **trophozoite** is pleomorphic and has a poorly defined cell membrane.

  2. The **cyst** is round to ovoid and measures 5–8 μm in diameter. It contains 2–8 sporozoites, which mature and are released as trophozoites when the cyst ruptures.

  3. *P. carinii* can be cultured in a variety of laboratory animals. The organism recently has been cultivated in tissue cell lines.

B. Epidemiology

  1. Distribution. Serologic studies suggest a **worldwide** distribution of pulmonary pneumocystosis. *P. carinii* appears to exist predominantly in a latent form; antibodies are detected in about 65% of the population.

  2. Incidence. Overt clinical disease occurs as sporadic cases and as epidemics. The disease is seen most frequently in patients receiving cancer chemotherapy, in immunosuppressed organ transplant recipients, and in homosexual men with AIDS.

  3. Many species of wild and domestic animals are infected with *P. carinii*, but the protozoa are sparse and difficult to detect in tissue samples.

C. Clinical disease. *P. carinii* shows a low degree of virulence for healthy, immunocompetent persons. The protozoan produces a progressive pneumonic syndrome in the compromised host.

  1. Infants generally have an insidious onset of pneumocystosis. Fever is either low-grade or completely absent. The disease lasts 3–4 weeks.

  2. Older children and adults receiving corticosteroid therapy comprise most of the patients.

    a. These patients generally experience an abrupt onset of disease that has a short duration.

    b. Fever in the range of 38°–40° C is characteristic of the pneumonia.

  3. The **disease in all patient groups** is characterized by progressive tachypnea and dyspnea. A nonproductive cough is prevalent in about 50% of patients, and hypoxia and cyanosis are common problems.

  4. Infiltrates are observed on radiologic examination of the lungs, but clinical signs of pneumonia usually are absent. The infiltrates are alveolar in character, and the disease usually involves the whole lung after beginning in the hilus.

  5. Pneumocystosis is a lethal disease in the immunocompromised patient; the mortality rate is 100%. Death usually is a result of progressive asphyxia.

D. Diagnosis

1. The finding of organisms (i.e., trophozoites or cysts) in the appropriate specimens from suspected patients generally is considered definitively diagnostic.

2. Bronchial lavage and endobronchial brush biopsy provide the best specimens without being too invasive.

3. Serologic tests for antibody detection are not sensitive or specific enough for diagnostic use.

E. Therapy

1. Co-trimoxazole reduces the mortality rate to 30% in most patients except those with AIDS. Long-term, low-dose therapy appears to lower the incidence of disease in high-risk populations.

2. Pentamidine also can be used in the treatment of pneumocystosis. Solubility difficulties and a wide spectrum of side effects limit the usefulness of this agent.

STUDY QUESTIONS

Directions: Each question below contains five suggested answers. Choose the **one best** response to each question.

1. Chagas' disease is caused by which of the following protozoa?

(A) *Toxoplasma gondii*
(B) *Trypanosoma brucei*
(C) *Leishmania donovani*
(D) *Trypanosoma cruzi*
(E) *Leishmania tropica*

2. Malabsorption syndrome is the primary characteristic of disease caused by which of the following protozoa?

(A) *Trypanosoma brucei*
(B) *Trypanosoma cruzi*
(C) *Entamoeba histolytica*
(D) *Giardia lamblia*
(E) *Toxoplasma gondii*

Directions: Each question below contains four suggested answers of which **one or more** is correct. Choose the answer

A if **1, 2, and 3** are correct
B if **1 and 3** are correct
C if **2 and 4** are correct
D if **4** is correct
E if **1, 2, 3, and 4** are correct

3. *Toxoplasma gondii* may be transmitted to humans by

(1) the fecal-oral route
(2) transplacental passage
(3) improperly cooked meat
(4) handling infected cats

4. Central nervous system involvement is characteristic of infections caused by which of the following protozoa?

(1) *Acanthamoeba* species
(2) *Giardia lamblia*
(3) *Naegleria fowleri*
(4) *Leishmania braziliensis*

5. Conditions typically seen in AIDS patients with primary toxoplasmosis include

(1) necrotizing encephalitis
(2) myocarditis
(3) pneumonitis
(4) hepatic abscess

6. Disseminated visceral leishmaniasis is caused by

(1) *Leishmania braziliensis*
(2) *Leishmania mexicana*
(3) *Leishmania tropica*
(4) *Leishmania donovani*

Directions: The group of questions below consists of lettered choices followed by several numbered items. For each numbered item select the **one** lettered choice with which it is **most** closely associated. Each lettered choice may be use once, more than once, or not at all.

Questions 7–11

Match the characteristic with the appropriate protozoan.

(A) *Pneumocystis carinii*
(B) *Entamoeba histolytica*
(C) *Naegleria fowleri*
(D) *Toxoplasma gondii*
(E) *Trichomonas vaginalis*

7. Oocysts in cat feces are transmitted to humans by the fecal-oral route

8. Etiologic agent of fulminant meningoencephalitis causing death in 3–5 days

9. Etiologic agent of a sexually transmitted disease

10. Etiologic agent of dysentery and hepatic abscess

11. Etiologic agent of pneumonia in immunocompromised people

ANSWERS AND EXPLANATIONS

1. The answer is D. [*III D 3*] *Trypanosoma cruzi* is the etiologic agent of American trypanosomiasis, or Chagas' disease. The disease is transmitted to humans by true bugs in the family Reduviidae. Most cases are asymptomatic, but an acute inflammatory, febrile disease occurs in children. Chronic febrile disease occurs in adults and is characterized by involvement of the heart or gastrointestinal tract with end-stage organ destruction. *Toxoplasma gondii* is the etiologic agent of lymphadenitis in immunocompetent people and of disseminated disease in fetuses and immunosuppressed patients. *Trypanosoma brucei* is the etiologic agent of African trypanosomiasis, or sleeping sickness. *Leishmania donovani* causes disseminated visceral leishmaniasis, and *Leishmania tropica* causes localized cutaneous leishmaniasis.

2. The answer is D. [*III B 4 a*] Malabsorption syndrome is the primary characteristic of disease caused by the intestinal flagellate *Giardia lamblia*. The absorption of lipids, carbohydrates, and vitamins is impaired by large numbers of parasites coating the intestinal epithelium and damaging the villi. The mechanism of the malabsorption is not completely understood. Heavy loads of other parasites occasionally cause malabsorption of nutrients. *Trypanosoma brucei* and *Trypanosoma cruzi* are the etiologic agents of sleeping sickness and Chagas' disease, respectively. *Entamoeba histolytica* causes intestinal disease typically characterized by diarrheal syndromes. *Toxoplasma gondii* normally produces a disease characterized by lymphadenitis.

3. The answer is E (all). [*I B 1 a–b, 2 c 3 c, 5 c*] *Toxoplasma gondii* is an intestinal flagellate that can be transmitted to humans in several ways. Generally, the parasite is transmitted by the fecal-oral route when the definitive host excretes oocysts in feces that are ingested by any warm-blooded host, including humans. The definitive host is the domestic cat and other felines. People also acquire toxoplasmosis by handling infected cats; oocysts on the hands are subsequently ingested. The flesh of warm-blooded animals becomes infected when macrophages ingest *T. gondii* and carry the sporozoites to any organ system. Improperly cooked meat from these animals becomes a source of infection. Toxoplasmosis contracted during pregnancy can result in congenital infection when the parasite crosses the placental membrane.

4. The answer is B (1, 3). [*II B 1 c (1), 2 c (1)*] *Acanthamoeba* species and *Naegleria fowleri* are etiologic agents of primary amoebic meningoencephalitis. These free-living amoeba usually are acquired by the respiratory route, and they penetrate the cribriform plate to cross into the brain. Primary amoebic meningoencephalitis is a serious disease typically causing death in 3–5 days. Infection caused by *Acanthamoeba* species is a milder form of disease. *Giardia lamblia* is an intestinal parasite, particularly of children, that occasionally may disseminate. *Leishmania braziliensis* is the etiologic agent of mucocutaneous leishmaniasis.

5. The answer is A (1, 2, 3). [*I B 3 b (2)*] Toxoplasmosis is a serious, life-threatening disease in neonates and severely immunosuppressed people such as AIDS patients. In immunocompromised people, the disease is characterized by necrotizing encephalitis, myocarditis, and pneumonitis. Disseminated toxoplasmosis usually is fatal. Hepatic abscess occurs in about 5% of people with amoebiasis caused by *Entamoeba histolytica*.

6. The answer is D (4). [*III C 5*] *Leishmania donovani* is the etiologic agent of disseminated visceral leishmaniasis. The disease, also known as kala-azar, is characterized by irregular fever, diarrhea, and malabsorption syndrome. It occurs in the tropical and subtropical areas of every continent except Australia. Other *Leishmania* species typically are associated with different disease syndromes. *Leishmania braziliensis* causes mucocutaneous leishmaniasis; *Leishmania mexicana* and *Leishmania tropica* cause localized cutaneous leishmaniasis.

7–11. The answers are: 7-D, 8-C, 9-E, 10-B, 11-A. [*I B 1 a, b, 2 c; II A 5 b, c, B 1 c (2); III A; IV*] Domestic cats and other felines are the primary definitive hosts of *Toxoplasma gondii*. Infectious oocysts are excreted in the feces of these animals, and humans become infected by the fecal-oral route.

Naegleria fowleri is the etiologic agent of primary amoebic meningoencephalitis. This fulminant disease typically causes death in 3–5 days. The free-living amoeba appears to be acquired from water and dust by the respiratory route.

Trichomonas vaginalis is the etiologic agent of a sexually transmitted disease frequently seen in women. It is estimated that 3 million women acquire the disease annually in the United States. Men who acquire the infection tend to be asymptomatic.

Entamoeba histolytica is the etiologic agent of amoebic dysentery; the disease usually occurs in debilitated people and during pregnancy. It is characterized by abdominal pain, fever, and profuse bloody stools. Hepatic abscess is seen in about 5% of patients with clinically overt amoebiasis caused by *E. histolytica*.

Pneumocystis carinii is the etiologic agent of pneumonia in immunosuppressed people. The disease generally is rare, but it has become an important cause of fatalities in AIDS patients.

25
Medically Significant Helminths

Gerald E. Wagner

I. INTESTINAL NEMATODES. Many nematodes parasitize the human gastrointestinal tract. These parasites are cylindrical and fusiform in shape and are covered with a protective cuticle. Six nematodes commonly infect humans; together they parasitize more than 25% of the world population.

A. Enterobiasis is caused by the pinworm **Enterobius vermicularis**. The infection is common in children.

1. **Morphology**
   a. **Size.** The female worm is about 10 mm long. The infrequently seen male is about 3 mm long.
   b. **Structure.** The pinworm is cream-colored. The female has a pointed tail and fan-like ridges at the anterior end.
   c. **Eggs.** The eggs of *E. vermicularis* are clear, ovoid, and flattened on one side. They measure about 50 μm in length and 25 μm in diameter.

2. **Epidemiology.** Enterobiasis affects about 200 million people worldwide and probably 30–40 million in the United States. The infectious eggs of *E. vermicularis* are relatively resistant to environmental conditions.
   a. Children are more susceptible than adults, but all age-groups are infected. Familial outbreaks are common.
   b. The incidence is higher in whites than in blacks.

3. **Clinical disease.** Enterobiasis rarely is a serious disease. The most common symptom is pruritus ani.
   a. **Pruritus ani** is most prominent at night and is associated with the migration of the pinworm. Irritability and other minor complaints are common.
   b. **Severe infestation** may induce scratching that can result in **secondary bacterial infections**. The worm can enter the female genital tract and cause vaginitis, endometritis, and salpingitis. The worm frequently is found in the appendix but probably plays no role in appendicitis.

4. **Diagnosis** of enterobiasis is suggested by the clinical manifestations and is confirmed by recovering and identifying eggs from the anal mucosa.

5. **Therapy** uses one of several agents, including **pyrantel pamoate** and **mebendazole**.
   a. **Asymptomatic cases.** Enterobiasis does not require treatment in the absence of symptoms. Many physicians, however, recommend simultaneous treatment of family members of symptomatic patients.
   b. **Symptomatic cases.** The cure rate with pyrantel pamoate or mebendazole treatment is high, but reinfection is common.

B. Ascariasis is caused by the short-lived roundworm **Ascaris lumbricoides** and is one of the most common infections in the world.

1. **Morphology**
   a. **Size.** *A. lumbricoides* is the largest intestinal nematode, measuring 150–500 mm in length. The male worm is somewhat smaller than the female.
   b. **Structure.** Like most ascarids, *A. lumbricoides* is pointed at the ends and has a firm, cream-colored cuticle.

    c. Eggs are ellipsoid, measure about 55×35 μm, and have a rough, albumin-like coat over a thin, chitinous shell.

        (1) Over 200,000 eggs are produced daily, even in the absence of fertilization.

        (2) The eggs can survive outside the intestinal tract for long periods of time.

2. Epidemiology. Ascariasis primarily affects children in warm climates.

    a. Incidence. More than 4 million Americans are infected, and it is estimated that over 1 billion people worldwide have the disease.

    b. Infectious, fertilized eggs can be ingested or inhaled. They are resistant to normal environmental conditions.

3. Clinical disease. Manifestations of disease are seen as a result of intestinal parasitism or of larval migration through the lungs. The worm load also affects the clinical picture.

    a. Gastrointestinal infestation

        (1) Asymptomatic infection. Infestation is asymptomatic if the worm load is small.

            (a) Worms usually are noted when they are vomited or passed in the stool.

            (b) Fever, although low grade, is common and appears to stimulate motility and migration of the worm.

        (2) Symptomatic infection. Heavy loads of 50 worms are common, and as many as 2000 worms have been recovered from a single child.

            (a) Abdominal pain and malabsorption are common, and **intestinal blockage** can occur.

                (i) Blockage results from the formation of a bolus of 100 worms or more.

                (ii) Blockage occurs in about 2 of every 1000 infected children.

            (b) The **mortality rate** from severe gastrointestinal infestation is 3%.

    b. Lung involvement. Pulmonary symptomatology is related to hypersensitivity in the previously exposed patient and to the existing number of parasites.

        (1) Fever, cough, wheezing, and shortness of breath are typical. Laboratory studies reveal hypoxia, eosinophilia, and pulmonary infiltration.

        (2) No specific mortality rate is associated with pulmonary ascariasis. Death occasionally occurs.

4. Diagnosis

    a. Gastrointestinal infestation usually is diagnosed by identifying eggs in the stool.

    b. Pulmonary disease is diagnosed by finding larvae and eosinophilia in the sputum.

5. Therapy. Pyrantel pamoate and **mebendazole** are effective therapeutic agents. These drugs are administered at 6-month intervals for communitywide control efforts.

C. Trichuriasis is caused by the whipworm *Trichuris trichiura*, which is a common cosmopolitan parasite.

1. Morphology

    a. Size. The adult whipworm is 30–50 mm in length.

    b. Structure. Whipworms have a thread-like anterior end and a bulbous posterior. The male has a coiled tail; the female's tail is straight.

    c. Eggs. The ovoid eggs are brown, thick-shelled structures with translucent knobs at the ends. Between 3,000 and 10,000 eggs are produced each day. They are about 50×25 μm in size.

2. Epidemiology. The whipworm is less widely distributed than helminths such as *E. vermicularis*, but it infects over 1 billion people.

    a. *T. trichiura* is seen primarily in areas with warm, moist climates and poor sanitary conditions. In the United States the disease occurs principally in the southeast.

    b. The attack rate in the tropics is much higher than in temperate areas; it has been reported to be 80%.

3. Clinical disease. Worm loads as high as 800 per infected person have been reported.

    a. Asymptomatic infestation. Patients with light worm loads usually are asymptomatic.

    b. Symptomatic disease

        (1) Typically, damage to the intestinal mucosa causes abdominal pain and diarrhea.

        (2) Heavy worm loads parasitize the entire colon and cause bleeding and anemia.

4. **Diagnosis** of symptomatic infection is made by identifying the eggs in the stool. There may be as many as 10,000 eggs per gram of feces.

5. **Therapy** is reserved for symptomatic cases. **Mebendazole** is the drug of choice, but the cure rate is only about 70%.

D. **Hookworm infection** is caused by ***Necator americanus* and *Ancylostoma duodenale***. It has been estimated that these two species consume 7 million L of blood per day from their 700 million victims.

1. **Morphology**
   a. **Size.** Both species measure about 10 mm in length.
   b. **Structure.** Hookworms are light pink to pinkish white.
      (1) The head and body usually are curved in opposite directions. The males have a unique opercular bursa, which is fan-shaped.
      (2) The species are differentiated by the structure of their oral cavity.
   c. **Eggs.** Thin-shelled eggs measure about 60 × 40 μm. Fertilized females release 10,000–20,000 eggs daily.

2. **Epidemiology.** Hookworms are found between latitudes 45° N and 30° S.
   a. **Transmission.** Disease is transmitted by skin contact with **filariform larvae**, which develop in warm, moist, shaded soil.
   b. **Infectivity.** Infection rates are highest in closed, heavily populated communities.
   c. **Distribution.** *N. americanus* is seen in Africa, Asia, and the Americas; *A. duodenale* occurs in China, Japan, India, the Middle East, and the Mediterranean area.

3. **Clinical disease**
   a. **Asymptomatic infestation.** In most cases of hookworm infestation, the worm load is small and the disease is asymptomatic.
   b. **Symptomatic disease.** Manifestations of symptomatic disease vary according to the original site of larval penetration, the migratory path of the larvae, and the presence or absence of adult worms in the gut.
      (1) **Rash.** A pruritic erythematous rash develops at the site of penetration, which usually is between the toes.
         (a) The rash probably is an **immunologic reaction** caused by prior sensitization to hookworm antigens.
         (b) There is some swelling at the site of larval penetration, and the rash may persist for months.
         (c) This phenomenon is seen most frequently in infections with *N. americanus*.
      (2) **Lung involvement.** Pulmonary migration of the larvae induces cough, wheezing, and low-grade fever. The symptoms usually are mild.
      (3) **Gastrointestinal infection.** Epigastric pain and abdominal peristalsis are typical symptoms of heavy gastrointestinal infestation.
         (a) Anemia and hypoalbuminemia occur because of chronic blood loss.
         (b) The manifestations of symptomatic gastrointestinal infestation by hookworms may be mild to severe.

4. **Diagnosis** is made by finding eggs in smears of stool. The worm usually is not identified specifically.

5. **Therapy. Pyrantel pamoate** and **mebendazole** are highly effective chemotherapeutic agents. Supportive therapy, such as iron supplementation and blood transfusions, may be necessary.

E. **Strongyloidiasis** is caused by the filarial parasite ***Strongyloides stercoralis***, which is a more evolutionarily advanced organism than the other nematodes.

1. **Morphology**
   a. **Size.** *S. stercoralis* is the smallest intestinal nematode; it is only 2 mm long.
   b. **Structure.** The parasite goes through a rhabditiform larval stage, a filarial stage, and an adult stage. The larvae are about 0.2 mm long.

2. **Epidemiology**
   a. **Transmission.** *S. stercoralis* is transmitted to humans both by direct skin contact with larvae in soil and by ingestion of filaria in food. Direct person-to-person transmission and autoinfection can occur.
   b. **Distribution.** Filaria have the same distribution as hookworms but are more prevalent in the tropics.

3. **Clinical disease**
   a. **Asymptomatic infestation.** Gastrointestinal infections with light worm loads usually are asymptomatic.
   b. **Symptomatic disease**
      (1) **Intestinal disease**
          (a) Heavy worm loads generally result in epigastric pain and tenderness, which are aggravated by the ingestion of food.
          (b) Cases of severe intestinal involvement may exhibit abdominal peristalsis, vomiting, diarrhea; and malabsorption.
      (2) **Lung involvement.** Pulmonary symptoms resemble those of ascariasis and hookworm infections. The inflammatory hypersensitivity response seen in hookworm infection is absent.
      (3) **Autoinfection.** Lesions on the lower back and buttocks can occur through autoinfection by the fecal route.
          (a) The raised erythematous lesions are transient and represent larval invasion of the skin.
          (b) The condition can recur over a period of decades if the infection is not treated.
      (4) **Hyperinfection.** Immunocompromised patients are susceptible to hyperinfection.
          (a) Severe enterocolitis and disseminated disease are seen in immunosuppressed persons.
          (b) The heart, lungs, and central nervous system (CNS) may become involved.

II. TISSUE NEMATODES.
Nematodes that invade tissues and the lymphohematogenous system of humans are a heterologous group. Humans may be incidental hosts of some tissue nematode infections, or they may be hosts for part of the life cycle of other nematodes.

A. **Toxocariasis** is caused by *Toxocara canis*, which is a natural parasite of dogs. Humans are incidental hosts; the disease also is known as **visceral larva migrans**.

1. **Morphology**
   a. **Size.** *T. canis* is a large ascarid, 150–350 mm long.
   b. **Structure.** The morphologic characteristics resemble those of *A. lumbricoides*.
   c. **Eggs.** The female *T. canis* produces up to 200,000 thick-walled eggs daily.

2. **Epidemiology**
   a. **Incidence in dogs.** Dogs in the United States have a high rate of toxocariasis; about 80% of puppies and 20% of adult dogs are affected.
   b. **Incidence in humans.** Serologic surveys indicate that 4%–20% of the human population in the United States have ingested viable eggs. Children 1–6 years old have the highest incidence of clinical disease.

3. **Clinical disease.** Any organ of the body can be invaded by *T. canis*. Hemorrhaging, necrosis, and the formation of eosinophilic granulomas may occur.
   a. **General characteristics**
      (1) The lungs, heart, liver, skeletal muscle, brain, and eyes are the most frequently involved tissues.
      (2) The severity of the disease is associated with the quantity and location of the lesions and the degree of immunologic response to larval antigens.
   b. **Disease in children.** Toxocariasis in children ranges from asymptomatic to severe. A persistent eosinophilia is characteristic of the disease.
      (1) **Moderate disease** manifests as fever and tender hepatomegaly.
      (2) **Severe disease** can present as splenomegaly, skin rash, pulmonary infiltration, asthma, behavior changes, focal neurologic defects, and convulsions.

       (3) Prognosis. The infection can persist for weeks to months. Death may occur as a result of cardiac dysfunction, respiratory failure, or brain damage.

    c. Disease in adults

        (1) Older children and adults who are immunocompetent rarely develop systemic manifestations.

        (2) Granulomatous endophthalmitis causing unilateral squinting and loss of visual acuity is the most common manifestation.

  4. Diagnosis

    a. Presumptive diagnosis of toxocariasis is made on the basis of the clinical syndrome and eosinophilic leukocytosis.

    b. Confirmed diagnosis is made by observing larvae in a liver biopsy and usually is not done until autopsy. Stool samples are of no benefit, because the parasite seldom attains the adult stage in humans.

  5. Therapy. The benefits and efficacy of antihelminthic agents are not certain. Corticosteroid therapy may prevent death in serious cases.

B. Trichinosis usually is caused by the ingestion of improperly cooked pork that contains *Trichinella spiralis*. The parasite becomes encapsulated within the muscles of its host and can remain viable for many years.

  1. Morphology

    a. Size. The male parasite measures about 1.5 mm in length, and the female is more than twice as long.

    b. Structure. The morphologic characteristics are consistent with those of other roundworms. The larvae become encysted in muscle tissue, and calcification can occur over a period of months.

    c. Eggs. Intrauterine embryogenesis occurs; therefore, second-stage larvae rather than eggs are released. About 1500 larvae measuring 100 μm in length are produced over a 6-week period. The male dies after copulation.

  2. Epidemiology. Human infection occurs worldwide except in Asia and Australia.

    a. Transmission from animals to humans. Most carnivores can be infected with *T. spiralis*.

        (1) Swine have the highest infection rate, and they are most frequently involved in the transmission of disease to humans.

        (2) In the United States most cases are due to the ingestion of improperly cooked contaminated pork. About 10% of cases are attributed to consumption of bear meat.

    b. Incidence. The incidence of cysts of *T. spiralis* in the diaphragm at autopsy has declined from 16% to 4% over the past 30 years in the United States.

        (1) It is estimated that 150,000–300,000 Americans are infected annually. Up to 1.5 million Americans carry *T. spiralis* in their muscles.

        (2) Most cases of trichinosis are asymptomatic; only about 100 cases are recognized clinically each year.

  3. Clinical disease. The lesions associated with trichinosis can be attributed to the invasion of striated muscle, the heart, and the CNS by the larvae. CNS involvement occasionally causes death.

    a. Initial infection

        (1) Symptoms appear 1–2 days after ingesting contaminated meat. The initial symptoms may be mild to severe.

        (2) The adult worms invade the intestinal mucosa. Abdominal pain, nausea, and diarrhea occur.

    b. Second stage

        (1) A second phase of the disease begins about 1 week later, with the larval invasion of muscle.

            (a) A load of 10 or fewer parasites per gram of muscle usually is asymptomatic.

            (b) A load of 100 or more parasites per gram of muscle produces significant disease.

            (c) A parasite load of 1000–5000 larvae per gram of muscle produces severe disease that may terminate in death.

        (2) Symptomatology

            (a) The most common symptoms are fever, weakness, and muscle pain.

(b) Eyelid edema, a maculopapular rash, and petechial hemorrhages in the conjunctiva and beneath the nails may develop.

(3) Complications

(a) Hemoptysis and pulmonary consolidation are common complications of trichinosis.

(b) Serious complications can include congestive heart failure, delirium, psychosis, paresis, and coma.

(4) Mortality rates range from 1% in mildly symptomatic cases to 10% when the CNS is involved.

4. Diagnosis

a. Blood count. Eosinophilic leukocytosis is the most common abnormality. Differential counts range from 15% to 50% during the second week of disease.

b. Serologic testing. Complement fixation (CF), indirect immunofluorescent antibody tests, and flocculation tests frequently are used in serologic diagnosis.

c. Biopsy. Cysts can be seen in muscle biopsy specimens, usually during the third week of illness.

5. Therapy

a. Corticosteroid treatment is beneficial in cases with severe edema, myocardial infection, pulmonary involvement, and CNS invasion.

b. The efficacy of **mebendazole** is questionable, but it may shorten the course of the disease.

C. Lymphatic filariasis is an acute inflammatory disease causing lymphatic blockage and, occasionally, elephantiasis. The syndrome may be caused by a variety of lymphatic parasites—most commonly **Wuchereria bancrofti** and **Brugia malayi**. Humans serve as definitive hosts during the life cycle of these parasites, which are transmitted by mosquitoes.

1. Morphology and life cycle

a. Size. The adult female *W. bancrofti* is about 100 mm long, and the male is about 40 mm long. Female and male *B. malayi* are about half these sizes. Male and female worms lie coiled together in the lymphatic system for up to 10 years.

b. Structural stages

(1) Eggs. Large numbers of embryonated eggs are produced, and at oviposition the embryos uncoil to become microfilaria.

(2) Microfilaria. Large numbers of microfilaria are produced. They are 200–300 μm long.

(a) The eggshell elongates to form a thin protective sheath for the microfilaria.

(b) The length, staining characteristics, and internal structures of microfilaria differentiate the two species.

(3) Larvae. The microfilaria develop into larvae that can migrate into the bloodstream and be taken by a mosquito during a blood meal.

(a) The larvae become filariform after going through a rhabditiform stage in the mosquito. **The filariform stage is infectious for humans.**

(b) These transformations occur in **specific mosquito vectors** for the two species.

2. Epidemiology. Lymphatic filariasis affects about 250 million inhabitants of Asia, Africa, Latin America, and the Pacific islands.

a. Approximately 75% of the infections occur in Asia.

b. *W. bancrofti* is cosmopolitan, and humans are the only known host.

c. *B. malayi* is seen in rural coastal areas, and it may infect other vertebrates as well as humans.

3. Clinical disease. Pathologic processes usually are confined to the lymphatic system. The disease can be both acute and chronic.

a. Acute disease. An acute syndrome is the result of the host response to dead or dying adult worms and the molting of younger worms.

(1) Acute lymphangitis with thrombus formation and hyperplastic changes in the lymphatic endothelium is accompanied by infiltration of lymphocytes, plasma cells, and eosinophils.

(2) Low-grade fever, lymphadenitis, and tender lymph nodes are typical findings. High fever (to 40° C), chills, muscle pain, and other systemic manifestations may occur.

(3) The symptomatology develops 8–12 months after exposure. The symptoms frequently subside spontaneously and then recur a few weeks later.

b. Chronic disease. Permanent lymphatic obstruction characterizes a chronic syndrome.

(1) Lymphatic blockage is the result of granuloma formation and fibrosis during repressed infection.

(2) Pulmonary and joint effusions, edema, hydrocele, and ascites are common.

4. Diagnosis

   a. Presumptive diagnosis. Eosinophilia and the clinical manifestations are strong evidence of disease in endemic areas, although mild cases may go unnoticed.

   b. Confirmed diagnosis requires the observation of microfilaria in the blood or in lymphatic, ascitic, or pleural fluids.

5. Therapy

   a. Diethylcarbamazine is effective in killing or suppressing microfilaria and adult worms.

   b. Corticosteroids and antihistamines are used to suppress the sometimes severe inflammatory response elicited by dead parasites.

D. Onchocerciasis is caused by *Onchocerca volvulus*, a cutaneous filarial parasite that spends part of its life cycle in humans. The disease, also known as **river blindness**, is transmitted to humans by an insect vector.

1. Morphology

   a. Size. Adult worms of *O. volvulus* are 20–50 mm long.

   b. Structure. Adult worms lie coiled in subcutaneous fibrous tissue. The microfilaria undergo transformation to filariform larvae within the insect vector.

   c. Embryos. About 2000 microfilaria are released daily by fertilized females for as long as 15 years.

2. Epidemiology. Onchocerciasis principally is a disease of the African tropics, although foci of infection occur in Yemen and Latin America.

   a. Transmission. The disease is transmitted to humans by **black flies** belonging to the genus *Simulium*, which live along rapidly flowing streams.

   b. Incidence. An estimated 50 million people are affected by onchocerciasis; 5% of victims are blind.

3. Clinical disease. Onchocerciasis is characterized by subcutaneous nodules, which can appear anywhere on the body.

   a. Nodule formation

   (1) The nodules are firm but movable, with diameters of 1–3 cm. They usually are painless unless they are located over a joint.

   (2) Typically, 10 or fewer nodules are found in each patient, but several hundred have been detected in severe cases.

   b. Immunologic response

   (1) Sensitization to antigens of the microfilaria results in a chronic inflammatory response.

   (2) The cutaneous manifestation is a papular erythematous rash with severe itching.

   (3) The skin eventually thickens, and subdermal elasticity is lost, resulting in the formation of large wrinkles and folds.

   c. Ocular involvement

   (1) Invasion of the eye results in punctate keratitis, iritis, and chorioretinitis. Visual acuity is lost, and eventually total blindness occurs.

   (2) In some African communities, 85% of the inhabitants have ocular lesions and 50% of the men are blind.

4. Diagnosis

   a. A confirmed diagnosis is made by demonstrating the microfilaria in a skin biopsy from an affected area.

   b. The microfilaria may be seen in the anterior chamber of the eye when ocular involvement develops.

5. Therapy. The microfilaria are killed by **diethylcarbamazine**. Treatment is initiated with low doses to prevent an intense inflammatory response to the dead parasites. Suramin or surgical excision must be used to eradicate adult worms.

E. Loiasis is a filarial disease of western Africa caused by *Loa loa*. The parasite spends part of its life cycle in the human host.

1. **Epidemiology.** The **deer fly** of the genus *Chrysops* is the vector of human loiasis. Sheathed microfilaria are found in the blood of infected people during daylight hours.

2. **Clinical disease.** The adult parasite migrates throughout the human body at a rate of about 1 cm/hr.
   a. **Immunologic response.** Localized allergic inflammation results in egg-sized lesions, which persist for 2–3 days.
   b. **Clinical symptomatology.** The patient usually experiences fever, itching, urticaria, and pain.
   c. **Ocular involvement.** The adult worm sometimes migrates across the subconjunctiva, causing intense tearing and pain but no blindness.

3. **Diagnosis** is made by recovering the adult parasite from the eye or by demonstrating microfilaria in the blood or in the lesions. Eosinophilia is a common laboratory finding.

4. **Therapy.** Microfilaria and adult worms are eradicated by **diethylcarbamazine**. The drug is administered at low doses initially to avoid a damaging allergic response.

III. CESTODE PARASITES.
The cestodes are long, segmented parasites commonly referred to as **tapeworms**. Infestation of the human bowel is a common, worldwide problem.

A. Morphology. Cestodes absorb nutrients across a complex cuticle, and the internal structures are encased in a solid parenchyma. The adult parasite consists of the scolex (or head), the neck, and the segmented body.

1. **Size.** Adult cestodes of different species range from 2 cm to 10 m in length.

2. **General structure and function**
   a. **Scolex.** The scolex typically is about 2 mm in diameter and equipped with sucking disks.
      (1) Usually, four muscular **disks** attach the parasite to the intestinal mucosa.
      (2) The **rostellum** is a retractable phalanx that aids in attachment.
   b. **Neck.** The neck is composed of segments known as **proglottids**.
      (1) Each proglottid is a hermaphroditic reproductive unit.
      (2) Proglottids are generated one at a time to constitute the body of the tapeworm.
         (a) The proglottids are joined by a common cuticle, nerve trunk, and excretory canals.
         (b) Mature, gravid proglottids release eggs by rupturing or disintegrating, or they pass them through a uterine pore.
            (i) *Taenia* species have a fully developed embryo in a solid shell at the time of release.
            (ii) *Diphyllobothrium latum* releases an opercular egg containing an immature embryo.

B. Life cycle. Most of the cestodes require one or more intermediate hosts to complete their life cycles.

1. Typically, eggs are passed to soil in the stool of the primary host.

2. The fertile eggs are ingested by an intermediate host, in which they hatch. The larvae enter the tissue and encyst.

3. The cycle is completed when the primary host ingests the cysts in the flesh of the intermediate host.

C. Clinical disease. The effect of cestodes on humans depends upon whether the affected person is the primary or the intermediate host.

1. **General considerations**
   a. **Primary host infection.** The adult parasite is limited to the lumen of the intestine when a human is the primary host.
      (1) The clinical manifestations of the infection usually are minor.
      (2) Humans are primary hosts for *Taenia saginata* and *D. latum*.
   b. **Intermediate host infection.** When the human is an intermediate host, larvae invade tissue.
      (1) Invasion of the larvae and their migration through organ systems usually results in the development of serious disease.

      (2) Humans are intermediate hosts for *Echinococcus granulosus*.

   c. Unique host relationships. *Taenia solium* and *Hymenolepsis nana* have the unique characteristic of using humans as both primary and intermediate hosts.

2. Specific tapeworm infections

   a. Beef tapeworm infection in humans is caused by *T. saginata*. Cattle are intermediate hosts.

      (1) Morphology. This cestode reaches a length of 4–6 m in the lumen of the human intestine.

      (2) Epidemiology

         (a) Incidence. Less than 1% of examined cattle in the United States are infected with *T. saginata*. Sanitary disposal of human feces and federal inspection of beef have contributed to this low incidence.

         (b) Infection is prevalent in the Middle East, Kenya, Ethiopia, Yugoslavia, parts of the Soviet Union, and parts of South America.

      (3) Clinical manifestations

         (a) Asymptomatic infestation. Most cases of disease are asymptomatic. The patient usually becomes aware of the infection through the spontaneous passage of proglottids.

         (b) Symptomatic infection. Some patients suffer from epigastric pain, diarrhea, nausea, irritability, and weight loss. The appendix, biliary duct, and pancreatic duct occasionally are obstructed.

      (4) Diagnosis is made by finding eggs or proglottids in the stool.

         (a) Examination of a proglottid is necessary to differentiate *T. saginata* from *T. solium*.

         (b) Collecting eggs from the perianal area by the cellophane tape method allows detection in 85%–95% of cases.

         (c) Direct examination of a stool specimen successfully diagnoses infection in only 50% of cases.

      (5) Therapy. Niclosamide is the drug of choice. A single dose is effective in killing the worms.

   b. Pork tapeworm infection is caused by *T. solium*. Like other species of the genus that cause human infections, it inhabits the jejunum.

      (1) Morphology and structure. *T. solium* rarely exceeds 5 m in length and usually consists of fewer than 1000 proglottids.

      (2) Epidemiology. *T. solium* is widely distributed throughout most of the world.

         (a) Human infection is common in eastern Europe, Africa, Asia, and Latin America.

         (b) Human disease **no longer occurs in the United States**.

      (3) Clinical manifestations

         (a) Primary host infection. Primary infection of humans by adult worms results in symptomatology similar to that caused by *T. saginata*.

         (b) Intermediate host infection. Viable encysted forms are seen in the brain, heart, eye, lung, subcutaneous tissue, and muscle when humans serve as intermediate hosts.

         (c) Immunologic response. Death of the parasite results in an inflammatory response with fever, muscle pain, and eosinophilia.

         (d) Neurologic involvement. CNS lesions cause meningoencephalitis, epilepsy, and other neurologic problems as well as psychiatric conditions.

      (4) Diagnosis

         (a) Microscopy. Microscopic diagnosis of primary human infection is made by demonstrating eggs or larvae in the stool.

         (b) Radiologic evidence. In endemic areas, larval infection is suspected in patients who have neurologic symptoms; radiologic examination may reveal calcified cycts.

         (c) Biopsy. Diagnosis is confirmed by biopsy, usually of subcutaneous nodules.

         (d) Serology. Serologic tests are nonspecific.

      (5) Therapy

         (a) Niclosamide eradicates the adult worm with a single dose.

         (b) There is no approved chemotherapy for the larval infection. Surgery may be used to remove ocular and cerebral cycts.

   c. Fish tapeworm infection is caused by *D. latum*.

      (1) Morphology and structure

         (a) The adult worm is 3–10 m long and contains 300–4000 proglottids, which are wider than they are long.

(b) The worm attaches to the intestinal mucosa by means of sucking grooves (i.e., **bothria**).

(c) The long-lived female releases more than 1 million eggs daily.

(2) Epidemiology. Human disease can be found in areas where raw, pickled, and improperly cooked fish are consumed.

(a) Route of transmission. Fish become infected from freshwater sources that have been contaminated by eggs from human feces.

(b) Distribution. The disease has been reported in Scandinavia, eastern Europe, the Baltic countries, central Europe, the Far East, and South America. It also is seen in the United States in Alaska, Florida, and the midwestern states.

(3) Clinical manifestations

(a) Intestinal symptomatology. Most fish tapeworm infections are asymptomatic, but epigastric pain, abdominal cramps, vomiting, and weight loss occasionally occur.

(b) Vitamin deficiency. The parasite competes for vitamin B_{12}, absorbing 80%–100% of this ingested vitamin from the upper jejunum.

(i) Low serum levels of the vitamin are seen in 40% of *D. latum* carriers.

(ii) Macrocytic anemia may occur.

(4) Diagnosis is confirmed by demonstrating eggs in the stool. Concentrating procedures usually are not necessary because of the large number of ova produced by the tapeworm.

(5) Therapy. Niclosamide is the drug of choice. **Vitamin supplementation** also can be beneficial.

d. Hydatid disease, or **echinococciasis**, usually is caused by *E. granulosus*. Human disease occasionally is caused by *Echinococcus multilocularis*. The adult worm inhabits the intestinal tract of canines.

(1) Morphology. There are three proglottids, and the eggs are identical to those of *T. saginata* and *T. solium*.

(2) Epidemiology. The pastoral and sylvatic forms of echinococciasis are distinct. The pastoral form is most common.

(a) Pastoral echinococciasis occurs in Australia, New Zealand, eastern and southern Africa, the Middle East, central Europe, and South America.

(i) About 200 cases are reported annually in the United States. They are acquired outside of the country, or they occur in shepherds in California, Utah, and southern Indiana.

(ii) Dogs are contaminated by feeding on the raw viscera of slaughtered sheep; humans acquire eggs from the fur of dogs.

(iii) Eggs on the hands contaminate human food, and ingestion begins the disease cycle.

(b) Sylvatic echinococciasis occurs in Alaska and western Canada.

(i) Wolves usually are the definitive hosts. Moose and caribou are intermediate hosts.

(ii) A cycle has been seen in California, where hunting dogs eat the infected offal of slain deer. A pastoral cycle is then established.

(3) Clinical manifestations. The symptomatology of human hydatid disease is the result of mechanical damage in tissue produced by enlarging cycts.

(a) In the pastoral form of disease, about 60% of the cysts are found in the liver, and 25% are found in the lungs.

(b) In sylvatic disease, 65% of the cysts are in the lungs, and the remainder are in the liver.

(c) Most cases of disease are asymptomatic; 5–20 years usually pass between infection and diagnosis.

(4) Diagnosis

(a) Radiologic evidence. Hydatid disease usually is diagnosed by finding pulmonary lesions (i.e., cysts) on routine radiologic examination of the chest.

(b) Serology. Confirmation of disease usually is based on serologic tests, although these are not totally satisfactory because they lack sensitivity and specificity.

(5) Therapy. The only available definitive therapy is surgical excision or drainage and sterilization of the lesion.

(a) Surgery is necessary only in symptomatic disease or when cysts become excessively enlarged.

(b) High-dose **mebendazole** therapy when surgery is contraindicated is in experimental stages.

IV. TREMATODES. The **flukes** survive in human tissue and in the vascular system for decades. They resist immunologic mechanisms of control and cause progressive damage to vital organs. The three major categories of trematodes seen in human infections are lung flukes, liver flukes, and blood flukes.

A. Morphology

1. **Size.** The trematodes vary in length from a few millimeters to several centimeters.
2. **Structure.** The worms are bilaterally symmetrical.
   a. There are two **suckers**, one surrounding the oral cavity and the other on the ventral surface. They are used for attachment and for an inchworm type of mobility.
   b. The flukes have a closed digestive tract and an excretory system consisting of flame cells and a posterior excretory pore.

B. Life cycle

1. **Reproductive systems.** Variations in the reproductive system divide the parasite into hermaphroditic species and schistosomes.
   a. **Hermaphroditic species** produce opercular eggs. **Schistosomes** deposit nonopercular offspring.
   b. The life cycles of the two groups of trematodes are similar.
2. **Miracidia development**
   a. Eggs are excreted in human feces and hatch in fresh water to form ciliated larvae known as **miracidia.**
   b. The miracidia of each species invade a specific **intermediate snail host**.
3. **Cercaria development**
   a. Asexual reproduction within the snail produces thousands of tail-bearing larvae known as **cercariae**.
   b. These larvae are released from the snail over a period of weeks. They swim about in search of their next host.
4. **Completion of the cycle**
   a. **Schistosomes** complete their cycle in the human. They attach to the skin, lose their tails, and invade the tissue.
   b. **Hermaphroditic trematodes** excyst on or in aquatic plants and animals and become infective **metacercariae**. Their life cycle is completed when this second intermediate host is ingested by a human.

C. Clinical disease

1. **Lung fluke infection**, or **paragonimiasis**, is caused by several species of *Paragonimus*. *Paragonimus westermani* is the most frequent cause of human disease.
   a. **Morphology and structure.** The reddish brown adult is about 10 × 5 mm and is found characteristically encapsulated in the pulmonary parenchyma.
   b. **Epidemiology**
      (1) **Distribution.** Paragonimiasis typically is seen in the Far East, but recently it has been described in India, Africa, and Latin America.
      (2) **Transmission.** Metacercariae are transmitted to humans in the Far East by ingestion of infected crabs.
         (a) The bodily fluid of crabs (i.e., crab juice) is ingested for medicinal purposes. Lightly cooked or pickled crabs also are a source of the parasite.
         (b) Other foods can become contaminated while the crabs are being prepared for cooking.
   c. **Clinical manifestations**
      (1) **Pulmonary disease.** Lung involvement resembles tuberculosis.
         (a) The adult worm causes an eosinophilic inflammatory response in the lung.

   (b) A fibrous capsule forms around the parasite and enlarges, eventually eroding into a bronchiole.
   (c) Eggs, blood, and inflammatory exudate are expectorated.
   (2) Abdominal disease. Intestinal and mesenteric involvement results in pain, bloody diarrhea, and occasionally palpable masses.
   (3) Brain involvement is a serious complication that occurs in about 1% of Oriental patients.
 d. Diagnosis usually is not as simple as with most helminthic diseases.
   (1) Microscopic examination
      (a) Eggs usually are not found in the sputum during the first 3 months of overt disease, but eventually 75% of patients develop positive sputa.
      (b) Pleural effusions, when they are present, should be checked for eggs.
      (c) Stool samples from children who swallow their sputa also may be positive.
   (2) Radiologic examination reveals that about 50% of patients with brain involvement have calcified lesions.
   (3) Serology. Serologic detection of circulating antibodies aids in diagnosis but is not specific.
 e. Therapy and prevention
   (1) Bithionol and **praziquantel** are effective chemotherapeutic agents for treating paragonimiasis.
   (2) Proper preparation and cooking of shellfish can prevent human infection.

2. Liver fluke infection, or **clonorchiasis**, is caused by species in the genera *Fasciola*, *Opisthorchis*, and *Clonorchis*. These flukes colonize and may obstruct the human biliary tract.
 a. Morphology. Liver flukes are lance shaped and measure 5–20 mm in length by 5 mm in diameter. These long-lived parasites discharge about 2000 tiny eggs each day.
 b. Epidemiology. *Clonorchis sinensis* (i.e., the **Chinese liver fluke**) is the most important species.
   (1) Transmission. Clonorchiasis is acquired by ingesting contaminated raw, dried, frozen, pickled, or salted fish.
   (2) Endemicity. The disease is endemic in the Far East, where human feces traditionally have been used to fertilize fish ponds; better disposal of human waste has decreased transmission of the disease.
   (3) Incidence. The parasite survives for more than 50 years in the human biliary tract. The incidence of disease in adults is decreasing slowly.
 c. Clinical manifestations. Light-load infections usually are asymptomatic. Heavy loads of 500–1000 flukes can be the result of frequent reinfection.
   (1) General symptomatology
      (a) Migration of the larvae from the duodenum to the bile duct frequently is associated with fever, chills, hepatomegaly, mild jaundice, and eosinophilia.
      (b) Inflammation and fibrosis of the smaller bile ducts, epithelial hyperplasia, and adenoma formation occur in the presence of adult flukes.
   (2) Heavy fluke loads. Clinically overt disease only occasionally occurs and is due to heavy fluke loads.
      (a) Bile stones and cholangiocarcinoma occasionally form.
      (b) Calculus formation during clonorchiasis frequently is associated with carriage of *Salmonella typhi* and other bacteria.
   (3) Biliary obstruction. The common biliary duct may be obstructed by dead worms.
      (a) Secondary bacterial infection accompanied by bacteremia, endotoxin shock, and hypoglycemia occurs.
      (b) Pancreatitis and pancreatic duct obstruction sometimes accompany biliary obstruction resulting from *C. sinensis* infection.
 d. Diagnosis requires identification of eggs in the stool or duodenal aspirates.
   (1) Blood analysis. Leukocytosis, eosinophilia, and elevated alkaline phosphatase levels are characteristic of clonorchiasis.
   (2) Physical findings. Abnormal radioisotopic and ultrasonographic liver findings are diagnostic.
 e. Therapy and prevention
   (1) Praziquantel appears to be an effective chemotherapeutic agent as indicated by experimental clinical use.

      (2) Transmission of the disease is controlled through proper cooking of freshwater fish and sanitary disposal of human feces.

3. Blood fluke infection, or **schistosomiasis**, is caused by a group of closely related blood flukes. Five species of *Schistosoma* are known to infect humans.

    a. Morphology. The 2-cm-long male remains in a lifelong copulatory embrace with the longer female, and 300–3000 eggs are deposited daily.

    b. Epidemiology

      (1) Distribution

        (a) *Schistosoma mansoni*, *Schistosoma haematobium*, and *Schistosoma japonicum* are the primary causes of disease worldwide.

        (b) *Schistosoma intercalatum* occurs in a limited area in Africa, and *Schistosoma mekongi* is found in a limited area in southeast Asia.

      (2) Incidence. The schistosomes cause more than 200 million human infections.

      (3) Transmission

        (a) Eggs hatch in fresh water and enter their species-specific **intermediate snail host**.

        (b) Over a 1- to 2-month period, the snail releases infectious **cercariae**, which infect humans when they come into contact with the skin.

    c. Clinical manifestations. Schistosomiasis can be divided into three stages.

      (1) In the **early stage**, the schistosomes penetrate and migrate.

        (a) **Immediate and delayed hypersensitivity responses** are characteristic. They are caused by the death of schistosomes that have penetrated the skin.

        (b) An **intensely pruritic skin rash** occurs. The viable larvae begin migrating to the liver, and the skin rash disappears.

        (c) The patient experiences abdominal pain, headache, and fever for 1–2 weeks.

      (2) The **intermediate stage** begins with oviposition.

        (a) This stage of schistosomiasis is an acute febrile illness resembling serum sickness.

        (b) Cough, urticaria, arthralgia, splenomegaly, lymphadenopathy, abdominal pain, and diarrhea are common signs and symptoms.

        (c) The intermediate stage occurs 1–2 months after initial exposure. The symptomatology may persist for 3 months.

      (3) The **chronic stage** is initiated by inflammation and scarring of the bowel and bladder caused by retained eggs.

        (a) About half of the deposited eggs are retained. They induce a T cell–mediated eosinophilic granuloma and fibroblast proliferation as a result of soluble antigens from the eggs.

        (b) Blockage of free blood flow is characteristic of chronic schistosomiasis.

    d. Diagnosis is made by the identification of eggs in the urine, stool, or biopsy specimens.

      (1) Biopsy. Cystoscopy and bladder biopsy may be needed for the diagnosis of mild infections.

      (2) Serology. Skin tests and serologic tests lack sensitivity and specificity, and cannot be used to confirm or exclude active infection.

    e. Therapy

      (1) Praziquantel kills adults of all *Schistosoma* species. Other agents are available for eliminating specific species.

      (2) There is no chemotherapy to eliminate schistosomal dermatitis, but **antihistamine** and **corticosteroid** therapy may help alleviate some of the symptoms.

STUDY QUESTIONS

Directions: Each question below contains five suggested answers. Choose the **one best** response to each question.

1. Visceral larval migrans is caused by

(A) *Toxocara canis*
(B) *Ancylostoma duodenale*
(C) *Loa loa*
(D) *Wuchereria bancrofti*
(E) *Ascaris lumbricoides*

2. The finding of encysted larvae in striated muscle is characteristic of

(A) onchoceriasis
(B) hydatid disease
(C) toxocariasis
(D) trichinosis
(E) paragonimiasis

3. Which of the following parasitic infections is transmitted to humans by the black fly and can cause blindness?

(A) Filariasis
(B) Toxocariasis
(C) Strongyloidiasis
(D) Trichuriasis
(E) Onchoceriasis

4. Distinct pastoral and sylvatic forms of disease are seen in

(A) echinococciasis
(B) pork tapeworm infection
(C) clonorchiasis
(D) whipworm infection
(E) schistosomiasis

5. Cholangiocarcinoma has been associated with infection by

(A) *Paragonimus westermani*
(B) *Clonorchis sinensis*
(C) *Loa loa*
(D) *Schistosoma haematobium*
(E) *Taenia saginata*

Directions: Each question below contains four suggested answers of which **one or more** is correct. Choose the answer

A   if **1, 2, and 3** are correct
B   if **1 and 3** are correct
C   if **2 and 4** are correct
D   if **4** is correct
E   if **1, 2, 3, and 4** are correct

6. Complications of ascariasis include

(1) intestinal obstruction
(2) blindness
(3) malabsorption
(4) elephantiasis

7. Anemia and hypoalbuminemia are characteristic manifestations of gastrointestinal infection caused by

(1) *Ancylostoma duodenale*
(2) *Enterobius vermicularis*
(3) *Necator americanus*
(4) *Ascaris lumbricoides*

8. Elephantiasis can be produced by

(1) *Loa loa*
(2) *Brugia malayi*
(3) *Onchocerca volvulus*
(4) *Wuchereria bancrofti*

9. Humans are both primary and intermediate hosts for

(1) *Taenia saginata*
(2) *Hymenolepis nana*
(3) *Diphyllobothrium latum*
(4) *Taenia solium*

10. Hydatid disease is characterized by a predominance of cysts in the

(1) lungs
(2) heart
(3) liver
(4) kidneys

11. Clinical manifestations of acute lymphatic filariasis include

(1) lymphangitis
(2) lymphadenitis
(3) thrombosis
(4) hyperplasia

ANSWERS AND EXPLANATIONS

1. The answer is A. [*II A*] *Toxocara canis* is a natural parasite of dogs; humans are incidental hosts. Toxocariasis in humans also is known as visceral larval migrans because of the nematode's potential to migrate into any organ system of the body. *Ancylostoma duodenale* is the etiologic agent of hookworm infection; *Loa loa* causes a filarial disease transmitted by the deer fly in Africa; *Wuchereria bancrofti* is a common cause of lymphatic filariasis that may lead to elephantiasis; and *Ascaris lumbricoides* is a roundworm that causes one of the most common parasitic diseases in the world.

2. The answer is D. [*II B 1 b, 3*] Trichinosis is characterized by invasion of striated muscle by roundworm larvae. The larvae encyst in the muscle tissue and may become calcified over a period of months. Human disease typically is associated with the ingestion of improperly cooked pork contaminated with *Trichinella spiralis*. Onchoceriasis is characterized by subcutaneous nodules throughout the body. Hydatid disease is characterized by cysts in the lungs, whereas paragonimiasis caused by the lung fluke can result in calcified lesions in the brain. Toxocariasis, also known as visceral larval migrans, is characterized by tissue damage due to migrating larvae.

3. The answer is E. [*II D 2 a, 3 c*] Onchoceriasis is primarily a disease of the African tropics; it affects about 50 million people. Approximately 5% of affected people are blind. The etiologic agent, *Onchocerca volvulus*, is transmitted to humans by black flies that live along rapidly flowing streams. The disease is characterized by subcutaneous nodules that may form in the eye and cause punctate keratitis, iritis, and chorioretinitis; this symptomatology may result in blindness. Filariasis is an acute inflammation of the lymphatics that may result in blockage. Toxocariasis produces pulmonary and gastrointestinal symptomatology and can cause loss of visual acuity with granulomatous eye involvement. Strongyloidiasis and trichuriasis produce primarily gastrointestinal disease.

4. The answer is A. [*III C 2 d (2)*] Echinococciasis, or hydatid disease, is caused by the tapeworm *Echinococcus granulosus*. Humans acquire the pastoral form of disease by ingesting eggs obtained from the contaminated hair of dogs that have eaten the raw viscera of slaughtered sheep. The disease occurs in a variety of countries, and a few cases are seen annually in the United States in shepherds in California, Utah, and Indiana. The sylvatic form of disease occurs in Alaska and western Canada, where wolves are the definitive hosts and moose and caribou are the intermediate hosts. Pork tapeworm is widely distributed throughout the world; humans may be both primary and intermediate hosts. Clonorchiasis and schistosomiasis are caused by flukes that infect humans after undergoing part of their life cycles in one or more intermediate aquatic hosts. Whipworm infection is a common cosmopolitan disease acquired by the fecal-oral route; the parasite spends part of its life cycle in soil.

5. The answer is B. [*IV C 2 c (2) (a)*] Human infection by the Chinese liver fluke *Clonorchis sinensis* has been associated with cholangiocarcinoma. *C. sinensis* is the most important species of liver fluke, although it rarely causes clinically overt disease. Biliary blockage and bile stones occasionally occur. Cholangiocarcinoma may be the result of chronic irritation caused by the trematode. *Paragonimus westermani*, the lung fluke, causes pulmonary disease that may resemble tuberculosis. *Loa loa* is a filarial disease characterized by localized allergic inflammation. *Schistosoma haematobium*, one of the blood flukes, causes hypersensitivity reactions during the early stage of infection, acute febrile disease during the intermediate stage, and inflammation and scarring of the bowel during the chronic stage. *Taenia saginata* is the etiologic agent of beef tapeworm.

6. The answer is B (1, 3). [*I B 3 a (2) (a)*] Ascariasis is a common helminthic infection caused by the roundworm *Ascaris lumbricoides*. This parasite is the largest intestinal nematode, and infection typically occurs in children. Ascariasis is asymptomatic when the worm load is light, but heavy loads of up to 2000 worms can cause serious complications, including intestinal blockage (which occurs in about 2 in 1000 infections) and malabsorption. Blindness is associated with onchocerciasis; elephantiasis occasionally is caused by the etiologic agents of lymphatic filariasis.

7. The answer is B (1, 3). [*I D 3 b (3) (a)*] *Ancylostoma duodenale* and *Necator americanus* are the etiologic agents of hookworm infection. Gastrointestinal infection by these helminths typically produces epigastric pain and abdominal peristalsis. Characteristic findings include anemia and hypoalbuminemia as a result of blood loss from the intestinal mucosa. These clinical manifestations of hookworm infection vary from mild to severe. Pruritis ani is the most prominent clinical manifestation of infestation with *Enterobius vermicularis*. Heavy loads of *Ascaris lumbricoides* in the intestines frequently result in abdominal pain and malabsorption syndrome.

8. The answer is C (2, 4). [*II C*] *Brugia malayi* and *Wuchereria bancrofti* are the most common causes of lymphatic filariasis. The inflammatory disease can result in chronic blockage of the lymphatic vessels. This condition occasionally leads to elephantiasis. *Loa loa* is the etiologic agent of a filarial disease of western Africa characterized by localized allergic inflammation resulting in egg-sized lesions. *Onchocerca volvulus* is a filarial parasite that causes river blindness.

9. The answer is C (2, 4). [*III C 1 c*] *Taenia solium*, the pork tapeworm, and *Hymenolepis nana*, a tapeworm of humans and rodents, have the unique ability to use humans as both primary and intermediate hosts. The beef tapeworm, *Taenia saginata*, and the fish tapeworm, *Diphyllobothrium latum*, use humans as primary hosts.

10. The answer is B (1, 3). [*III C 2 d (3) (a), (b)*] Hydatid disease is caused primarily by *Echinococcus granulosus* and occasionally by *Echinococcus multilocularis*. The clinical manifestations of human disease are the result of the enlarging cysts that cause mechanical damage in tissues. In the pastoral form of hydatid disease, about 60% of the cysts occur in the liver and 25% in the lungs. In the sylvatic form, approximately 65% of the cysts occur in the lungs and the remainder in the liver. The heart and kidneys may be involved, particularly in the pastoral form, but only a few cysts occasionally occur in these organs.

11. The answer is E (all). [*II C 3 a (1), (2)*] Lymphatic filariasis is most commonly caused by *Wuchereria bancrofti* and *Brugia malayi*. The acute disease of the lymph nodes and lymphatic vessels is inflammatory in nature and is due to the host response to dead and dying adult worms and the molting of young worms. Acute lymphangitis with thrombus formation and hyperplastic changes in the lymphatic endothelium is associated with the infiltration of lymphocytes, plasma cells, and eosinophils. Lymphadenitis, tender lymph nodes, and low-grade fever also are typical findings. High fever, muscle pain, chills, and other systemic manifestations occasionally occur.

26
Bacteremia and Septicemia

Gerald E. Wagner

I. DEFINITIONS. The terms bacteremia and septicemia often are used interchangeably, particularly in the United States. There are, however, distinct differences between these two clinical conditions.

A. Bacteremia is the condition of having bacteria in the blood. In Europe, bacteremia is differentiated from septicemia by the absence of significant clinical symptomatology during bacteremia.

1. **Sources.** Bacteria may be introduced into the circulation by physical means (e.g., trauma), or they may be shed into the bloodstream from a focus of infection.

2. **Transient bacteremia** is a temporary condition. Recent studies in the United States have indicated that bacteria such as *Staphylococcus epidermidis, Bacteroides melaninogenicus,* and *Clostridium perfringens* frequently can be recovered from the blood of apparently healthy persons.

B. Septicemia is a condition in which bacteria actually grow and reproduce in the bloodstream, usually producing systemic toxic effect (i.e., toxemia).

1. **Sources.** The etiologic agent of septicemia usually is introduced into the bloodstream by dissemination from an infectious focus. Traumatic inoculation also occurs.

2. **Persistence.** Septicemia tends to persist until effective therapy is administered or the patient dies.

II. BACTEREMIC AND SEPTICEMIC INFECTIONS

A. Bacteremia. Most bacteria have the potential to cause bacteremia.

1. **Gram-negative bacteremia**
   a. **Etiology.** The term gram-negative bacteremia usually is reserved for bacteremia caused by members of the families Enterobacteriaceae and Pseudomonadaceae.
      (1) *Escherichia coli* is the most common etiologic agent of gram-negative bacteremia, accounting for 35% of cases.
      (2) *Klebsiella, Enterobacter,* and *Serratia* **species** account for an additional 27% of cases.
      (3) *Proteus* **species** cause about 11% of cases of gram-negative bacteremia.
      (4) *Pseudomonas aeruginosa* is the etiologic agent in 12% of diagnosed cases of gram-negative bacteremia.
   b. **Origin.** The gastrointestinal tract, urinary tract, and skin frequently are the sites of origin of the bacteria.
   c. **Clinical manifestations** of gram-negative bacteremia are identical regardless of the species of gram-negative bacteria causing the disease. The severity of the symptomatology, however, may vary considerably from patient to patient.
      (1) Typical symptoms include high fever, shaking chills, prostration, and, occasionally, nausea and vomiting.
      (2) In the classic case, the symptoms occur 1–2 hours after the bacteria enter the bloodstream. Endotoxin shock or hypotension may become apparent within 3–8 hours after the initial symptoms.
      (3) Classic symptoms occur in only about 30%–40% of patients with gram-negative bacteremia.

(a) Most cases of the disease have more subtle clinical manifestations.

(b) Fever may be the only symptom of bacteremia in patients who have undergone genitourinary or gastrointestinal procedures.

d. Predisposing factors. Surgery, urinary tract manipulations (e.g., catheterization), and underlying diseases are regarded as major precipitating factors in the development of gram-negative bacteremia.

e. Incidence. The incidence of gram-negative bacteremia in the United States is estimated at 71,000–300,000 cases per year. The infection occurs in about 1% of all patients admitted to hospitals.

2. Staphylococcal bacteremia

a. Etiology. Diagnosed staphylococcal bacteremia usually is caused by coagulase-positive **Staphylococcus aureus**. Coagulase-negative staphylococci (i.e., **S. epidermidis** and **Staphylococcus saprophyticus**) occasionally cause bacteremia.

b. Origin. Staphylococcal bacteremia may be primary in origin, with the major source of infection being the skin, or it may be secondary to an infectious focus.

(1) Cutaneous abscesses are an important source of the staphylococci, and simple palpation of such abscesses may seed the bloodstream.

(2) Virtually all cases of bacteremia due to coagulase-negative staphylococci result from contaminated intravenous catheters, cannulae, hydrocephalic shunts, or the paraphernalia used in intravenous drug abuse.

c. Clinical manifestations of staphylococcal bacteremia correlate with the source of the bacteria, the primary foci of infection, and dissemination of the bacteria.

(1) The disease usually progresses slowly with no specific symptomatology. Shock is less common than in cases of gram-negative bacteremia.

(2) Bacteremia caused by *S. aureus* occasionally is a fulminating disease characterized by sudden onset of fever, chills, and tachycardia. The patient's condition rapidly deteriorates into shock, which may be fatal in 24–48 hours.

(3) Bacteremia caused by *S. epidermidis* is more insidious and less dramatic than that caused by *S. aureus*. Transient, asymptomatic bacteremia caused by *S. epidermidis* probably is a daily occurrence.

d. Predisposing factors. Mechanical manipulations involving the skin, such as insertion of intravenous catheters and surgery, contribute to the risk of staphylococcal bacteremia. Coagulase-negative staphylococcal bacteremia virtually always occurs in this manner.

e. Incidence. Staphylococci are the second most common cause of bacteremia. *S. epidermidis* and *S. saprophyticus* now are the suspected causes of many undiagnosed cases of bacteremia. These coagulase-negative staphylococci account for about 0.1% of diagnosed cases, but their actual clinical significance is controversial.

3. Enteric (typhoid) fever

a. Etiology. Enteric fever usually is caused by **Salmonella typhi**. Other salmonellae that occasionally cause enteric fever include *Salmonella enteritidis* serotype paratyphi A, *S. enteritidis* serotype paratyphi B, and *S. enteritidis* serotype typhimurium.

b. Origin. *Salmonella* species colonize the gastrointestinal tracts of a wide variety of animals. Enteric fever typically is transmitted by the fecal-oral route.

c. Clinical manifestations. The primary characteristic of enteric fever is bacteremia, occurring after invasion of intestinal mucosa by the ingested bacteria. Symptomatology may last 4 weeks or longer.

(1) The classic symptomatology begins with fever that generally increases in a stepwise fashion. Sustained or intermittent fever currently is seen most commonly.

(2) Nonspecific symptomatology includes headache, anorexia, malaise, lethargy, and myalgia. Diarrhea, constipation, nonproductive cough, bronchitis, and bradycardia also may occur.

(3) The patient often looks toxic and has facial flushing, dilated pupils, dry skin, and a dull, detached appearance. Total prostration may occur, especially in children.

d. Predisposing factors. The exact nature of predisposition is not clear. Ingestion of improperly cooked animal products increases the risk of developing disease. Sickle cell trait appears to predispose to serious disease.

e. Incidence. Enteric fever has declined steadily in the United States and western Europe. Most cases now are imported in young adults who travel to endemic areas. It remains an important disease in developing nations, where the mortality rate approaches 50%.

4. Gonococcal bacteremia (gonococcemia)

  a. Etiology. Gonococcemia is caused by strains of **Neisseria gonorrhoeae** capable of disseminated disease.

  b. Origin. The site of primary infection usually is the urethra, but it may be the anal mucosa or other primary foci. The strains capable of dissemination are sensitive to penicillin, resistant to the bactericidal action of serum, and have specific nutritional requirements.

  c. Clinical manifestations. Gonococcemia typically occurs in patients with asymptomatic gonorrhea (i.e., primarily women).

   (1) Gonococcemia leads to disseminated gonococcal disease; syndromes range from low-grade fever or dermatitis to severe metastatic joint infection or meningitis.

   (2) Polyarthritis, dermatitis, and fever are the most frequent clinical manifestations of gonococcemia; shock seldom occurs. Gonococcemia rarely is fatal.

   (3) Gonococcemia is an important cause of pelvic inflammatory disease and salpingitis leading to sterility.

  d. Predisposing factors. Asymptomatic gonorrhea in people carrying strains of *N. gonorrhoeae* capable of disseminating typically leads to gonococcemia. Patients with complement deficiency are at high risk for developing disease.

  e. Incidence. The incidence of gonococcemia corresponds to the number of people carrying strains of gonococci capable of disseminating. Geographic variation is seen; a disseminated gonococcal infection (DGI) biotype has emerged in some areas of the world, although its incidence still is rare.

5. Meningococcal bacteremia (meningococcemia)

  a. Etiology. **Neisseria meningitidis** serogroups A, B, and C appear to be the most common causes of meningococcemia. Serogroup C currently appears to predominate in epidemics.

  b. Origin. Humans appear to be the only reservoir of *N. meningitidis*. The carriage rate normally is 1%–30%, but during epidemics it may be as high as 95%.

  c. Clinical manifestations. *N. meningitidis* invades the bloodstream from the nasopharynx and produces a wide range of clinical symptomatology. Meningococcemia may be a mild, transient bacteremia or a fulminating and rapidly fatal disease. Typically, meningococcemia leads to meningitis.

   (1) In the mildest presentation, the patient has fever and malaise; these symptoms subside spontaneously in 1–2 days.

   (2) Acute meningococcemia is more serious; patients present with fever, malaise, weakness, nausea, vomiting, chills, headache, myalgia, and arthralgia. Petechial hemorrhages are the most characteristic clinical manifestation of acute meningococcemia.

   (3) Serious, fulminating meningococcemia with a high mortality rate is known as **Waterhouse-Friderichsen syndrome**.

    (a) Patients present with an abrupt onset of fever as high as 40.6° C, chills, nausea, vomiting, myalgia, weakness, and headache. Widespread petechial hemorrhages appear suddenly.

    (b) The symptomatology of shock may predominate, but meningitis is rare.

  d. Predisposing factors. Meningococcemia principally is seen in children and young adults. High rates of nasopharyngeal carriage usually precede epidemics.

  e. Incidence. Meningococcal disease occurs sporadically throughout the world; it is most common in the spring. Waterhouse-Friderichsen syndrome occurs in 5%–15% of patients with meningococcemia.

B. Septicemia. When the virulence of the bacteria is sufficiently high or when host factors predispose to sepsis, septicemia occurs. Most of the etiologic agents of septicemia also can cause bacteremia.

1. Bacteroides septicemia

  a. Etiology. Most cases of bacteroides septicemia are caused by **Bacteroides fragilis**, although **B. melaninogenicus** is occasionally the etiologic agent.

  b. Origin. Generally, bacteroides septicemia originating from the oropharynx is caused by *B. melaninogenicus*; septicemia originating from sites below the diaphragm is caused by *B. fragilis*.

   (1) Septicemia caused by bacteroides commonly results from an abscess or other focus of infection seeding the bloodstream.

   (2) Sepsis also may develop as the result of minor trauma to the oral cavity, genitourinary tract, gastrointestinal tract, or skin.

    c. Clinical manifestations. Bacteroides septicemia tends to be more protracted and less precipitous in onset than other gram-negative septicemias.

        (1) Usually, the septicemia appears as a prostrating febrile illness characterized by toxemia. It develops over a period of several days to a few weeks.

        (2) Peripheral leukocytosis (sometimes exceeding 40,000/mm^3), chills, spiking fever, and the persistence of bacteria in the bloodstream are characteristic.

        (3) Jaundice, thrombophlebitis, and infective embolism in the lung, liver, or brain are seen in a significant percentage of septicemic patients.

        (4) Septic shock occurs in 25%–35% of cases but differs from gram-negative endotoxin shock in that it occurs late in the course of the disease.

        (5) A few patients may demonstrate only low-grade fever and malaise.

    d. Predisposing factors. Peritoneal abscess, septic abortion, gynecologic surgery or manipulation, lower back or hip trauma, and infected decubitus ulcer are predisposing conditions.

    e. Incidence. Septicemia caused by *Bacteroides* species occurs most frequently in debilitated patients or in persons with underlying disease.

        (1) Hospitalized patients recovering from abdominal or gynecologic surgery frequently develop bacteremia that can progress to septicemia.

        (2) Nonhospitalized patients commonly are over 40 years of age and are recovering from recent surgery or are receiving prophylactic antibiotics in preparation for abdominal surgery.

        (3) Acute illness typically is seen in young women following septic abortion.

2. Clostridial septicemia

    a. Etiology. Any of the more than 70 species of *Clostridium* potentially are etiologic agents of bacteremia and septicemia.

        (1) *Clostridium perfringens* causes more than 90% of clostridial septicemia.

        (2) Less virulent species, *Clostridium septicum* and *Clostridium ramosum*, can produce shock or fatal illness in the newborn, in patients with malignant disease, and in immunocompromised patients.

    b. Origin. Clostridial septicemia originates from the colon, biliary tract, and uterus. A clinically apparent site of origin is identified in only about half the cases; intestinal, biliary tract, and gynecologic sites of origin are assumed in the remaining cases.

    c. Clinical manifestations

        (1) Clostridial septicemia originating from the colon or biliary tract presents as an acute, serious, febrile illness. High fever, extensive intravascular hemolysis, acute renal tubular necrosis, and hypotension are characteristic.

        (2) Septicemia originating from the uterus often occurs 24–72 hours after an attempted abortion and is characterized by fever and chills.

            (a) Symptomatology also may include severe myalgia, malaise, headache, abdominal cramps, sharp gastrointestinal pain, nausea, vomiting, diarrhea, and foul bloody or brown vaginal discharge.

            (b) The septicemia may progress dramatically within a few hours, causing oliguria, hypotension, jaundice, intravascular hemolysis, and shock.

        (3) Chills and fever may be the predominant clinical presentation in some patients; intravascular hemolysis is observed in 30%–50% of cases.

        (4) Patients with malignant disease, especially those receiving radiation or cytotoxic therapy, may develop rapidly fatal septicemia originating from minor foci of infection.

    d. Predisposing factors. Transient clostridial bacteremia now is thought to be common, but abdominal and gynecologic surgery can predispose to septicemia. Most patients also have underlying medical problems.

    e. Incidence. Clostridia are identified in 1%–2.5% of positive blood cultures; in 30%–40% of blood cultures positive for clostridia, these anaerobes are mixed with other anaerobic bacteria, aerobic bacteria, or both.

C. Bacteremia and septicemia of miscellaneous etiology. Most bacteria have the potential to cause transient bacteremia; many can cause septicemia, partricularly in compromised hosts. The only criterion for either condition is the seeding of the bloodstream with bacteria by trauma or from infectious foci.

1. Bacteremia or septicemia can be an integral part of many infectious disease processes. For example, intravascular dissemination takes place during pneumococcal pneumonia, bartonellosis, Rocky Mountain spotted fever, and other rickettsial diseases, syphilis, and leptospirosis.

2. Bacteremia or septicemia produces clinical symptomatology characteristic of some diseases.

3. The most predictable clinical features of bacteremia and septicemia, regardless of etiology, are fever and chills.

III. COMPLICATIONS

A. Endotoxin shock is one of the most common complications of bacteremia or septicemia. Gram-negative enteric bacteria are the most frequent etiologic agents of bacteremia and septicemia. Endotoxin shock manifests as a broad spectrum of symptomatology.

B. Disseminated foci of infection can develop regardless of the etiologic agent of bacteremia or septicemia.

1. Meningitis or other syndromes involving the central nervous system generally are considered the most serious complications of intravascular infection.

2. Patients particularly susceptible to hematogenous dissemination of bacteria include those with underlying diseases, natural or therapeutic immunosuppression, or prosthetic implants.

IV. DIAGNOSIS. A confirmed diagnosis of either bacteremia or septicemia is made by isolating bacteria from the blood. A variety of factors influence the successful recovery of bacteria from blood.

A. Serial blood cultures

1. Precautions. The skin at the site of venipuncture and the fingers used to palpate the vein should be thoroughly disinfected with an iodophor and 70% alcohol.

2. Volume of sample. In adult patients, a minimum of 10 ml of blood should be taken; lesser quantities should be taken from children, depending upon their size.

3. Microscopic examination. Bacteria rarely are seen in direct microscopic examination of blood smears. *Borrelia recurrentis* is one of the few bacteria that can be observed in blood smears.

4. Culturing success. Studies indicate that a total of three blood cultures taken over a 24-hour period allows for successful recovery of bacteria in 99% of patients with bacteremia or septicemia.
  a. The recovery rate is 80% when only one blood culture is obtained.
  b. If the patient's status prevents obtaining cultures over a 24-hour period, three blood cultures taken over several hours increases the chance of isolating the etiologic agent.
  c. Blood cultures from patients receiving antimicrobial therapy may not be positive during a 24-hour period, and additional samples must be obtained over a longer time period.

B. Culturing technique. A relatively standardized procedure is used for culturing blood samples.

1. Media for blood culture include trypticase soy broth, thioglycolate broth, or some other appropriate bacteriologic medium containing an anticoagulant-anticomplement-antiphagocytic compound such as sodium polyanetholsulfonate. The medium typically is dispensed in 100-ml volumes.

2. Culture vessel. Each blood culture uses a set of two blood-culture bottles that can be vented.

3. Inoculation. Each bottle is inoculated with 5 ml or less blood; one bottle is vented to recover aerobic and facultatively anaerobic bacteria, and the unvented second bottle is used to recover strict anaerobes.

4. Inspection for growth. The culture bottles are incubated at 35° C and inspected daily for turbidity for the first 7 days; if no growth is present, they are inspected again on day 14.
  a. After 24 hours of incubation, it is standard procedure to prepare a Gram stain and subculture samples of culture medium onto blood agar plates from bottles showing no turbidity.
    (1) The Gram stain may detect small numbers of bacteria when growth is insufficient to impart gross turbidity to the culture medium.
    (2) Subculturing significantly enhances the recovery of bacteria from blood cultures.

> **b.** Unless fastidious bacterial species are suspected, blood culture bottles usually are discarded as negative after 2 weeks.
>
> **5. Commercially prepared systems** that lyse the patient's erythrocytes and recover the bacteria on a substrate during centrifugation of the sample have been introduced.

V. THERAPY. Immediate aggressive therapy is mandatory when bacteria are isolated from the bloodstream of a patient exhibiting any clinical symptomatology.

> **A. Selection of a chemotherapeutic agent.** The choice of a therapeutic agent may be based on a Gram stain of purulent material from any focus of infection (e.g., an abscess) that is a potential source of bacteremia or septicemia.
>
> **B. Administration.** Intravenous infusion of the selected chemotherapeutic agent achieves the highest attainable serum level in the shortest period of time.
>
> **C. Duration of therapy.** Therapeutic serum levels of an efficacious antimicrobial agent should be maintained for a minimum of 10–14 days. Intravenous antimicrobial therapy can be continued for 6 weeks or longer to achieve eradication of the causative bacteria.
>
> **D. Supplemental therapy.** Supportive therapy based on clinical symptomatology frequently is needed for the patient with intravascular infection. Oral administration of an antimicrobial agent may be prescribed following a successful intravenous course of therapy.

STUDY QUESTIONS

Directions: Each question below contains five suggested answers. Choose the **one best** response to each question.

1. What organism most frequently causes gram-negative bacteremia?

(A) *Pseudomonas aeruginosa*
(B) *Bacteroides fragilis*
(C) *Escherichia coli*
(D) *Neisseria meningitidis*
(E) *Clostridium perfringens*

2. What is the best means of diagnosing bacteremia?

(A) Gram stain
(B) Agar subculture
(C) Centrifugation
(D) Serial blood culture
(E) Complement fixation

3. What organism is the most comon etiologic agent of clostridial septicemia?

(A) *Clostridium tetani*
(B) *Clostridium septicum*
(C) *Clostridium difficile*
(D) *Clostridium ramosum*
(E) *Clostridium perfringens*

4. Sickle cell trait appears to be a predisposing factor in serious intravascular infection caused by what organism?

(A) *Salmonella typhi*
(B) *Clostridium septicum*
(C) *Bacteroides melaninogenicus*
(D) *Neisseria gonorrhoeae*
(E) *Staphylococcus saprophyticus*

5. What organism is the etiologic agent of Waterhouse-Friderichsen syndrome?

(A) *Clostridium perfringens*
(B) *Neisseria meningitidis*
(C) *Staphylococcus aureus*
(D) *Bacteroides fragilis*
(E) *Pseudomonas aeruginosa*

Directions: Each question below contains four suggested answers of which **one or more** is correct. Choose the answer

    A   if **1, 2, and 3** are correct
    B   if **1 and 3** are correct
    C   if **2 and 4** are correct
    D   if **4** is correct
    E   if **1, 2, 3, and 4** are correct

6. The most characteristic symptomatology of bacteremia or septicemia includes

(1) chills
(2) shock
(3) fever
(4) nausea

7. Complications of bacteremia and septicemia include

(1) meningitis
(2) disseminated infection
(3) endotoxin shock
(4) petechial hemorrhages

8. Characteristics of gonococci capable of causing bacteremia include which of the following?

(1) They are killed by serum factors
(2) They have special nutritive requirements
(3) They belong to a single serotype
(4) They are sensitive to penicillin

9. Symptomatology that is characteristic of meningococcemia includes

(1) petechial hemorrhages
(2) arthralgia
(3) malaise
(4) jaundice

ANSWERS AND EXPLANATIONS

1. The answer is C. [*II A 1 a (1)*] The term gram-negative bacteremia generally is reserved for infections by agents in the Enterobacteriaceae and Pseudomonaceae families. *Escherichia coli* is the most common cause of gram-negative bacteremia. Frequently the bacteremia is transient. *Pseudomonas aeruginosa* less frequently causes gram-negative bacteremia, typically following direct inoculation of the bacteria into the bloodstream. Meningococcal bacteremia (i.e., meningococcemia) is characteristic of disease caused by *Neisseria meningitidis*; the meningococcemia accounts for much of the symptomatology of meningococcal disease. *Bacteroides fragilis* and *Clostridium perfringens* typically produce clinically obvious intravascular infection in the form of septicemia.

2. The answer is D. [*IV*] Diagnosis of bacteremia or septicemia is made by isolating the etiologic agent from blood. Serial blood cultures, when performed properly by a relatively standardized method, successfully isolate the bacteria from 99% of patients. The Gram stain rarely reveals the presence of bacteria in blood samples taken during bacteremia because of the dilution factor. Subculturing negative blood culture bottles is performed routinely in an attempt to isolate small populations of bacteria. Lysis of erythrocytes and harvesting of bacteria by centrifugation is used in a commercial isolation kit. The detection of complement-fixing antibodies, or any other type of antibodies, does not indicate bacteremia or septicemia.

3. The answer is E. [*II B 2 a (1)*] *Clostridium perfringens* is the etiologic agent of more than 90% of diagnosed cases of clostridial septicemia. Any of the more than 70 species of clostridia, however, is capable of causing either bacteremia or septicemia. *Clostridium septicum* and *Clostridium ramosum* frequently have been reported as the cause of shock and death in neonates, patients with malignancy, and immunocompromised people. *Clostridium difficile* is the etiologic agent of antibiotic-induced pseudomembranous colitis, but it can disseminate and cause septicemia. *Clostridium tetani* and *Clostridium botulinum* most commonly produce toxemia but, under the appropriate conditions, can cause bacteremia or septicemia.

4. The answer is A. [*II A 3 d*] Sickle cell trait appears to predispose to development of serious enteric fever caused by *Salmonella typhi*; bacteremia is a critical step in the pathogenesis of this disease. The predisposition to serious bacteremia caused by *S. typhi* is not understood clearly. *Clostridium septicum* can cause both bacteremia and septicemia; immune deficiency and underlying disease are predisposing factors. Predisposing factors for intravascular infection caused by *Bacteroides melaninogenicus* include abscesses and trauma, particularly in the oropharyngeal area. Predisposing factors for gonococcemia include asymptomatic carriage of a strain of *Neisseria gonorrhoeae* capable of dissemination and complement deficiency. The primary predisposing factor for staphylococcal bacteremia is trauma to the skin.

5. The answer is B. [*II A 5 c (3)*] Serious, fulminant meningococcemia is known as Waterhouse-Friderichsen syndrome. The disease has a high mortality rate and is characterized by abrupt onset, high fever, chills, nausea, vomiting, myalgia, weakness, headache, petechial hemorrhages, and shock; meningitis is rare. *Pseudomonas aeruginosa* is the etiologic agent of serious bacteremia that has a high mortality rate; *Staphylococcus aureus* also causes bacteremia. *Clostridium perfringens* and *Bacteroides fragilis* are causes of septicemia.

6. The answer is B (1, 3). [*II C 3*] The most predictable characteristics of bacteremia or septicemia are fever and chills, regardless of the etiologic agent. Fever and chills are caused by the release of endogenous pyrogen or by direct effects of bacterial components that act as pyrogens. Shock usually occurs in serious bacteremic or septicemic infections and typically, although not exclusively, is caused by the release of lipopolysaccharide endotoxin of gram-negative bacteria. Some patients, however, can develop tolerance to lipopolysaccharide. Nausea is a frequent systemic response to disseminated infection.

7. The answer is A (1, 2, 3). [*III A, B 1*] Complications of intravascular infections tend to be the same for bacteremia and septicemia. Most of them can develop regardless of the etiologic agent. Endotoxin shock is a common complication because gram-negative enteric rods, particularly *Escherichia coli*, are the most common causes of bacteremia. Endotoxin shock is manifested by a variety of signs and symptoms in different patients. Disseminated foci of infection frequently occur as the bacteria are entrapped by phagocytic cells or lodge in various tissues. Meningitis is considered one of the most serious disseminated infections resulting from bacteremia or septicemia. Petechial hemorrhages are a sign of some bacterial infections and may be caused by endotoxin, vasculitis, or blockage of capillaries. Petechial hemorrhages are the most common clinical sign in meningococcemia.

8. The answer is C (2, 4). [*II A 4 b*] Gonococcemia is a bacteremia caused by strains of *Neisseria gonorrhoeae* capable of disseminating from the primary site of infection, usually the urethra. The strains that are capable of disseminating are highly sensitive to penicillin, resist the bactericidal action of serum, and have special nutritive requirements; the growth requirements are arginine, uracil, and hypoxanthine. In some geographic regions, disseminated gonococcal infection has been associated with a specific biotype, but its incidence is rare.

9. The answer is A (1, 2, 3). [*II A 5 c 1, 2, 3*] Meningococcemia can present in several forms ranging from mild to severe. Patients with mild meningococcemia typically experience fever and malaise; this symptomatology generally subsides spontaneously. Acute meningococcemia is a more severe disease characterized, in part, by arthralgia and petechial hemorrhages. Patients with severe meningococcemia experience malaise, petechial hemorrhages, and many other signs and symptoms. Jaundice is seen in a significant percentage of patients with bacteroides septicemia.

27

Pneumonia

Gerald E. Wagner

I. BACTERIAL PNEUMONIA, caused by a variety of bacteria, is acute and abrupt in onset. The inflammatory process in the lung is characterized by an outpouring of fibrinous edema fluid, erythrocytes, and leukocytes into the alveoli. The result is consolidation of the lung.

A. Pneumococcal pneumonia is the most common form of bacterial pneumonia; it accounts for about 80% of cases.

    1. Etiology. Any of the more than 80 serotypes of **_Streptococcus pneumoniae_**, commonly referred to as the **pneumococci,** can cause pneumonia, although 14 capsular types cause 80% of pneumococcal disease in the United States. The pneumococci are indigenous flora of the upper respiratory tract in 30%–40% of people.

    2. Epidemiology. The incidence of pneumococcal pneumonia is low in healthy people; at least 75% of patients with overt disease have an underlying illness.

      a. Predisposing factors

        (1) Disease occurs most frequently in the very young, the elderly, and people with compromised immunologic defenses.

        (2) Patients who have trouble handling upper respiratory secretions because of increased volume of the secretions or poor laryngeal reflexes during sleep also are more susceptible to infection.

      b. The **mortality rate** for pneumococcal pneumonia is approximately 10% even among patients who receive adequate chemotherapy based on in vitro sensitivity testing; most deaths occur among patients with underlying disease.

    3. Pathogenesis. The pathogenesis of pneumococcal pneumonia is not understood completely.

      a. Infection

        (1) It is postulated that infection occurs when upper respiratory tract secretions containing pneumococci are aspirated rather than when bacteria are inhaled into the lungs.

        (2) Whether virulent pneumococci are spread by aerosolization or by direct contact has not been determined satisfactorily.

      b. The pneumonic process

        (1) Pneumonia is initiated by the outpouring of fibrinous edema fluid and erythrocytes into alveolar spaces as a response to pneumococcal multiplication.

          (a) Increased capillary permeability is accompanied by migration of neutrophils into the alveolar spaces, and the pneumococci are phagocytosed.

          (b) The process spreads as pneumococci rapidly proliferate at the leading edge of infection.

          (c) Consolidation occurs at the center of the process due to leukocyte activity and coagulating exudates.

        (2) Healing occurs as macrophages remove alveolar debris, dead pneumococci and neutrophils, erythrocytes, and strands of fibrin.

        (3) Pulmonary function returns to normal within a few weeks despite the intense inflammatory response; necrosis of the lung is rare.

    4. Clinical manifestations. A history of rhinorrhea, pharyngitis, and cough of several days' duration is common in patients with pneumococcal pneumonia.

    a. Symptomatology
      (1) Disease is characterized by sudden onset of fever (typically 39.8° C), pleuritic chest pain, and a hard, shaking chill accompanied by teeth chattering; repeated rigors are uncommon.
      (2) Cough and sputum production intensify with time, and the sputum classically is rust colored due to alveolar hemorrhage.
      (3) Symptomatology is severe and usually is accompanied by prostration within 48 hours of onset.
    b. Physical findings
      (1) The patient typically is perspiring and warm with tachycardia and tachypnea.
      (2) Cool, vasoconstricted skin is a sign of imminent cardiovascular collapse.
      (3) Fine rales are present even before consolidation is evident.

5. Diagnosis is made by establishing the symptomatology of pneumonia and showing that *S. pneumoniae* is the etiologic agent.
    a. Differential diagnosis
      (1) Physical and roentgenographic findings confirm the diagnosis of pneumonia. Proving that the pneumococcus is the causative agent may be more difficult.
      (2) Pneumonia caused by other bacteria, pulmonary embolism, and carcinoma of the lung may simulate pneumococcal pneumonia and should be considered in the differential diagnosis.
    b. Culture. Positive cultures of expectorated sputum and nasopharyngeal swabs are difficult to interpret because of the prevalence of pneumococci as indigenous flora. Other contaminating microbial flora may obscure the presence of pneumococci.
      (1) The best specimen for culture is blood, but only about 30% of patients with pneumococcal pneumonia are bacteremic.
      (2) Transtracheal aspirates may be useful, but false-positive cultures are obtained in 30% of patients.
    c. Stains. Gram-stained preparations of expectorated sputum are useful; large numbers of lancet-shaped cocci in pairs in conjunction with macrophages and neutrophils are strongly supportive of the diagnosis.

6. Therapy
    a. Antimicrobial chemotherapy
      (1) Pneumococcal pneumonia usually is treated with oral or parenteral penicillin.
      (2) Tetracycline is contraindicated as 5%–40% of isolates of *S. pneumoniae* are resistant.
    b. Immunoprophylaxis. Immunization with polyvalent vaccine consisting of capsular polysaccharide of the 14 most virulent strains of *S. pneumoniae* significantly reduces the incidence of disease in susceptible adult populations. The vaccine is less efficacious in children and generally ineffective in infants under 2 years of age.

B. Pneumonia caused by other pyogenic cocci

    1. Etiology. *Staphylococcus aureus*, *Streptococcus pyogenes*, other *Streptococcus* species, and *Neisseria meningitidis* occasionally cause pneumonia.

    2. Epidemiology. These pneumonias usually are secondary to other infections, such as influenza, and to noninfectious conditions, such as intravenous drug abuse.

    3. Clinical manifestations of pneumonia caused by these pyogenic cocci vary little from those of pneumococcal pneumonia. Unlike pneumococcal pneumonia, these infections—particularly staphylococcal pneumonia—cause considerable destruction of lung tissue.

    4. Definitive diagnosis is made by culture and identification of the etiologic agent in conjunction with strong clinical evidence of its causative role.

    5. Therapy consists of appropriately selected antimicrobial agents and supportive treatment.

C. Gram-negative bacillary pneumonia

    1. Etiology
      a. *Klebsiella pneumoniae*, *Escherichia coli*, and *Pseudomonas aeruginosa* are the principal causes of gram-negative bacillary pneumonia.

b. Occasionally, other gram-negative rods may cause pneumonia that generally is indistinguishable clinically and epidemiologically from the pneumonia caused by the most frequently isolated etiologic agents.

    (1) *Haemophilus influenzae* is a common cause of pneumonia in children under 5 years of age and in adults more than 60 years of age.

    (2) *Legionella pneumophila* has been increasingly recognized as a cause of progressive toxic pneumonia since establishment of its causative role in Legionnaires' disease.

2. Epidemiology. Gram-negative bacillary pneumonia accounts for about 10% of all cases of bacterial pneumonia.

  a. The **incidence** of gram-negative bacillary pneumonia is highest among compromised people who are likely to have their upper respiratory tract colonized by gram-negative aerobic rods; these patients have acute or chronic illness.

    (1) Nearly 50% of hospitalized patients have oropharyngeal colonization, in contrast to only 5% of healthy individuals.

    (2) About 25% of colonized patients develop pneumonia; lung infection probably is caused by aspiration of oropharyngeal secretions.

  b. Predisposing factors. Alcoholics, drug abusers, and other people subject to periods of unconsciousness and decreased laryngeal reflexes are susceptible.

  c. The **mortality rate** is about 50%, even with chemotherapy. Debilitated patients with pneumonia caused by *P. aeruginosa* have a mortality rate of 80%.

3. Clinical manifestations are similar to those of pneumococcal pneumonia: fever, chills, and a productive cough.

  a. Highly mucoid and bloody sputum—**"currant jelly" sputum**—may be seen when *K. pneumoniae* is the etiologic agent.

  b. Cavitation of the lung, hypotension, delirium, and shock are common.

  c. Physical findings include variable radiologic evidence of nonspecific lung involvement ranging from peribronchial effusion to dense unilobar consolidation of the upper lobe.

4. Diagnosis is difficult except in the approximately 25% of patients with positive blood cultures.

  a. Culture and identification of the etiologic agent confirm the diagnosis, but differentiation between bacteria colonizing the upper respiratory tract and those causing pneumonia may be difficult.

  b. Transtracheal or transthoracic aspirates and transbronchoscopic lung biopsy are the ideal specimens when obtained aseptically.

5. Therapy. Antimicrobial agents and supportive measures are necessities in the treatment of gram-negative bacillary pneumonia.

  a. Combined therapy with cephalosporins and aminoglycosides usually is beneficial and should be based on in vitro sensitivity testing.

  b. Quinolone antibiotics are proving useful in treating infection caused by *P. aeruginosa* and *Proteus* species.

D. Mycoplasma pneumonia, formerly referred to as **atypical pneumonia,** is highly infectious and frequently occurs in epidemics.

1. Etiology. *Mycoplasma pneumoniae* is a small, membrane-bound, pleomorphic bacterium.

2. Epidemiology. Mycoplasma pneumonia is distributed worldwide and has been recognized in every country where specific testing has been conducted.

  a. The lungs are the primary target organ in about 10% of infected people.

  b. In close environments, as exists in families, 50% of infected people develop clinical pneumonia as a consequence of dissemination of contaminated respiratory droplets.

3. Clinical manifestations. The clinical syndrome of mycoplasma pneumonia develops slowly.

  a. Initial symptomatology usually is a nonproductive cough, which eventually produces a watery to mucoid sputum that rarely is tinged with blood. The cough lasts about 2 weeks in untreated patients.

  b. Fever of 37.8°–39.4°C and headache are complaints of virtually all patients. Chills, rhinorrhea, chest pain, malaise, and generalized myalgia also are common.

  c. Physical findings, radiology, and laboratory values resemble those found in viral pneumonia.

  d. Most patients recover without complications or sequelae.

4. Diagnosis is made by culturing *M. pneumoniae* on special media and identifying it by metabolic characteristics and growth neutralization tests using antibody-impregnated disks.

   a. Nonspecific cold agglutinins develop in about half the patients. A high titer of cold agglutinins is good presumptive evidence of infection and justifies appropriate antibiotic therapy.

   b. Type-specific antibodies develop during convalescence and are useful only for retrospective confirmation of the disease.

5. Therapy

   a. Tetracycline or erythromycin in 4 equally divided oral doses amounting to 2 g/day for 10 days is the treatment of choice.

   b. Antibiotic therapy decreases the severity and duration of disease but does not immediately eliminate the bacteria from the upper respiratory tract.

II. VIRAL PNEUMONIA typically has an insidious onset. The disease may be a predisposing factor to, a complication of, or concomitant with bacterial pneumonia.

A. Etiololgy. A variety of viruses are capable of causing inflammation of lung parenchyma.

   1. Common etiologic agents include respiratory syncytial virus, adenovirus, parainfluenza, influenza A, and influenza B.

   2. Less common etiologic agents include herpesviruses, rhinoviruses, rubeola, echoviruses, coronaviruses, and coxsackieviruses.

B. Epidemiology. The incidence of viral pneumonia is difficult to determine because it is not a reportable disease, and some patients never seek medical attention.

   1. Viral pneumonia is estimated to affect approximately 0.1% of the population each year. Children less than 5 years old have an incidence more than four times that of the overall population; respiratory syncytial virus is the most common etiologic agent in this age-group.

   2. Predisposing factors. The physical and metabolic condition of the host influences the etiology of viral pneumonia. Age and immunologic status of the host and the physical features of the environment play important roles in the type and severity of disease.
      a. The incidence is highest in patients with underlying disease that affects cell-mediated immunity.
      b. Hospitalized patients, organ transplant recipients, diabetics, and the elderly are prime candidates for infection.

   3. Dissemination. Viral pneumonia is spread by droplet inhalation; coughing and sneezing are primary means of transmission. Direct contact with respiratory secretions also appears to be important in some infections.

   4. Seasonal variation. Viral pneumonia is most common in midwinter and spring, in part because of "population closeness."

C. Pathogenesis

   1. Primary foci of infection occur in the cells of the upper respiratory tract.

   2. Dissemination into the lungs requires cell-to-cell transmission. The infection spreads via contaminated mucous secretions and hematogenous and lymphatic dissemination.

   3. Viral pneumonia also may occur secondary to the more classic or disseminated viral syndromes.
      a. Cytomegalovirus (CMV) pneumonia typically is seen in patients with leukemia, lymphoma, or tissue transplantation. Primary CMV pneumonia occurs in neonates and immunocompromised adults.
      b. Varicella (chickenpox) pneumonia occurs in children secondary to the skin rash; 90% of primary cases of disease in adults occur in individuals more than 20 years old.

D. Clinical manifestations

1. **General features**
   a. Symptomatology initially includes nasal congestion, rhinorrhea, coryza, and eye discomfort.
   b. Variable symptoms of sore throat and hoarseness occur followed by headache, fever, chills, malaise, myalgia, and a nonproductive cough.
   c. Nonspecific symptomatology becomes more severe as the pneumonia progresses; prostration may occur suddenly.
   d. Physical examination usually is not specific or remarkable; the principal benefit is in developing a differential diagnosis.
   e. Viral pneumonia complicated by bacterial superinfection is the major cause of morbidity and mortality.

2. **Specific features**
   a. **Influenza A virus pneumonia** is the most common pneumonia in adults.
      (1) **Symptomatology**
         (a) Disease begins abruptly with fever of 39°–40° C, other systemic symptomatology, and prostration.
         (b) Coughing generally is nonproductive, but small quantities of bloody sputum may be produced as disease progresses.
         (c) Chest pain is nonpleuritic and substernal; rales and rhonchi are typical auscultatory findings.
      (2) **Laboratory findings**
         (a) Chest roentgenograms show diffuse bronchopneumonia; it often is seen bilaterally.
         (b) White blood cell counts are elevated, frequently to more than 10,000/mm³, with a marked left shift.
         (c) Blood gases and pH changes are important nonspecific clues to the severity of infection; hypoxemia and hypercarbia are prognostic.
   b. **Respiratory syncytial virus pneumonia** is most serious in the very young, although it also occurs in the elderly.
      (1) **Symptomatology.** The most prominent clinical manifestations are fever, dyspnea, and wheezing with intercostal retraction.
         (a) Children less than 1 year old tend to develop bronchiolitis.
         (b) Children 4–5 years old commonly develop bronchopneumonia.
         (c) Rales and rhonchi are present; cough is variable and sputum is scanty.
      (2) **Laboratory findings**
         (a) Chest radiographs show bilateral bronchopneumonia.
         (b) White blood cell counts and differential counts are not specific.
   c. **Adenoviral pneumonia** usually occurs in military recruits and children; it seldom occurs in civilian adults.
      (1) **Symptomatology.** The pneumonic syndrome produced by adenovirus types 1, 2, 3, and 7 is indistinguishable from pneumococcal pneumonia with respect to symptomatology and physical findings, although the incidence of some features varies.
         (a) In children the disease is severe and may be fatal.
         (b) Complications of bronchiectasis and chronic pulmonary disease are seen only in children.
         (c) Pharyngitis, rhinitis, rales, nausea, and vomiting are common features of adenoviral pneumonia.
      (2) **Laboratory findings.** White blood cell counts are 500–30,000 cells/mm³, but two-thirds of patients have counts less than 10,000/mm³.
   d. **Rubeola (measles) pneumonia** occurs in 7%–50% of patients with rubeola. Pneumonia develops within 5 days of appearance of the rash, and the disease usually is seen in children under 6 years old.
      (1) Rales and rhonchi on chest examination generally are the only abnormal findings in rubeola pneumonia.
      (2) Radiographic evidence of interstitial pneumonia is common.
      (3) Primary rubeola pneumonia principally is a fatal disease of immunologically compromised people.

III. *PNEUMOCYSTIS CARINII* PNEUMONIA is a major cause of morbidity and mortality in certain patient populations.

    A. Etiology. The protozoan *P. carinii* has been found in humans and other animals with symptomatic and asymptomatic infection. Its life cycle is unknown.

    B. Epidemiology

        1. *P. carinii* is the most common cause of sporadic interstitial pneumonia in immunosuppressed patients, affecting 25%–50% of patients; 50% of AIDS patients develop the disease.

        2. Disease occasionally is seen as endemic or intermediately epidemic illness in premature or malnourished infants. Infection appears to be common during the first 2 years of life.

        3. Predisposing factors include prolonged use of corticosteroids for lymphoproliferative malignancy, transplantation, and collagen-vasculature disease. Latent disease can be activated by corticosteroid therapy.

    C. Clinical manifestations vary greatly, in part because of infection with other agents; 25% of patients also have CMV infection. Most cases progress rapidly over a period of a few days; however, worsening pneumonia may develop gradually. Respiratory distress is indicated by tachypnea, nasal flaring, and eventually cyanosis.

        1. Symptomatology.
            a. Symptomatology includes fever of 39°–40° C and shaking chills. A nonproductive cough develops in fewer than half the patients.
            b. Shortness of breath may be the only patient complaint. Chest auscultation is remarkable for absence of findings.

        2. Laboratory findings
            a. Chest radiographs typically show diffuse granular infiltrates in both lungs resembling those seen in pulmonary edema.
            b. The most consistent finding in blood chemistry is hypoxemia. Lymphopenia and hypoalbuminemia often are present.

    D. Diagnosis requires demonstration of *P. carinii* in the lung.

        1. Stains and culture
            a. Samples often are obtained by needle aspiration of the lung, usually through the axilla.
            b. Smears usually are stained with Giemsa stain or toluidine O; silver stains also may be used. Standard histopathologic preparation also should be performed.
            c. Cultures for bacteria, mycobacteria, fungi, and viruses should be used to determine other complicating infectious processes.

        2. Serology. The absence of antibody formation in many cases makes serodiagnosis difficult. Immunofluorescent stains have been employed to detect the organism in lung tissue.

    E. Therapy. Pentamidine and co-trimoxazole (i.e., trimethoprim-sulfamethoxazole) are the most widely used antimicrobial agents. Treatment is difficult in the severely immunosuppressed patient, and new therapies are being investigated.

IV. PNEUMONIC PROCESSES IN OTHER DISEASES. Pneumonia characterized by specific histopathologic findings in the lung occurs in a variety of disease syndromes that are not considered in this chapter. For example, certain stages of tuberculosis and primary pulmonary mycosis produce pneumonia. Other diseases in which pneumonia may be part of the prodrome include psittacosis, melioidosis, Q fever, and legionellosis.

STUDY QUESTIONS

Directions: Each question below contains five suggested answers. Choose the **one best** response to each question.

Questions 1–4

A 79-year-old man who is bedridden with a broken hip suddenly develops a fever of 39.6° C and worsening cough that produces blood-tinged sputum. The patient complains of chilliness and chest pain. Sputum is obtained for microscopic examination and culture.

1. The most probable etiologic agent is

(A) *Legionella pneumophila*
(B) *Klebsiella pneumoniae*
(C) *Streptococcus pneumoniae*
(D) Adenovirus type 4
(E) *Mycoplasma pneumoniae*

2. Gram stain of the sputum reveals a predominance of gram-positive cocci in pairs; antipneumococcal chemotherapy is initiated. The treatment of choice is

(A) penicillin
(B) tetracycline
(C) pentamidine
(D) co-trimoxazole
(E) erythromycin

3. About 40 hours after onset of fever, the patient's symptomatology is severe. The nurse notes that the patient now feels cool to the touch rather than warm and perspiring; the attending physician notes general vasodilation. Which of the following conditions appears imminent for the patient?

(A) A hard, shaking chill
(B) Endotoxin shock
(C) Renal failure
(D) Vascular collapse
(E) Bacteremia

4. The patient dies within 72 hours of the onset of overt symptomatology. A major factor in the death probably is

(A) ineffective chemotherapy
(B) patient age
(C) underlying disease
(D) poor circulation
(E) pulmonary embolism

5. "Currant jelly" sputum is characteristic of pneumonia caused by

(A) *Streptococcus pneumoniae*
(B) *Klebsiella pneumoniae*
(C) *Mycoplasma pneumoniae*
(D) *Neisseria meningitidis*
(E) *Pneumocystis carinii*

Directions: Each question below contains four suggested answers of which **one or more** is correct. Choose the answer

A   if **1, 2, and 3** are correct
B   if **1 and 3** are correct
C   if **2 and 4** are correct
D   if **4** is correct
E   if **1, 2, 3, and 4** are correct

6. Prominent physical findings on chest auscultation in patients with pneumonia caused by *Pneumocystis carinii* include

(1) rales
(2) bilateral consolidation
(3) diffuse infiltrates and pulmonary edema
(4) an absence of findings

7. Cytomegalovirus pneumonia typically is seen in

(1) neonates
(2) leukemics
(3) transplant recipients
(4) lymphoma patients

ANSWERS AND EXPLANATIONS

1. The answer is C. [*I A 2 a (1), (2), 4 a, b*] Pneumococcal pneumonia—most often caused by 14 of the more than 80 capsular serotypes of *Streptococcus pneumoniae*—accounts for about 80% of all cases of bacterial pneumonia. The disease often occurs in debilitated elderly patients who have diminished immunologic function; the relative immobilization of the patient affects his or her ability to handle respiratory secretions and predisposes to pneumococcal infection. The sudden onset of fever, a worsening cough producing rust-colored (i.e., blood-tinged) sputum, chest pain, and chilliness make up the classic symptomatology of pneumococcal pneumonia. *Legionella pneumophila* produces a severe, progressively toxic pneumonia. *Klebsiella pneumoniae* produces a syndrome initially resembling pneumococcal pneumonia, but the sputum usually is very bloody and takes on a classic "currant jelly" appearance because of the destruction of lung tissue. Although it resembles pneumococcal pneumonia, adenovirus pneumonia rarely occurs in the adult civilian population. *Mycoplasma pneumoniae* produces a less severe pneumonia; sputum production is scanty.

2. The answer is A. [*I A 6 a (1)*] The treatment of choice for pneumococcal pneumonia is orally or parenterally administered penicillin, or one of its derivatives. Tetracycline is contraindicated because approximately 40% of the clinical isolates of *Streptococcus pneumoniae* are resistant. Tetracycline and erythromycin are the antibiotics of choice for mycoplasma pneumonia. Pentamidine and co-trimoxazole (i.e., trimethoprim-sulfamethoxazole) are widely used in the treatment of pneumonia caused by *Pneumocystis carinii*.

3. The answer is D. [*I A 4 d (2)*] Patients with pneumococcal pneumonia generally feel warm and perspiring to the touch. A change in this physical finding to cool skin and vasodilation indicates the imminence of vascular collapse and a poor prognosis for the patient. A single hard, shaking chill with teeth chattering is characteristic of pneumococcal pneumonia and usually signals recovery as the patient's cell-mediated immune system begins clearing the bacteria. The capsular polysaccharide of *Streptococcus pneumoniae* has a very low toxicity, and the microbe does not possess lipopolysaccharide (LPS). Renal failure is not a normal complication of pneumococcal pneumonia; the disease resolves with few if any residual side effects. Bacteremia occurs and isolation of the bacteria from the blood is confirmation of the etiologic agent; blood cultures are positive in only about 30% of patients.

4. The answer is B. [*I A 2 b*] The mortality rate associated with pneumococcal pneumonia is 10%, even when apparently effective chemotherapy is administered. Fatalities occur most commonly in patients with underlying disease that affects the immune system. In this case, the most prominent predisposing factor would be the patient's age. Immobilization of the patient because of the hip fracture might predispose to acquisition of the disease but probably does not significantly affect the prognosis. This patient, however, probably has diminished immunologic defenses because of his age. Poor circulation in the elderly usually affects the extremities; there is no indication that this is a problem in this patient. Pulmonary embolism is a condition that is included in the differential diagnosis before confirmatory evidence of the etiologic agent is obtained.

5. The answer is B. [*I C 3 a*] *Klebsiella pneumoniae* produces a highly destructive pneumonia that can result in considerable scarring to the lungs. A highly mucoid and bloody sputum, referred to as "currant jelly" sputum, is characteristic of the infection. *Streptococcus pneumoniae* classically causes production of rust-colored sputum when it is the etiologic agent of pneumonia. *Mycoplasma pneumoniae* causes pneumonia characterized by a scanty, watery sputum. Occasional cases of pneumonia caused by *Neisseria meningitidis* resemble pneumococcal pneumonia. Fewer than half the patients with pneumonia caused by *Pneumocystis carinii* develop a nonproductive cough.

6. The answer is D (4). [*III C 4*] Most cases of pneumonia caused by *Pneumocystis carinii* progress rapidly, but a variety of symptomatologies may develop. Chest auscultation is remarkable in that it yields no physical findings; shortness of breath may be the only patient complaint although hypoxemia is common. Radiographic examination of the lung shows diffuse infiltrates resembling those seen in pulmonary edema. Rales and consolidation are not typical findings.

7. The answer is E (all). [*II C 3 a*] Cytomegalovirus (CMV) pneumonia usually is secondary to a more classic disseminated viral syndrome. People most likely to develop CMV pneumonia are those with leukemia, lymphoma, and organ transplants. Primary CMV pneumonia occurs most frequently in neonates.

I. SUSCEPTIBILITY TO INFECTION. The incidence of infection in surgical patients is influenced by a variety of endogenous and exogenous factors. Endogenous factors include the immunologic status and indigenous flora of the patient; exogenous factors include environmental cleanliness and the staff-to-patient ratio, which affects care of wounds.

A. Compromised host defenses

1. **Skin and mucous membranes** are the body's initial defense against microbial invasion. They are compromised by trauma and by disruption of specialized functions of cells.

   a. **Incisions** disrupt the integrity of skin, disturb chemical barriers to infection such as perspiration, and can alter beneficial indigenous flora.

   b. **Anesthesia** depresses the ciliary action of bronchial epithelium and inhibits normal body secretions that entrap and remove pathogens by washing action.

2. **Underlying metabolic disorders** such as diabetes, diseases affecting the immune system, obesity, and malnutrition greatly increase the risk of developing a postoperative infection.

B. Foreign bodies.
The ability of foreign bodies to predispose to infection is not understood in detail. Foreign bodies appear to provide colonization sites for microorganisms that are unable to adhere to normal tissue.

1. **Prosthetic implants**, whether derived from synthetic materials or animal tissue, provide a suitable surface for microbial colonization. The inherent avascularity of these materials prevents delivery of immunologically important substances and cells as well as chemotherapeutic agents to the colonized implant.

2. **Indwelling devices** such as intravenous and urinary catheters, stomach tubes, wound drains, sutures, and other foreign bodies employed on a temporary, supportive basis for surgical patients can be colonized by microorganisms.

C. Damaged tissue.
Tissue injury occurs at the site of surgical incisions.

1. Disruption of normal vascularization by surgery may inhibit the ability of tissue in localized areas to mount an inflammatory response.

2. Surface characteristics of tissue and the formation of protective clots aid the colonization and survival of infecting microorganisms.

3. The ability of opportunistic pathogens to colonize damaged tissue is exemplified by the nearly exclusive predilection of *Salmonella choleraesuis* for vascular grafts.

D. Antimicrobial therapy.
The use of antiseptics to disinfect operative sites and the administration of broad-spectrum antimicrobial agents for therapeutic and prophylactic purposes predispose to superinfection by opportunistic pathogens.

1. Elimination of indigenous flora permits the colonization of tissue by less desirable microorganisms.

2. Indigenous bacterial flora usually is replaced by either drug-resistant bacteria or opportunistic fungal pathogens.

II. TYPES OF INFECTIONS. There are two major categories of surgical infections—localized infection, characterized by incisional wound colonization and abscess formation, and systemic infection, characterized by sepsis.

A. Localized infection

1. **Wound infections.** An infected surgical wound typically discharges pus.
   a. **Categorization.** Wounds incurred during surgery are assigned to one of four categories.
      (1) **Clean wounds** are wounds that do not enter the gastrointestinal tract or respiratory tract and that do not involve inflammation or a break in aseptic technique.
         (a) Wounds resulting from hysterectomies, cholecystectomies, and incidental appendectomies are included in this category if no initial inflammation of the tissue is present.
         (b) The incidence of infection in clean surgical wounds is as high as 5.3%.
      (2) **Clean-contaminated wounds** are those involving a hollow muscular area without significant spillage of indigenous flora.
         (a) Respiratory tract and gastrointestinal tract incisions with a minimal number of microbes contaminating normally sterile tissue are considered to be clean-contaminated wounds.
         (b) Up to 10.8% of clean-contaminated wounds become infected.
      (3) **Contaminated wounds** are those involving tissue with acute inflammation but no pus or those with gross spillage from a hollow, muscular organ.
         (a) Contaminated wounds also include those involving a major break in aseptic technique.
         (b) The incidence of infection in contaminated wounds is 15.2%–21.9%.
      (4) **Dirty wounds** are those involving tissue with pus and a perforated viscus.
         (a) Old traumatic wounds such as decubitous ulcers are included in this category.
         (b) The incidence of infection is 22%–40% in dirty wounds.
   b. **Monitoring.** Inflamed or erythematous wounds are monitored. Only when inflammation resolves are they classified as not infected.
   c. **Diagnosis.** Incisional wound infections usually are obvious in the monitored postoperative patient. Careful, regular visual examination of the incisional wound and its dressing provides early evidence of infection.
      (1) Signs of classic inflammatory response—heat, redness, edema, and pain—indicate an infectious process. These signs precede the appearance of a purulent discharge.
      (2) Heavy wound colonization is characterized by bacterial or fungal growth on the incision and dressing; colonization does not always lead to disease.
         (a) When bacterial or fungal growth is visible, the risk of infection is increased.
         (b) Pus or other material from a wound with persistent heavy colonization should be Gram stained and cultured; the isolated microorganism should be identified and tested for antimicrobial susceptibility.

2. **Abscess formation.** Abscesses form as localized infections in skin and soft tissues after surgical or accidental trauma.
   a. Abscesses are localized collections of purulent exudate, but the pus generally is not visible through the skin.
      (1) One or more drainage tracts (i.e., sinuses) to the skin may develop as the abscess progresses.
      (2) Localized pain and inflammation usually are present.
   b. **Monitoring.** Patients must be followed closely because early symptomatology of abscess formation often is absent. Unless the abscess is very deep in the tissue, it can be detected by palpation.
   c. **Diagnosis.** Knowledge of indigenous flora is important in specimen collection and interpretation of results.
      (1) Early diagnosis of an abscess is difficult because it is walled off, and local or systemic symptomatology may not be obvious.
      (2) A subcutaneous abscess at the wound site is apparent when it becomes large enough to visibly displace tissue or when it begins to drain.
      (3) A deep abscess, such as one occurring after abdominal or thoracic surgery, is difficult to diagnose; persistent fever and leukocytosis in the patient should be regarded as a potential sign of deep abscess.

(4) Roentgenographic examination is useful in locating the space-occupying lesion.

(5) Abscesses should be aspirated. Pus should be Gram stained and cultured for aerobic and anaerobic bacteria, fungi, or parasites when appropriate.

B. Systemic infection. Disseminated infections in the surgical patient develop as a complication of microorganisms in the bloodstream.

1. Types

    a. Bacterial infection

        (1) Clinically, disease caused by bacteria in the bloodstream (i.e., bacteremia or septicemia) is a serious, life-threatening complication of surgery.

        (2) Sepsis may result from direct inoculation of bacteria into the bloodstream during surgery, dissemination from a wound infection or abscess, or colonization of indwelling foreign bodies with subsequent seeding of the circulatory system.

    b. Fungal infection. Fungemia is relatively rare as a direct complication of surgery, but when it occurs it usually has a fatal outcome.

        (1) *Candida albicans* or other *Candida* species most frequently cause postoperative fungemia. The source usually is endogenous, and the yeast can directly invade the bloodstream from wounds or it can colonize indwelling foreign bodies.

        (2) Opportunistic fungal infections from environmental sources cause disease when the patient's cell-mediated defenses have been compromised.

    c. Viral infections. Viremia, usually of unknown origin, can be inapparent or life-threatening, depending upon the immune status of the surgical patient.

2. Diagnosis

    a. Sepsis in the surgical patient is diagnosed by the same methods employed for any patient with bacteremia or septicemia (see Chapter 26, section IV).

        (1) Serial blood cultures should be obtained from postoperative patients with persistent fever and other systemic symptomatology of infection.

        (2) The prophylactic use of broad-spectrum antibiotics can make isolation of the etiologic agent difficult.

    b. Fungemia is diagnosed by blood cultures. Isolation from peripheral blood may be difficult; special culture media and conditions are required.

    c. Viremia usually is diagnosed on the basis of clinical symptomatology and the elimination of bacterial and fungal etiologies.

III. ETIOLOGY OF INFECTION

A. Sources. The etiology of a particular surgical infection is dependent primarily upon the operative site. Table 28-1 lists the most common etiologic agents of surgical infections and their distribution among various clinical services.

1. The respiratory, urogenital, and gastrointestinal tracts are the endogenous sources of the most common etiologic agents of postoperative infections.

2. Infections originating from the environment or from indigenous flora of the skin account for about 2% of surgical infections. They are easily controlled through the use of disinfectants and antiseptics.

3. Viruses and fungi have been associated with surgical wounds but are uncommon as the primary etiologic agents of infections.

B. Clinically significant surgical infections. Some of the more important infections and their clinical characteristics and treatment are described here.

1. Staphylococcal infection

    a. Etiology

        (1) *Staphylococcus aureus* is the most common cause of incisional wound infection; it causes a purulent discharge from about 10% of all surgical incisions.

        (2) *Staphylococcus epidermidis*, once considered nonpathogenic, and coagulase-negative *S. aureus* are found with increasing frequency.

Table 28-1. Distribution of Etiologic Agents of Surgical Infections with Respect to Medical Service

Etiologic Agent	Distribution as to Service Percentage						Percentage of All Surgical Infections
	Surgery	Gyne-cology	Medicine	Obstetrics	Pediatrics	Newborn	
Escherichia coli	78.6	13.6	2.9	3.6	0.8	0.4	18.7
Staphylococcus aureus	85.1	5.6	5.5	1.7	1.4	0.7	18.6
Pseudomonas aeruginosa	88.5	3.6	5.6	0.6	1.6	0.1	8.8
Proteus mirabilis	78.5	13.3	3.5	3.9	0.5	0.1	5.6
Bacteroides species	76.4	15.7	2.7	4.3	0.5	0	3.8
Proteus species	74.2	16.1	5.1	4.0	0.6	0.2	3.2
Hemolytic streptococci	72.6	17.3	3.6	4.7	0.7	0.5	1.6
Group A streptococci	79.1	8.8	7.3	2.4	2.4	0	1.0
Clostridium perfringens	91.3	2.3	3.9	2.0	0.6	0	1.0
Other pathogens	79.4	10.7	4.6	3.9	0.8	0.5	37.6

Adapted from Cruse PJE: Wound infections: epidemiology and clinical characteristics. In *Surgical Infectious Diseases.* Edited by Simmons RL, Howard RJ. New York, Appleton-Century-Crofts, 1982, p. 439. Data from the National Nosocomial Infections Study, January 1970 to August 1973.

b. Clinical characteristics
    (1) Localized infections caused by *S. aureus* become apparent 4–6 days after surgery.
        (a) Abscess formation is common, and hematogenous dissemination can occur.
        (b) Certain toxin-producing strains of *S. aureus* are responsible for occasional cases of wound-associated toxic shock syndrome (TSS).
    (2) Infections caused by *S. epidermidis* usually are mild and occur after the patient has left the hospital.
        (a) The subacute nature of infection by *S. epidermidis* in the areas of prosthetic devices is a major clinical concern.
        (b) Coagulase-negative *S. aureus* formerly was identified as *S. epidermidis*; disease is identical to that caused by *S. epidermidis*.
c. Therapy
    (1) Surgery. Staphylococcal wound infections or abscesses should be opened, irrigated, and debrided if needed. The cavity should be packed with antiseptic gauze.
    (2) Chemotherapy. Definitively localized infections generally do not require antibiotic therapy.
        (a) Penicillinase-resistant antistaphylococcal antibiotics should be administered if an attempt to debride or evacuate the wound is contemplated.
        (b) Intravenous antistaphylococcal antibiotics are administered if cellulitis, lymphangitis, lymphadenitis, or septicemia is suspected.

2. Gram-negative bacillary infection. Surgical wound infections caused by gram-negative enteric rods primarily are seen in debilitated, elderly patients.
    a. Etiology. *Escherichia coli* and *Enterobacter*, *Klebsiella*, *Proteus*, and *Pseudomonas* species produce purulent incisional wound infections, often in conjunction with anaerobic streptococci and *Bacteroides fragilis*.
    b. Clinical characteristics
        (1) Incisional infections caused by gram-negative bacilli occur 1–2 weeks after surgery.
        (2) Cryptogenic fever, tachycardia, and other systemic symptomatology of sepsis generally develop in untreated or undiagnosed cases. Cellulitis, edema, erythema, and pain are relatively rare.
        (3) Bacteremia often is diagnosed before the localized infection is discovered.
            (a) Classic symptomatology of endotoxin shock may not develop.
            (b) Hypertriglyceridemia, hyperglycemia, and hypertension may occur.

(4) Persistent systemic symptomatology following chemotherapy may indicate deeper sub-fascial abscess or intra-abdominal source of infection.

  c. Therapy

    (1) Surgery. The infected surgical wound should be opened and debrided and the pus evacuated.

    (2) Chemotherapy

      (a) Systemic antimicrobial therapy should be initiated immediately to prevent hematogenous dissemination of the infection.

      (b) Combined therapy with an aminoglycoside, clindamycin, and ampicillin is recommended when Gram's stain of purulent material reveals mixtures of gram-negative rods and gram-positive cocci.

3. Group A streptococcal infection is either endogenous in origin or derives from people in contact with the patient.

  a. Etiology. Group A beta-hemolytic *Streptococcus pyogenes* is found on the skin and in the nasopharynx of carriers.

  b. Clinical characteristics

    (1) Group A streptococcal wound infection is characterized by a fulminant course.

    (2) Diffuse cellulitis, lymphangitis, and lymphadenitis with a large, blood-filled bleb surrounding the primary focus of infection are seen. A thin, watery, purulent discharge typically occurs within 3 days after surgery.

    (3) Septicemia frequently occurs with fever, chills, sweats, tachycardia, prostration, and other symptomatology of toxemia.

  c. Therapy

    (1) Surgery

      (a) Wounds should be opened and drained if there is accumulation of pus or necrotic tissue, or if the wound becomes undermined or gangrenous.

      (b) Tissue destruction and scarring can be substantial. Skin grafts ultimately may be needed to cover the wound.

    (2) Chemotherapy. High-dose parenteral penicillin or an alternative antibiotic is sufficient to control some group A streptococcal wound infections.

4. Other streptococcal infections. Non-group A streptococci most often cause surgical wound infections in conjunction with other bacteria.

  a. Etiology

    (1) *Streptococcus faecalis* and other group D streptococci (i.e., the enterococci), microaerophilic streptococci, peptostreptococci, and other anaerobic streptococci cause surgical wound infections.

    (2) Enterococci most frequently are found in mixed infections with gram-negative enteric rods, and the microaerophilic streptococci nearly always are found in synergistic processes with *S. aureus* or *Proteus* species.

  b. Clinical characteristics. Non-group A streptococcal infections vary according to the etiologic agent and range from mild, noninvasive infection to severe infection with bacteremia.

    (1) Enterococci are less invasive than group A streptococci.

      (a) Colonization of surgical wounds can occur without systemic manifestations.

      (b) Enterococci also cause severe infection and sepsis alone or, more frequently, with other bacteria.

    (2) Microaerophilic streptococci produce an intense inflammatory response in surgical wounds 10–14 days after the operation.

      (a) Massive cellulitis develops, and a characteristic purplish lesion with an ulcerated center forms and is surrounded by gangrenous skin.

      (b) The violaceous region of the characteristic lesion is extremely sensitive and painful.

    (3) Anaerobic streptococci, such as the peptostreptococci, can produce a variety of severe postoperative infections, particularly after surgery involving the urogenital, gastrointestinal, or respiratory tract.

      (a) Peptostreptococci frequently are found in incisional wound infections and deep abscesses.

      (b) Pus characteristically is thick and grayish with a fetid, anaerobic odor.

      (c) Bacteremia occasionally occurs.

c. Therapy

(1) **Surgery.** Lesions caused by microaerophilic streptococci generally are excised to remove gangrenous and necrotic tissue, and abscesses are drained.

(2) **Chemotherapy.** Postoperative streptococcal infections require specific antimicrobials administered in combination with other antibiotics to eradicate the typical mixed etiology.

5. **Myonecrosis and clostridial cellulitis.** Contamination of the surgical wound with soil or feces can lead to serious, life-threatening clostridial infections.

a. **Etiology.** *Clostridium perfringens* is the most important etiologic agent of clostridial wound infections, although other gangrene-causing *Clostridium* species may be involved.

b. **Clinical characteristics.** Clostridial infection presents as spreading crepitant or noncrepitant cellulitis without the manifestations of the systemic gas gangrene syndrome or as progressive myonecrosis with systemic toxemia.

(1) Clostridial cellulitis is a relatively benign disease of the fascia and subcutaneous tissue.

(2) Myonecrosis typically is a rapidly spreading, destructive disease of skeletal muscle.

c. **Therapy**

(1) **Surgery**

(a) Clostridial cellulitis requires opening the wound and debriding all necrotic tissue.

(b) Fulminant myonecrosis is treated by amputation when possible.

(2) **Chemotherapy.** Parenteral administration of appropriate chemotherapy is mandatory in all clostridial infections. Penicillin, chloramphenicol, and clindamycin as combined therapy is recommended because not all clostridia are susceptible to each antibiotic alone.

6. **Rare causes of surgical infection.** Numerous microorganisms occasionally cause surgical infections.

a. The "other" pathogens (see Table 28-1) account for about 38% of wound infections.

b. Infections include tuberculosis of the wound following surgery on tuberculous lesions; actinomycotic infection following oral, esophageal, gastric, thoracic, genital, or colonic surgery; mycotic wound infection and fungemia, usually caused by *C. albicans* or other *Candida* species; and surgical diphtheria.

IV. CHEMOPROPHYLAXIS.
Prophylactic use of antibiotics is controversial because evidence of real benefit to the patient often is lacking. Additionally, no antibiotic regimen completely eliminates the body's microbial flora. Infections subsequent to chemoprophylaxis usually are caused by bacteria with multiple drug resistance. Conservative chemoprophylaxis is recommended.

A. Administration sequence

1. **Initiation.** Prophylactic administration of antibiotics should be initiated only when its value is known. Chemoprophylaxis is beneficial for contaminated or dirty wounds and in cases of vascular surgery.

2. **Preadministration.** Prophylaxis should not be initiated too early. Replacement of indigenous flora with antibiotic-resistant microorganisms is avoided by administration of the antimicrobial agent 1–3 hours before surgery.

3. **Postadministration.** Prophylaxis should not be used any longer than necessary. In the absence of infection, 1–3 doses of antimicrobials is considered adequate.

B. Dosage.
Appropriate doses of antimicrobial agents used for prophylaxis must be administered. The tendency is to administer too small a dose, which provides selective pressure for drug-resistant strains of microorganisms.

STUDY QUESTIONS

Directions: Each question below contains five suggested answers. Choose the **one best** response to each question.

1. A violaceous lesion with an ulcerated center and surrounded by gangrenous skin is characteristic of a surgical wound infection caused by

(A) gram-negative rods
(B) anaerobic streptococci
(C) coagulase-negative staphylococci
(D) microaerophilic streptococci
(E) clostridia

Directions: Each question below contains four suggested answers of which **one or more** is correct. Choose the answer

> **A**   if **1, 2, and 3** are correct
> **B**   if **1 and 3** are correct
> **C**   if **2 and 4** are correct
> **D**   if **4** is correct
> **E**   if **1, 2, 3, and 4** are correct

Questions 2 and 3

A 63-year-old white man has abdominal surgery to repair an intestinal hernia. During surgical manipulation, the bowel is pinched and requires a suture for repair. Following surgery, the patient develops an inflamed incision and cellulitis.

2. Cellulitis is a characteristic finding in surgical infections caused by

(1) gram-negative rods
(2) microaerophilic streptococci
(3) staphylococci
(4) clostridia

3. The patient develops lymphangitis and lymphadenitis in conjunction with the diffuse cellulitis. A large, blood-filled vesicle also develops. Gram-positive cocci in short chains are observed in smears of a thin, watery, purulent discharge from the wound. Probable sources of the infection include

(1) the skin of patient or staff
(2) the patient's damaged intestine
(3) the nasopharynx of a carrier
(4) the oral mucosa of patient or staff

4. Common endogenous sources of microbes that cause surgical wound infections include the

(1) urogenital tract
(2) gastrointestinal tract
(3) respiratory tract
(4) skin

5. Anesthesia inhibits natural immune defenses by

(1) drying and cracking the skin
(2) altering bodily secretions
(3) slowing migration of neutrophils
(4) depressing ciliary action of cells

Directions: The group of questions below consists of lettered choices followed by several numbered items. For each numbered item select the **one** lettered choice with which it is **most** closely associated. Each lettered choice may be use once, more than once, or not at all.

Questions 6–10

Match the characteristic symptomatology of surgical infections with the etiologic agent.

(A) *Escherichia coli*
(B) *Streptococcus pyogenes*
(C) *Peptostreptococcus micros*
(D) *Clostridium perfringens*
(E) *Candida albicans*

6. Fungemia

7. Diffuse cellulitis and thin, watery, purulent discharge

8. Fulminant myonecrosis of skeletal muscles

9. Thick, grayish pus with a fetid odor

10. Hyperglycemia, hypertriglyceridemia, and hypertension

ANSWERS AND EXPLANATIONS

1. The answer is D. [*III B 4 b (2) (a)*] Microaerophilic streptococci produce massive cellulitis 10–14 days after surgery. A purplish lesion characteristically forms. It has an ulcerated center and is surrounded by an area of gangrenous skin. The violaceous lesion is extremely sensitive and painful. Gram-negative enteric rods cause inflammation and may produce a nonclassical endotoxin shock syndrome characterized by hyperglycemia, hypertriglyceridemia, and hypertension. Coagulase-negative staphylococci typically cause a mild inflammation and purulent discharge from the wound site. Clostridial infection of surgical wounds causes cellulitis and myonecrosis.

2. The answer is C (2, 4). [*III B 4 b (2) (a), 5 b*] Cellulitis most commonly is a characteristic of wound infections caused by microaerophilic streptococci, clostridia, and group A streptococci. Gram-negative rods usually cause cryptogenic fever, tachycardia, and other systemic symptomatology; there is little cellulitis. Staphylococci usually produce inflammation of the incisional wound with a purulent discharge; abscess formation is common.

3. The answer is B (1, 3). [*III B 3 a, b (2)*] The clinical description of the wound infection suggests a classical etiology of group A streptococci—*Streptococcus pyogenes.* The usual source of infectious *S. pyogenes* is the skin and nasopharynx of carriers. The patient himself or other people in contact with the patient, such as hospital staff, would be likely sources of the bacteria. *S. pyogenes* does not normally inhabit the oral mucosa, and it cannot survive in the intestine for any length of time.

4. The answer is A (1, 2, 3). [*III A 1*] The urogenital tract, gastrointestinal tract, and respiratory tract are the sources of most surgical wound infections. Although the skin's indigenous flora contains many microbes capable of causing wound infections, the skin is relatively easily disinfected prior to surgery.

5. The answer is C (2, 4). [*I A 1 b*] A variety of situations during surgery affect the development of incisional wound infections. Endogenous factors such as the indigenous flora and immune status of the patient are important contributory factors in the development of surgical wound infection. Exogenous factors such as environmental cleanliness and staff-to-patient ratios also are important. A high ratio provides more intensive wound care for the patient. Anesthesia is an exogenous factor that affects the natural defense mechanisms of the body. It alters bodily secretions that eliminate pathogenic microbes by mechanical washing action and depresses the ciliated cell activity of specialized respiratory cells.

6–10. The answers are 6-E, 7-B, 8-D, 9-C, 10-A. [*III B 2 b (3) (b), 3 b (2), 4 b (3) (b), 5 b (2), 6 b*] Although most surgical wound infections can present with a wide range of symptomatology, certain signs and symptoms are relatively characteristic of specific etiologic agents. Fungi in the bloodstream (i.e., fungemia) most frequently is caused by *Candida albicans* or other *Candida* species. Fungemia is a relatively rare result of surgical wound infections.

Diffuse cellulitis with a thin, watery, purulent discharge from a surgical wound indicates infection by *Streptococcus pyogenes.* A large, blood-filled bleb, lymphangitis, and lymphadenitis also are seen in *S. pyogenes* incisional infections.

Fulminant myonecrosis of skeletal muscles is a serious, life-threatening complication of clostridial wound infections. *Clostridium perfringes* is the most common etiologic agent of these infections. The clostridia inhabit soil and the gastrointestinal tract of most mammals.

Anaerobic streptococci such as *Peptostreptococcus* species frequently are associated with wound infections contaminated by indigenous flora of the gastrointestinal tract, urogenital tract, or respiratory tract. They also are found in deep abscesses in infections of mixed etiology. They characteristically produce thick, grayish pus with fetid, anaerobic odor.

Escherichia coli and other gram-negative enteric rods often contaminate surgical wounds involving the urogenital tract and gastrointestinal tract. Instead of classic symptomatology associated with endotoxin shock, they may produce hyperglycemia, hypertriglyceridemia, and hypertension in elderly, debilitated patients.

29
Urogenital Infections
Gerald E. Wagner

I. URINARY TRACT INFECTION (UTI). Understanding the epidemiology and etiology of UTI is essential to rapid, accurate diagnosis and to the development of an effective therapeutic regimen.

A. Epidemiology. Most cases of UTI occur in ambulatory, nonhospitalized people who are otherwise healthy, although a variety of factors predispose to the condition.

1. **Gender.** Females are 30 times more likely to develop UTI than males, in part because of their shorter urethra.
   a. The incidence of UTI in females appears to increase linearly with age. UTI is diagnosed in about 1% of school-aged girls and the incidence peaks at about 10% in women over 60 years of age.
   b. Males rarely develop UTI; the incidence peaks at about 1% in men over 60 years of age.
      (1) During the first week after birth, males are most likely to develop UTI as the result of hematogenous dissemination.
      (2) Males also are infected more commonly in areas where schistosomiasis is endemic; the incidence is as high as 5%.

2. **Anatomic factors**
   a. Abnormalities in anatomic structure of the urinary tract, especially those resulting in obstruction or the reflux and incomplete voiding of urine, predipose to UTI.
   b. Frequent, recurrent, and chronic UTI suggest an anatomic abnormality.

3. **Physical factors**
   a. Manipulations with mechanical devices such as catheters, probes, and swabs may introduce microbes into the urethra and bladder, causing UTI.
   b. Improper use of tampons, douches, and vaginal swabs may inoculate the tip of the urethra with indigenous vaginal flora.

4. **Metabolic factors**
   a. Disorders such as diabetes significantly increase the risk of UTI. Diabetic males show about the same incidence of infection as healthy females (i.e., 30 times higher than that of healthy males).
   b. Kidney stones and other obstructive entities resulting from metabolic disorders also increase the risk of UTI.

5. **Hospitalization.** The hospitalized patient is at a greater risk for developing UTI than is the ambulatory patient.
   a. Patients with indwelling urinary catheters inevitably develop UTI in spite of precautions with aseptic techniques and disinfection.
   b. The risk of infection increases with the length of time the indwelling urinary catheter is in place.

B. Etiology

1. **Common etiologic agents**
   a. **Enterobacteriaceae** that are indigenous flora of the gastrointestinal tract cause the majority of cases of UTI.
      (1) *Escherichia coli* accounts for 50%–85% of UTIs, and *Klebsiella pneumoniae* is responsible for another 8%–13%.

       (2) In hospitalized patients the etiology of UTI changes radically; *Proteus, Serratia, Aeromonas,* and *Pseudomonas* species frequently are causes of infection.

   b. *Streptococcus faecalis* is the only gram-positive enteric bacterium commonly causing UTI.

2. Infrequent etiologic agents

   a. *Staphylococcus aureus,* **corynebacteria,** and **lactobacilli** occasionally are isolated from infected urines but often are secondary to other syndromes.

   b. *Trichomonas vaginalis, Mycoplasma hominis, Ureaplasma urealyticum,* and *Chlamydia trachomatis* frequently are causes of urethritis but do not appear to cause true UTI.

   c. *Candida albicans* **and other** *Candida* **species** may be isolated from the urine of diabetic females and from patients with indwelling catheters.

       (1) The presence of the yeast usually represents only saprophytic colonization of the urethra.

       (2) Ascending candidiasis of the urinary tract is rare; repeated isolation of this yeast could indicate pyelonephritis secondary to disseminated candidiasis.

3. Viruses play a poorly defined role in UTI except in certain immune complex diseases, in which viruses are deposited in glomeruli of the kidneys.

   a. Adenovirus type 2 is the apparent cause of acute hemorrhagic cystitis in children.

   b. Cytomegaloviruses (CMVs) are implicated in some cases of acute rejection of renal transplants.

   c. Viruses frequently are found in large numbers in urine during the course of viral disease.

C. Clinical manifestations. Symptomatology in UTI varies depending upon whether the infection is in the lower urinary tract (urethritis and cystitis) or in the upper urinary tract (acute nonobstructive pyelonephritis). Individual variation in patients also is common.

1. Urethritis

   a. Characteristic symptomatology includes dysuria (i.e., pain or burning during urination), frequency of urination, and urgency of urination.

   b. This symptomatology also occurs in women with vulvitis or vaginitis, but in true UTI the pain is described as internal dysuria, in contrast to labial discomfort.

2. Cystitis

   a. Characteristic symptomatology includes suprapubic pain and tenderness, frequency of urination caused by diminished bladder capacity, and occasional hematuria.

   b. Although urethrocystitis may be asymptomatic, it usually causes malodorous urine and incontinence, especially in older women.

3. Acute pyelonephritis

   a. Characteristic symptomatology includes flank pain, renal tenderness upon palpation, fever and chills, and hematuria. The symptomatology of lower UTI usually is present.

   b. Nonspecific symptomatology such as nausea, vomiting, diarrhea, and constipation may confuse the diagnosis.

   c. The clinical manifestations of acute pyelonephritis may resolve spontaneously in the absence of therapy.

4. Complications

   a. UTI is the major source of **gram-negative rods in the bloodstream.**

       (1) Bacteremia and septic shock are serious and frequently fatal complications of UTI, especially in patients with urinary tract abnormalities and underlying disease.

       (2) Urologic procedures on patients with UTI may introduce the infecting microorganism into the circulation.

   b. Severe kidney damage resulting from papillary necrosis has been described in patients recovering from acute pyelonephritis.

       (1) Renal and perirenal abscess may occur and lead to bacteremia.

       (2) End-stage chronic pyelonephritis as the result of UTI is rare.

D. Diagnosis of UTI can be made with assurity only by demonstrating the etiologic agent, usually bacteria, in the urine.

1. **Collection of urine samples**
   a. **General considerations**
      (1) Care must be taken to minimize contamination by the indigenous microbiota of the urethra or vagina
      (2) Urine samples must be transported rapidly to the clinical laboratory to prevent overgrowth of the sample by contaminating microbes. Refrigeration of the sample may affect results.
   b. **Collection techniques**
      (1) **The midstream clean-catch technique** is the most commonly employed urine collection method.
         (a) The glans penis or urethral meatus should be thoroughly cleaned with soap and water and dried with a sterile sponge. Disinfectant solutions should not be used because they can inhibit the growth of the etiologic agent.
         (b) After voiding is initiated, the urine sample is collected in midstream.
      (2) **Catheterization or suprapubic aspiration** may be necessary when the patient is unable to cooperate.

2. **Quantitative urine cultures** have been proposed as a means of distinguishing between contamination and actual infection.
   a. **Significant bacteriuria,** indicating UTI, is defined as 10^5 or more gram-negative rods/ml of properly collected and transported urine.
      (1) Fewer bacteria cause bacteriuria in patients infected with bacteria other than Enterobacteriaceae or in patients with a known predisposition to UTI.
      (2) Symptomatology is more important than the numbers of bacteria in UTI; any degree of bacteriuria should be evaluated in patients with urinary tract symptomatology.
      (3) Contamination frequently is indicated by large numbers of more than one species of bacteria.
   b. **Techniques for quantitative culture**
      (1) **Medium**
         (a) A differential/selective medium such as MacConkey or eosin-methylene blue (EMB) agar is used.
         (b) Blood agar often is inoculated to obtain a total bacteria count.
      (2) **Inoculation**
         (a) The **pour plate technique** is the most accurate method for obtaining evenly distributed colonies.
         (b) The **calibrated loop technique** is the fastest and easiest method of inoculation.
         (c) **Commercial dip slides** generally yield interpretable results.
      (3) **Counts.** Automated, semiautomated, or manual counts of colonies are made after 18–24 hours of incubation.
   c. **Microscopic examination** of uncentrifuged urine is a rapid means of estimating the number of bacteria present in the specimen.
      (1) Observation of at least one bacterium per high-power field ($\times$ 1000) is considered equivalent to 10^5 bacteria/ml of urine.
      (2) Leukocytes or erythrocytes in the sample is consistent with, but not diagnostic for, UTI.
      (3) Squamous epithelial cells of vaginal origin in urine indicate improper collection of the sample.

E. **Therapy.** Treatment of UTI is aimed at the prevention of complications and at early recurrence of infection.

   1. **General considerations**
      a. Urinary levels of most antimicrobial agents used to treat UTI are 10–100 times greater than peak serum levels; disk diffusion susceptibility tests are based on average serum levels of the drug.
      b. Susceptibility testing may not be necessary in simple UTI, but it is important in pyelonephritis and cases of recurrent chronic infections.

   2. **Therapeutic regimens**
      a. **Urethritis and cystitis.** The proper dose of the antimicrobial agent and the length of therapy for lower UTI are controversial.
         (1) Conventional therapy consists of minimal to moderate doses of a drug for 10–15 days.

   (2) A single dose of an effective drug frequently is sufficient to clear uncomplicated urethritis or cystitis.

  b. Acute pyelonephritis. Antimicrobial therapy must achieve rapid sterilization to prevent relapse or serious complications.

   (1) Combined therapy with synergistic drugs is important in preventing the emergence of resistant populations of bacteria.

   (2) β-Lactam antibiotics and aminoglycosides in synergistic combinations may achieve sterility faster and more reliably than either drug alone.

II. PROSTATITIS is a term often used to refer to noninfectious symptom complexes as well as infections of the prostate gland.

A. Etiology

 1. Acute and chronic bacterial prostatitis usually are caused by the most common etiologic agents of UTI.

  a. Enterobacteriaceae (*E. coli*, *Klebsiella* species, and *Proteus* species) and ***Pseudomonas* species** are the most common causes.

  b. *S. faecalis* and *Staphylococcus epidermidis* are less frequent causes of prostatitis.

 2. Nonbacterial prostatitis, in some cases is a misnomer; anaerobic bacteria, mycoplasmas, and chlamydiae may be the etiologic agents.

  a. Parasites such as *T. vaginalis* and fungi such as *C. albicans* rarely cause prostatitis.

  b. There is no compelling evidence that viruses cause prostatitis, although many cases of symptomatic disease occur following viral infections of the upper respiratory tract.

B. Clinical manifestations

 1. Acute bacterial prostatitis has a sudden onset. The patient presents with fever, chills, and other symptomatology of systemic involvement.

  a. The **characteristic symptomatology** of lower UTI is present; nonspecific complaints such as anorexia and myalgia also are common.

  b. Physical findings. The prostate is swollen, firm, indurated, warm to the touch, and very tender.

 2. Chronic bacterial prostatitis is more subtle in clinical presentation. The patient usually has a history of recurrent UTI.

  a. A unique characteristic of this syndrome is the absence of specific symptomatology until the patient complains of cystitis.

  b. The patient has significant bacteriuria and pyuria during the symptomatic phase of disease.

  c. Physical findings. The prostate usually is small and firm; it may be tender during the symptomatic phase.

 3. Nonbacterial prostatitis describes a syndrome consistent with the inflammatory reaction in bacterial prostatitis.

  a. Symptomatology includes variable degrees of perineal ache, back pain, groin pain that may radiate to the inner aspect of the thigh, musculoskeletal soreness, and irritation on voiding (i.e., frequency, urgency, and dysuria).

  b. Ejaculatory pain may or may not be a complaint. Urethral discharge is uncommon.

 4. Complications

  a. Pyelonephritis is a common complication of bacterial prostatitis when antimicrobial therapy is delayed.

  b. Other complications are the same as those of any severe UTI.

  c. No direct evidence supports a role for bacterial prostatitis in carcinoma.

C. Diagnosis of prostatitis is made on the basis of clinical manifestations and cultural results. Accurate diagnosis of the type of prostatitis is essential to effective therapy.

 1. Acute and chronic bacterial prostatitis generally present no diagnostic dilemma.

  a. Acute bacterial prostatitis. Typically, urinalysis reveals acute UTI (i.e., significant bacteriuria) and a large number of white cells.

    b. Chronic bacterial prostatitis. Microscopic characteristics of expressed prostatic secretions are variable but may be useful in making the diagnosis.

  2. Nonbacterial prostatitis
    a. Prostatic examination generally is inconclusive.
    b. Microscopic examination of expressed prostatic secretions reveals inflammatory characteristics consistent with infection.
    c. Cultures usually are negative unless specialized techniques for nonbacterial etiologic agents are used.

D. Therapy

  1. Acute bacterial prostatitis. An antimicrobial agent chosen on the basis of in vitro susceptibility testing produces dramatic results even though the agent normally does not penetrate the prostatic membrane.

  2. Chronic bacterial prostatitis. Antimicrobial therapy is unsatisfactory, but prevention of cystitis and treatment of symptoms are recommended courses of therapy.

  3. Nonbacterial prostatitis. Effective treatment can be formulated only when the etiology of this condition is known.

III. NONVENEREAL INFECTIONS OF THE FEMALE GENITALIA. Infections of the female genitalia that are not acquired by sexual contact may be pyogenic or granulomatous.

A. General considerations

  1. Pyogenic and granulomatous infections of the female genitalia usually are caused by indigenous flora of the vagina.

  2. Inferior aseptic techniques in obstetric and gynecologic procedures may result in exogenously acquired infections caused by *S. aureus, Streptococcus pyogenes,* or other microbes.

  3. A few etiologic agents, such as *T. vaginalis* and *C. albicans,* may be present in small numbers as indigenous flora of the vagina, or they may be acquired from asymptomatic sexual partners or from fomites.

B. Clinical syndromes

  1. Pyogenic infections
    a. Epidemiology. Infection generally occurs in poorly oxygenated spaces or in cavities containing blood, amniotic fluid, and dead fetal tissue.
    b. Etiology
      (1) Strict anaerobes, such as *Bacteroides* and *Peptostreptococcus* species, and the microaerophilic streptococci usually are found in large numbers.
      (2) Fastidious bacteria such as *Actinomyces israelii, Propionibacterium* species, *Clostridium* species, and *Gardnerella vaginalis* are found less frequently in large numbers.
      (3) Facultative anaerobes such as *E. coli, Proteus mirabilis, S. epidermidis, Streptococcus agalactiae,* and *S. faecalis* may be found in small numbers alone or in mixed infections with strict anaerobes.
    c. Clinical manifestations. Fever, chills, and other systemic symptomatology are present; abscesses are common.
    d. Therapy
      (1) Chemotherapy. Immediate, effective antimicrobial therapy is essential.
      (2) Invasive procedures. Surgical excision of affected tissues, debridement, or drainage of abscesses frequently is necessary.

  2. Vulvovaginitis
    a. Epidemiology. Vulvovaginitis may develop from the overgrowth of indigenous flora, from sexual transmission, or from recently contaminated fomites.
    b. Etiology
      (1) *C. albicans* and *T. vaginalis* are the most frequent etiologic agents.
      (2) *G. vaginalis, U. urealyticum,* and *M. hominis* are less frequent causes of the syndrome.

    c. Clinical manifestations
- **(1)** An abnormal increase in vaginal secretions and inflammation of the vaginal mucosa and vulvar skin characterize vulvovaginitis.
- **(2)** Severe discomfort is common and a malodorous discharge frequently occurs.

    d. Therapy. Specific and nonspecific topical antimicrobial agents usually are beneficial.

3. Toxic shock syndrome (TSS)

    a. Epidemiology. TSS typically occurs during the menstrual period of previously healthy women and is associated with improper use of expandable tampons of synthetic materials that occlude the vagina.

    b. Etiology
- **(1)** *S. aureus* has been cultured from the vagina in 98% of women with TSS; these bacteria rarely are cultured from the vagina of healthy women.
- **(2)** **Staphylococcal toxin** is implicated as the cause of TSS, although the pathway by which the toxin enters the circulation is speculative.

    c. Clinical manifestations
- **(1)** TSS is an acute febrile illness characterized by severe hypotension and other symptomatology of shock.
- **(2)** An erythematous macular skin rash is common.
- **(3)** Death sometimes occurs from shock.

    d. Diagnosis. The general shock-type syndrome along with the rash may make diagnosis difficult when Rocky Mountain spotted fever, Kawasaki disease, or Reye's syndrome must be considered.

    e. Therapy
- **(1)** **Supportive therapy** for shock is the major course of treatment for TSS.
- **(2)** **Chemotherapy.** Antistaphylococcal β-lactamase–resistant antibiotics have been used, but their therapeutic value is questionable.

4. Granulomatous infection

    a. Epidemiology. Infection usually is caused by hematogenous spread of microbes from primary pulmonary or intestinal infection to the fallopian tubes.
- **(1)** Upward dissemination of the infection produces localized pelvic peritonitis in about 50% of patients and perioophoritis in 30% of patients.
- **(2)** Downward spread produces endometritis in almost 60% of patients.

    b. Etiology. Granulomatous infection of the female genitalia usually is caused by *Mycobacterium tuberculosis*.

    c. Clinical manifestations. Systemic manifestations of chronic infection are rare; patients appear healthy in the absence of progressive pulmonary tuberculosis.
- **(1)** Mild pelvic pain and abnormal vaginal bleeding are the most common complaints.
- **(2)** About 85% of patients are nulliparous, and sterility is the chief complaint of 50% of the patients.

    d. Diagnosis. Granulomatous infection of the female genitalia is suspected on the basis of symptomatology in patients with primary tuberculosis. Confirmation of disease is by biopsy.

    e. Therapy. Effective treatment is achieved with antituberculosis drugs.

IV. GONORRHEA AND NONGONOCOCCAL URETHRITIS (NGU) are sexually transmitted diseases with identical symptomatology. These diseases may coexist; NGU often becomes prominent following effective treatment of gonorrhea.

A. Etiology

1. *Neisseria gonorrhoeae*, a fastidious aerobic gram-negative coccus usually occurring in pairs, is the only etiologic agent of gonorrhea.

2. *Chlamydia trachomatis*, an obligate intracellular bacterium, is the usual cause of NGU, formerly referred to as nonspecific urethritis.

    a. *U. urealyticum, M. hominis, T. vaginalis, C. albicans*, and herpes simplex virus also are implicated as etiologic agents of NGU.

    b. Approximately 20% of cases of NGU do not yield a likely pathogenic microorganism when urethral exudate is cultured.

B. Epidemiology

1. Gonorrhea

a. Transmission normally is by direct sexual contact. The bacteria rapidly die when exposed to the environment. Humans are the natural reservoir.

b. The typical patient is a sexually promiscuous, young, urban male.

    (1) Reported cases in men outnumber those in women about 3 to 1; this appears to reflect the greater frequency of asymptomatic disease among women.

    (2) Most cases occur in the sexually active age-group, with a peak in incidence in those who are 20–25 years old.

c. About 3 million cases are reported yearly in the United States; estimates of unreported cases range as high as 30 million.

2. NGU is prevalent in the same population and is as common as gonorrhea in the United States. In England, NGU is the most common venereal disease reported.

C. Pathogenesis. The pathogenesis of gonorrhea and genital chlamydial infections is not understood completely.

1. Gonorrhea. Susceptible columnar epithelium of the urethra, endocervix, and deeper genital tissue are invaded.

a. Mechanical and chemical defenses of the genital microenvironment eliminate unattached gonococci.

b. Dissemination up the genital tract probably involves attachment to sperm and twitching motility.

c. A major inflammatory response occurs in the submucosa.

    (1) Intact mucosal cells are penetrated and an intense neutrophil response is induced.

    (2) Overlying mucosa of the tissues is destroyed.

d. Virulence factors

    (1) Gonococcal toxin, probably endotoxin, is a potent inhibitor of ciliary activity, preventing removal of the organisms from fallopian tubes.

    (2) An enzyme capable of destroying secretory immunoglobulin A (IgA) is produced by strains of pathogenic gonococci.

2. NGU

a. Macrophages and lymphocytes containing typical inclusion bodies collect in subepithelial cells of the urethra.

b. Chlamydial urethritis has a follicular appearance that is distinguished from the deep mucosal erosion of gonorrhea.

D. Clinical manifestations

1. Gonorrhea. Clinical manifestations vary depending upon the gender of the patient and the site of infection.

a. Genital gonorrhea in men

    (1) Acute anterior gonococcal urethritis occurs 2–5 days after exposure and reaches a symptomatic peak in 2–3 weeks.

    (2) Symptomatology occasionally lasts for months, and relapse may occur.

        (a) Urethritis classically is characterized by severe dysuria and purulent discharge.

        (b) Frequency and urgency of urination, genital itching, inguinal adenitis, and fever are less common symptoms.

    (3) All portions of the urogenital tract commonly become infected in untreated patients.

    (4) Asymptomatic infections recently have been recognized as an important reservoir of infection.

b. Genital gonorrhea in women

    (1) Site of infection

        (a) The endocervix is the primary site of infection.

        (b) Urethral involvement typically is secondary to endocervical infection.

        (c) The paraurethral Skene's gland and Bartholin's duct also may become infected.

    (2) Symptomatology

        (a) Endocervical infection causes increased vaginal secretions, purulent discharge, or menorrhagic or intermenstrual bleeding.

       (b) Urethral infection produces dysuria and frequency and urgency of urination.

       (c) Pelvic inflammatory disease (PID; salpingitis) is the most common complication, occurring in about 15% of women.

     (3) Asymptomatic genital gonorrhea is much more frequent than in men, and women are considered a major reservoir of disease.

   c. Anorectal gonorrhea. Infection of the anal mucosa of women and homosexual men occurs by penoanal contact; secondary spread from the endocervix also is common.

     (1) Most patients remain asymptomatic and are at risk of disseminated gonococcal disease.

     (2) Symptomatic manifestations

       (a) Columnar epithelium of the rectum becomes inflamed and friable, and mucopurulent or bloody discharge may occur.

       (b) Acute symptomatology of proctitis, such as burning, tenesmus, and purulent discharge, occurs in less than 5% of patients.

       (c) Mild symptomatology, such as itching, mucoid discharge, painful defecation, and constipation, occurs in less than 10% of patients.

   d. Gonococcal pharyngitis. Pharyngitis from oral-genital contact occurs in about 20% of homosexual men and 10% of heterosexuals with gonorrhea at other sites. Disease usually is asymptomatic.

     (1) Sore throat, tonsillitis, and gingivitis have been attributed to gonococci.

     (2) Hematogenous dissemination from the pharynx is more common than from other sites.

   e. Gonococcal conjunctivitis is the most common form of neonatal conjunctivitis. Infection in both children and adults may be initiated by autoinoculation, sexual activity, laboratory accident, or use of contaminated urine as folk therapy for other forms of conjunctivitis.

     (1) Infection usually becomes evident 3–7 days postpartum.

     (2) Bilateral, profuse, and purulent conjunctivitis is characteristic.

     (3) Progressive disease may lead to keratitis or panophthalmitis.

   f. Gonococcal perihepatitis. A gonococcal etiology has never been proven by culture of the peritoneum.

     (1) Clinical manifestations. Acute, severe, pleuritic, right upper quadrant abdominal pain has been associated with salpingitis and cervical gonorrhea.

       (a) Early stage syndrome is characterized by exudative peritonitis.

       (b) Late stage developments include "violin string" adhesions.

     (2) Therapy. The disease manifestations are alleviated by antigonococcal chemotherapy.

 2. NGU. The clinical manifestations of NGU caused by bacteria (i.e., *C. trachomatis, U. urealyticum, M. hominis*) are indistinguishable from those of genital gonorrhea. Disease is characterized by urethritis and a purulent discharge.

E. Diagnosis

 1. Gonorrhea generally is easily diagnosed.

   a. Stains

     (1) Gram-negative pairs of typical cocci inside neutrophils are strong presumptive evidence of gonorrhea.

       (a) Gram stains correlate well with urethral cultures from men.

       (b) Gram stains often are falsely negative with endocervical specimens.

       (c) The stain is of no value in rectal and pharyngeal disease.

     (2) Fluorescent antibody techniques are diagnostic, but they are complicated and usually not practical for the routine laboratory.

   b. Culture and biochemical identification of *N. gonorrhoeae* are required for confirmation of the diagnosis.

 2. NGU may be distinguished from gonorrhea by the lack of microorganisms on Gram stains of urethral discharge and other specimens and by culture of pathogens other than gonococci.

F. Therapy

 1. Genital gonorrhea

   a. Penicillins remain the drugs of choice in gonococcal infection, although there has been increasing resistance by plasmid-mediated penicillinase-producing strains of *N. gonorrhoeae* (i.e., PPNG).

      (1) Procaine penicillin is administered intramuscularly at a minimum dose of 4.8 million U.

      (2) Orally administered probenecid (1.0 g) is a standard adjunct.

    b. Spectinomycin, tetracycline, and a cefoxitin are alternative chemotherapeutic agents.

      (1) Spectinomycin is used to assure treatment of PPNG in walk-in patients.

      (2) Tetracycline assures treatment of concurrent NGU.

    c. Patient compliance and reliability are important considerations in therapy.

2. Neonatal gonococcal conjunctivitis is successfully treated prophylactically by instilling antibiotic solutions or 1% silver nitrate into the eyes of the neonate at birth.

3. Intravenous administration of chemotherapy is preferable in some cases of **complicated gonorrhea** such as salpingitis.

4. NGU generally responds, at least symptomatically, to oral tetracycline or erythromycin therapy. Optimal doses and duration of therapy have not been established.

V. SYPHILIS

V. SYPHILIS is a serious, potentially debilitating veneral disease expressed in three distinct phases—primary, secondary, and tertiary. Congenital syphilis is a nonvenereally transmitted syndrome. Only primary syphilis and congenital syphilis are associated with the urogenital tract.

A. Primary syphilis

1. Etiology. Primary syphilis is contracted when ***Treponema pallidum*** enters the body through small breaks in the squamous or mucosal epithelium during intimate sexual contact.

2. Clinical manifestations. *T. pallidum* proliferates in regional lymph nodes and disseminates hematogenously.

    a. Chancre formation at the site of inoculation is the principal manifestation.

      (1) A painless red pimple initially appears 2–3 weeks after contact.

      (2) The chancre develops into an ulcerative lesion within a few days.

      (3) Chancres usually appear on the penis, labia, fourchette, or cervix, although any mucocutaneous tissue contacted during sexual activity may develop lesions.

    b. Inguinal lymph nodes are enlarged, usually bilaterally, in about 80% of patients with a genital chancre; the lymph nodes are tender in most of these patients.

    c. Clinical manifestations of primary infection disappear spontaneously within 3–6 weeks.

3. Diagnosis

    a. Clinical diagnosis of syphilis should be considered in any patient with painless, ulcerative genital lesions. Infection caused by herpesvirus or *Hemophilus ducreyi* can resemble syphilis, as can trauma during sexual intercourse, a fixed drug reaction, or erosion of the cervix.

    b. Microbiologic diagnosis

      (1) **Dark-field microscopy** of exudate obtained directly from a suspected lesion is used to observe live treponemes.

        (a) Characteristic treponemal morphology and motility confirm the diagnosis.

        (b) 'Most chancres are positive for *T. pallidum*, but bacterial superinfection and prior treatment with ointments may make treponemes scarce or absent.

      (2) *T. pallidum* has not been cultured in synthetic medium.

    c. Serologic diagnosis

      (1) **The Venereal Disease Research Laboratory (VDRL) and rapid plasma reagin (RPR) tests** are used for presumptive screening.

        (a) Results depend upon induction of nonspecific Wassermann antibodies during the course of syphilis.

        (b) Leprosy, lupus erythematosus, mononucleosis, and other disease syndromes also induce positive tests.

      (2) **Specific serologic diagnosis** is made most commonly by the **fluorescent treponemal antibody-absorbed serum (FTA-AbS) test.**

        (a) The test uses patient's serum, lyophilized *T. pallidum*, and fluorescently labeled anti-human gammaglobulin.

        (b) About 90% of patients develop a positive test within 2 weeks of the appearance of the primary syphilitic chancre.

4. Therapy for primary syphilis generally follows the guidelines proposed by the United States Public Health Service.

  a. A single parenteral dose of 2.4 million U of benzathine penicillin, or 600,000 U of oral procaine penicillin for 8 days, is administered. Patient reliability and compliance must be considered in determining which regimen to use.

  b. Tetracycline or erythromycin, 500 mg four times daily for 15 days, is the alternative treatment for patients with penicillin allergy.

B. Congenital syphilis

  1. Etiology. *T. pallidum* is capable of crossing the placental membrane and infecting the fetus. Fetuses aborted as early as the ninth or tenth week of gestation have been infected with the treponeme.

  2. Clinical manifestations

   a. Early consequences of congenital syphilis include abortion, stillbirth, or a variety of manifestations of disseminated syphilis in the infant.

    (1) Splenomegaly and hepatomegaly occur in nearly all affected infants.

    (2) Skeletal abnormalities also are seen in nearly all patients upon radiologic examination.

    (3) Enlarged lymph nodes are seen in about 50% of infected infants.

    (4) Generally, symptomatology appears during weeks 2–6 of life.

     (a) Snuffles, a syndrome resembling a head cold, is the initial syndrome.

     (b) Skin and mucocutaneous tissue develop lesions.

     (c) The syndrome typically is fatal if not treated.

   b. Late manifestations of congenital syphilis occur after 2 years of age and are of two types.

    (1) Stigmata are structural abnormalities that become apparent with the development of teeth and long bones.

     (a) These manifestations are caused by early infection producing tissue damage.

     (b) Stigmata can be prevented by antimicrobial treatment before 3 months of age.

    (2) Manifestations of obscure pathogenesis, such as keratitis, skin lesions, gummas of nasal and facial bone, periostitis, and central nervous system (CNS) disease, are the second syndrome seen late in congenital syphilis.

  3. Therapy depends upon whether the cerebrospinal fluid (CSF) is normal or abnormal.

   a. Treatment for patients with normal CSF is a single dose of 500,000 U of benzathine penicillin per kilogram of body weight.

   b. In patients with abnormal CSF, a 10-day course of procaine penicillin, 500,000 U per kilogram of body weight per day, is the recommended therapy.

   c. Treatment of syphilitic pregnant women with penicillin prevents congenital syphilis.

VI. GENITAL HERPES. Herpes genitalis is an acute inflammatory disease of the genital tract caused by **herpes simplex virus (HSV)**. The disease syndrome results from either primary or recurrent infection.

  A. Etiology. The etiologic agent of genital herpes is **HSV type 2 (HSV-2)**. HSV-2 is related closely to HSV-1, the cause of fever blisters and other lesions above the waist.

  1. The two types share 40% of their DNA base sequences, have many common antigens, and produce identical lesions.

  2. HSV-2 can be differentiated from HSV-1 on the basis of specific antigenic, biologic, and biochemical characteristics.

  B. Epidemiology. HSV-2 is introduced into the genital mucosa by sexual contact with either symptomatic or asymptomatic genital infection.

  1. The virus replicates in cells of the stratum spinosum epidermidis.

   a. A variety of cytopathic effects are produced, including Cowdry type A intranuclear inclusion bodies and multinucleated giant cells.

   b. Viral replication reaches a peak in 3–4 days.

  2. Latent infection occurs when HSV-2 invades local sensory nerve endings and ascends the axon.

C. Clinical manifestations. The development of herpetic lesions (i.e., vesicles) is the most characteristic feature of the disease. Clinical manifestations, however, differ for primary and recurrent infections.

1. **Lesions**
   a. Initially, an erythematous papule appears as the result of intracellular edema, inflammation, and capillary dilation.
   b. Vesicles form with the influx of edematous fluid as the infection and inflammation progress, lifting uninvolved stratum corneum.
   c. Vesicle fluid may be clear or yellowish and often becomes cloudy with the influx of inflammatory cells. The lesions tend to rupture easily.
   d. Drying vesicles that are not secondarily infected with bacteria develop a reddish brown crust.
   e. Healed lesions may leave temporarily hypopigmented areas.

2. **Primary genital herpes,** a severely painful disease, is expressed in 2–7 days following sexual contact.
   a. **Symptomatology.** Initial symptomatology characteristically is local tenderness or burning. Typical herpetic vesicles appear subsequently.
   b. **Lesions** may be widespread.
      (1) In women, lesions can involve the labia, vaginal mucosa, cervix, clitoris, urethra, perianal skin, buttocks, and thighs.
      (2) In men, lesions usually occur on the penis; the scrotum, thighs, and buttocks are less frequently involved.
   c. **Local lymph node infection** often occurs, but the disease normally is self-limited.
   d. **Pain** begins to subside in 10–14 days, and healing occurs without residue in 3–5 weeks.
   e. **Vesicles**
      (1) Ruptured vesicles leave very tender ulcers covered with grayish yellow exudate and surrounded by a red areola.
      (2) Vesicles may coalesce to form large bullae. Severe dysuria sometimes occurs if the urethra becomes involved.

3. **Recurrent genital herpes** tends to involve the same sites as primary infection, but the symptomatology usually is less severe and the lesions more circumscribed.
   a. Pain, burning, itching, or tingling occurs at the sites where vesicles eventually form in clusters.
   b. Vesicles evolve more rapidly but generally persist for a shorter period of time than do primary lesions.

4. **Complications and sequelae** may take a variety of clinical forms, although they are rare in normal people.
   a. **Complications** include bacterial or fungal superinfection, extragenital infection or aberrant viral behavior in apparently normal hosts, and disseminated HSV in immunocompromised hosts.
   b. **Neonatal herpes.** Transmission of HSV-2 in utero to the fetus results in serious disseminated disease.
      (1) **The risk of infection** varies according to the time in pregnancy when the mother is infected.
         (a) Transplacental transmission is rare when primary infection occurs during the first trimester.
         (b) The risk is about 10% if the mother is symptomatic after 32 weeks gestation.
         (c) The risk is 50% if virus is present at delivery.
      (2) **Incidence.** Estimates of the incidence of neonatal infection range from 1 in 7500 to 1 in 20,000 deliveries.
      (3) **Mortality** is about 80%.

D. Diagnosis

1. **Differential diagnosis** of genital herpes should include syphilis, chancroid, candidiasis, lymphogranuloma venereum, granuloma inguinale, varicella zoster, contact dermatitis, and other vesicular diseases.

2. **Presumptive diagnosis** often is made on the basis of clinical symptomatology, history, and the appearance of lesions.

3. **Virus isolation and microscopic studies** can be useful in confirming the diagnosis and in identifying HSV-2.
   a. Virus may be isolated from vesicle fluid by inoculating cell cultures such as primary human embryonic kidney. Cells are examined for typical cytopathic effects.
   b. Viral isolates may be typed as HSV-2 by a variety of serologic techniques or by electrophoretic analysis of viral DNA and proteins. These definitive techniques are not performed in the routine clinical laboratory.

4. **Microscopic examination of scrapings** from the base of a vesicle (i.e., the Tzanck smear) is a simple, rapid means of diagnosis.
   a. Smears are examined microscopically for typical intranuclear inclusion bodies and multi-nucleated giant cells, indicating HSV or varicella zoster.
   b. Fluorescent antibody staining of Tzanck smears yields more specific identification, but the accuracy depends on the quality of the staining reagents.
   c. Material from the vaginal and cervical walls can be examined in the Papanicolaou (Pap) smear.

E. **Therapy** for genital herpes infection does not effect a cure despite many claims to the contrary in popular press advertisements.

1. **Chemotherapy**
   a. Topical application of acycloguanosine (acyclovir) and 2-deoxy-D-glucose is the best method for inhibiting spread of the lesions.
   b. There is controversy about the effects of drugs in recurrent genital herpes.

2. **Supportive therapy**
   a. Routinely used analgesic agents may provide minimal relief of the severe pain of genital herpes.
   b. Counseling may alleviate some of the psychosocial stress of recurrent disease.

3. **Surgery.** Cesarean section generally prevents infection of the neonate during the birth process.

VII. MISCELLANEOUS INFECTIONS OF THE UROGENITAL TRACT. Several other clinically significant bacterial infections are transmitted venereally. Although relatively rare in the United States and Europe, these diseases are endemic in other countries, particularly in tropical and subtropical areas.

A. **Chancroid (soft chancre)**

1. **Etiology.** *H. ducreyi*, a gram-negative pleomorphic rod, is the cause of chancroid.

2. **Clinical characteristics**
   a. Chancroid is characterized by a painful genital ulceration.
      (1) The primary lesion begins as a small papule that develops into a pustule and then ulcerates within 2 days.
      (2) The chancre is soft and tender, in contrast to the firm and painless chancre of syphilis.
   b. Inguinal lymphadenitis usually is present.
   c. There is no systemic symptomatology.

3. **Diagnosis.** The only means of definitive diagnosis is isolation of *H. ducreyi*. Presumptive diagnosis is made on the basis of clinical manifestations.

4. **Therapy.** The clinical disease generally responds to sulfonamides or tetracyclines.

B. **Lymphogranuloma venereum (LGV)**

1. **Etiology.** *C. trachomatis* serotypes L1, L2, and L3 are the etiologic agents of LGV.

2. **Clinical characteristics**
   a. LGV is characterized by a transient lesion on skin or mucous membranes of the genitalia or rectum.

      (1) The lesion is small and has been described as a papule, pustule, or ulcer.

      (2) The painless lesion is transient and often goes unnoticed.

    b. Regional lymphadenitis progresses to bubo formation.

    c. Systemic symptomatology of fever, chills, headache, myalgias, and hyperglobulinemia develops as the disease progresses.

    d. Bacteremia also may develop.

  3. Diagnosis. The diagnosis of LGV usually is made on the basis of clinical evaluation and serology; it is confirmed by isolating specific LGV strains of *C. trachomatis.*

  4. Therapy consists of surgical drainage of buboes and 4–6 weeks of tetracycline.

C. Granuloma inguinale

  1. Etiology. *Calymmatobacterium granulomatis* is the etiologic agent of this sexually transmitted disease.

  2. Epidemiology. Granuloma inguinale is the rarest venereal disease in North America and Europe, but it is endemic to tropical and subtropical climates.

  3. Clinical characteristics

    a. The lesion appears as an indurated papule that ulcerates over days to weeks.

    b. Pink to red hypertrophic granulomatous-looking tissue forms. Purulent exudate and inflammation of surrounding tissue is minimal or absent; true granulomas do not form.

    c. Lesions occur on external genitalia, perianal skin, and inguinal areas in more than 90% of patients.

  4. Diagnosis. Suspected cases are confirmed by identification of Donovan's bodies in fresh lesion scrapings or in crushed biopsy tissue stained by the Wright or Giemsa method. The bacterium is difficult to culture.

  5. Therapy. Tetracycline is the antibiotic of choice for treating the disease.

STUDY QUESTIONS

Directions: Each question below contains five suggested answers. Choose the **one best** response to each question.

1. The most common cause of granulomatous infection of the female genitalia is

(A) *Trichomonas vaginalis*
(B) *Candida albicans*
(C) Herpes simplex virus
(D) *Actinomyces israelii*
(E) *Mycobacterium tuberculosis*

2. Initial symptomatology of congenital syphilis typically occurring 2–6 weeks after birth is referred to as

(A) Burton agammaglobulinemia
(B) snuffles
(C) Waterhouse-Friderichsen syndrome
(D) whitlow
(E) tabes dorsalis

Directions: The group of questions below consists of lettered choices followed by several numbered items. For each numbered item select the **one** lettered choice with which it is **most** closely associated. Each lettered choice may be use once, more than once, or not at all.

Questions 3–8

Match each statement regarding prostatitis with the most appropriate response.

(A) Acute prostatitis
(B) Chronic prostatitis
(C) Both
(D) Neither

3. Antibacterial chemotherapy generally is unsatisfactory

4. Urinary tract symptomatology and findings closely resemble those of simple urinary tract infection (UTI)

5. Common etiologic agents include the Enterobacteriaceae and *Pseudomonas* species

6. Initial symptomatology usually is absent

7. The disease has been associated with prostatic carcinoma

8. Patients usually have a history of recurrent UTI

Questions 9–12

Match each statement regarding urethritis with the most appropriate response.

(A) Gonococcal urethritis
(B) Nongonococcal urethritis
(C) Both
(D) Neither

9. Diagnosis is confirmed by culture and identification of an etiologic agent

10. A purulent urethral exudate is characteristic of disease

11. Tetracycline is an effective chemotherapeutic agent

12. Spectinomycin is an alternative chemotherapeutic agent

ANSWERS AND EXPLANATIONS

1. The answer is E. [*III B 4 b*] *Mycobacterium tuberculosis* is the most common etiologic agent of granulomatous infections of the female genitalia. Disease usually is the result of hematogenous dissemination of tubercle bacilli from a primary focus of infection in the lung. Endometritis, localized pelvic peritonitis, and perioophoritis are the most frequent syndromes. *Actinomyces israelii* most often causes a pyogenic infection of the genitalia, but it also has been associated with granulomatous disease in women using intrauterine contraceptive devices. *Trichomonas vaginalis* and *Candida albicans* are etiologic agents of vulvovaginitis.

2. The answer is B. [*V B 2 a (4) (a)*] Snuffles, a syndrome resembling a common cold, is the initial symptomatology of congenital syphilis typically occurring 2–6 weeks after birth. Burton agammaglobulinemia is a genetic disorder that causes immunosuppression. Waterhouse-Friderichsen syndrome is a fulminant form of meningococcemia that has a high mortality rate. Whitlow is an infection of the finger caused by herpes simplex virus. Tabes dorsalis is a serious complication of tertiary syphilis.

3–8. The answers are: 3-B, 4-A, 5-C, 6-B, 7-D, 8-B. [*II A 1 a, B 1 a, 2 a, 4 c, D 2*] Antibacterial chemotherapy for chronic prostatitis is unsatisfactory, although antibacterial agents can prevent the development of complications such as cystitis.

The clinical manifestations of acute prostatitis include systemic symptomatology, such as fever and chills, and urinary tract symptomatology and findings resembling those found in simple urinary tract infection (UTI), such as frequency and urgency of urination, burning pain on urination, and many white cells in the urine.

The most common etiologic agents of acute and chronic prostatitis include the Enterobacteriaceae (e.g., *Escherichia coli*, *Klebsiella pneumoniae*, *Proteus* species) and *Pseudomonas* species.

A unique characteristic of chronic prostatitis is the absence of clinical symptomatology; cystitis, which occurs with progression of the disease, frequently is the initial symptomatology.

No direct evidence has supported any association between prostatitis of any type and prostate carcinoma.

Patients with chronic prostatitis usually have a history of recurrent UTI of bacterial etiology.

9–12. The answers are: 9-C, 10-C, 11-C, 12-A. [*IV D 1 a (1) (a), 2, E 1 b, 2, F 1 b, 4*] The confirmed diagnosis of both gonococcal urethritis and nongonococcal urethritis (NGU) is made on the basis of culture and identification of the etiologic agent. In approximately 20% of cases of NGU an obvious pathogenic etiologic agent is not cultured from the purulent exudate.

Purulent exudate from the urethra is characteristic of both gonococcal urethritis and NGU, particularly in the male. The absence of gram-negative cocci in pairs on a Gram stain of a smear of urethral discharge is a presumptive differentiation between gonococcal urethritis and NGU.

Tetracycline is the drug of choice in treating NGU, and it generally is an effective alternative chemotherapeutic agent for treating gonococcal urethritis in patients with penicillin allergy.

Spectinomycin is the primary antigonococcal therapy in cases where the etiologic agent is a penicillinase-producing gonococcus.

30
Central Nervous System Infections
Gerald E. Wagner

I. MENINGITIS. Inflammation of the meninges may result from direct invasion of the tissue or subarachnoid space by microorganisms or from seeding of the central nervous system (CNS) via the bloodstream. Clinical signs of meningitis are relatively consistent regardless of the etiologic agent; laboratory findings are significant and usually definitive.

A. Bacterial meningitis is an acute, life-threatening infection. The fatality rate is approximately 15% even with appropriate antimicrobial therapy. The incidence of disease is associated inversely with age; the frequency with which a particular etiologic agent is isolated also is related to the patient's age.

1. **Etiology.** Encapsulated bacteria are most commonly associated with bacterial meningitis.
   a. More than 80% of cases are caused by *Haemophilus influenzae*, *Neisseria meningitidis*, and *Streptococcus pneumoniae*.
      (1) *H. influenzae* **type B**, a gram-negative, pleomorphic rod, is the most frequent cause of **bacterial meningitis**.
         (a) **Epidemiology**
            (i) **Incidence in children.** Purulent meningitis caused by *H. influenzae* is found most frequently in infants 6–12 months old. The incidence of disease decreases with each year after 5 years of age.
            (ii) **Incidence in adults.** Adults have a low incidence of meningitis caused by *H. influenzae*, but it has increased since the widespread use of antibiotics.
         (b) **Immunity.** Maternally passed antibodies probably are protective during the first 6 months of life. Subclinical infections during childhood induce formation of protective anticapsular antibodies.
      (2) *N. meningitidis* is the etiologic agent of **epidemic meningitis.** Four serogroups (A, B, C, and Y) based on the polysaccharide capsule cause most cases of meningococcal meningitis.
         (a) **Epidemiology.** Meningococcal meningitis occurs primarily among young adults.
            (i) Incidence of disease is highest during childhood.
            (ii) Military recruits, presumably because of crowding, have been subject to outbreaks.
         (b) **Immunity.** Nasopharyngeal colonization induces the formation of protective antibodies.
      (3) *S. pneumoniae* produces a large polysaccharide capsule that is responsible for virulence. There are more than 80 recognized capsular serotypes.
         (a) **Epidemiology.** *S. pneumoniae* causes meningitis in all age-groups; 18 serotypes are responsible for 90% of cases of meningitis.
            (i) Factors that predispose to pneumococcal meningitis include splenectomy, sickle cell anemia, alcoholism, and head trauma.
            (ii) Children with untreated otitis media caused by *S. pneumoniae* risk developing meningitis.
            (iii) The pneumococcus is the most frequent cause of meningitis among the **elderly**.
         (b) **Immunity.** Capsular polysaccharide induces specific protective antibodies.
   b. **Other etiologic agents.** The remaining 20% of cases are caused by *Listeria monocytogenes*, gram-negative enteric rods, *Staphylococcus aureus*, *Streptococcus pyogenes*, *Staphylococcus epidermidis*, *Pasteurella multocida*, *Acinetobacter calcoaceticus*, *Mycobacterium tuberculosis*, or other pathogenic bacteria.

2. **Clinical manifestations.** Bacterial meningitis among adults follows a characteristic pattern although the progression of symptomatology may vary. The symptomatology of neonatal meningitis is highly variable.
   a. **Duration.** The duration of the disease is short, but it is affected by the etiologic agent, size of the inoculum, and the immune and physical status of the patient.
   b. **Symptomatology.** Nearly all patients have fever, headache, and nuchal rigidity, although the severity of these symptoms varies. The clinical pattern may be atypical with absence of headache and nuchal rigidity.
      (1) The most common physical finding is a stiff neck; it is characterized by pain and resistance on flexion.
      (2) Lethargy and drowsiness are common signs; confusion, agitated delirium, and stupor occur less frequently.
      (3) Coma is infrequent; it is a poor prognostic sign.
      (4) Other neurologic findings such as positive Brudzinski's and Kernig's signs may be present. Irritability on movement is a frequent sign, and seizures are common.
      (5) In neonates, bulging of the fontanelle, indicating increased intracranial pressure, may occur.

3. **Diagnosis.** Laboratory findings assist in the diagnosis of bacterial meningitis. They indicate the degree of meningeal inflammation but have no prognostic value. Changes are seen in cerebrospinal fluid (CSF) and vary depending upon the stage of the disease.
   a. **Pressure.** The CSF pressure is moderately elevated in most patients, but it may be very high in the severely ill.
   b. **Cell count.** The CSF contains many white blood cells and appears turbid. The white blood cell count typically is 1000–10,000 cells/mm^3 (normal count is 0–5); neutrophils are prominent.
   c. **Protein and glucose.** Increased total protein in proportion to the white cell count and decreased CSF glucose to less than 40% of the simultaneous blood glucose level are characteristic.
   d. **Culture.** CSF should be sent for laboratory culture and sensitivity testing. Meningitis due to *M. tuberculosis* or to *L. monocytogenes* typically yields negative cultures.

4. **Therapy**
   a. **Chemotherapy**
      (1) **Administration.** Antimicrobial agents generally are administered intravenously. It is important to select an antimicrobial that is effective against the specific etiologic agent. It is essential to obtain adequate concentrations of the antimicrobial in the CSF.
         (a) Some chemotherapeutic compounds, such as the aminoglycoside antibiotics, are not employed except intraventricularly because they do not effectively cross the blood–CSF barrier.
         (b) Meningitis caused by gram-negative enteric rods frequently requires intrathecal administration of chemotherapeutic agents.
      (2) **Chemotherapeutic agents**
         (a) **Penicillin G** is used as initial therapy for community-associated bacterial meningitis when examination of a CSF smear reveals no microorganisms.
         (b) **Chloramphenicol** is used to treat meningitis in both children and adults.
            (i) It has both excellent capacity for penetrating the CNS and broad-spectrum bactericidal activity.
            (ii) Chloramphenicol should not be used in neonatal meningitis because of toxic side effects.
         (c) **Cephalosporin** antibiotics are contraindicated in meningitis because they are partially inactivated in CSF.
   b. **Supportive therapy** during bacterial meningitis prevents water and electrolyte imbalance, prevents seizures, and lowers fever.
   c. **Surgery.** Chronic, recurrent bacterial meningitis caused by anatomic defects requires corrective surgery.

5. **Prevention.** Immunization has been partially effective in preventing bacterial meningitis through the use of pneumococcal, meningococcal, and *Haemophilus influenzae* type b (Hib) vaccines.

    a. The purified capsular polysaccharide in these vaccines is not highly immunogenic.
    b. Immunization is effective in adults but not in children, particularly those under the age of 2 years, who have the highest incidence of bacterial meningitis.

B. Viral meningitis and encephalitis. The term **aseptic meningitis** often has been used synonymously with viral meningitis. As certain bacterial infections produce so-called aseptic meningitis, the term is applied correctly only to nonviral etiologies.

  1. Etiology. Enteroviruses and mumps virus are the primary causes of viral meningitis and encephalitis in the United States.
    a. Coxsackievirus B, echovirus, mumps virus, coxsackievirus A, and poliovirus are the most frequently isolated etiologic agents.
    b. Herpesvirus, rubella, vaccinia, cytomegalovirus (CMV), rabies virus, the arboviruses, and many other viral agents also may cause meningitis and encephalitis.

  2. Epidemiology
    a. Enterovirus infections commonly occur in late summer and early fall; mumps virus infections typically occur in late winter and early spring.
    b. Viral meningitis principally is a disease of the young; it rarely occurs after age 40.

  3. Clinical manifestations
    a. Meningitis. Signs and symptoms of viral meningitis become apparent gradually.
      (1) Patients usually complain of fever, anorexia, malaise, myalgia, and a sore throat. Symptomatology includes lethargy, vomiting, severe headache, and a stiff neck.
      (2) Definite evidence of meningeal irritation and inflammation is seen within a few days to a week.
      (3) Physical findings include fever and nuchal rigidity, with or without a positive Brudzinski's or Kernig's sign.
      (4) The prognosis in viral meningitis is good; complications, if any, are mild and usually reversible.
    b. Encephalitis
      (1) Encephalitis is an infection of the brain substance; it has a much poorer prognosis than viral meningitis.
      (2) Dramatic signs of cerebral dysfunction are the major differentiating manifestations of viral encephalitis.
    c. Mortality. Whereas the mortality rate of uncomplicated viral meningitis is relatively low, the mortality rate of encephalitis is 20%–70%.

  4. Laboratory diagnosis. The etiologic agent varies relative to the season in which the infection occurs.
    a. Examination of CSF obtained by lumbar puncture is most diagnostic.
      (1) Opening pressure is elevated.
      (2) The number of leukocytes is increased typically to 100–500 cells/mm^3; counts in excess of 1000 rarely are observed.
      (3) Within the first 48 hours of disease, segmented neutrophils may predominate; later, more than 50% of the total cells are mononuclear cells.
    b. Other laboratory findings have limited value in the diagnosis of viral meningitis.
      (1) Peripheral blood analyses are within normal limits or occasionally are slightly elevated.
      (2) CSF glucose concentration usually is normal, although about 10% of patients with mumps virus meningitis have a low glucose level.
      (3) Protein concentrations in the CSF are moderately elevated in viral meningitis.
    c. Culture. Identification by cell culture of a specific virus as the etiologic agent is successful in about 25%–30% of cases.

  5. Therapy for most cases of viral meningitis is symptomatic and supportive. In an otherwise healthy individual, hospitalization is not required when the diagnosis of viral meningitis is unequivocal.

C. Fungal meningitis. Most fungi that cause the systemic and opportunistic mycoses may disseminate and affect the CNS; a few show a distinct predilection for meningeal tissue.

1. **Etiology.** The most frequent etiologic agents of fungal meningitis are *Cryptococcus neoformans* and *Coccidioides immitis*. Both infect the CNS by hematogenous dissemination from primary foci of infection in the lungs.
   a. *C. neoformans* is the only encapsulated yeast of medical importance.
   b. *C. immitis* is a filamentous fungus that forms highly infectious arthrocomidia.

2. **Epidemiology**
   a. *C. neoformans* is distributed worldwide. Most people, particularly in urban areas, are exposed to this opportunistic yeast without developing overt disease. More than 50% of patients who develop cryptococcal meningitis have underlying disease affecting cell-mediated immunity.
   b. *C. immitis* is endemic to Senorran life zones of the southwestern United States, Mexico, and South America. More than 80% of the population in these areas have been exposed to the fungus, as indicated by positive skin tests. However, fewer than 1% of primary cases develop coccidioidal meningitis.

3. **Clinical manifestations** of fungal meningitis vary with the extent of meningeal involvement. Symptomatology differs from that of acute bacterial meningitis.
   a. **Cryptococcal meningitis** is subacute or chronic.
      (1) Patients develop symptomatology of either meningoencephalitis or a space-occupying lesion in addition to those of meningeal irritation.
         (a) The most frequent symptom is frontal headache of increasing severity.
         (b) Additional symptomatology may include fever, weight loss, nausea, vomiting, and mental and visual aberrations.
         (c) Physical findings include cranial nerve dysfunction, positive Brudzinski's and Kernig's signs, and increased intracranial pressure.
         (d) Skin lesions, bone lesions, lymphadenitis, and other visceral lesions occur in about 10% of patients.
      (2) The disease course can last from a few days in rapid, precipitous illness to years. Progressive deterioration of the patient over a period of weeks to months normally is seen.
   b. **Coccidioidal meningitis** is a granulomatous disease of the CNS that usually is fatal if untreated.
      (1) Meningitis may be the first recognizable evidence of coccidioidomycosis because primary disease usually is asymptomatic and undiagnosed. Meningitis occurs 6 months or more after primary infection.
         (a) Headache is the most common symptom.
         (b) Fever, weakness, confusion, mental and behavioral abnormalities, stiff neck, diplopia, ataxia, and vomiting are variable symptomatology.
         (c) A frequent physical finding is skin lesions at the nasolabial fold.
      (2) Relapses are relatively common; they occur most frequently 1–2 years after apparent resolution of disease.

4. **Diagnosis.** Early diagnosis of fungal meningitis improves the prognosis.
   a. **Differential diagnosis**
      (1) **Cryptococcal meningitis** always is considered as a diagnosis in people with compromised cell-mediated immunity who develop the symptomatology of meningitis.
      (2) **Coccidioidal meningitis** should be suspected both in patients with the appropriate symptomatology who have past or concurrent coccidioidomycosis at other sites and in patients with a history of travel to areas where coccidioidomycosis is endemic or of exposure to fomites from such areas. Exclusion of other etiologic agents of meningitis often assumes a major role in the diagnosis.
   b. **Laboratory findings**
      (1) **Cell count.** Mononuclear cells are the predominant inflammatory cell type in CSF; total white cell counts are 50–500 cells/mm^3.
      (2) Protein levels are elevated.
      (3) The glucose level may be only 20%–50% of simultaneous blood glucose concentrations.
      (4) **Microscopic examination of the CSF**
         (a) **Cryptococcal meningitis.** A negative stain showing encapsulated yeast in CSF is diagnostic evidence of disease and warrants initiation of chemotherapy. The negative stain is positive in 50%–70% of patients with disease.
         (b) **Coccidioides meningitis.** Visualization of *C. immitis* in CSF is rare.

c. Culture and identification

(1) **Cryptococcal meningitis.** Confirmation of the diagnosis is made by physiologic and biochemical identification of *C. neoformans* cultured from the CSF.

(2) **Coccidioides meningitis.** About 70% of patients have negative CSF cultures. Cultures of *C. immitis* are identified on the basis of microscopic and colonial morphology.

d. Serodiagnosis

(1) **Cryptococcal meningitis.** Detection of capsular antigen in CSF by a latex agglutination test can be made in about 94% of patients with disease. Rheumatoid factor can yield false-positive results.

(2) **Coccidioides meningitis.** Serodiagnosis can be made by demonstrating complement-fixing (CF) antibodies in CSF.

    (a) About 70% of patients with coccidioidal meningitis have detectable CF titers on the first examination; in some cases the disease may be difficult to diagnose initially.

    (b) As the disease progresses, nearly all patients develop CF antibodies in the CSF.

5. Therapy. Treatment of fungal meningitis is limited by the number of chemotherapeutic agents available.

a. Amphotericin B traditionally has been the drug of choice in fungal meningitis, but nephrotoxicity frequently compromises a full therapeutic regimen.

b. Cryptococcal meningitis

(1) 5-Fluorocytosine (5-FC; flucytosine) as the sole chemotherapeutic agent frequently causes emergence of resistant strains of *C. neoformans*.

(2) Combined therapy with 5-FC and low-dose amphotericin B has proven effective in treating cryptococcal meningitis.

c. Coccidioides meningitis

(1) The new imidazole antifungal agents, miconazole and ketoconazole, are alternatives to amphotericin B therapy.

(2) Intrathecal or intraventricular administration of drugs may be necessary to achieve therapeutic levels in the CSF.

II. BRAIN ABSCESS.
Focal, suppurative bacterial infections of brain substance include extradural abscess, subdural abscess, and septic cortical thrombophlebitis.

A. Predisposing factors. Brain abscess is most likely to result from chronic cerebral anoxia, chronic osteomyelitis of bones adjacent to the brain, septic embolization from other foci of infection, or direct implantation of bacteria by accidental or surgical trauma.

B. Etiology

1. Indigenous anaerobic gram-positive streptococci, *Bacteroides* species, and capnophilic aerotolerant streptococci are isolated most frequently from brain abscesses.

2. Facultatively anaerobic enteric rods are etiologic agents of brain abscess in neonates and debilitated elderly patients.

3. Staphylococci and aerobic streptococci are associated with trauma-induced abscesses.

4. Mixtures of aerobic and anaerobic (or facultatively anaerobic) bacteria are seen as etiologies of brain abscesses.

5. *H. influenzae* is found only in children 1–5 years old.

6. *Actinomyces israelii, Nocardia asteroides*, and *Haemophilus aphrophilus* occasionally are implicated as etiologic agents of a brain abscess.

C. Epidemiology. Brain abscesses are rare, even during sustained bacterial meningitis and bacteremia. A high incidence of brain abscesses in children with congenital cyanotic heart disease, but a low incidence with cerebral atherosclerosis, suggests that clearing of bacteria from the lungs is an important feature in preventing brain abscesses.

D. Clinical manifestations of brain abscesses are associated with systemic reactions to the infection, increased intracranial pressure, and damage to or destruction of brain tissue.

1. **Systemic reactions** generally are fever and chills; this symptomatology usually is associated with septicemia, acute sinusitis, or acute otitis.
   a. Examination of the patient may reveal other suppurative foci of infection.
   b. Low-grade fever may be the only symptom of brain abscess in chronically ill patients with underlying disease, and in some patients there are no systemic manifestations.

2. **Increased intracranial pressure causing headache** is the most common symptomatology of brain abscess. It is seen in nearly all patients, and it contributes significantly to morbidity and mortality.
   a. Headache typically becomes progressively more severe and is not alleviated by analgesic agents. The pain may be localized to the side of the head where the abscess is located. Patients accustomed to headache may note a change in the character of the pain.
   b. Vomiting, coma, and papilledema may occur in severe and advanced cases. Rupture of the abscess causes meningeal irritation and usually is fatal.

3. **Neurologic signs often indicate the location** of the abscess in the brain.
   a. **Temporal lobe abscess** causes expressive aphasia and homonymous, contralateral, upper-quadrant visual field defect. Temporary paralysis of the ipsilateral third and sixth cranial nerves occurs because of increased pressure in the temporal lobe.
   b. **Cerebellar abscess** causes nystagmus, ipsilateral poor coordination, and hypotonia. Temporary paralysis of the ipsilateral sixth and seventh cranial nerves may be caused by pressure.
   c. **Parietal lobe abscess** causes contralateral hemisensory defects.
   d. **Frontal lobe abscess** leads to contralateral motor defects and impaired consciousness.
   e. **Occipital lobe abscess** may cause contralateral homonymous hemianopia.

4. **Focal neurologic signs** are the result of either tissue destruction or inflammation and edema.
   a. Symptoms of increased intracranial pressure and inflammation usually precede focal neurologic signs by days or weeks.
   b. Fever, vision abnormalities, vomiting, and drowsiness are common.

E. Diagnosis. Early diagnosis of brain abscess is imperative. Classical symptomatology such as persistent headache may indicate the need for an exhaustive neurologic examination.

1. **Radiologic examination**
   a. The **computed tomography (CT) scan** is the best radiologic technique for detection of brain abscesses. It can locate areas where abscesses are not completely formed.
   b. **Routine skull roentgenographs** are of limited diagnostic value but may indicate suppurative involvement of regional bones.
   c. **Technetium 99m (^{99m}Tc) radioisotopic scans** are an excellent screen for localizing inflammatory processes.

2. **Culture and staining**
   a. Pus obtained from the abscess should be submitted immediately for anaerobic and aerobic culture; gram-stained smears should be examined as soon as possible.
   b. Lumbar puncture for obtaining CSF is contraindicated when brain abscess is suspected.

F. Therapy. The treatment of bacterial brain abscess conceptually is straightforward. The principal reason for treatment failure is delay in diagnosis, which predisposes to development of intolerable intracranial pressure and eventual destruction of brain tissue.

1. **Preferred therapy** is drainage of pus and administration of antimicrobial agents. Although surgical excision is not complicated and antibiotic-resistant bacteria are not a problem, the morbidity and mortality rates are high.

2. **Successful treatment** is dependent upon three features of the therapeutic regimen.
   a. **Timing.** The timing and selection of neurologic procedures can alleviate reaccumulation of pus in the abscess.
      (1) CT scans permit localization of single or multiple lesions that can be excised.
      (2) Surgical intervention is advisable if the patient's status is good and a well-encapsulated abscess is located in a nonvital region of the brain.
         (a) Encapsulation generally requires as long as 3 weeks. The patient who survives this period without serious deterioration usually can withstand surgery.

    (b) Typically, however, increasing intracranial pressure causes degeneration of the patient's status as indicated by loss of consciousness.

        (i) It is imperative that pressure be relieved; this is accomplished by craniotomy and aspiration of pus.

        (ii) Follow-up CT scans are useful in evaluating the effectiveness of surgery.

  b. Antimicrobial therapy. The mortality rate has not significantly declined since initiation of the complimentary use of antibiotics in surgical management of brain abscess.

    (1) Ignorance of etiologic agents of brain abscesses and of the pharmacokinetics of antibiotics in the CNS, as well as failure to initiate effective therapy early in disease, contributes to chemotherapeutic failure.

    (2) By the time specimens can be obtained for culture and sensitivity testing, the effectiveness of antimicrobial therapy is compromised.

    (3) There is no established time period for antimicrobial treatment of brain abscesses, but 4–6 weeks of parenteral administration is recommended.

    (4) Understanding associations between certain bacterial species and the origin of the brain abscess, as well as the use of new antimicrobial agents in synergistic combinations, should decrease the mortality rate.

3. Monitoring. The patient's recovery generally is monitored by radiography or CT scans.

  a. Physical examination often is unreliable because the patient may have residual defects and relapses may occur.

  b. Healing of excised or drained abscesses can be monitored by simple head roentgenography if a radiopaque substance is injected into the space left by the surgery.

  c. ^{99m}Tc and CT scans are used to monitor recovery in patients managed without surgery.

III. MISCELLANEOUS INFECTIONS OF THE CNS.
A variety of infectious diseases involve the CNS. The primary focus of infection generally is in some other tissue, and CNS involvement (the etiology of which may be difficult to establish) is an infrequent complication.

A. Pott's disease (spinal tuberculosis)

  1. Etiology. Pott's disease is caused by hematogenous dissemination of *M. tuberculosis* from either a separate primary focus of infection or from reactivated disease.

  2. Clinical manifestations. *M. tuberculosis* colonizes the vertebral body and destroys the disk. Paraspinal or epidural abscesses form and cause compression of the spinal cord, eventually producing paraplegia.

  3. Diagnosis. The disease is indistinguishable from blastomycosis of the vertebrae. A definitive diagnosis can be made only by biopsy and culture of the causative microorganism.

B. Neurosyphilis. Three major forms of neurosyphilis are meningovascular neurosyphilis, general paresis of the insane, and tabes dorsalis, but mixtures of the manifestations are common.

  1. Epidemiology. CNS involvement generally occurs in the tertiary stage of syphilis. About 8% of patients with primary syphilis eventually develop signs of CNS involvement.

  2. Clinical syndromes

    a. Early manifestations of neurosyphilis occur in at least 2% of patients with secondary syphilis.

      (1) Symptomatology. The syndrome is characterized by aseptic meningitis with headache, vomiting, and irritability occurring either during or within a few months of secondary syphilis.

      (2) Diagnosis is easy when signs of meningitis accompany the rash characteristic of secondary syphilis.

        (a) CSF chemistry. Increased protein levels and a normal glucose concentration are detected in the CSF.

        (b) Cell count. An average of 500 white cells/mm^3 with a predominance of lymphocytes is found in the CSF.

        (c) Serology. The Venereal Disease Research Laboratory (VDRL) test is positive.

      (3) Therapy. Patients with early manifestations of neurosyphilis respond rapidly to adequate antimicrobial therapy.

      b. Late manifestations of neurosyphilis occur.years after primary infection.
        (1) Symptomatology. Mixed symptomatology of meningovascular disease, general paresis, and tabes dorsalis is seen. It resembles that of a variety of neurologic diseases; that is, partial stroke, adult-onset seizures, brain trauma secondary to gummas, isolated optic atrophy, abdominal or joint pain, incapacitating facial pain, and partial or progressive dementia.
        (2) Diagnosis
          (a) CSF chemistry. The CSF is abnormal; the protein concentration is elevated.
          (b) Cell counts tend to be low in meningovascular disease and long-standing tabes dorsalis. In general paresis cell counts are high, and mononuclear cells predominate.
          (c) Serology. The VDRL test generally is positive except that in cases of tabes dorsalis, 10%–20% are negative.
      c. Atypical manifestations of neurosyphilis have increased since the widespread use of penicillin.
        (1) Two-thirds of all patients with CNS involvement demonstrate atypical symptomatology. A third of patients with primary and secondary syphilis show abnormalities in their CSF and are considered asymptomatic neurosyphilitics.
        (2) Diagnosis usually is based on clinical manifestations and may be difficult.

C. Mycotic invasion and abscess. A variety of opportunistic fungi can infect the CNS, particularly in patients with known underlying disease or immunosuppression such as AIDS.

  1. Zygomycosis
    a. Etiology. Ubiquitous species in the genera *Mucor, Rhizopus,* and *Absidia* cause zygomycosis.
    b. Pathogenesis and clinical manifestations. The etiologic agent colonizes the nasal passages and invades the brain by direct extension. Persistent headache and ocular changes are prominent manifestations. The disease progresses rapidly and usually is fatal in 48–72 hours.
    c. Diagnosis and therapy. Early diagnosis is difficult, and by the time a diagnosis is made the patient's prognosis is poor. Aggressive surgery and chemotherapy rarely are successful.

  2. Aspergilloma of the brain
    a. Etiology. *Aspergillus fumigatus, Aspergillus niger,* and *Aspergillus flavus* are the most frequent etiologic agents, although any of the more than 100 species of *Aspergillus* can cause disease.
    b. Pathogenesis and clinical manifestations. Aspergilloma or invasive aspergillosis of the brain usually is secondary to disseminated disease. The effects on the CNS may mimic a variety of neurologic diseases.
    c. Diagnosis and therapy. Differentiation of aspergillosis of the brain from other neurologic diseases may be difficult or impossible. Excision of the aspergilloma, chemotherapy, and control of underlying disease play a role in treatment.

D. Parasitic disease. A number of parasitic infections, particularly malaria, toxoplasmosis, amebiasis, giardiasis, and those caused by other protozoa, may involve the CNS. These infections, including meningoencephalitis and abscess formation, occur rarely in patients with localized disease.

  1. Etiology. *Naegleria fowleri,* a free-living ameba, causes primary meningoencephalitis.

  2. Diagnosis is made on the basis of neurologic symptomatology accompanying that of more generalized infection.

  3. The **morbidity and mortality** rates are high. Primary amebic meningoencephalitis has a mortality rate of 100% in 48–72 hours.

STUDY QUESTIONS

Directions: Each question below contains five suggested answers. Choose the **one best** response to each question.

Questions 1 and 2

A patient on immunosuppressive therapy for a renal transplant develops a frontal headache that increases in severity over a period of 1 week. Lumbar puncture is performed, and the cell count is 100 cells/mm^3, and mononuclear cells are the prominent type. Protein is elevated, and glucose is 30 mg/dl (blood glucose 145 mg/dl).

1. The probable diagnosis based upon history, clinical manifestations, and laboratory findings is

(A) phycomycosis
(B) mumps virus encephalitis
(C) aspergilloma
(D) primary amebic meningoencephalitis
(E) cryptococcal meningitis

2. Serodiagnosis in this case is most likely made by detecting capsular polysaccharide with

(A) counterimmunoelectrophoresis
(B) latex agglutination
(C) enzyme-linked immunosorbent assay
(D) Ouchterlony radial diffusion
(E) complement fixation

3. What is the etiologic agent of Pott's disease?

(A) *Actinomyces israelii*
(B) *Haemophilus influenzae*
(C) *Treponema pallidum*
(D) *Mycobacterium tuberculosis*
(E) *Coccidioides immitis*

Directions: Each question below contains four suggested answers of which **one or more** is correct. Choose the answer

A if **1, 2, and 3** are correct
B if **1 and 3** are correct
C if **2 and 4** are correct
D if **4** is correct
E if **1, 2, 3, and 4** are correct

4. Lumbar puncture is contraindicated in

(1) cryptococcal meningitis
(2) aspergilloma
(3) early stage neurosyphilis
(4) staphylococcal brain abscess

5. Chemotherapeutic agents used in the treatment of fungal meningitis include

(1) 5-fluorocytosine
(2) miconazole
(3) amphotericin B
(4) ketoconazole

6. True statements about tabes dorsalis include which of the following?

(1) The VDRL test is negative in a significant percentage of patients
(2) It is one of three major forms of neurosyphilis
(3) It is a late manifestation of neurosyphilis, occurring years after primary infection
(4) CSF shows a slightly elevated protein level but otherwise is normal

Directions: The group of questions below consists of lettered choices followed by several numbered items. For each numbered item, select the **one** lettered choice with which it is **most** closely associated. Each lettered choice may be used once, more than once, or not at all. Choose the answer

- **A**   if the item is associated with **(A) only**
- **B**   if the item is associated with **(B) only**
- **C**   if the item is associated with **both (A) and (B)**
- **D**   if the item is associated with **neither (A) nor (B)**

Questions 7–11

Match the listed characteristic with the appropriate form of meningitis.

(A) Bacterial meningitis
(B) Viral meningitis
(C) Both
(D) Neither

7. Disease course of short duration
8. Designated aseptic meningitis
9. High mortality rate of 15%
10. Peripheral blood analysis often is normal
11. Characterized by nuchal rigidity

ANSWERS AND EXPLANATIONS

1. The answer is E. [*I C 2 a, 3 a (1) (a), 4 b*] Cryptococcal meningitis is characterized by a subacute or chronic course. The most characteristic symptom of overt disease is a frontal headache of increasing severity with time. The CSF shows an elevated cell count of 50–500 cells/mm^3; protein concentration usually is elevated, and the glucose level is only 20%–50% of a simultaneous serum concentration. Phycomycosis is characterized by direct invasion of the brain by *Mucor, Absidia*, or *Rhizopus* species; a frontal headache initially is prominent, but the disease is fatal in 48–72 hours. Mumps virus encephalitis is characterized by dramatic neurologic manifestations and a poor prognosis. Aspergilloma is a mycotic abscess, and lumbar puncture is contraindicated. Primary amebic meningoencephalitis is a fulminant disease characterized by severe neurologic symptomatology and death in 48–72 hours.

2. The answer is B. [*I C 4 e (1)*] The capsular polysaccharide of *Cryptococcus neoformans* can be detected in CSF by latex agglutination. This serodiagnostic test is positive in 94% of patients with cryptococcal meningitis. Counterimmunoelectrophoresis can be used for rapid diagnosis of staphylococcal disease, and complement fixation is used to detect antibodies against coccidioidal antigens. Enzyme-linked immunosorbent assay (ELISA) has been developed for a number of viral infections. Ouchterlony radial diffusion generally is not used in serodiagnosis.

3. The answer is D. [*III A 1*] Pott's disease is caused by an infection of the spinal column by *Mycobacterium tuberculosis*. Tubercle bacilli colonize the vertebral body and destroy the disk, causing compression of the spinal column and eventual paraplegia. *Actinomyces israelii* is an occasional cause of bacterial meningitis by extension of infection from cervicofacial, thoracic, or gastrointestinal foci. *Haemophilus influenzae* is the most common etiologic agent isolated from patients with acute bacterial meningitis, particularly children 6–12 months old. *Treponema pallidum* is the etiologic agent of syphilis; approximately 8% of patients with primary syphilis develop central nervous system involvement. *Coccidioides immitis* is an etiologic agent of fungal meningitis; disease is endemic to the semiarid southwestern United States.

4. The answer is C (2, 4). [*II E 2 b*] The invasive procedure of lumbar puncture is contraindicated in patients with brain abscess. Laboratory examination of the CSF provides little useful information in making a diagnosis of staphylococcal brain abscess or aspergilloma (i.e., fungal abscess). In contrast, microscopic examination of CSF and other laboratory findings such as cell counts, protein concentrations, and glucose concentrations are useful in the diagnosis of all forms of meningitis.

5. The answer is E (all). [*I C 5*] Meningitis caused by fungi is life threatening and must be treated aggressively. Chemotherapeutic agents include 5-fluorocytosine (5-FC), amphotericin B, and imidazoles, such as miconazole and ketoconazole. Combined chemotherapy with 5-FC and low-dose amphotericin B is recommended for cryptococcal meningitis to decrease the risk of severe nephrotoxicity. Amphotericin B, miconazole, and ketoconazole are used in treating coccidioidal meningitis.

6. The answer is A (1, 2, 3). [*III B 2 b*] Tabes dorsalis is one of three major forms of neurosyphilis; the other two forms are meningovascular neurosyphilis and general paresis of the insane. It is a late manifestation of neurosyphilis and usually occurs up to 20 years after primary infection. The Venereal Disease Research Laboratory (VDRL) flocculation test is negative in 10%–20% of patients with tabes dorsalis; it is positive in the other forms of neurosyphilis. The CSF in patients with tabes dorsalis is abnormal. There is an elevated protein concentration and a predominence of mononuclear cells.

7–11. The answers are: 7-A, 8-D, 9-A, 10-B, 11-C. [*I A 2 a, b; B 3 a (1), (3); 4 b (2) (a)*] Bacterial meningitis typically is a disease of short duration. The mortality rate is high (approximately 15%), particularly in the very young and elderly. Characteristic symptomatology varies in its progression but virtually always includes fever, headache, and nuchal rigidity. Peripheral blood typically shows an elevated white blood cell count and, depending upon the etiologic agent, decreased glucose and other abnormalities.

Viral meningitis typically is more insidious in onset and has a longer course than bacterial meningitis. For many years, viral meningitis was referred to as aseptic meningitis, but with the realization that some bacterial diseases cause meningitis-like symptomatology without actually infecting the tissue, the term now is correctly applied only to meningitis of nonviral origin. Symptomatology of viral meningitis includes fever, headache, nuchal rigidity, nausea and vomiting, and lethargy. Peripheral blood usually is normal, or values may be slightly elevated. The mortality rate for viral meningitis is relatively low, although the mortality rate for viral encephalitis is 20%–70%.

31
Hospital-Acquired Infections and Management of the Immunocompromised Patient

Gerald E. Wagner

I. HOSPITAL-ACQUIRED INFECTIONS

A. Epidemiology. Debilitation of hospitalized patients due to trauma and underlying disease increases their susceptibility to infections and in particular to certain infectious diseases.

  1. Psychologic effects. Factors such as psychological stress undoubtedly play a role in the incidence of nosocomial infections among relatively healthy patients who are hospitalized for elective surgery, mental disorders, and treatment of other non-life-threatening conditions.

  2. Physical manipulation. Invasive procedures such as intravenous (IV) cannulation, urinary catheterization, and surgery frequently result in iatrogenic infections.

  3. Incidence. The incidence of nosocomial infections increases in proportion to the time the patient remains hospitalized; as many as 50% of infections treated in hospitals are nosocomial in origin.

B. Etiology. Staphylococci, pneumococci, gram-negative enteric rods, pseudomonads, and anaerobes are frequent etiologic agents of disease. Some common opportunistic pathogens and the conditions that predispose hospitalized patients to them are listed in Table 31-1.

  1. Drug-resistant strains
    a. Bacteria isolated from people with nosocomial infections usually are more virulent and exhibit a higher degree of drug resistance than species isolated from the general population.
    b. The widespread and frequent use of therapeutic and prophylactic antimicrobial agents provides selective pressure for the proliferation of drug-resistant microorganisms.

  2. Sources of infection. The patient's indigenous flora, other patients, environmental contaminants, and hospital personnel are sources of etiologic agents of nosocomial infections.

C. Clinical manifestations. The site of infection and the size of inoculum regulate to some degree the clinical manifestations of nosocomial infections.

  1. General types of manifestations. Localized or systemic manifestations result from invasive diagnostic, corrective, and maintenance procedures.

  2. Initial symptomatology. The onset of infection may be more rapid and the symptomatology more dramatic in the hospitalized patient than in the ambulatory patient with an analogous infection.
    a. Fever often is the initial indication of infection, whether it is localized wound colonization or sepsis.
    b. The inflammatory response and other classic manifestations of infection may be masked or absent in patients with underlying disease or immunosuppression.

D. Diagnosis. Rapid diagnosis of nosocomial infection is important to a favorable prognosis for the patient.

Table 31-1. Predisposing Conditions to Nosocomial Infections and Common Etiologic Agents

Condition	Etiologic Agents
Presence of a urinary catheter	*Serratia marcescens* *Pseudomonas aeruginosa* *Proteus* species
Presence of foreign bodies (e.g., IV cannulas, catheters, prostheses)	*Staphylococcus epidermidis* *Staphylococcus aureus* *Propionibacterium acnes* *Candida* species *Aspergillus* species
Surgery	*S. epidermidis* *S. aureus* *Bacteroides* species *Clostridium perfringens* *P. aeruginosa* Other aerobic, facultative, and anaerobic bacteria
Burns	*P. aeruginosa*
Splenectomy	*Streptococcus pneumoniae*
Diabetes mellitus	*S. aureus* *Candida albicans* *P. aeruginosa* Phycomycetes
Hematoproliferative disorders	*Cryptococcus neoformans* Varicella zoster virus Cytomegalovirus *Listeria monocytogenes*
Alcoholism	*S. pneumoniae* *Klebsiella pneumoniae* *L. monocytogenes*
Cortisone therapy	*S. epidermidis* *S. aureus* *Mycobacterium tuberculosis* Fungi Viruses

1. A sudden rise in the patient's temperature following an invasive procedure is strong presumptive evidence of iatrogenic infection. Examination of the site often reveals further evidence of infection, and appropriate cultures should be obtained.

2. Underlying disease increases the difficulty of diagnosing nosocomial infection.
   a. Symptomatology may be vague or atypical, and alert, careful monitoring is essential.
   b. Suspicion of infection merits a complete physical examination.
   c. Cultures should be taken regularly and from multiple sites. Meticulous aseptic culture techniques are critical; any microorganism is a potential etiologic agent.

3. Serologic testing is useful when a specific disease is suspected, and the differential diagnosis can be narrowed significantly by a process of elimination.

E. Therapy. The hospitalized patient is at increased risk of infection, making prophylaxis an important form of specific therapy.

1. **Supportive therapy**
   a. Frequent changes of dressings, irrigation, and débridement decrease the incidence of wound infections by as much as 95%.
   b. Maintaining a clean, disinfected environment reduces or eliminates patient exposure to large inocula of opportunistic pathogens. Sterilization of living areas is neither feasible nor necessary, even for immunosuppressed patients.
   c. Physicians and other hospital personnel who are shown to be carriers of potential pathogens, such as *Staphylococcus aureus*, should be treated with specific chemotherapy to eliminate the carrier state.

2. **Antimicrobial chemotherapy**
   a. The benefits of broad-spectrum prophylactic antimicrobial chemotherapy to some patients, particularly those with underlying disease, may surpass the risks of selecting for drug-resistant strains of bacteria.
   b. Antimicrobial susceptibility testing should be performed for microorganisms cultured from patients with suspected iatrogenic or nosocomial infection.

II. THE IMMUNOCOMPROMISED PATIENT is at high risk for developing infections, particularly those caused by opportunistic microorganisms that are ubiquitous in the environment or are indigenous flora.

A. **Epidemiology.** The incidence of these infections has increased dramatically in recent years as a result of increased survival of compromised persons. A variety of physical, metabolic, and therapeutic conditions compromise the immune system and predispose to infection.

1. **General factors.** Immunosuppression frequently is the consequence of a noninfectious underlying disease or a therapeutic regimen.
   a. **Breakdown of physical barriers** such as skin, mucous membranes, and the ciliary action of bronchial epithelium compromises the host's defense against infection.
   b. **Immunologic cell dysfunction.** Reduction in the number of segmented neutrophils, inadequacy of macrophage function, suppressed cellular immune reactions, and decreased antibody synthesis or function result in immunosuppression.

2. **Specific factors**
   a. **Foreign bodies** such as catheters and prosthetic implants compromise the integrity of the body's barriers against infection.
      (1) The complex interactions between implanted foreign bodies and invading microorganisms are not understood.
      (2) The infectious processes initially are localized, and abscess formation is common.
      (3) Hematogenous dissemination of the etiologic agent leads to systemic manifestations.
   b. **Extensive skin burns** provide easy entry for microorganisms into the body and disturb electrolyte and fluid balances.
      (1) Delayed-type hypersensitivity (DTH) responses are markedly decreased after second- and third-degree burns.
      (2) The chemotactic activity of mononuclear leukocytes is decreased in patients with burns over more than 20% of their bodies.
         (a) The dysfunction is manifest 15–45 days after thermal injury.
         (b) A chemotactic inhibitor has been identified in the serum.
      (3) The number of circulating suppressor mononuclear cells is increased in severely burned patients with sepsis. These cells inhibit the phytohemagglutinin responsiveness of peripheral blood leukocytes and suppress mixed leukocyte reactions.
   c. **Altered splenic function or splenectomy** causes a decreased production of immunoglobulin M (IgM) antibodies.
      (1) IgM is synthesized to a large extent by splenic lymphoid tissue and accounts for opsonization of encapsulated microorganisms.
      (2) Patients with altered or absent splenic function are at an increased risk for developing pneumonia, bacteremia, and meningitis.
   d. **Bone marrow failure** causes a rapid decline in the number of circulating segmented neutrophils, which have a short biologic half-life.

      (1) Leukopenia may progress to the point of almost total absence of segmented neutrophils in the blood (i.e., agranulocytosis).
      (2) Patients with agranulocytosis are susceptible to myriad microbial infections; pneumonia, bacteremia, and urinary tract infections (UTIs) are common.
  e. Malignancies of all types produce some degree of impaired immunity.
      (1) Patients with solid epithelial tumors and chronic lymphocytic hematoproliferative disease have suppressed cell-mediated immune responses.
      (2) Patients with acute leukemia usually have a reduced number of segmented neutrophils in the circulation.
      (3) Therapeutic use of irradiation and cytotoxic drugs for malignant diseases further suppresses immunologic capabilities.
      (4) Bacteremia, pneumonia, fungemia, and UTI frequently are diagnosed in patients with malignant disease.
  f. Allograft recipients require immunosuppressive and anti-inflammatory therapy to prevent rejection of the donor tissue.
      (1) Antithymocyte serum, azathioprine, cyclosporine, and prednisone commonly are used to depress or abolish the cell-mediated immune response of allograft recipients.
      (2) These immunosuppressed patients are particularly susceptible to disseminated viral and fungal diseases, pneumonia, bacteremia, and UTI.
  g. Genetic disorders that cause impaired immunity include sex-linked recessive (Bruton's) agammaglobulinemia and a variety of other syndromes showing some familial patterns.
      (1) Patients with other genetically transmitted diseases, such as diabetes and cystic fibrosis, also appear to be more susceptible to infections.
      (2) The exact links between the genetic disorders and deficiencies in the immune response are not understood clearly.
  h. Acquired immune deficiency syndrome (AIDS) is seen predominantly in promiscuous homosexual males and intravenous drug abusers.
      (1) Human immunodeficiency viruses (HIVs) infect a subclass of T cells and produce severe immunosuppression.
      (2) AIDS patients inevitably succumb to fatal infections by a variety of opportunistic pathogens.

B. Etiology. Immunocompromised patients appear to be uniquely susceptible to some infections as well as to any of the classic infectious diseases, such as tuberculosis. Some important etiologic agents of infection in immunosuppressed patients are listed here.

  1. Staphylococci and streptococci are responsible for localized skin infections and disseminated disease associated with foreign bodies and trauma.
    a. Aggressive invasion of tissue characterizes these pyogenic cocci once they are introduced through the skin.
    b. *S. aureus* is a major cause of localized and deep abscesses.
    c. Streptococci, particularly *Streptococcus pyogenes*, cause skin and soft-tissue infections alone or in mixed etiologies with staphylococci.

  2. Streptococcus pneumoniae poses a serious threat to patients who are unable to produce sufficient opsonizing antibody.
    a. *S. pneumoniae* is a normal inhabitant of the oropharynx, especially in children.
    b. Patients with impaired splenic function are at high risk for developing pneumococcal pneumonia, bacteremia, and meningitis.
    c. Patients with multiple myeloma often are deficient in antibody production and have an increased incidence of pneumococcal pneumonia.

  3. Pseudomonas aeruginosa causes significant tissue destruction once it gains entrance through the skin.
    a. It is a major cause of fatal sepsis in burn victims.
    b. Diabetics and patients with reduced numbers of segmented neutrophils frequently are infected.
      (1) *P. aeruginosa* causes osteomyelitis, UTI, and disseminated disease in immunocompromised patients.
      (2) The bacteria have the ability to invade epithelial cells of blood vessels, causing repeated seeding of the bloodstream.

   c. *P. aeruginosa* routinely colonizes the respiratory tract of patients with cystic fibrosis; the pathogenesis of the bacteria in relation to the disease symptomatology is not clear.

   d. *Pseudomonas* species display multiple drug resistance and frequently are resistant to standard antimicrobial therapy; the new quinolone antibiotics appear to be effective therapy.

4. *Serratia marcescens* once was considered a harmless saprobe.

   a. The species is a leading cause of bacteremia in immunosuppressed patients with indwelling IV and urinary catheters.

   b. Most strains of *S. marcescens* are resistant to standard antimicrobial agents; a few strains have been resistant to all clinically available antibacterial compounds.

5. *Listeria monocytogenes* occurs almost exclusively in patients with some degree of impaired immunity, including neonates and the elderly.

   a. Septicemia and meningitis are diagnosed most commonly in immunosuppressed patients with T-cell dysfunction.

   b. *L. monocytogenes* is difficult to cultivate and often is overlooked in clinical specimens because of its resemblance to diphtheroids.

6. *Nocardia asteroides* is an opportunistic pathogen principally infecting patients receiving long-term immunosuppressive therapy for allograft maintenance.

   a. Nocardiosis usually presents as multiple lung abscesses, but the disease may disseminate hematogenously and cause metastatic abscesses in any organ system.

   b. Skin abscesses usually are indicative of disseminated disease.

7. **Mycobacteria** in general, and *Mycobacterium tuberculosis* in particular, cause disease in patients with defects in T cell function.

   a. Latent tuberculosis may develop into active disease during long-term administration of corticosteroids.

   b. *Mycobacterium avium*, *Mycobacterium intracellulare*, and other mycobacteria are significant causes of morbidity and mortality among AIDS and other immunocompromised patients.

8. *Cryptococcus neoformans* is the only encapsulated yeast of medical importance, and infection occurs almost exclusively in immunosuppressed individuals.

   a. Primary and disseminated cryptococcosis is common in patients with Hodgkin's disease and other forms of malignant lymphoma and in renal allograft recipients.

   b. Some evidence suggests that cryptococcosis in the immunosuppressed host results from the reactivation of primary, asymptomatic pulmonary infection.

   c. *C. neoformans* shows a predilection for the central nervous system, but any organ, including the kidneys, spleen, and bones, can become involved.

9. *Aspergillus species* are some of the most ubiquitous microorganisms in the environment.

   a. *Aspergillus fumigatus* and other species frequently are the etiololgic agents of pulmonary and disseminated disease in patients with malignancies and therapeutic immunosuppression.

   b. These tenacious fungi contaminate prosthetic devices, inhalation therapy equipment, parenteral fluids, and most other medical supplies.

10. *Candida albicans*, a member of the indigenous flora, causes serious mucocutaneous and disseminated disease in patients with underlying hematoproliferative malignancy and in individuals receiving corticosteroid or long-term, broad-spectrum antimicrobial therapy.

   a. Diabetics have a higher incidence of skin colonization by *C. albicans* than other people, and disseminated disease may occur with implantation of foreign bodies or transient disturbances in the immunologic status of the patient.

   b. Disseminated disease often is difficult to diagnose because the fungus is not easily recovered from the bloodstream even in cases of endocarditis.

11. *Toxoplasma gondii* is the cause of serious, often fatal, disseminated infection in AIDS patients and in people receiving immunosuppressive therapy to prevent allograft rejection.

   a. Toxoplasmosis apparently occurs through reactivation of latent infection in most patients.

   b. There is some evidence that the protozoan can be transmitted through donor tissues.

12. ***Pneumocystis carinii*** is an unculturable protozoan.
   a. *P. carinii* causes clinical disease exclusively in immunosuppressed individuals; it is the leading cause of fatal pneumonia in AIDS patients.
   b. Clinical disease usually presents as bilateral pneumonia; subclinical colonization is thought to occur early in life in many individuals.
   c. Predisposing factors include depressed cell-mediated immunity, hematoproliferative malignancy, AIDS, and malnutrition in the very young.

13. **Herpes simplex virus, varicella zoster virus, and cytomegalovirus (CMV)** are opportunistic DNA viruses that cause disease in immunosuppressed patients.
   a. Each of these viruses is capable of producing disseminated disease with pulmonary involvement.
   b. Extensive skin lesions are a conspicuous indication of disseminated herpes simplex and varicella zoster virus infection.
   c. CMV can be transmitted to allograft recipients as a latent infection in transplanted tissue. CMV appears to be an immunosuppressive agent; superinfection by bacteria, fungi, and other viruses causes life-threatening disease.

C. **Diagnosis.** Aggressive diagnostic approaches must be initiated at the earliest suspicion of infection in the immunosuppressed patient.

1. Immunosuppression often prevents or delays an exudative cellular response to infection, causing atypical symptomatology.

2. Physical findings may be absent in the severely immunosuppressed patient, or they may be masked by underlying disease.

3. All available microbiologic, serologic, cultural, roentgenographic, and radioisotopic procedures are essential for early diagnosis.

D. **Therapy** for infectious disease in the immunosuppressed patient is designed to eradicate completely or inhibit the growth of the causative agent and to ameliorate the compromising condition when possible.

1. **Antibacterial therapy.** Synergistic antibacterial agents accelerate the clearance of infecting microorganisms and discourage the in vivo emergence of drug-resistant microbial populations.

2. **Antiviral therapy.** The recent availability of compounds that inhibit the replication of DNA viruses has provided chemotherapeutic regimens for herpesvirus infections.

3. **Antifungal therapy.** New imidazole compounds are alternatives to amphotericin B in treating opportunistic mycoses.

4. **Immunologic potentiation.** The use of modulators of the immune response, such as interferon and thymosin, has shown potential in treating certain types of infectious diseases and is the subject of active clinical research.

5. **Prophylaxis** for immunosuppressed patients may include reducing the number of environmental and indigenous microbial flora, immunization, and the use of gamma globulin and long-term antimicrobial therapy.

STUDY QUESTIONS

Directions: Each question below contains five suggested answers. Choose the **one best** response to each question.

Questions 1-3

A 32-year-old volunteer fireman receives second-degree burns over approximately 30% of his body and third-degree burns over another 40%. Intravenous fluid and electrolyte therapy is initiated on site, and the patient is transported to the burn ward of a large metropolitan hospital. Approximately 2½ weeks after thermal injury, the patient becomes septic.

1. The most probable etiologic agent is

(A) *Serratia marcescens*
(B) *Nocardia asteroides*
(C) *Staphylococcus aureus*
(D) *Pseudomonas aeruginosa*
(E) *Streptococcus pneumoniae*

2. The patient's sepsis possibly developed because of

(A) inability of B cells to produce opsonins
(B) inhibition of chemotactic activity of mononuclear leukocyte
(C) lack of circulating segmented neutrophils in the bloodstream
(D) inhibition of the production of immunoglobulin M
(E) induction of Burton's agammaglobulinemia by the severe thermal injury

3. The most promising chemotherapeutic agent for treatment of this patient belongs to which of the following categories?

(A) Imidazole antibiotics
(B) Immune modulators
(C) Quinolone antibiotics
(D) Thymosin fractions
(E) Amphotericin B

4. Human immunodeficiency virus (HIV) causes immunosuppression by infecting

(A) T cells
(B) segmented neutrophils
(C) mononuclear leukocytes
(D) bone marrow cells
(E) suppressor cells

5. Infections acquired by mechanical manipulation during medical procedures are referred to as

(A) nosocomial infections
(B) opportunistic infections
(C) fomite-induced infections
(D) superinfections
(E) iatrogenic infections

Directions: The group of questions below consists of lettered choices followed by several numbered items. For each numbered item select the **one** lettered choice with which it is **most** closely associated. Each lettered choice may be use once, more than once, or not at all.

Questions 6–10

Match each form of immunosuppression with the most likely clinical setting.

(A) Bone marrow failure
(B) Splenectomy
(C) Allograft surgery
(D) Genetic disease
(E) Extensive skin burns

6. Inhibited delayed-type hypersensitivity
7. Deficiency of immunoglobulin M
8. Agammaglobulinemia
9. Decreased populations of T cells
10. Agranulocytosis

ANSWERS AND EXPLANATIONS

1. The answer is D. [*II B 3 a*] *Pseudomonas aeruginosa* is a major cause of sepsis in burn patients. The bacterium is ubiquitous in the environment and is a member of the indigenous flora of some people. Once it gains entry through the physical barrier of the skin, *P. aeruginosa* causes significant tissue damage. It is capable of invading the epithelium of blood vessels, leading to repeated seeding of the bloodstream. *Serratia marcescens* is a leading cause of sepsis and urinary tract infections in patients with indwelling catheters. *Nocardia asteroides* is seen primarily in patients receiving long-term antirejection therapy. *Staphylococcus aureus* causes localized skin infections and deep abscesses associated with foreign bodies. *Streptococcus pneumoniae* causes pneumonia in patients with underlying diseases affecting antibody production and cell-mediated immunity; patients with splenectomy are particularly at risk.

2. The answer is B. [*II A 2 b (2)*] Patients with burns over more than 20% of their body demonstrate a lack of chemotactic activity of mononuclear leukocytes 15–45 days after the thermal injury. Inhibition of activity corresponds to the appearance of a chemotactic inhibitor in the serum. An inability to produce opsonin antibodies and a decrease in the production of immunoglobulin M (IgM) are associated with splenectomy. A decrease in the number of circulating segmented neutrophils occurs during bone marrow failure. Burton's agammaglobulinemia is a sex-linked genetic dysfunction.

3. The answer is C. [*II B 3 d*] *Pseudomonas aeruginosa* and other pseudomonads frequently display multiple drug resistance. The mortality rate in pseudomonas septicemia approaches 80%. A new group of antibiotics, the quinolones, have shown good in vitro activity against *P. aeruginosa*; the number of cases of disease successfully treated with the quinolones is increasing. Imidazole antibiotics and amphotericin B are antifungal agents; the imidazoles are now considered alternatives to amphotericin B therapy in some mycoses. Immune modulators, such as thymosin and interferon, have shown potential in treating infections controlled by the cell-mediated immune system.

4. The answer is A. [*II A 2 h (1)*] Human immunodeficiency virus (HIV) is the etiologic agent of acquired immune deficiency syndrome (AIDS). The syndrome is characterized by progressive immunosuppression, and the patient usually develops fatal opportunistic infections or malignancy. The virus initially infects a specific set of T cells (i.e., CD4+ cells) and proliferates within the cells during blastogenesis. The activity of segmented neutrophils is affected by bone marrow failure; the bone marrow dysfunction causes agranulocytosis. The chemotactic activity of mononuclear leukocytes is significantly decreased in patients who are burned over more than 20% of their body. Severely burned patients with sepsis also have been shown to have increased numbers of suppressor mononuclear cells.

5. The answer is E. [*I A 2*] Iatrogenic infections are acquired by the physical manipulation of diagnostic or therapeutic devices in patients. Most of these infections occur in hospitalized patients. Nosocomial (i.e., hospital-acquired) infections, therefore, may include iatrogenic infections. Opportunistic infections are those caused by etiologic agents that normally do not cause disease in healthy, immunocompetent people. Infections acquired from inanimate objects, or fomites, may affect hospitalized and nonhospitalized patients and do not necessarily involve the use of the fomite in medical procedures. Superinfections are overwhelming infectious processes that normally occur in immunocompromised patients.

6–10. The answers are: 6-E, 7-B, 8-D, 9-C, 10-A. [*II A 2 b (1), c–g*] Patients with extensive second- and third-degree burns show a marked decrease in the ability to mount a delayed-type hypersensitivity response.

Much of the immunoglobulin M (IgM) responsible for opsonizing encapsulated microorganisms is produced in the spleen. For example, splenectomy makes patients susceptible to pneumococcal infection.

A sex-linked genetic disorder is responsible for a form of agammaglobulinemia known as Burton's agammaglobulinemia. A variety of other immunosuppressive conditions are thought to have genetic causes.

Decreased populations of T cells are desired to prevent tissue rejection in allograft recipients. This is accomplished with antithymocyte serum and with cytotoxic compounds.

Agranulocytosis occurs in patients with bone marrow failure. There is a rapid decline in the numbers of circulating segmented neutrophils. This leukopenia can be so severe that there is almost a total lack of segmented neutrophils in the blood; the condition is known as agranulocytosis.

Challenge
Exam

Introduction

One of the least attractive aspects of pursuing an education is the necessity of being examined on what has been learned. Instructors do not like to prepare tests, and students do not like to take them.

However, students are required to take many examinations during their learning careers, and little if any time is spent acquainting them with the positive aspects of tests and with systematic and successful methods for approaching them. Students perceive tests as punitive and sometimes feel that they are merely opportunities for the instructor to discover what the student has forgotten or has never learned. Students need to view tests as opportunities to display their knowledge and to use them as tools for developing prescriptions for further study and learning.

A brief history and discussion of the National Board of Medical Examiners (NBME) examinations (i.e., Parts I, II, and III and FLEX) are presented in this preface, along with ideas concerning psychological preparation for the examinations. Also presented are general considerations and test-taking tips, as well as ways to use practice exams as educational tools. (The literature provided by the various examination boards contains detailed information concerning the construction and scoring of specific exams.)

National Board of Medical Examiners Examinations

Before the various NBME exams were developed, each state attempted to license physicians through its own procedures. Differences between the quality and testing procedures of the various state examinations resulted in the refusal of some states to recognize the licensure of physicians licensed in other states. This made it difficult for physicians to move freely from one state to another and produced an uneven quality of medical care in the United States.

To remedy this situation, the various state medical boards decided they would be better served if an outside agency prepared standard exams to be given in all states, allowing each state to meet its own needs and have a common standard by which to judge the educational preparation of individuals applying for license.

One misconception concerning these outside agencies is that they are licensing authorities. This is not the case; they are examination boards only. The individual states retain the power to grant and revoke licenses. The examination boards are charged with designing and scoring valid and reliable tests. They are primarily concerned with providing the states with feedback on how examinees have performed and with making suggestions about the interpretation and usefulness of scores. The states use this information as partial fulfillment of qualifications upon which they grant licenses.

Students should remember that these exams are administered nationwide and, although the general medical information is similar, educational methodologies and faculty areas of expertise differ from institution to institution. It it unrealistic to expect that students will know all the material presented in the exams; they may face questions on the exams in areas that were only superficially covered in their classes. The testing authorities recognize this situation, and their scoring procedures take it into account.

Scoring the Exams

The diversity of curriculum necessitates that these tests be scored using a criteria-based normal curve. An individual score is based not only on how many questions were answered correctly by a specific student but also on how this one performance relates to the distribution of all scores of the criteria group. In the case of NBME, Part I, the criteria group consists of those students who have completed 2 years of medical training in the United States and are taking the test for the first time and those students who took the test during the previous four June sittings.

Since this test has been constructed to measure a wide range of educational situations, the mean, or average, score generally can be achieved by answering 64% to 68% of the questions correctly. Passing the exam requires answering correctly 55% to 60% of the questions. The competition for acceptance into medical school and the performance levels necessary to stay in school are so high that many students who have always achieved these high levels naturally assume they must perform in a similar fashion and attain equivalent scores on the NBME exams. This is not the case. In fact, among students who are accustomed to performing at levels exceeding 80% to 90%, fewer than 4% taking these tests perform at that high level. Unrealistically high personal expectations leave students psychologically unprepared for these tests, and the anxiety of the moment renders them incapable of doing their best work.

Actually, **most students have learned quite well**, but they fail to display this learning when they are tested because they do not understand the construction, purpose, or scoring procedures of board exams. It is imperative that they understand that they are **not** expected to score as well as they have in the past and that the measurement criteria is group performance, not only individual performance.

While preparing for an exam, it is important that students learn as much as they can about the subject they will be tested on, as well as prepare to discover just how much they may not know. Students should study to acquire knowledge, not just to prepare for tests. **For the well-prepared candidate, the chances of passing far exceed the chances of failing.**

Materials Needed for Test Preparation

In preparation for a test, many students collect far too much study material only to find that they simply do not have the time to go through all of it. They are defeated before they begin because either they cannot get through all the material, leaving areas unstudied, or they race through the material so quickly that they cannot benefit from the activity.

It is generally more efficient for the student to use materials already at hand; that is, class notes, one good outline to cover or strengthen areas not locally stressed and to quickly review the whole topic, and one good text as a reference for looking up complex material needing further explanation.

Also, many students attempt to memorize far too much information, rather than learning and understanding less material and then relying on that learned information to determine the answers to questions at the time of the examination. Relying too heavily on memorized material causes anxiety, and the more anxious students become during a test, the less learned knowledge they are likely to use.

Positive Attitude

A positive attitude and a realistic approach are essential to successful test taking. If concentration is placed on the negative aspects of tests or on the potential for failure, anxiety increases and performance decreases. A negative attitude generally develops if the student concentrates on "I must pass" rather than on "I can pass." "What if I fail?" becomes the major factor motivating the student to **run from failure rather than toward success**. This results from placing too much emphasis on scores rather than understanding that scores have only slight relevance to future professional performance.

The score received is only one aspect of test performance. Test performance also indicates the student's ability to use information during evaluation procedures and reveals how this ability might be used in the future. For example, when a patient enters the physician's office with a problem, the physician begins by asking questions, searching for clues, and seeking diagnostic information. Hypotheses are then developed, which will include several potential causes for the problem. Weighing the probabilities, the physician will begin to discard those hypotheses with the least likelihood of being correct. Good differential diagnosis involves the ability to deal with uncertainty, to reduce potential causes to the smallest number, and to use all learned information in arriving at a conclusion.

This same thought process can and should be used in testing situations. It might be termed **paper-and-pencil differential diagnosis**. In each question with five alternatives, of which one is correct, there are four alternatives that are incorrect. If deductive reasoning is used, as in solving a clinical problem, the choices can be viewed as having possibilities of being correct. The elimination of wrong choices increases the odds that a student will be able to recognize the correct choice. Even if the correct choice does not become evident, the probability of guessing correctly increases. Just as differential diagnosis in a clinical setting can result in a correct diagnosis, eliminating incorrect choices on a test can result in choosing the correct answer.

Answering questions based on what is incorrect is difficult for many students since they have had nearly 20 years experience taking tests with the implied assertion that knowledge can be displayed only by knowing what is correct. It must be remembered, however, that students can display knowledge by knowing something is wrong, just as they can display it by knowing something is right. **Students should begin to think in the present as they expect themselves to think in the future.**

Paper-and-Pencil Differential Diagnosis

The technique used to arrive at the answer to the following question is an example of the paper-and-pencil differential diagnosis approach.

> A recently diagnosed case of hypothyroidism in a 45-year-old man may result in which of the following conditions?

> (A) Thyrotoxicosis
> (B) Cretinism
> (C) Myxedema
> (D) Graves' disease
> (E) Hashimoto's thyroiditis

It is presumed that all of the choices presented in the question are plausible and partially correct. If the student begins by breaking the question into parts and trying to discover what the question is attempting to measure, it will be possible to answer the question correctly by using more than memorized charts concerning thyroid problems.

- The question may be testing if the student knows the difference between "hypo" and "hyper" conditions.
- The answer choices may include thyroid problems that are not "hypothyroid" problems.
- It is possible that one or more of the choices are "hypo" but are not "thyroid" problems, that they are some other endocrine problems.
- "Recently diagnosed in a 45-year-old man" indicates that the correct answer is not a congenital childhood problem.

- "May result in" as opposed to "resulting from" suggests that the choices might include a problem that **causes** hypothyroidism rather than **results from** hypothyroidism, as stated.

By applying this kind of reasoning, the student can see that choice **A**, thyroid toxicosis, which is a disorder resulting from an overactive thyroid gland ("hyper") must be eliminated. Another piece of knowledge, that is, Graves' disease is thyroid toxicosis, eliminates choice **D**. Choice **B**, cretinism, is indeed hypothyroidism, but it is a childhood disorder. Therefore, **B** is eliminated. Choice **E** is an inflammation of the thyroid gland—here the clue is the suffix "itis." The reasoning is that thyroiditis, being an inflammation, may **cause** a thyroid problem, perhaps even a hypothyroid problem, but there is no reason for the reverse to be true. Myxedema, choice **C**, is the only choice left and the obvious correct answer.

Preparing for Board Examinations

1. **Study for yourself.** Although some of the material may seem irrelevant, the more you learn now, the less you will have to learn later. Also, do not let the fear of the test rob you of an important part of your education. If you study to learn, the task is less distasteful than studying solely to pass a test.

2. **Review all areas.** You should not be selective by studying perceived weak areas and ignoring perceived strong areas. This is probably the last time you will have the time and the motivation to review **all** of the basic sciences.

3. **Attempt to understand, not just to memorize, the material.** Ask yourself: To whom does the material apply? When does it apply? Where does it apply? How does it apply? Understanding the connections among these points allows for longer retention and aids in those situations when guessing strategies may be needed.

4. Try to **anticipate questions that might appear on the test.** Ask yourself how you might construct a question on a specific topic.

5. **Give yourself a couple days of rest before the test.** Studying up to the last moment will increase your anxiety and cause potential confusion.

Taking Board Examinations

1. In the case of NBME exams, be sure to **pace yourself** to use the time optimally. As soon as you get your test booklet, go through and circle the questions numbered 40, 80, 120, and 160. The test is constructed so that you will have approximately 45 seconds for each question. If you are at a circled number every 30 minutes, you will be right on schedule. A 2-hour test will have 150–170 questions and a 2½-hour test will have approximately 200 questions. You should use all of your allotted time; if you finish too early, you probably did so by moving too quickly through the test.

2. **Read each question and all the alternatives carefully** before you begin to make decisions. Remember the questions contain clues, as do the answer choices. As a physician, you would not make a clinical decision without a complete examination of all the data; the same holds true for answering test questions.

3. **Read the directions for each question set carefully.** You would be amazed at how many students make mistakes in tests simply because they have not paid close attention to the directions.

4. It is not advisable to leave blanks with the intention of coming back to answer the questions later. Because of the way board examinations are constructed, you probably will not pick up any new information that will help you when you come back, and the chances of getting numerically off on your answer sheet are greater than your chances of benefiting by skipping around. If you feel that you must come back to a question, mark the best choice and place a note in the margin. Generally speaking, it is best not to change answers once you have made a decision, unless you have learned new information. Your intuitive reaction and first response are correct more often than changes made out of frustration or anxiety. **Never turn in an answer sheet with blanks.** Scores are based on the number that you get correct; you are not penalized for incorrect choices.

5. **Do not try to answer the questions on a stimulus-response basis.** It generally will not work. Use all of your learned knowledge.

6. **Do not let anxiety destroy your confidence.** If you have prepared conscientiously, you know enough to pass. Use all that you have learned.

7. **Do not try to determine how well you are doing as you proceed.** You will not be able to make an objective assessment, and your anxiety will increase.

8. **Do not expect a feeling of mastery** or anything close to what you are accustomed to. Remember, this is a nationally administered exam, not a mastery test.

9. **Do not become frustrated or angry** about what appear to be bad or difficult questions. You simply do not know the answers; you cannot know everything.

Specific Test-Taking Strategies

Read the entire question carefully, regardless of format. Test questions have multiple parts. Concentrate on picking out the pertinent key words that might help you begin to problem solve. Words such as "always," "all," "never," "mostly," "primarily," and so forth play significant roles. In all types of questions, distractors with terms such as "always" or "never" most often are incorrect. Adjectives and adverbs can completely change the meaning of questions—pay close attention to them. Also, medical prefixes and suffixes (e.g., "hypo-," "hyper-," "-ectomy," "-itis") are sometimes at the root of the question. The knowledge and application of everyday English grammar often is the key to dissecting questions.

Multiple-Choice Questions

Read the question and the choices carefully to become familiar with the data as given. Remember, in multiple-choice questions there is one correct answer and there are four distractors, or incorrect answers. (Distractors are plausible and possibly correct or they would not be called distractors.) They are generally correct for part of the question but not for the entire question. Dissecting the question into parts aids in discerning these distractors.

If the correct answer is not immediately evident, begin eliminating the distractors. (Many students feel that they must always start at option A and make a decision before they move to B, thus forcing decisions they are not ready to make.) Your first decisions should be made on those choices you feel the most confident about.

Compare the choices to each part of the question. **To be wrong**, a choice needs to be incorrect for only part of the question. **To be correct**, it must be **totally** correct. If you believe a choice is partially incorrect, tentatively eliminate that choice. Make notes next to the choices regarding tentative decisions. One method is to place a minus sign next to the choices you are certain are incorrect and a plus sign next to those that potentially are correct. Finally, place a zero next to any choice you do not understand or need to come back to for further inspection. Do not feel that you must make final decisions until you have examined all choices carefully.

When you have eliminated as many choices as you can, decide which of those that are left has the highest probability of being correct. Remember to use paper-and-pencil differential diagnosis. Above all, be honest with yourself. If you do not know the answer, eliminate as many choices as possible and choose reasonably.

Multiple True-False Questions

Multiple true-false questions are not as difficult as some students make them. These are the questions in which you must mark:

> **A** if **1, 2, and 3** are correct,
> **B** if **1 and 3** are correct,
> **C** if **2 and 4** are correct,
> **D** if only **4** is correct, or
> **E** if **all** are correct.

Remember that the name for this type of question is multiple true-false and then use this concept. Become familiar with each choice and make notes. Then concentrate on the one choice you feel is definitely incorrect. If you can find one incorrect alternative, you can eliminate three choices immediately and be down to a fifty-fifty probability of guessing the correct answer. In this format, if choice 1 is incorrect, so is choice 3; they go together. Alternatively, if 1 is correct, so is 3. The combinations of alternatives are constant; they will not be mixed. You will not find a situation where choice 1 is correct, but 3 is incorrect.

After eliminating the choices you are sure are incorrect, concentrate on the choice that will make your final decision. For instance, if you discard choice 1, you have eliminated alternatives A, B, and E. This leaves C (2 and 4) and D (4 only). Concentrate on choice 2, and decide if it is true or false. Rereading and concentrating on choice 4 only wastes time; choice 2 will be the decision maker. (Take the path of least resistance and concentrate on the smallest possible number of items while making a decision.) Obviously, if none of the choices is found to be incorrect, the answer is E (all).

Comparison-Matching Questions

Comparison-matching questions are also easier to address if you concentrate on one alternative at a time. Choose option:

> **A** if the question is associated with **(A) only,**
> **B** if the question is associated with **(B) only,**
> **C** if the question is associated with **both (A) and (B),** or
> **D** if the question is associated with **neither (A) nor (B).**

Here again, the elimination of obvious wrong alternatives helps clear away needless information and can help you make a clearer decision.

Single Best Answer-Matching Sets

Single best answer-matching sets consist of a list of words or statements followed by several numbered items or statements. Be sure to pay attention to whether the choices can be used more than once, only once, or not all. Consider each choice individually and carefully. Begin with those with which you are the most familiar. It is important always to break the statements and words into parts, as with all other question formats. **If a choice is only partially correct, then it is incorrect.**

Guessing

Nothing takes the place of a firm knowledge base, but with little information to work with, even after playing paper-and-pencil differential diagnosis, you may find it necessary to guess at the correct answer. A few simple rules can help increase your guessing accuracy. Always guess consistently if you have no idea what is correct; that is, after eliminating all that you can, make the choice that agrees with your intuition or choose the option closest to the top of the list that has not been eliminated as a potential answer.

When guessing at questions that present with choices in numerical form, you will often find the choices listed in an ascending or descending order. It is generally not wise to guess the first or last alternative, since these are usually extreme values and are most likely incorrect.

Using the Challenge Exam to Learn

All too often, students do not take full advantage of practice exams. There is a tendency to complete the exam, score it, look up the correct answers to those questions missed, and then forget the entire thing.

In fact, great educational benefits can be derived if students would spend more time using practice tests as learning tools. As mentioned earlier, incorrect choices in test questions are plausible and partially correct or they would not fulfill their purpose as distractors. This means that it is just as beneficial to look up the incorrect choices as the correct choices to discover specifically why they are incorrect. In this way, it is possible to learn better test-taking skills as the subtlety of question construction is uncovered.

Additionally, it is advisable to go back and attempt to restructure each question to see if all the choices can be made correct by modifying the question. By doing this, four times as much will be learned. By all means, look up the right answer and explanation. Then, focus on each of the other choices and ask yourself under what conditions they might be correct. For example, the entire thrust of the sample question concerning hypothyroidism could be altered by changing the first few words to read:

> "Hyperthyroidism recently discovered in. . . ."
> "Hypothyroidism prenatally occurring in. . . ."
> "Hypothyroidism resulting from. . . ."

This question can be used to learn and understand thyroid problems in general, not only to memorize answers to specific questions.

The Challenge Exam that follows contains 180 questions and explanations. Every effort has been made to simulate the types of questions and the degree of question difficulty in the various licensure and qualifying exams (i.e., NBME Parts I, II, and III and FLEX). While taking this exam, the student should attempt to create the testing conditions that might be experienced during actual testing situations. Approximately 1 minute should be allowed for each question, and the entire test should be finished before it is scored.

Summary

Ideally, examinations are designed to determine how much information students have learned and how that information is used in the successful completion of the examination. Students will be successful if these suggestions are followed:

- Develop a positive attitude and maintain that attitude.
- Be realistic in determining the amount of material you attempt to master and in the score you hope to attain.
- Read the directions for each type of question and the questions themselves closely and follow the directions carefully.
- Guess intelligently and consistently when guessing strategies must be used.
- Bring the paper-and-pencil differential diagnosis approach to each question in the examination.
- Use the test as an opportunity to display your knowledge and as a tool for developing prescriptions for further study and learning.

National Board examinations are not easy. They may be almost impossible for those who have unrealistic expectations or for those who allow misinformation concerning the exams to produce anxiety out of proportion to the task at hand. They are manageable if they are approached with a positive attitude and with consistent use of all the information the student has learned.

Michael J. O'Donnell

STUDY QUESTIONS

Directions: Each question below contains five suggested answers. Choose the **one best** response to each question.

1. A very painful, spreading, cutaneous edematous erythema is clinically descriptive of

(A) erysipeloid
(B) diphtheria
(C) Pontiac fever
(D) listeriosis
(E) nocardiosis

2. All of the following statements about genetic mutations are true EXCEPT

(A) many mutations are spontaneous
(B) many mutations are simple changes of one nucleotide base for another
(C) mutations can result from either insertion or deletion of nucleotides
(D) missense mutations result in termination of peptide chain growth
(E) microdeletions lead to a change in the reading frame

3. If forbidden clones are not deleted during T cell development, a person may develop

(A) hypogammaglobulinemia
(B) type I hypersensitivity to exogenous antigens
(C) autoimmune disease
(D) tolerance to autoantigens
(E) none of the above

4. All of the following combinations of viruses can participate in pseudotyping EXCEPT

(A) murine leukemia virus and influenza virus
(B) influenza virus and rabies virus
(C) parainfluenza virus and murine leukemia virus
(D) rabies virus and poliovirus
(E) vesicular stomatitis virus and parainfluenza virus

5. The acid-fast staining characteristic of mycobacteria appears to be due to which of the following cell wall constituents?

(A) Mycolic acid
(B) Lipopolysaccharide
(C) Lipid A
(D) *N*-Acetylmuramic acid
(E) *N*-Glycolylmuramic acid

6. The most common etiologic agents of bacterial endocarditis are

(A) staphylococci
(B) group A streptococci
(C) pneumococci
(D) anaerobic gram-positive cocci
(E) viridans streptococci

7. Which of the following structures commonly seen in yeast cells also is found in bacteria?

(A) Nuclear membrane
(B) Mitochondria
(C) Lysosomes
(D) Discrete chromosomes with histones
(E) Semipermeable cytoplasmic membrane

8. Which of the following statements regarding infectious mononucleosis is correct?

(A) The disease occurs primarily in persons from lower socioeconomic groups in developed countries
(B) The viral agent principally affects the activity of $CD4^+$ T cells
(C) It has disappeared from the United States because of a highly effective vaccine
(D) The incubation period varies from 10 to 50 days, depending on patient age
(E) The diagnosis is made by positive identification of Epstein-Barr virus on culture

9. A common infection in American children that is characterized by pruritus ani is caused by

(A) *Necator americanus*
(B) *Ascaris lumbricoides*
(C) *Enterobius vermicularis*
(D) *Trichuris trichiuria*
(E) *Onchocerca volvulus*

10. The capsomer constitutes which of the following structural features of viruses?

(A) The unique 5′ terminal structure found on minus-strand viruses
(B) The protein and lipid that enclose the nucleocapsid
(C) The glycosylated appendage protruding from the viral envelope
(D) The protein subunits that form the capsid
(E) The internal core protein of the virus

11. A 27-year-old geologist who had been exploring caves in the Ohio River valley developed an influenza-like syndrome that lasted several weeks, but no medical attention was sought. Several months later, during a routine physical examination, coin lesions were seen on chest x-ray. This clinical picture is most suggestive of primary infection caused by

(A) *Blastomyces dermatitidis*
(B) *Coccidioides immitis*
(C) *Cryptococcus neoformans*
(D) *Aspergillus fumigatus*
(E) *Histoplasma capsulatum*

12. What bacterial species would an oxidase-positive, glucose-fermenting, gram-negative rod suggest?

(A) *Vibrio* species
(B) *Erwinia* species
(C) *Shigella* species
(D) *Acinetobacter* species
(E) *Pseudomonas* species

13. Multiplication of the microbe in host tissue primarily depends upon the microbe's ability to

(A) produce antiphagocytic virulence factors
(B) adhere tenaciously to host cells
(C) compete with host cells for nutrients
(D) elaborate membrane-destroying enzymes
(E) penetrate the intact skin of the host

14. To which infectious processes is delayed-type hypersensitivity skin testing most applicable?

(A) Exotoxin-caused diseases
(B) Granulomatous infections
(C) Intracellular bacterial infections
(D) Extracellular bacterial infections
(E) Viral diseases

15. Which of the following diseases produces symptomatology resembling that of infectious mononucleosis?

(A) Trichomoniasis
(B) Trypanosomiasis
(C) Leishmaniasis
(D) Toxoplasmosis
(E) Giardiasis

16. Which of the following procedures is useful for diagnostic evaluation of infectious disease patients, regardless of the infecting agent?

(A) Bacteriologic culture
(B) Skin testing
(C) Biopsy for culture and staining
(D) Inoculation of tissue cultures
(E) History and physical examination

17. The most common form of actinomycosis is

(A) cervicofacial actinomycosis
(B) pelvic actinomycosis
(C) thoracic actinomycosis
(D) pulmonary actinomycosis
(E) abdominal actinomycosis

18. The only known human RNA tumor viruses have been isolated from cases of

(A) malignant melanoma
(B) cervical carcinoma
(C) T cell leukemia
(D) multiple myeloma
(E) nasopharyngeal carcinoma

19. Which of the following statements best explains the relatively high success rates associated with organ transplantation between identical twins?

(A) T cells can be suppressed by drugs used during the procedure
(B) The transplanted organ shares 50% of the histocompatibility antigens with the cells of the recipient
(C) The transplanted organ is regarded as "self" by the lymphocytes of the recipient
(D) Tolerance to the new organ can be induced with passive transfer of antibody
(E) Suppressor T cells are selectively transferred with the organ

20. Viral sensitivity to extraction by organic solvents (e.g., ether, chloroform) is a reflection of which of the following structural effects?

(A) Denaturation of viral matrix proteins
(B) Denaturation of viral glycoproteins
(C) Denaturation of viral nucleic acids
(D) Solvent-induced cross-linking of nucleic acids
(E) Solvent solubility of lipids in viral envelope

21. Banana-shaped gametocytes are seen in malaria caused by which etiologic agent?

(A) *Plasmodium falciparum*
(B) *Plasmodium vivax*
(C) *Plasmodium malariae*
(D) *Plasmodium ovale*
(E) None of the above

22. Immunosuppressed diabetics in ketoacidosis appear to be particularly susceptible to which of the following mycoses?

(A) Blastomycosis
(B) Zygomycosis
(C) Sporotrichosis
(D) Rhinosporidiosis
(E) Chromoblastomycosis

23. Several cultures of the nasopharynx of a 35-year-old man are obtained over a 1-year period, and in 40% of the cultures the same phage type of *Staphylococcus aureus* is isolated. This man is best described as a

(A) prodromal carrier
(B) transient carrier
(C) intermittent carrier
(D) successive carrier
(E) persistent carrier

24. All of the following statements describe properties of the human wart virus EXCEPT

(A) it has a papovavirus morphology
(B) it replicates well in human and monkey tissue cultures
(C) it causes the most common small DNA virus infection of humans
(D) its incubation period is 1–6 months
(E) the virus is spread by direct contact or autoinoculation

25. Which of the following illnesses is a frequent cause of blindness?

(A) Inclusion conjunctivitis
(B) Q fever
(C) Lymphogranuloma venereum
(D) Trachoma
(E) Psittacosis

26. All of the following statements describe the characteristics of all viruses EXCEPT

(A) viruses are obligate intracellular parasites
(B) viruses contain active enzymes
(C) viruses divide by binary fission
(D) a virus contains only one type of nucleic acid
(E) only one viral particle is required for infection

Questions 27 and 28

A 30-year-old, juvenile-onset diabetic pierces his left foot with a large nail while jogging near a construction site. Surgery is performed a week after the injury to remove bone chips from the damaged third metatarsal and to débride the wound. A rapidly spreading, crepitant necrosis subsequently develops, and the patient becomes toxemic.

27. What is the probable etiologic agent of this infection?

(A) *Candida albicans*
(B) *Pseudomonas aeruginosa*
(C) *Actinomyces israelii*
(D) *Clostridium perfringens*
(E) *Staphylococcus aureus*

28. What is the recommended treatment for this infectious process?

(A) Amputation of the affected limb
(B) Combined antimicrobial chemotherapy
(C) Surgical opening and draining
(D) Repeated débridement of necrotic tissue
(E) Hyperbaric oxygen and chemotherapy

29. Which of the bacteria listed below serve as recipients in F-factor-mediated conjugation?

(A) Hfr bacteria
(B) F^+ bacteria
(C) F' bacteria
(D) F^- bacteria
(E) None of the above

30. Currently, the major cause of transfusion-associated hepatitis in the United States is

(A) hepatitis A virus
(B) hepatitis B virus
(C) hepatitis C virus
(D) cytomegalovirus
(E) Epstein-Barr virus

31. The most striking feature of herpes simplex virus (HSV) infection is its propensity to recur. All of the following statements regarding recurrent infection by HSV are true EXCEPT

(A) the lesions tend to recur at the site of primary infection
(B) a recurrent infection may be induced by sunlight
(C) the virus appears to be latent in the epithelium at the site of recurrence
(D) circulating antibodies play a small role in the control of recurrent infections
(E) the virus appears to be latent in cervical ganglia

32. Most cases of Rocky Mountain spotted fever in the United States occur in

(A) Middle Atlantic states
(B) northeastern states
(C) western states
(D) southwestern states
(E) Rocky Mountain states

33. Which of the following is the most frequently isolated intestinal parasite in the United States?

(A) *Ascaris lumbricoides*
(B) *Giardia lamblia*
(C) *Enterobius vermicularis*
(D) *Entamoeba histolytica*
(E) *Trichomonas vaginalis*

34. Which technique currently is used for the serodiagnosis of AIDS?

(A) Radioimmunoassay
(B) Enzyme-linked immunosorbent assay
(C) Polymerase chain reaction
(D) Two-dimensional immunoelectrophoresis
(E) Hemagglutination inhibition

35. Rabies is a member of which of the following virus groups?

(A) Arbovirus
(B) Rhabdovirus
(C) Myxovirus
(D) Togavirus
(E) Arenavirus

36. All of the following enzymes can be seen in viruses EXCEPT

(A) neuraminidase
(B) RNA-dependent RNA polymerase
(C) collagenase
(D) protease
(E) RNA-dependent DNA polymerase

37. What is the most rapid response to gram-negative endotoxin?

(A) Neutropenia
(B) Hypotension
(C) Leukocytosis
(D) Fever
(E) Hypoferrinemia

38. Rhinoviruses have all of the following characteristics EXCEPT

(A) they are closely related to enteroviruses
(B) there are many serologic types
(C) they replicate only in primate cells
(D) they demonstrate optimal replication at 33° C and are inhibited at 37° C
(E) they contain a defective neuraminidase in the viral envelope

39. The most selective and active antiviral purine and pyrimidine analogs share which of the following features?

(A) All contain iodine in the 5 position of the uracil ring in place of the methyl group of thymine
(B) All are based on substituted cytosine
(C) All are based on substituted thymidine
(D) All are selectively phosphorylated by virus-induced kinases
(E) All interact directly with viral DNA polymerase

40. The most common form of human pasteurellosis is

(A) septicemia
(B) localized wound infection
(C) gastrointestinal disease
(D) fulminating toxic pneumonia
(E) lymphadenopathy

41. Virulent strains of *Mycobacterium tuberculosis* characteristically produce

(A) mycolic toxin
(B) lipopolysaccharide
(C) tuberculin
(D) cord factor
(E) niacin

42. Which of the following skin test reagents is prepared from the tissue-phase growth of a dimorphic fungus?

(A) Coccidioidin
(B) Histoplasmin
(C) Spherulin
(D) Blastomycin
(E) Trichophytin

43. Influenza virus has all of the following properties EXCEPT

(A) it is a minus-strand RNA virus
(B) it contains a fragmented genome
(C) it consists of three immunologic types based on the nucleocapsid protein (NP)
(D) it lacks a neuraminidase
(E) it will grow in cells derived from several different species

44. What condition is marked by formation of a malignant capsule?

(A) Enteritis necroticans
(B) Lockjaw
(C) Cutaneous anthrax
(D) Pseudomembranous colitis
(E) Woolsorter's disease

45. True statements regarding bacterial growth include all of the following EXCEPT

(A) growth is exponential, not arithmetic
(B) growth begins immediately after bacteria are placed in fresh growth medium
(C) growth rates are regulated by the availability of nutrients
(D) environmental temperature affects the rate of growth
(E) growth can be limited by enzyme inhibition

46. The incubation period for hepatitis B is

(A) less than 15 days
(B) 15–40 days
(C) 40–60 days
(D) 60–160 days
(E) more than 160 days

47. The specific sites on antigens with which antibodies react are called

(A) isotopes
(B) immunogens
(C) epitopes
(D) carriers
(E) haplotypes

48. All of the following statements about collaboration between T cells and B cells are true EXCEPT

(A) T cells may stimulate antibody production by B cells
(B) T cells may depress antibody production by B cells
(C) T cells participating in carrier recognition must recognize the same antigenic epitope as the cooperating B cell
(D) collaboration between T cells and B cells is not required for all antibody responses
(E) antigen processing by macrophages may be required to optimize T cell effects

Questions 49–51

A 23-year-old man who is otherwise healthy is brought to the emergency room with lower right quadrant pain, nausea, and an elevated white blood cell count. The diagnosis is appendicitis, and the patient is prepared for routine appendectomy. Surgery is uneventful.

49. To what category does the surgical wound belong?

(A) Dirty
(B) Contaminated
(C) Nonsterile
(D) Clean-contaminated
(E) Clean

50. Approximately 4 days after surgery the incision becomes inflamed, and a purulent discharge is noted. The patient has no systemic symptomatology. Which of the following conditions is the most probable diagnosis?

(A) Clostridial infection
(B) Staphylococcal infection
(C) Gram-negative bacillary infection
(D) Group A streptococcal infection
(E) Anaerobic streptococcal infection

51. Coagulase-positive, gram-positive cocci are isolated from the purulent discharge. A possible life-threatening condition of this localized wound infection is

(A) myonecrosis
(B) cellulitis
(C) toxic shock syndrome
(D) subcutaneous abscess
(E) endotoxin shock

Questions 52–55

A 68-year-old woman who is a volunteer at a church-sponsored kindergarten suddenly develops a fever of 38.2° C and a severe headache one evening. The following morning she also experiences a stiff neck and uncharacteristic drowsiness. At the emergency room, her temperature is 38.8° C, and there is pain and resistance on flexion of her neck. The patient is noted to be mentally competent although lethargic. A CSF sample is obtained by lumbar puncture.

52. On the basis of the history and physical examination of this patient, what is the most probable diagnosis?

(A) Viral meningitis
(B) Fungal meningitis
(C) Bacterial meningitis
(D) Viral encephalitis
(E) Brain abscess

53. On the basis of the patient's age, the probable etiologic agent is

(A) *Staphylococcus aureus*
(B) *Haemophilus influenzae*
(C) *Actinomyces israelii*
(D) *Neisseria meningitidis*
(E) *Streptococcus pneumoniae*

54. Opening pressure on lumbar puncture was slightly elevated, and the diagnosis was acute bacterial meningitis based upon the finding of gram-positive cocci in pairs in the CSF. The cell count was elevated. The prominent cell type most likely was

(A) mononuclear cells
(B) neutrophils
(C) lymphocytes
(D) red cells
(E) segmented neutrophils

55. Protein and glucose concentrations in the CSF probably were

(A) both elevated
(B) elevated and low, respectively
(C) low and elevated, respectively
(D) both low
(E) unaffected

Questions 56–60

A 19-year-old, recently married woman visits her family doctor complaining of frequency and urgency of urination with burning pain in the vaginal area. Questioning of the patient does not reveal any recent history of systemic symptomatology; physical examination reveals a well-developed, well-nourished, healthy person. The presumptive diagnosis is uncomplicated urinary tract infection (UTI).

56. The most probable etiologic agent is

(A) *Staphylococcus aureus*
(B) *Klebsiella pneumoniae*
(C) *Gardnerella vaginalis*
(D) *Escherichia coli*
(E) *Serratia marcescens*

57. What is the best means of collecting a urine sample for cultures from this patient?

(A) Suprapubic aspiration
(B) Early morning voiding
(C) Midstream clean catch
(D) Catheterization
(E) Mantoux technique

58. What is the most accurate technique for determining the concentration of bacteria in the urine sample?

(A) Calibrated loop
(B) Pour plates
(C) Dip slide
(D) Spectrophotometry
(E) Microscopy

59. What minimal number of gram-negative enteric rods indicates significant bacteriuria?

(A) 10^3/ml
(B) 10^4/ml
(C) 10^5/ml
(D) 10^6/ml
(E) 10^7/ml

60. A rapid means of estimating gram-negative rod bacteriuria is the microscopic examination of uncentrifuged urine. What is the minimum average number of gram-negative rods per high-power microscopic field that indicates significant bacteriuria?

(A) 0.5
(B) 1
(C) 1.5
(D) 2
(E) 3

Directions: Each question below contains four suggested answers of which **one or more** is correct. Choose the answer

A if **1, 2, and 3** are correct
B if **1 and 3** are correct
C if **2 and 4** are correct
D if **4** is correct
E if **1, 2, 3, and 4** are correct

61. AIDS patients show a particular susceptibility to infection by

(1) *Mycobacterium avium*
(2) *Toxoplasma gondii*
(3) cytomegalovirus
(4) *Pneumocystis carinii*

62. Enterotoxin causing increased adenyl cyclase activity is produced by

(1) *Vibrio cholerae* type 0:1
(2) *Shigella dysenteriae* type 1
(3) enterotoxigenic *Escherichia coli*
(4) *Salmonella enteritidis* var. *typhimurium*

63. Primary cutaneous lesions may occur without disseminated disease in which of the following systemic mycoses?

(1) Coccidioidomycosis
(2) Paracoccidioidomycosis
(3) Histoplasmosis
(4) Blastomycosis

64. Anaerobic gram-positive cocci that have been isolated in pure culture from disease processes include

(1) *Peptostreptococcus* species
(2) *Peptococcus* species
(3) *Sarcina* species
(4) *Streptococcus* species

65. The potential for most bacteria to cause infection depends upon their ability to

(1) adhere to host cells
(2) colonize the host tissue
(3) multiply on or within host tissue
(4) penetrate host tissue

66. Viral proteins required for cell transformation by SV40 or polyoma viruses include which of the following?

(1) Small T antigen
(2) Viral protein 1 (VP1)
(3) Large T antigen
(4) Viral protein 2 (VP2)

67. Sequelae of group A β-hemolytic streptococcal (i.e., *Streptococcus pyogenes*) infection include

(1) pharyngitis
(2) migratory arthritis
(3) endocarditis
(4) acute glomerulonephritis

68. Immunologic activities that would be expected to occur immediately after primary immunization include

(1) production of IgM
(2) proliferation of a restricted set of B cells
(3) production of low-affinity IgG
(4) degradation of antigen by thymic epithelial cells

69. Draining sinus tracts discharging granular colonies of bacteria characterize

(1) listeriosis
(2) nocardiosis
(3) erysipeloid
(4) actinomycosis

70. The adsorption of interferon to cell surfaces induces an antiviral state. True statements about this process include which of the following?

(1) It requires synthesis of mRNA
(2) It spreads from interferon-treated cells to resistant cells
(3) It is associated with production of new cellular proteins
(4) It immediately follows interferon treatment

71. Infections may be acquired by

(1) human-to-human transmission
(2) aerosolization of respiratory droplets
(3) contact with fomites
(4) contact with arthropod vectors

72. A few days after returning from a 2-week hunting trip on the Texas plains, a 32-year-old white man develops a serious illness with symptomatology characteristic of bacteremia. Axillary lymph nodes become enlarged and tender and eventually break open and drain. The differential diagnosis should include

(1) brucellosis
(2) tularemia
(3) pasteurellosis
(4) plague

73. Giardiasis usually is characterized by

(1) fecal-oral transmission
(2) invasion of intestinal mucosa
(3) sudden, explosive diarrhea
(4) elevated serum IgE

74. True statements concerning the epidemic spread of influenza include

(1) the disease must become established in a nonimmune or partially immune population
(2) respiratory IgA is the most effective immunoglobulin in the early halt of infection
(3) the predominant mode of transmission is through infectious aerosols
(4) the high rate of infection in nursing homes is due to direct spread via contaminated bed sheets and dishes

75. True statements about the Coombs' test include

(1) the test is used primarily to detect agglutinins of pathogenic bacteria
(2) the test is used to detect particulate antigens that are too large to agglutinate
(3) the test is dependent upon the activation of complement and its lysis of sensitized red blood cells
(4) the test uses antiglobulin to produce a lattice for the agglutination of red blood cells

76. Virulence factors of *Streptococcus pneumoniae* include

(1) streptolysin O
(2) protein A
(3) pili
(4) a capsule

77. Occupational hazards for people working with animals include

(1) inclusion conjunctivitis
(2) psittacosis
(3) louse-borne typhus
(4) Q fever

78. Important considerations in preventing nosocomial infections in hospitalized patients include

(1) elimination of carrier states in hospital personnel
(2) sterilization of the patient's environment
(3) careful monitoring of wounds
(4) discouraging the use of prophylactic antibiotics

79. Bacteremia or septicemia is a characteristic part of the infectious process in which of the following diseases?

(1) Venereal syphilis
(2) Bartonellosis
(3) Rocky Mountain spotted fever
(4) Pneumococcal pneumonia

80. Legionnaires' disease and Pontiac fever differ in

(1) attack rate
(2) initial symptomatology
(3) mortality rate
(4) etiologic agent

81. Tertiary stage lesions are characteristic of

(1) syphilis
(2) pinta
(3) yaws
(4) Vincent's angina

		SUMMARY OF DIRECTIONS		
A	**B**	**C**	**D**	**E**
1,2,3 only	1,3 only	2,4 only	4 only	All are correct

82. Important factors in immunity to parainfluenza viruses include

(1) circulating serotype-specific IgG
(2) high titers of circulating antibodies to the F protein
(3) high titers of group-specific IgG
(4) serotype-specific secretory IgA

83. Species that cause at least two distinct syndromes of food poisoning include

(1) *Bacillus cereus*
(2) *Salmonella typhi*
(3) *Clostridium perfringens*
(4) *Staphylococcus aureus*

84. Diarrheal syndrome is associated with the ingestion of food or fluids contaminated with

(1) *Bacteroides melaninogenicus*
(2) *Campylobacter fetus*
(3) *Acinetobacter anitratus*
(4) *Vibrio parahaemolyticus*

85. A painless lesion on the genitalia is characteristic of

(1) chancroid
(2) lymphogranuloma venereum
(3) genital herpes
(4) syphilis

86. Effective, commercially available vaccines have been produced from the capsules of

(1) *Streptococcus pneumoniae*
(2) *Neisseria meningitidis*
(3) *Haemophilus influenzae*
(4) *Bacillus anthracis*

87. True statements regarding infections by varicella-zoster virus (VZV) include which of the following?

(1) Primary infection usually occurs in epidemic form
(2) Primary disease is most common in the summer months
(3) Transmission is by respiratory droplets or direct contact with lesions
(4) Recurrent infection is most common in the winter months

88. Etiologic agents of human tuberculosis include

(1) *Mycobacterium tuberculosis*
(2) *Mycobacterium bovis*
(3) *Mycobacterium africanum*
(4) *Mycobacterium fortuitum*

89. Syphilis is transmissible by

(1) sexual contact during the primary stage
(2) transplacental passage during pregnancy
(3) contact with lesions during the secondary stage
(4) transfusion during the secondary stage

90. Clinical effects of infection with toxigenic *Corynebacterium diphtheriae* include

(1) changes in myocardial structure and function
(2) demyelination of cranial nerves
(3) reversible paralysis
(4) necrotic cutaneous lesions

91. The growth of clinical isolates of pathogenic *Neisseria* species is enhanced by

(1) an atmosphere of increased (i.e., 4%–8%) carbon dioxide
(2) addition of soluble starch as a detoxifying agent
(3) inhibition of indigenous flora with antibacterial and antifungal compounds
(4) addition of tellurite as a differential factor

92. The pathogenesis and clinical manifestations of primary tuberculosis are characterized by

(1) a cell-mediated immune response with macrophages and lymphocytes infiltrating the focus of infection
(2) lymphatic and hematogenous spread of mycobacteria to a variety of organ systems
(3) a typical clinical presentation of mild, influenza-like symptomatology
(4) the persistence of viable mycobacteria in tubercles for as long as 20 years

93. All interferons are alike in which of the following ways?

(1) All are small proteins
(2) All are induced by similar mechanisms both in cultured cells and in vivo
(3) All are species specific and not virus specific
(4) All are directly antiviral

94. The classic symptomatology caused by rickettsial organisms includes

(1) headache
(2) fever
(3) vasculitis
(4) rash

Questions 95–97

A 19-year-old, recently married woman reports to the emergency room complaining of severe abdominal pain and fever. She is subsequently diagnosed as having pelvic inflammatory disease (PID).

95. Potential etiologic agents of PID in this patient include

(1) *Mycoplasma hominis*
(2) *Chlamydia trachomatis*
(3) *Ureaplasma urealyticum*
(4) *Neisseria gonorrhoeae*

96. Cultures taken from the cervical os and anal mucosa of the patient are examined microscopically and show gram-negative cocci in pairs. These bacteria might be

(1) *Neisseria sicca*
(2) *Branhamella catarrhalis*
(3) *Neisseria gonorrhoeae*
(4) *Chlamydia trachomatis*

97. The isolated bacterium is further identified as *Neisseria gonorrhoeae* by biochemical tests. If this infection is not treated effectively, the patient might develop

(1) lymphogranuloma venereum
(2) ascending arthritis
(3) granulomatosis infantiseptica
(4) sterility

Questions 98 and 99

A 23-year-old intravenous drug abuser who was in labor was admitted to a large urban hospital. Her admission blood workup indicated a positive VDRL test.

98. Probable conditions of the neonate at the time of birth include

(1) hepatosplenomegaly
(2) blindness
(3) skeletal abnormalities
(4) skin lesions

99. The clinical manifestations of congenital syphilis could be prevented by

(1) cesarean section
(2) chemotherapy for the neonate
(3) vaccination of the mother
(4) penicillin therapy during pregnancy

Directions: The groups of questions below consist of lettered choices followed by several numbered items. For each numbered item, select the **one** lettered choice with which it is **most** closely associated. Each lettered choice may be used once, more than once, or not at all. Choose the answer

A   if the item is associated with **(A) only**

B   if the item is associated with **(B) only**

C   if the item is associated with **both (A) and (B)**

D   if the item is associated with **neither (A) nor (B)**

Questions 100–104

Match each characteristic with the appropriate toxin or toxins.

(A) Tetanospasmin
(B) Botulinum toxin
(C) Both
(D) Neither

100. Interferes with the release of acetylcholine

101. Causes neurotoxicity

102. Is clinically neutralized with specific anti-toxin

103. Induces immunity at the time of intoxication

104. Blocks transmitter release at synapses

Questions 105–108

Match the characteristics below with the appropriate microbe or microbes.

(A) *Haemophilus influenzae*
(B) *Haemophilus ducreyi*
(C) Both
(D) Neither

105. Requires X and V factors for growth

106. Polysaccharide capsule is a vaccine

107. Primarily an animal pathogen

108. Major cause of meningitis in children

Questions 109–111

Match each description with the appropriate type or types of lymphocytes.

(A) T cells
(B) B cells
(C) Both
(D) Neither

109. The surface of these lymphocytes is coated with immunoglobulin

110. These lymphocytes are separated from peripheral blood by density gradient centrifugation

111. These lymphocytes form rosettes when mixed with sheep erythrocytes

Directions: The groups of questions below consist of lettered choices followed by several numbered items. For each numbered item select the **one** lettered choice with which it is **most** closely associated. Each lettered choice may be use once, more than once, or not at all.

Questions 112–115

Match each statement regarding a mechanism of genetic recombination in bacteria with the process that is most likely to play a major role.

(A) Recombination involving the rec A gene
(B) Recombination involving rec B and rec C genes
(C) Site-specific recombination
(D) Illegitimate recombination
(E) Molecular cloning

112. The best example of this type of recombination is the integration of bacteriophage λ

113. This type of recombination uses neither extensive homology nor specific sites

114. The loss of this mechanism leads to an almost complete loss of generalized recombination

115. This type of recombination is unaffected by the species or sequence of the introduced DNA

Questions 116–120

Match each description with the most appropriate virus.

(A) Herpes simplex virus
(B) Measles virus
(C) Rubella virus
(D) Adenovirus
(E) Coronavirus

116. Closely associated with "the common cold"

117. Closely associated with acute respiratory disease

118. Often spread via contaminated swimming pools

119. A highly infectious paramyxovirus

120. A plus-strand RNA virus

Questions 121–125

Match the description with the appropriate etiologic agent.

(A) *Streptococcus pneumoniae*
(B) *Mycoplasma pneumoniae*
(C) Adenovirus
(D) Influenza A virus
(E) *Pneumocystis carinii*

121. Major cause of interstitial pneumonia in immunosuppressed patients

122. Probably the most highly infectious agent of pneumonia

123. Etiologic agent of the most common pneumonia in adults

124. Accounts for most cases of bacterial pneumonia

125. Common cause of pneumonia in military recruits but rare in other adults

Questions 126–130

Match each viral property with the virus most likely to exhibit that characteristic.

(A) Human immundeficiency virus
(B) Poliovirus
(C) Influenza virus
(D) Adenovirus
(E) Vaccinia virus

126. Contains a single molecule of plus-strand RNA

127. Carries the genetic information for its own membrane

128. All viral gene products are made in equimolar amounts

129. Contains two copies of its nucleic acid genome

130. May induce tumor formation in newborn hamsters

Questions 131–135

Match each common site of infection with the appropriate helminth.

(A) *Paragonimus westermani*
(B) *Diphyllobothrium latum*
(C) *Clonorchis sinensis*
(D) *Brugia malayi*
(E) *Schistosoma mansoni*

131. Lymph nodes and lymphatic vessels

132. Common bile duct

133. Lungs

134. Blood vessels

135. Small intestines

Questions 136–141

Match each disease below with its etiologic agent.

(A) *Treponema pallidum*
(B) *Treponema pertenue*
(C) *Treponema carateum*
(D) *Leptospira interrogans*
(E) *Borrelia recurrentis*

136. Relapsing fever

137. Bejel

138. Pinta

139. Syphilis

140. Fort Bragg fever

141. Yaws

Questions 142–146

Match each drug listed below with its appropriate mode of action.

(A) Inhibits cell wall synthesis
(B) Alters cell membrane function
(C) Inhibits protein synthesis by affecting the 30S ribosomal subunit
(D) Inhibits protein synthesis by affecting the 50S ribosomal subunit
(E) Alters nucleic acid synthesis or function

142. Erythromycin

143. Chloramphenicol

144. Polymyxin

145. Metronidazole

146. Cycloserine

Questions 147–150

Match each statement describing a feature of bacterial energy metabolism with the most appropriate metabolic process.

(A) Mixed acid fermentation
(B) Glycolysis
(C) Complete aerobic respiration
(D) Autotrophic metabolism
(E) Entner-Doudoroff pathway

147. This process uses inorganic salts as a source of terminal electron acceptors

148. This process yields 38 mol of ATP for each mol of glucose used

149. This is a common form of energy metabolism among members of the Enterobacteriaceae family

150. This is the major hexose-reducing pathway in pseudomonads

ANSWERS AND EXPLANATIONS

1. The answer is A. [*Chapter 16 V B 1*] The etiologic agent of erysipeloid is *Erysipelothrix rhusiopathiae*, a bacterium widely distributed in the environment. Human erysipeloid is clinically described as a slowly spreading cutaneous erythema. Cutaneous edema is characteristic, and the disease is very painful. Cutaneous diphtheria is characterized by a necrotic lesion sometimes associated with insect bites; the bite apparently provides the break in the skin through which toxigenic *Corynebacterium diphtheriae* enter the tissue. Cutaneous nocardiosis is characterized by draining sinus tracts discharging purulent exudate-containing granules. Pontiac fever and listeriosis do not have cutaneous manifestations.

2. The answer is D. [*Chapter 3 III A 1 d (2)*] The mutations labeled as missense do not introduce nucleotide changes that signal chain termination but rather change the amino acid assignment for that codon, thereby producing a polypeptide with an altered primary sequence. These polypeptides frequently are partially functional. Mutations are both induced (by mutagens such as ultraviolent light) or spontaneous; the latter result from errors in replication or repair of DNA. Spontaneous mutations occur randomly in DNA. In some cases mutations are not apparent or have only a very small impact. Mutations in which one base is changed to another may have no impact on the amino acid sequence, may be conservative (the replacement of an amino acid by an almost identical chemical), or may result in a major chemical change in the replaced amino acid. These single-base changes are the most common spontaneous mutations. In insertion and deletion mutagenesis, the size of the DNA change plays a major role in the impact of the mutation.

3. The answer is C. [*Chapter 5 V B*] The presence of forbidden clones provides a supply of cells that carry recognition of self antigens and that will serve as cells to stimulate both humoral and cell-mediated immune responses leading to autoimmunity. B cell deficiency—primarily a lack of circulating B cells—is a primary cause of hypogammaglobulinemia. Type I hypersensitivity (anaphylactic hypersensitivity) is mediated by humoral antibodies, which result from a normal immune response to exogenous antigens. The deletion of forbidden clones leads to tolerance to autoantigens —the opposite of autoimmunity.

4. The answer is D. [*Chapter 19 III B 1 c; Chapter 22 III C*] Pseudotyping refers to the exchange of membrane proteins between enveloped viruses. Among the viruses listed, only poliovirus—a picornavirus—is a naked capsid virus; therefore, poliovirus could not participate in pseudotyping. All of the other viral combinations mentioned are capable of exchanging envelope proteins during viral maturation when both are within the same cell.

5. The answer is A. [*Chapter 14 I A 2 a*] The cell wall of mycobacteria differs significantly from that of other bacterial species. Lipid compounds account for as much as 60% of the dry weight of the mycobacterial cell, resulting in a highly hydrophobic structure. In addition, the cell wall of mycobacteria contains N-glycolylmuramic acid rather than N-acetylmuramic acid. The high lipid content of the mycobacterial cell wall does not allow the penetration of basic aniline dyes, and special staining techniques must be employed. Mycobacteria have the characteristic of being acid fast, apparently as a result of the mycolic acid content of the cell wall. The Ziehl-Neelsen stain is the most commonly employed acid-fast staining technique. Lipid A is a component of the lipopolysaccharide component characteristic of all gram-negative bacteria.

6. The answer is E. [*Chapter 8 II C 2 b*] The viridans streptococci account for more than 50% of all cases of bacterial endocarditis. *Staphylococcus aureus* and *Staphylococcus epidermidis* are important causes of acute and subacute endocarditis, respectively, especially following thoracic surgery. *Streptococcus pyogenes* (a group A streptococcus) and *Streptococcus pneumoniae* occasionally are the etiologic agents of endocarditis. Anaerobic gram-positive cocci, such as *Peptococcus* species, rarely have been associated with endocarditis.

7. The answer is E. [*Chapter 2 II A 2*] There are several differences in organization between eukaryotic yeast cells and prokaryotic bacterial cells. Most of these differences involve the organization of the genetic material within the eukaryotic nucleus and the prokaryotic nucleoid, and the localization of the electron transport system of eukaryotic cells within the mitochondria and of prokaryotic cells within the cytoplasmic membrane. All living cells share one common feature—their external barrier is a semipermeable membrane.

8. The answer is D. [*Chapter 21 III D 2 a (1)*] Infectious mononucleosis remains a disease of middle- and upper-class teenagers in developed countries. The major target cell of the virus is the B cell—not the T cell lines as is the case in diseases such as AIDS. No vaccination program is in place for prevention of this disease, and vaccine development is unlikely in the near future; live or killed virus vaccines would not be used because of the oncogenic potential of Epstein-Barr virus (EBV). The incubation period in young children is 10–40 days and in adults is 30–50 days. Diagnosis relies on the identification of specific antibodies against the virus proteins, as EBV is not easily cultured.

9. The answer is C. [*Chapter 25 I A 2 a, 3 a*] *Enterobius vermicularis* is the etiologic agent of pinworm, a common disease of American children, although enterobiasis occurs in all age-groups. Pinworm infection is characterized by pruritus ani, which is associated with migration of the pinworm outside the anal canal at night. Gastrointestinal infection with the hookworm *Necator americanus* typically produces epigastric pain and abdominal peristalsis. Gastrointestinal ascariasis caused by *Ascaris lumbricoides* is characterized by abdominal pain and malabsorption syndrome. *Trichuris trichiuria*, the whipworm, is a common cosmopolitan parasite that causes abdominal pain and diarrhea. *Onchocerca volvulus*, a tissue nematode, is the etiologic agent of an ocular infection that normally does not occur in the United States.

10. The answer is D. [*Chapter 18 II B 2*] The capsomer is the basic building block that aggregates to form the viral coat, or capsid. The capsomer may be composed of several individual polypeptides or protomers. The protomers represent the product of a single viral gene, and in most viruses several different protomers associate to form a capsomer. For enveloped viruses, the protein and lipid that enclose the nucleocapsid constitute the viral envelope. The unique 5′ terminus on a minus-strand virus is referred to as a "cap." The surface glycoproteins of different viruses have different functions but generally appear as spikes in the electron microscope. The complex of nucleic acid and protein in an enveloped virus is the nucleocapsid.

11. The answer is E. [*Chapter 17 IV A 3 a*] Histoplasmosis—a systemic mycosis caused by *Histoplasma capsulatum*— is endemic to the midwestern United States, particularly the Ohio River valley, where greater than 50% of the residents are skin-test positive. Primary disease is characterized by an influenza-like syndrome that rarely requires medical attention and usually resolves spontaneously. Coin lesions may be seen on chest x-ray and are the healed, calcified focal lesions of infection. Common sources of infection include bat caves and bird roosts. Primary blastomycosis and coccidioidomycosis usually occur in the lungs and tend to be only mildly symptomatic. Healed primary pulmonary lesions, however, do not resemble classic coin lesions.

12. The answer is A. [*Chapter 11 II A 1 a (1)*] *Vibrio* species resemble Enterobacteriaceae biochemically except that they are oxidase positive. *Vibrio* species are gram-negative rods that may be curved. They ferment glucose and other carbohydrates. *Erwinia* and *Shigella* species belong to the Enterobacteriaceae family and are oxidase negative. *Pseudomonas* and *Acinetobacter* species are glucose nonfermenters.

13. The answer is C. [*Chapter 7 I A 3 b*] Multiplication of the microorganism primarily depends upon its ability to compete with the host cells for nutrients. The infecting microbe must be able to acquire sufficient nutrition to replicate at a rate that allows it to overcome the natural defense mechanisms of the host. Factors that are necessary for successful infection of the host include adherence to tissue cells, penetration of the host cell, and production of antiphagocytic capsules and other virulence factors. Whereas these factors aid the growth and survival of the microbe, successful competition for nutrients is an absolute requirement for multiplication.

14. The answer is B. [*Chapter 6 III D*] Delayed-type hypersensitivity (DTH) skin testing is most useful as an indicator of chronic granulomatous infection. It is used routinely to screen populations for infection by *Mycobacterium tuberculosis* and the systemic fungal agents. The test relies on the microorganism eliciting a DTH reaction (i.e., a cell-mediated immune response); humoral immunity generally provides no protection. Humoral immunity is important in the control of viral infections and diseases caused by exotoxins. Extracellular and intracellular bacterial pathogens elicit a variety of humoral and cell-mediated immune responses.

15. The answer is D. [*Chapter 24 I B 3 a (1)*] Otherwise healthy people with toxoplasmosis normally exhibit lymphadenopathy involving the cervical lymph nodes. Other symptomatology may include swollen nontender nodes, pharyngitis, fever, rash, hepatomegaly, splenomegaly, and atypical lympho-cytosis. The disease may resemble infectious mononucleosis. Trichomoniasis is a vaginitis with an odiferous discharge. Trypanosomiasis occurs in two forms: African trypanosomiasis is sleeping sickness, and American trypanosomiasis is an inflammatory, febrile disease that may involve the heart or gastrointestinal tract. Leishmaniasis can involve localized cutaneous tissues or mucocutaneous tissues or it can be a disseminated disease involving many organs. Giardiasis is an intestinal disease characterized by malabsorption syndrome.

16. The answer is E. [*Chapter 1 III B*] A complete history and physical examination enable the clinician to gauge the appropriate diagnostic tests to be performed for a definitive diagnosis. Not all pathogens are detected by culture or through staining, and some insight into symptoms, exposure, and physical findings will assist in determining the best diagnostic approach.

17. The answer is A. [*Chapter 16 VI A 5 a*] Cervicofacial actinomycosis is the most common form of actinomycosis; the etiologic agent is *Actinomyces israelii*. The disease typically occurs as the result of trauma to the mouth or jaw, including tooth extraction. It also may result from poor dental hygiene. *A. israelii* is a strict anaerobe that requires a deep wound with devitalized tissue or an anaerobic microenvironment to cause disease. Thoracic and abdominal actinomycoses are rare. Thoracic actinomycosis may occur by inhalation of the bacteria, causing pneumonia, or by invasive extension from cervicofacial or abdominal disease. Abdominal actinomycosis occurs as the result of trauma to the intestine from *A. israelii*, a member of the indigenous flora of the colon. Pelvic involvement occasionally occurs by invasive extension from the gastrointestinal tract. Pelvic infections also have been associated with the use of intrauterine contraceptive devices.

18. The answer is C. [*Chapter 23 II D 6*] Despite many years of research and many suggestive results, only a single class of human tumors has yielded a retrovirus. These tumors have been uniformly of a class of T cell lymphoma, and the virus is termed human T cell leukemia virus (HTLV). There are several identifiable serologic variants of HTLV.

19. The answer is C. [*Chapter 5 III A 2; V*] Because identical twins share identical surface antigens, their cells are looked upon as "self" when transferred from one twin to another. The transplanted organ shares 100% of the histocompatibility antigens with the cells of the recipient. Thus, no immunosuppressive drugs are required, and tolerance to the newly transplanted organ already exists.

20. The answer is E. [*Chapter 18 II D 1*] It is well known that lipids are soluble in organic solvents (e.g., ether, chloroform), and all lipid-containing membranes—cellular or viral—share this property. Proteins and nucleic acids generally are unaffected by organic solvents; harsh treatments such as phenol extraction may denature proteins but not nucleic acids.

21. The answer is A. [*Chapter 24 Table 24-1*] The etiologic agents of human malaria are differentiated in the venous blood of symptomatic patients by their morphology and the effects they produce on infected erythrocytes. *Plasmodium falciparum* is characterized by banana-shaped gametocytes in infected red blood cells. The other *Plasmodium* species display rounded gametocytes.

22. The answer is B. [*Chapter 17 V D 5*] Zygomycosis is an opportunistic mycosis most commonly caused by *Mucor* species. Immunosuppressed patients, and especially diabetics in ketoacidosis, are at particularly high risk for this infection. The increased capability of performing successful renal transplants on type I diabetics has increased the incidence of zygomycosis. In diabetic patients, zygomycosis most commonly presents as a rhinocerebral infection. Conidia germinate in the posterior nares, and the rapidly growing filamentous fungi penetrate the cribriform plate and enter the brain. The disease usually is fatal within 48–72 hours.

23. The answer is C. [*Chapter 7 I C 3*] Intermittent carriers of pathogenic microbes are found to have the same serotype, biotype, or phage type at the same site at different (i.e., intermittent) times. Intermittent and transient carriers are similar with the exception that, in transient carriage, different strains or different microbes are isolated at different times. Persistent carriers always carry the same species at a particular site; it may or may not be the same strain.

24. The answer is B. [*Chapter 21 I B 1 c*] The human wart virus is the most common papovavirus infection of humans. The virus is a typical papovavirus; however, it does not grow in tissue culture of any type. All of the information currently available about these viruses has been derived from viruses that were isolated from human wart tissue.

25. The answer is D. [*Chapter 15 I G 1 a (3)*] Trachoma caused by *Chlamydia trachomatis* serotypes A, B, Ba, and C is a major problem in some regions of Africa; it also is seen in native American populations in the southeastern United States. It is estimated that worldwide 400 million people are affected by trachoma, and approximately 20 million are blind. Inclusion conjunctivitis is caused by different serotypes of *C. trachomatis*, and it is a relatively mild disease of childhood. Lymphogranuloma venereum also is caused by specific serotypes of *C. trachomatis*, but the disease does not affect the eyes. Q fever and psittacosis are respiratory diseases caused by *Coxiella burnetii* and *Chlamydia psittaci*, respectively.

26. The answer is C. [*Chapter 18 I A; Chapter 19 I B*] All viruses are obligate intracellular parasites; they rely on the host cell for their replication functions. Viral replication occurs by somewhat of an "assembly line" process, not by duplication (the method used by bacteria); therefore, viruses do not divide by binary fission. Many viruses contain highly specialized enzymes, but these enzymes are involved only in specific steps in the replication cycle, not in general metabolic processes. Although the particle-to-infectivity ratio for animal viruses always exceeds 1, it is clear that only one viral particle is required for infection.

27. The answer is D. [*Chapter 28 III B 5 a, b*] *Clostridium perfringens* is the most common cause of crepitant necrosis in surgical wounds. Myonecrosis of skeletal muscles can progress until the patient is toxemic. *C. perfringens* is present in soil and feces. *Candida albicans* is the most common cause of fungemia; *Pseudomonas aeruginosa* frequently colonizes wounds and can lead to gram-negative sepsis; *Actinomyces israelii* is an occasional cause of wound infections in patients undergoing oral or gastrointestinal surgery; *Staphylococcus aureus* is a frequent cause of wound infection, but it does not cause crepitant necrosis.

28. The answer is A. [*Chapter 28 III B 5 c (1) (b)*] When possible, amputation is the recommended treatment for clostridial cellulitis and myonecrosis. The diabetic patient presents a special problem in that peripheral vascularization probably is poor. In early, uncomplicated clostridial cellulitis, combined chemotherapy with penicillin, clindamycin, and chloramphenicol is useful, along with opening and draining the wound. Regular débridement also may be beneficial in preventing spread of the infection. Hyperbaric oxygen generally has not proven effective in eliminating the infection, particularly in patients with poor circulation.

29. The answer is D. [*Chapter 3 II C 1*] Bacteria that carry the F factor appear to repel complex formation with other male bacteria and do not form conjugation pairs with them. This probably is a function of the presence of the F pilus, which is expressed on F^+, F', and Hfr bacteria; F^- bacteria are recipient cells. The proteins found in the F pilus are unique and are different from those in F^- organisms. These specific proteins act as receptors for some bacteriophages; these phages are considered to be male specific because they attach only to the F factor.

30. The answer is C. [*Chapter 23 I*] Because of the availability of good diagnostic tests for hepatitis B virus (HBV) and hepatitis A virus (HAV), most blood supplies and blood donors are screened carefully for these viruses. As a result, the incidence of transfusion-associated disease related to these two viruses is much lower than that related to non-A, non-B hepatitis virus (now called hepatitis C virus), which cannot be detected by normal screening procedures. Both cytomegalovirus (CMV) and EBV have been associated with cases of hepatitis but are rare in this disease.

31. The answer is C. [*Chapter 21 III B 3 b*] Herpes simplex virus (HSV) can be isolated from neurons of cervical or sacral ganglia in the general region of recurrent lesions. There is no evidence to support the theory that the virus may remain latent in local epithelial tissue. The exact mechanism of induction is unknown, but it is clear that such agents as sunlight, stress, hormonal changes, and menstruation are common inducers of recurrent infection. Patients with recurrent lesions often have high antibody titers to HSV; however, these antibodies do not suppress the recurrent lesions.

32. The answer is A. [*Chapter 15 II D 1 a*] Rocky Mountain spotted fever is the most common rickettsial disease in the United States; more than 1000 cases occur annually. Despite the name, most cases of Rocky Mountain spotted fever occur in Middle Atlantic states such as Maryland, Virginia, and the Carolinas. The disease is transmitted to humans by the bite of one of several varieties of tick.

33. The answer is B. [*Chapter 24 III B*] *Giardia lamblia* is the most frequently isolated intestinal parasite in the United States. Giardiasis is common in both children and adults. Most cases are asymptomatic. Symptomatic disease is characterized by malabsorption syndrome and diarrhea. *Ascaris lumbricoides* and *Enterobius vermicularis* are helminths that are common causes of infection. *E. vermicularis*, the pinworm, is the cause of a common infection in children in the United States, and the short-lived roundworm *A. lumbricoides* causes one of the most common parasitic infections worldwide. *Entamoeba histolytica* is the etiologic agent of gastrointestinal amoebiasis, amoebic dysentery, and hepatic abscess. The infection rate in the United States is 1%–5%. *Trichomonas vaginalis* is the etiologic agent of vaginitis, which affects about 3 million women annually in the United States.

34. The answer is B. [*Chapter 6 III F*] Circulating antibodies to the human immunodeficiency virus (HIV) currently are detected by the enzyme-linked immunosorbent assay (ELISA). Electrophoretic and agglutination techniques are not sensitive or specific enough for the serodiagnosis of AIDS. Radioimmunoassay is sensitive and specific, but the use of radioactive materials is avoided when possible. The polymerase chain reaction technique is a gene amplification procedure that has been suggested as a diagnostic tool for AIDS, but the testing procedure has not yet been completely developed and approved.

35. The answer is B. [*Chapter 22 IV A–B*] Rabies is a bullet-shaped virus that consists of a single helical strand of ribonucleoprotein containing a single, minus-strand RNA. The virus carries its own transcriptase and, therefore, clearly is a rhabdovirus. Arboviruses, myxoviruses, togaviruses, and arenaviruses all are RNA viruses, but they do not share the characteristic morphology of the rhabdoviruses.

36. The answer is C. [*Chapter 18 II E 2*] Although viruses neither contain nor encode enzymes that are involved in energy metabolism, virus particles contain several unique enzymes involved in nucleic acid synthesis as well as neuraminidase and proteases, which are involved in viral maturation. Most virion-associated enzymes perform critical steps in viral replication, which generally are not done by the host cell. For example, RNA-directed polymerases are not common in nature, and virion neuraminidase and protease carry out steps that must be done at the specific site of viral replication. There is no evidence of collagenase in a virus.

37. The answer is A. [*Chapter 11 I D 2 d*] Neutropenia occurs within minutes of exposure of the circulatory system to toxic gram-negative endotoxin. Leukocytosis follows the neutropenia, and eventually monoblasts and other immature cellular forms appear. Hypotension and fever typically occur about 30 minutes after exposure to the lipopolysaccharide. Hypoferrinemia is a metabolic alteration that typically shows up later in the endotoxin shock syndrome.

38. The answer is E. [*Chapter 22 III A, F*] The rhinoviruses are members of the picornavirus group as are the enteroviruses, to which they are closely related. These plus-strand RNA viruses have no viral envelope. The rhinoviruses replicate only in cells of primate origin, and their growth is naturally temperature sensitive, especially on primary isolation.

39. The answer is D. [*Chapter 20 II C 3*] The selective phosphorylation of purine and pyrimidine analog drugs by virus-induced enzymes leads to very favorable therapeutic indexes and allows the administration of effective antiviral doses without host toxicity. It is not completely clear why these drugs have a selective affinity for herpes-induced thymidine kinase, but their effectiveness is based on this affinity. It has become clear that no single substituted base is uniquely effective as an antiviral compound; specific configurations of cytosine, thymidine, and inosine have been effective.

40. The answer is B. [*Chapter 12 III D 2*] In humans, pasteurellosis caused by *Pasteurella multocida* most commonly occurs as a localized wound infection following the bite or scratch of a dog or cat. The bacteria are indigenous flora of the oropharynx of these and other animals. Pneumonia and septicemia occur in animals, usually as the result of stress or some other debilitating factor. Septicemia, meningitis, and chronic lung infections occasionally occur in humans. The localized wound infections in humans are purulent and rarely involve the lymph nodes, although lymph node involvement occurs in disseminated disease.

41. The answer is D. [*Chapter 14 II B 2 b*] The production of cord factor is characteristic of virulent strains of *Mycobacterium tuberculosis*. The glycolipid causes growing cells of *M. tuberculosis* to form bundles of entwined strands. The actual relationship, if any, of cord factor to virulence is unknown. A component of all mycobacterial cell walls, mycolic acid apparently gives the cells their acid-fast staining characteristic; mycolic acid has no specific toxic effects. Niacin is produced by *M. tuberculosis* and is used as an identifying characteristic, but it has no relationship to virulence. Tuberculin is a heterologous mixture of proteins derived from the culture filtrate of *M. tuberculosis* grown in liquid medium. Tuberculin is used as a skin-test reagent, and large quantities of it do not produce toxic effects on tissue; tuberculin injections do not convert a person to skin test positive. Lipopolysaccharide is not a component of mycobacterial cells.

42. The answer is C. [*Chapter 17 IV C 4 c*] Spherulin is a skin test antigen prepared from the tissue phase of the dimorphic fungus *Coccidioides immitis*. The spherules for use as a skin test reagent are grown in tissue culture. Spherulin appears to be more specific than coccidioidin, a skin test antigen prepared from the mycelial phase of *C. immitis*. Histoplasmin, blastomycin, and trichophytin all are skin test reagents prepared from mycelial-phase cultures. Histoplasmin skin testing provides useful epidemiologic information about histoplasmosis; blastomycin skin testing is not a useful screening tool for blastomycosis. Trichophyton testing, used for *Trichophyton* infection, is used primarily to test for anergy in men.

43. The answer is D. [*Chapter 22 I A 2*] Influenza is a large, enveloped, minus-strand virus with a fragmented genome. It has two important virion surface projections, a hemagglutinin and a neuraminidase. Both of these proteins are highly significant antigens in the development of immunity to influenza virus infection. The serologic differentiation of the major types of influenza virus is based on antigen found on the core nucleocapsid protein (NP). Influenza virus replicates in many different animals, ranging from humans to birds. This wide host range is reflected in the types of cells that are sensitive to viral infection in culture.

44. The answer is C. [*Chapter 10 I A 6 a*] A malignant pustule is a clinical manifestation of cutaneous anthrax. It occurs at the site of inoculation and is characterized by a black eschar at its base surrounded by an inflamed ring. Enteritis necroticans caused by *Clostridium perfringens* and pseudomembranous colitis caused by *Clostridium difficile* are diseases of the gastrointestinal tract that may be characterized by ulcerative lesions in the intestinal mucosa. Lockjaw is a lay name for tetanus; it refers to the muscle and neural spasms caused by the neurotoxin tetanospasmin. Woolsorter's disease is pulmonary anthrax—a diffuse, lethal, progressive pneumonia caused by the inhalation of spores of *Bacillus anthracis*.

45. The answer is B. [*Chapter 2 III A 2 a*] Bacteria increase in mass and number in an exponential manner. The rate of bacterial growth is determined by the availability of essential nutrients, temperature, and events that regulate the metabolic activities of the organism (e.g., end-product inhibition). Upon transfer to fresh growth medium, bacterial cells undergo a lag phase, which varies with the recent growth history of the organisms and the medium into which the cells have been transferred. Even exponentially growing cells transferred to fresh, identical medium require a brief adaptation period for such functions as carbon dioxide accumulation and pH adjustment.

46. The answer is D. [*Chapter 23 Table 23-1*] Hepatitis B has a long incubation period, which may extend as long as 160 days but rarely surpasses that length of time; it rarely is under 45 days. Hepatitis B generally is associated with instruments contaminated by infectious blood or blood products. The route of infection is via direct inoculation of the infectious material into the blood.

47. The answer is C. [*Chapter 5 II A 2*] The specific site of antibody reactivity on antigen is the epitope, or antigenic determinant. Immunogens may have many epitopes; the exact number is dependent upon the size of the molecule. Carrier molecules frequently are used to make specific epitopes (in the form of haptens) immunogenic. Carriers have many epitopes of their own in addition to the passenger epitopes. Haplotype is the term applied to genetic pattern of alleles of genes in the major histocompatibility complex. Isotopes refers to the different forms of an element, which vary only by atomic weight; isotopes may be stable or unsable (i.e., radioactive).

48. The answer is C. [*Chapter 5 IV B 2*] Individual subpopulations of T cells can promote or suppress antibody production. Some antigens, such as bacterial polysaccharides, may stimulate antibody production that does not depend on T cell help; such antigens are referred to as T-independent antigens. Macrophage processing of antigen with subsequent presentation of processed antigen to the T cell is an important mechanism for promoting T cell activity. In the collaboration between T cells and B cells, the T cell may recognize epitopes on the carrier portion of the antigen molecule, which are different from the epitopes to which the B cell is reacting.

49. The answer is E. [*Chapter 28 II A 1 a (1) (a)*] Routine appendectomy without complications produces a clean surgical wound. The gastrointestinal tract is not entered although the appendix is in the proximity of the intestines, and because the surgery in this patient was uneventful it can be assumed that there was no break in aseptic techniques. The fact that routine appendectomy does not involve a ruptured or highly inflamed appendix contributes to the categorization of the incision as a clean wound. Wounds are dirty when pus is encountered during the surgery, and a contaminated wound involves the finding of acute inflammation or the gross spillage of indigenous flora from a hollow, muscular organ. The categorization of clean-contaminated wound implies entrance into a hollow, muscular organ without significant spillage of indigenous flora. "Nonsterile" is not a surgical wound category.

50. The answer is B. [*Chapter 28 III B 1 a (1), b (1)*] Localized inflammation and pus developing 4–6 days after appendectomy indicate that the patient has probably developed a staphylococcal wound infection. Whether *Staphylococcus aureus* or *Staphylococcus epidermidis* is the etiologic agent cannot be determined by the case history. Clostridial and group A streptococcal wound infections generally produce cellulitis and frequently are fulminant. Anaerobic streptococci produce incisional wound infections and are found in deep abscesses, frequently in mixed etiologies. They produce a thick grayish pus with a distinctive fetid, anaerobic odor.

51. The answer is C. [*Chapter 28 III B 1 b (1) (b)*] The etiologic agent of the infection apparently is *Staphylococcus aureus.* Certain strains of S. aureus produce an exotoxin that causes toxic shock syndrome (TSS). This syndrome most commonly is associated with the improper use of highly absorbent tampons made of synthetic fibers, but wound TSS also may occur when the toxin is absorbed through an incisional wound. TSS usually is fatal. *S. aureus* also can produce subcutaneous abscesses, but these generally are localized infections. Myonecrosis and cellulitis are produced primarily by *Clostridium perfringens*; localized cellulitis also is caused by *Streptococcus pyogenes*, microaerophilic streptococci, and staphylococci. Localized cellulitis rarely is life threatening. Endotoxin shock is a potentially fatal complication of wound infections caused by gram-negative bacteria.

52. The answer is C. [*Chapter 30 I A 2 a, b*] The findings of fever, headache, nuchal rigidity, and lethargy with an acute onset and the lack of dramatic neurologic manifestations suggest acute bacterial meningitis. Viral meningitis causes much of the same symptomatology, but the onset typically is more insidious and the patient usually is less acutely ill. Patients with viral encephalitis display the same general symptomatology as those with viral meningitis, but encephalitis is differentiated by dramatic neurologic manifestations and a much poorer prognosis. Fungal meningitis is more chronic and frequently is seen with other systemic signs of mycotic disease. Brain abscess usually is seen with other foci of infection, and the patient typically has deficits that reflect the location of the lesion.

53. The answer is E. [*Chapter 30 I A 1 a (3) (a) (iii)*] *Streptococcus pneumoniae* is the most common cause of bacterial meningitis among the elderly. *Haemophilus influenzae* type b is the most common cause of bacterial meningitis overall. Its incidence is highest in children 6–12 months old and decreases with age; the incidence of meningitis caused by *H. influenzae* is low in adults. Meningococcal meningitis occurs primarily among young adults, and *Neisseria meningitidis* serogroups A, B, C, and Y cause most cases. *Staphylococcus aureus* occasionally causes meningitis but is a common cause of brain abscess. *Actinomyces israelii* is a rare cause of meningitis associated with trauma to the jaw and gingiva or gastrointestinal tract.

54. The answer is B. [*Chapter 30 I A 3 b*] The cell count in the CSF is elevated during acute bacterial meningitis to 1,000–10,000 cells/mm^3; neutrophils are the prominant cell type. Normal cell counts are

0–5 cells/mm³ CSF. Mononuclear cells are predominant in the CSF late in viral meningitis and in patients with late manifestations of neurosyphilis; lymphocytes are prominent in early manifestations of neurosyphilis. A few erythrocytes may contaminate the CSF during the lumbar puncture; many erythrocytes can indicate brain hemorrhage. Segmented neutrophils are predominant in the CSF during viral meningitis, but the count rarely exceeds 1,000 cells/mm³.

55. The answer is B. *[Chapter 30 I A 3 c]* CSF chemistry is an important tool in determining the general diagnosis of meningitis. During bacterial meningitis, CSF protein characteristically is increased in relation to the cell count; glucose levels are low with a concentration of about 40% of simultaneous serum glucose levels. Elevated protein and a very low glucose level usually are found in fungal meningitis. In viral meningitis, protein levels are moderately elevated and glucose concentrations are normal or slightly decreased.

56. The answer is D. *[Chapter 29 I B 1 a]* *Escherichia coli* is the most common etiologic agent of simple urinary tract infection (UTI) in unhospitalized ambulatory people who are otherwise healthy. This member of the Enterobacteriaceae family accounts for about 50% of UTIs in these persons. *Klebsiella pneumoniae* is the second most common cause of UTI in this population, accounting for 8%–13% of disease. *Staphylococcus aureus* is only an occasional cause of UTI; *Serratia marcescens* is a frequent cause of UTI in hospitalized and debilitated patients with indwelling urinary catheters. *Gardnerella vaginalis* is an occasional etiologic agent of vulvovaginitis.

57. The answer is C. *[Chapter 29 I D 1 a]* The midstream clean-catch technique of urine collection is the most widely used method of obtaining urine samples in suspected cases of UTI. The sample is collected in midstream of the voiding process after thoroughly cleaning the glans penis or urethral meatus with soap and water; disinfectant solutions are contraindicated. Suprapubic aspiration and catheterization are invasive procedures and should be used only as a last resort in collecting urine to diagnose significant bacteriuria. Collection of samples during early morning voiding has no particular advantage in determining significant bacteriuria. The Mantoux technique is a method for determining sensitivity to tuberculoproteins.

58. The answer is B. *[Chapter 29 I D 3 b (2) (a)]* The most accurate means of quantitative culture is the pour plate. This technique evenly distributes the bacteria within agar and permits accurate counting. Contamination of the agar by environmental microbes usually is revealed by surface colonies. The calibrated loop method of urine culture is the most widely used technique because it is more rapid and convenient. Commercially prepared dip slides also are used to determine significant bacteriuria, but this method generally gives only an estimation of the number of bacteria per milliliter of urine. Spectrophotometry is not useful because it determines the number of particles in the urine rather than the number of living bacteria and it can be affected by urine color. Microscopy of uncentrifuged urine provides only an estimation of the bacterial count.

59. The answer is C. *[Chapter 29 I D 3 a]* The presence of 10⁵ or more gram-negative enteric rods/ml of urine defines significant bacteriuria. Note that this number is specific for gram-negative enteric rods. The number of other bacteria, such as *Staphylococcus aureus* or enterococci, and fungi, such as *Torulopsis glabrata* or *Candida albicans*, indicating UTI usually is smaller.

60. The answer is B. *[Chapter 29 I D 2 a]* A rapid means of estimating a count of 10⁵ bacteria/ml of urine is by microscopic examination of uncentrifuged urine. The finding of at least 1 gram-negative rod per high-power microscopic field indicates significant bacteriuria.

61. The answer is E (all). *[Chapter 27 III B 1, C 1; Chapter 31 II B 7 b, 11, 12 a]* AIDS patients are predisposed to a variety of opportunistic infections. *Pneumocystis carinii* is the leading cause of fatal pneumonia in AIDS patients; 25% of the patients with *P. carinii* infection also have CMV infection. There also is an increased susceptibility to mycobacterial infections caused by the tubercle bacillus and the etiologic agents of tuberculosis syndromes. The sporozoan *Toxoplasma gondii* frequently is the cause of disseminated, usually fatal disease in AIDS patients.

62. The answer is B (1, 3). *[Chapter 11 I E 1 b (1) (a) (i); II A 1 c (2)]* Enterotoxigenic *Escherichia coli* and *Vibrio cholerae*, of which type 0:1 is one of the most virulent, produce an enterotoxin that stimulates the activity of adenyl cyclase in intestinal epithelium. The increased activity of the enzyme results in increased concentration of cyclic adenosine 3', 5'-monophosphate (cAMP). Increased cAMP causes the efflux of electrolytes and water, resulting in a diarrheal syndrome. The dysentery caused by *Shigella dysenteriae* is the result of enterotoxin and neurotoxin. Salmonella enteritidis produces diarrhea by invasion of intestinal mucosa and, perhaps, by the release of an enterotoxin. The enterotoxic effects of these two bacteria is not completely defined.

63. The answer is D (4). [*Chapter 17 IV B 3 a (3)*] The respiratory tract is the usual route of primary infection in the systemic mycoses; most cases occur following the inhalation of microconidia into the lungs. Blastomycosis, however, may occur as primary cutaneous lesions. The pathogenesis of blastomycosis is not understood completely, but primary cutaneous blastomycosis is thought to be acquired as a result of traumatic inoculation of *Blastomyces dermatitidis* by splinters of wood or other vegetation contaminated with the fungus. Cutaneous lesions seen in cases of histoplasmosis, coccidioidomycosis, and paracoccidioidomycosis usually are the result of disseminated disease.

64. The answer is E (all). [*Chapter 8 III B 1*] The anaerobic gram-positive cocci include *Peptococcus* species, *Peptostreptococcus* species, anaerobic *Streptococcus* species, and *Sarcina* species. Confusion over the taxonomy of these members of the indigenous flora has led to some doubt about their pathogenicity. However, these bacteria have been isolated in pure culture from meningitis, osteomyelitis, joint infection, brain and liver abscesses, breast infections, and other diseased tissues. Bones, joints, soft tissues, and vascular tissues are the most common sites of infection.

65. The answer is E (all). [*Chapter 7 I A*] Several steps are involved in the process by which most bacteria cause infection. They adhere to specific host cells as the result of both bacterial and host factors and, frequently, by attachment to specific receptors. Bacteria colonize the host tissue if they are capable of overcoming the host's natural defense mechanisms. In order to survive, they multiply if they can compete for nutrients. Finally, bacteria must penetrate host tissues in order to produce a pathologic process.

66. The answer is B (1, 3). [*Chapter 23 II A 3 b*] The bulk of genetic information indicates that the large T antigen is a multifunctional protein with a role in both establishing and maintaining transformation. The small T antigen has an overlapping amino acid sequence plus a unique region that is missing from the large T antigen. This protein appears to be essential for the maintenance of transformation. The late viral proteins, such as the capsid proteins, play no role in the transformation process.

67. The answer is C (2, 4). [*Chapter 8 II A 4 b (2), (3)*] Sequelae are nonsuppurative consequences of a previous infection. Upper respiratory tract and skin infections caused by group A β-hemolytic streptococci (e.g., *Streptococcus pyogenes*) may result in migratory arthritis, acute glomerulonephritis, and rheumatic fever as sequelae. Generally, the more frequently an individual has had *S. pyogenes* infection, the higher the risk of developing one of the sequelae. Several theories exist about the development of these sequelae, including cross-reacting antibodies, immune complex deposition, and autoimmune reactions. Pharyngitis and endocarditis are active infectious processes caused by *S. pyogenes*.

68. The answer is A (1, 2, 3). [*Chapter 5 II G 2*] Initial antibody production after primary immunization is largely of the IgM isotype. The IgG that is produced initially is of lower affinity than IgG produced after secondary boosting. Immunization results in proliferation of the subpopulation of B cells with surface receptors for that antigen, whereas other B cells remain quiescent. Initial antigen processing is carried out by macrophages, which present the processed antigen to T cells. The thymic epithelial cells are not involved in antigen processing; they function only as a stimulus for T cell differentiation prior to antigen exposure.

69. The answer is C (2, 4). [*Chapter 16 VI A 6 a, B 5 b (1)*] Actinomycosis and nocardiosis are two distinct clinical syndromes with distinct causes and clinical manifestations. However, cutaneous nocardiosis and actinomycosis have a common feature: Both diseases can cause draining sinus tracts with granular colonies in the discharge. The granules consist of bacteria and amorphous tissue exudate. Erysipeloid is a cutaneous disease characterized by a painful, spreading, edematous erythema. Listeriosis has no cutaneous manifestations.

70. The answer is A (1, 2, 3). [*Chapter 20 I C 1–3*] Since interferons themselves are not antiviral, they must interact with host cells to induce the antiviral state. Induction involves the derepression (or activation) of host genes coding for several proteins that are effectors of the antiviral state. Since the process of establishing the antiviral state requires both mRNA and protein synthesis, there is a considerable lag between interferon adsorption and the appearance of the antiviral state. In mixtures of cells that are sensitive and resistant to interferon, the antiviral state spreads from sensitive cells to resistant ones.

71. The answer is E (all). [*Chapter 7 I B*] Humans acquire infectious agents by nearly every conceivable route, although human-to-human transmission through inhalation of contaminated, aerosolized respiratory droplets probably is the most common means of acquisition. Human-to-human transmission by intimate contact, as occurs in venereal diseases, also accounts for many infections. Animal-to-human transmission of the etiologic agents of zoonoses can occur by inhalation of contaminated droplets, by direct contact with tissue or excrement, and by ingestion of contaminated tissue or products. Fomites are inanimate objects that become contaminated with an infectious agent, which then is transmitted by human contact with the fomite. Vectors, particularly arthropods and insects such as flies, transmit to humans a variety of diseases, such as the rickettsial diseases.

72. The answer is C (2, 4). [*Chapter 12 III A 6, B 5 a (1), C 4 a, D*] Plague and ulceroglandular tularemia cause bacteremia following the bite of a contaminated vector. Regional lymph nodes, such as those in the axillae, become enlarged and tender and characteristically form buboes in plague. These buboes normally rupture and drain. In ulceroglandular tularemia, the regional lymph nodes, typically in the axillae, become enlarged and tender and may break open and drain in serious cases. In brucellosis and pasteurellosis, bacteremia can occur and lymph nodes may be involved, but suppurated lymph nodes would be very uncharacteristic of these diseases.

73. The answer is B (1, 3). [*Chapter 24 III B 3 b, 4 c (2)*] Giardiasis is a common intestinal infection caused by the flagellate *Giardia lamblia*. The organisms colonize the duodenum and jejunum, where they attach to epithelial cells but do not invade the mucosa. Free parasites are carried into the colon, and oocysts are excreted in the feces. The fecal-oral route is the most common means of transmission of giardiasis. The disease is characterized by diarrhea, which can occur suddenly and explosively, particularly in children. Serologic tests for elevated immunoglobulins are of no value in the diagnosis.

74. The answer is A (1, 2, 3). [*Chapter 22 I E*] The initiation of an influenza epidemic requires that the infection be established in a portion of the population that does not have adequate immunity. The highly variable antigenic structure of the influenza virus is a predisposing factor to the establishment of such an epidemic spread. The presence of respiratory immunoglobulin A (IgA) halts the initial replication of the virus in the respiratory epithelium, which occurs following inhalation of infected droplets. Virtually all spread of influenza virus involves virus-laden droplets, not contaminated objects such as sheets and dishes. Hospitals offer a ready environment for the person-to-person transmission of infectious aerosols.

75. The answer is D (4). [*Chapter 6 I D 4*] The Coombs' test is a widely used technique for detecting nonagglutinating antibodies or amounts of antibodies that are too small to agglutinate red blood cells. The test, also known as the antiglobulin test, uses antiglobulin produced in a heterologous species to form a lattice for agglutination of red blood cells. Particulate antigens usually are agglutinated easily. Bacterial agglutinins are cellular components that, when combined with antibodies, cause an agglutination reaction with the bacterial cell as the detectable agglutinating particle. The Coombs' test does not use complement or the lysis of red blood cells.

76. The answer is D (4). [*Chapter 8 II D 3*] The polysaccharide capsule of *Streptococcus pneumoniae* appears to be the only significant virulence factor of the bacterium. It is antiphagocytic, and all virulent strains of the pneumococcus typically possess large capsules. Streptolysin O is an extracellular product of streptococci responsible, in part, for the hemolysis of erythrocytes, but its role as a virulence factor is unclear; pneumococci produce pneumolysin. Protein A is a virulence factor of *Staphylococcus aureus*, and pili aid in adherence of various microbes to tissue cells.

77. The answer is C (2, 4). [*Chapter 15 I H 1; II D 2 e (1) (c)*] Humans acquire psittacosis and Q fever from animals that are natural hosts of *Chlamydia psittaci* and *Coxiella burnetii*, respectively. *C. psittaci* frequently is the cause of a latent infection in wild and domesticated birds. Shipping and handling of the birds often activates the disease, and humans are infected by inhaling contaminated respiratory droplets from the bird. Q fever is primarily an occupational disease of people who handle animals or animal products. Humans normally acquire it also through the respiratory route. Inclusion conjunctivitis is a disease of neonates and children. It is typically acquired from a mother who is carrying *Chlamydia trachomatis* serotype A, B, Ba, or C in the cervix. Louse-borne typhus is transmitted from human to human by the body louse.

78. The answer is B (1, 3). [*Chapter 31 I E 1, 2 a*] Several types of supportive and specific therapy can be used to diminish the likelihood that nosocomial infections will develop. Physicians and other hospital personnel who are shown to carry potentially pathogenic microorganisms should be treated with effective antimicrobial therapy, and their contact with susceptible patients should be limited. Wounds should be monitored carefully for signs of infection; frequent dressing changes, irrigation, and débridement significantly decrease the incidence of wound infections. The patient's environment should be kept clean to decrease the number of microorganisms to which the patient is exposed; sterilization of the environment is impractical and unnecessary. The benefits of discriminating use of prophylactic antibiotics in hospitalized patients with underlying disease frequently outweigh the risks of selective pressures of the drugs on microbial populations.

79. The answer is E (all). [*Chapter 26 II C 1*] Many microbial diseases are characterized by bacteremia or septicemia as an integral part of the disease process. *Treponema pallidum* disseminates hematogenously from the site of inoculation, and this bacteremia is the cause of the gummas characteristic of secondary disease. Bartonellosis is a rare disease confined to certain Andean valleys; the etiologic agent *Bartonella bacilliformis* shows an affinity for vascular endothelium and erythrocytes. Rocky Mountain spotted fever and other rickettsial diseases are characterized by vasculitis caused by the hematogenously disseminating rickettsia. Pneumococcal pneumonia is characterized by bacteremia that can lead to meningitis.

80. The answer is B (1, 3). [*Chapter 16 II E 1, 3, F 1, 2*] Legionnaires' disease and Pontiac fever are two distinct clinical syndromes caused by the same etiologic agent, *Legionella pneumophila*. The initial, early symptomatology of both syndromes is characterized by headache and myalgia. However, following onset, the diseases differ greatly. Legionnaires' disease progresses to a severe pneumonia with a mortality rate as high as 60%. Pontiac fever develops into a mild flu-like syndrome; there have been no reported fatalities. The attack rate in Legionnaires' disease is less than 1%, whereas the attack rate in Pontiac fever is greater than 90%.

81. The answer is B (1, 3). [*Chapter 13 I A 5 a (3), C 2 c*] Syphilis and yaws characteristically have primary, secondary, and tertiary stages. The tertiary stage of both diseases is characterized by the presence of disseminated lesions. Lesions in tertiary syphilis are most serious when they occur in the heart or central nervous system. The tertiary stage lesions of yaws typically occur in the skin and bones and may cause gross disfigurement. Pinta is a disease primarily of children caused by *Treponema carateum*. Unlike syphilis and yaws, pinta does not appear to have a secondary or tertiary stage. Vincent's angina once was believed to be caused by *Treponema vincentii* in conjunction with *Bacteroides melaninogenicus*; the etiologic agent now is thought to be a herpesvirus.

82. The answer is D (4). [*Chapter 22 II C 1 b*] Parainfluenza viruses are important causes of human respiratory disease, especially in the fall and winter. Infection is initiated in the respiratory epithelium, which is not directly accessible to circulating antibodies of any type. Therefore, the only significant immunity is derived from local IgA and is type specific. Infection occurs in the presence of high titers of group-specific circulating IgG.

83. The answer is B (1, 3). [*Chapter 8 I B 3 d; Chapter 10 I B 1; II A 6 b; Chapter 11 II B 1 c, 2*] Food poisoning is a common disease caused by a variety of bacteria, a few of which are the etiologic agents of two distinct syndromes. *Bacillus cereus* produces an enterotoxin that causes severe nausea and vomiting; the syndrome has been incorrectly attributed to staphylococcal enterotoxins. A separate enterotoxin produced by *B. cereus* causes food poisoning characterized by abdominal cramps and diarrhea, similar to symptoms of clostridial food poisoning. *Clostridium perfringens* also is the etiologic agent of two distinct syndromes of food poisoning. The first is a mild form produced by type A strains and characterized by nausea, abdominal pain, and diarrhea. It often occurs in epidemic proportions in hospitalized or institutionalized people. Enteritis necroticans is a serious disease produced by the ingestion of type C strains of *C. perfringens*. *Salmonella typhi* is the etiologic agent of enteric fever, and *Staphylococcus aureus* produces food poisoning as the result of enterotoxin production; different types of these two bacteria produce analogous syndromes.

84. The answer is C (2, 4). [*Chapter 11 II A 2 a, B 2 a*] *Vibrio parahaemolyticus* and *Campylobacter fetus* most frequently cause gastroenteritis in humans as the result of ingesting contaminated food or fluids. *V. parahaemolyticus* is a marine vibrio that can be isolated from a variety of fish and shellfish. It has been associated most frequently with a severe enteritis brought on by the ingestion of improperly cooked crabs, although raw fish dishes are an increasingly important source in the United States. It is the most frequent single cause of food poisoning in Japan, where the ingestion of raw fish is a cultural characteristic. *C. fetus* normally causes gastroenteritis, especially in children, when food or fluids contaminated by animals is ingested. *Bacteroides melaninogenicus* occasionally causes human disease of the respiratory tract or other organs above the diaphragm. *Acinetobacter anitratus* has been associated with respiratory and urinary tract infections and wound infections.

85. The answer is C (2, 4). [*Chapter 29 V A 2 a (1); VII B 2 a (2)*] The chancre of primary syphilis is firm and painless; the lesion of lymphogranuloma venereum also is painless. The chancre of chancroid, in contrast to that of syphilis, is soft and tender. The lesions of genital herpes (i.e., vesicles) are extremely painful.

86. The answer is A (1, 2, 3). [*Chapter 8 II D 4; Chapter 9 III C; Chapter 10 I A 9; Chapter 12 I A 7*] Effective human vaccines have been produced from the polysaccharide capsules of *Streptococcus pneumoniae*, *Neisseria meningitidis*, and *Haemophilus influenzae*. These vaccines induce a high degree of immunity in adults, but they are relatively ineffective in children under the age of 2 years. The polypeptide capsule of *Bacillus anthracis* is a virulence factor, but it has not been used to produce a vaccine. At present, there is no anthrax vaccine available for use in humans.

87. The answer is B (1, 3). [*Chapter 21 III E*] The primary infection of varicella-zoster virus (VZV) is varicella, or chickenpox, which occurs in epidemic form in children, most commonly in the winter months. In children, chickenpox usually is a mild self-limited disease, but in adults it can be quite severe. Transmission is by respiratory droplets and direct contact. The recurrent form of the disease seen most often in older patients and in immunosuppressed patients is zoster, or shingles; that form of disease results from a reactivation of a latent infection and occurs with equal frequency throughout the year.

88. The answer is B (1, 3). [*Chapter 14 II A*] Several species of mycobacteria are capable of causing tuberculosis in humans. *Mycobacterium tuberculosis* is the most important etiologic agent of tuberculosis in developed countries. *Mycobacterium bovis* once was a major cause of gastrointestinal tuberculosis in the United States, but eradication of diseased herds of cattle and routine pasteurization of dairy products has significantly decreased the incidence of tuberculosis due to this agent. *Mycobacterium africanum* is similar to *M. tuberculosis* and is an important cause of human tuberculosis in Africa. *Mycobacterium fortuitum* is the occasional etiologic agent of abscesses at injection sites in drug abusers.

89. The answer is E (all). [*Chapter 13 I A 4, 5 a (1) (a), (2) (b), 6*] Syphilis is most contagious during the primary and secondary stages of the disease. The chancre at the site of infection contains many treponemes, and the usual means of transmission is by sexual contact. The disseminated skin lesions characteristic of secondary syphilis also contain many treponemes, and contact with these lesions can transmit the disease. Both primary and secondary stages of infection during pregnancy can lead to transplacental passage of *Treponema pallidum* and potentially fatal congenital syphilis in the fetus. Bacteremia during the secondary stage, and late in the primary stage, makes transmission during transfusion probable if the blood is not screened for *T. pallidum*.

90. The answer is E (all). [*Chapter 16 III E 2 b, 3 c (1), (2)*] Toxigenic *Corynebacterium diphtheriae* produces a potent exotoxin that inhibits protein synthesis. The toxin is responsible for the systemic symptomatology associated with diphtheria, including effects on the structure and function of myocardial cells and demyelination of peripheral and cranial nerves. The paralysis resulting from nerve demyelination usually is reversible as the toxin is metabolized and synthesis of the myelin sheath resumes. Pharyngeal diphtheria is characterized by the formation of a pseudomembrane that does not absolutely require toxin. Cutaneous diphtheria is characterized by a cutaneous lesion that may be a simple papule or may be a chronic necrotic lesion that does not heal.

91. The answer is A (1, 2, 3). [*Chapter 9 I B 4, 7*] The metabolic activity and growth requirements of pathogenic *Neisseria* species make it extremely difficult to isolate colonies from clinical specimens on routine bacteriologic media. Chocolate agar is the basic medium, and it should be supplemented with solubilized starch to detoxify fatty acids contained in the medium. The presence of selected antibacterial and antifungal compounds in the medium prevents the overgrowth of less fastidious indigenous flora. The growth of pathogenic neisseriae, especially upon initial isolation, is greatly enhanced by an atmosphere of increased (i.e., 4%–8%) carbon dioxide. Commercial isolation media incorporating these features are available. Tellurite is a component of differential media for *Corynebacterium diphtheriae.*

92. The answer is E (all). [*Chapter 14 II E 1 a–e*] Primary tuberculosis occurs upon initial infection with one of the etiologic agents of human tuberculosis, usually *Mycobacterium tuberculosis.* The mycobacteria usually are inhaled and establish a focus of infection in the periphery of the midzone of the lung, which is characterized by a cell-mediated immune response with the infiltration of macrophages and lymphocytes. The infecting mycobacteria are phagocytized by macrophages and carried to the cervical lymph nodes. The infected macrophages then spread the mycobacteria to any organ system via lymphatic and blood vessels. In most cases, the tubercles formed by the disseminating bacteria heal and sometimes calcify. The infected person develops a mild, influenza-like syndrome if there is any symptomatology at all. Everyone who has primary tuberculosis has a risk of reactivation tuberculosis, because the mycobacteria can remain viable within the healed tubercles for at least 20 years.

93. The answer is A (1, 2, 3). [*Chapter 20 I A, B, C*] Interferons are naturally produced proteins and glycoproteins that are induced both in cultured cells and in vivo by an identical set of inducers. Interferons are not virus specific but, instead, are host specific. They act by attaching to host cells and inducing an antiviral state mediated by other proteins rather than interacting directly with an infecting virus.

94. The answer is E (all). [*Chapter 15 II C 1*] The classic symptomatology associated with rickettsial diseases is headache, fever, and rash. The rash is the result of vasculitis characterized by focal infection causing hyperplasia and inflammation of the vascular epithelium. Although clinical manifestations vary from patient to patient and from disease to disease, most people with rickettsial disease experience one or more of these classic manifestations.

95. The answer is E (all). [*Chapter 9 II D 2; Chapter 15 I G 2 b (2); Chapter 16 I D 1, 2 b*] Pelvic inflammatory disease (PID) can be caused by a variety of bacteria, all of which usually are sexually transmitted. *Mycoplasma hominis, Chlamydia trachomatis,* and *Neisseria gonorrhoeae* are important causes of nongonococcal urethritis and PID in women. *Ureaplasma urealyticum* is more difficult to associate with specific disease, but it occasionally is the only microorganism isolated from cases of urethritis and PID in women.

96. The answer is A (1, 2, 3). [*Chapter 9 I A; IV; Chapter 15 I G*] All *Neisseria* species are gram-negative cocci usually occurring in pairs in clinical specimens; this characteristic is true for the pathogenic gonococci and meningococci as well as the nonpathogenic members of the indigenous flora, such as *Neisseria sicca, Neisseria flavescens,* and others. *Branhamella catarrhalis* is morphologically identical to the neisseriae and was formerly placed in the genus *Neisseria. Chlamydia trachomatis* is a small, round-to-oval bacterium with an outer membrane that resembles gram-negative bacteria, although the Gram stain is not used routinely to examine these cells as they grow intracellularly and form inclusion bodies.

97. The answer is C (2, 4). [*Chapter 9 II D 2; Chapter 15 I G 1 b (3); Chapter 16 IV C 1*] Disseminated gonorrhea may result from the failure to diagnose and treat asymptomatic gonococcal infection. PID can occur and result in additional systemic syndromes, such as ascending gonococcal arthritis and scarring of the fallopian tubes, causing permanent sterility. Lymphogranuloma venereum is caused by *Chlamydia trachomatis* serotypes L1, L2, and L3. It begins as a small ulcer on the genitalia and involves the local lymph nodes. Granulomatosis infantiseptica is the result of vaginal colonization by *Listeria monocytogenes.* The bacterium can cross the placenta, infecting the fetus and possibly resulting in an abortion or stillbirth.

98. The answer is B (1, 3). [*Chapter 29 V B 2 a (1), (2)*] A positive VDRL (Venereal Disease Research Laboratory) test is epidemiologically suggestive evidence of syphilis in this pregnant woman. Immediate manifestations of congenital syphilis include hepatomegaly, splenomegaly, and skeletal abnormalities on radiographic examination. These conditions occur in virtually all neonates born with disseminated syphilis. Skin lesions typically appear 2–6 weeks after birth, and blindness is a late manifestation of congenital syphilis.

99. The answer is C (2, 4). [*Chapter 29 V B 3*] Effective chemotherapeutic regimens for the neonate with congenital syphilis are based upon laboratory analysis of the neonatal CSF. A single dose of penicillin is effective in neonates with normal CSF, and a 10-day course of chemotherapy is recommended for those with abnormal CSF. In addition, congenital syphilis can be prevented by penicillin therapy for the pregnant woman. Cesarean section prevents neonatal infections, such as herpes, that are acquired by passage through the birth canal; congenital syphilis is acquired by transplacental passage of *Treponema pallidum.* No vaccine is available for syphilis.

100–104. The answers are: 100-B, 101-C, 102-C, 103-D, 104-A. [*Chapter 10 II B, C 4, 5*] The neurotoxin of *Clostridium botulinum* acts at the motor nerve endings of the neuromuscular junction. The toxin interferes with the release of acetylcholine.

Both tetanospasmin and botulinum toxin are potent neurotoxins. They bind to nerve cells and exert neurotoxic effects, usually resulting in death from paralysis of essential muscles such as the diaphragm.

The neurotoxins of both *Clostridium tetani* and *C. botulinum* are neutralized by specific antitoxins (i.e., antibodies). Tetanus usually is prevented by the administration of tetanus toxoid throughout life, inducing the body to produce anti-tetanospasmin antibodies. Specific neutralizing antitoxin produced in vitro is administered as a therapeutic agent to prevent progression of disease.

Patients who receive antitoxin therapy in the form of hyperimmune serum and recover from tetanus or botulism are not immune to subsequent intoxication. The lethal quantity of toxin in both diseases is probably too small to induce antibody formation.

Tetanospasmin of *C. tetani* blocks spontaneous and evoked transmitter release at both the central and peripheral synapses. The neurotoxin binds irreversibly to the nerve cells and exerts its influence. Convulsions, spasms, and paralysis are the frequently fatal results.

105–108. The answers are: 105-C, 106-A, 107-D, 108-A. [*Chapter 12 I A 3, 5, 6 b, 7 b, C 2*] The genus *Haemophilus* is characterized by special growth requirements; the X factor (hematin) and the V factor (nicotinamide-adenine dinucleotide; NAD) are absolute requirements for growth. They may be added to the culture medium as separate biochemical components, or they may be provided by the addition of freshly hemolyzed blood.

The antiphagocytic polysaccharide capsule of *Haemophilus influenzae* has been used as an effective human vaccine. Its primary drawback is that it is not effective in inducing protective antibodies in children younger than 2 years; this group is at the highest risk for developing meningitis or other serious disseminated disease.

H. influenzae and *Haemophilus ducreyi* are obligate human pathogens. Avirulent strains of *H. influenzae* are carried in the nasopharynx and can convert to virulent strains, although the mechanism of conversion is not clearly understood. *H. ducreyi* is the etiologic agent of chancroid, a sexually transmitted disease.

H. influenzae is an important cause of meningitis in children between the ages of 6 months and 3 years. *H. influenzae* and *Streptococcus pneumoniae* are the most frequent causes of meningitis in this age-group.

109–111. The answers are: 109-B, 110-C, 111-A. [*Chapter 6 II B*] B lymphocytes, or B cells have been shown to have up to 10^5 immunoglobulin molecules on their cell surface. They are attached to the cell membrane by the Fc portion of the molecule. The B cell synthesizes the immunoglobulin attached to its membrane, and there usually is only one heavy- or light-chain type per B cell.

Both T cells and B cells can be separated from peripheral blood by density gradient centrifugation. This technique is a high-speed centrifugation of a gradient of cesium chloride resulting in the separation of cells of differing densities. The gradient zones can be drained to yield a relatively pure mixture of each cell population.

Human T cells characteristically bind sheep erythrocytes to their surface to form a rosette structure. The binding sites on the T cells probably recognize either glycoprotein or carbohydrate receptors on the surface of the sheep erythrocytes. Immunoglobulin and complement receptors do not appear to be involved.

112–115. The answers are: 112-C, 113-D, 114-A, 115-E. [*Chapter 3 III B 1 c (1), 2 a, b; IV B*] Bacteriophage λ has a region of nucleotide homology with its host, and that region defines the normal integration site. Integration of λ into the chromosome involves a site-specific recombination and can occur effectively in the absence of a functional recombination system in the host. The major recombination system in bacteria is the rec system with rec A controlling the major part, and rec B and rec C controlling much smaller proportions. A small amount of recombination can occur without a high degree of sequence homology or the intervention of the enzymes of the rec system. This mechanism is not fully understood and is referred to as illegitimate recombination.

Molecular cloning has become an extraordinarily powerful tool for the study of gene structure and organization. One of the most useful characteristics of cloning is the fact that the DNA is manipulated in vitro, attached to a vector, and then introduced into the microorganism for replication, thereby making the source of the DNA insignificant.

116–120. The answers are: 116-E, 117-D, 118-D, 119-B, 120-C. [*Chapter 21 II C 1 a, b; Chapter 22 II E; V A; VIII D*] The common cold has several etiologies. The principal causes of this acute, afebrile disease of the upper respiratory tract are the rhinoviruses and the coronaviruses.

Adenoviruses often are associated with acute respiratory disease—a condition that does not resemble the syndrome known as the common cold. Acute respiratory disease generally occurs only in confined populations, such as military training camps. Adenoviruses also are associated with a pharyngoconjunctivitis that is spread via contaminated swimming pools, most often during the summer months.

Measles virus is the only paramyxovirus listed and is one of the most highly infectious viruses that infects humans. Rubella virus is a plus-strand RNA virus belonging to the rubivirus group of the togavirus family. It, too, is highly contagious.

121–125. The answers are: 121-E, 122-B, 123-D, 124-A, 125-C. [*Chapter 27 I A 1, D; II D 2 a, c; III B 1*] The protozoan *Pneumocystis carinii* is the most common cause of interstitial pneumonia in immunosuppressed patients. It is a major cause of morbidity and mortality among certain patient populations, such as AIDS patients.

The membrane-bound, pleomorphic bacterium *Mycoplasma pneumoniae* probably is the most highly infectious etiologic agent of pneumonia. In families and other "close environment" groups, approximately 50% of the people exposed develop disease.

Bacterial pneumonias account for only a small percentage of all pneumonias. Overall, influenza A virus is the etiologic agent of the most common pneumonia in the adult population.

Although bacterial pneumonia accounts for only a small proportion of all human pneumonia, the encapsulated bacterium *Streptococcus pneumoniae* is responsible for about 80% of all bacterial pneumonia cases. Only 14 of the more than 80 capsular serotypes account for most cases of pneumococcal pneumonia.

Adenovirus pneumonia occurs primarily in military recruits and children; it rarely occurs in the civilian adult population.

126–130. The answers are: 126-B, 127-E, 128-B, 129-A, 130-D. [*Chapter 19 II A 1, C 3 c (2), D; Chapter 21 II C 2*] Poliovirus, a picornavirus, contains a single RNA molecule of positive polarity. There is no regulatory control in poliovirus replication, and the single RNA is translated into a polyprotein that contains a single copy of each gene product; therefore, each is made in equimolar amounts. Each of the other viruses listed has some type of regulatory control over transcription, thus limiting the amount of each gene product.

Only one group of viruses—the poxviruses—encodes their own membranes. Vaccinia virus is the most well-studied member of this viral group.

The retroviruses are the only viruses that contain two copies of their nucleic acid genomes. HIV, a retrovirus, is a plus-strand virus. Influenza virus is a minus-strand RNA virus, and adenovirus and vaccinia virus are DNA viruses.

Several serotypes of adenovirus can induce tumor formation in newborn hamsters. None of the other viruses listed is oncogenic for hamsters or any other animal species. There has been speculation that some HIV genes encode oncogenic regions, which, under well-defined circumstances, may be involved in tumorigenesis in humans; however, this has not been clearly established.

131–135. The answers are: 131-D, 132-C, 133-A, 134-E, 135-B. [*Chapter 25 II C; III C 2 c (3); IV C 1, 2, 3 b (1) (a)*] *Brugia malayi* is one of the frequent causes of lymphatic filariasis that can cause blockage of the lymphatic system, resulting in elephantiasis.

Clonorchis sinensis, the Chinese liver fluke, inhabits the common bile duct and can cause blockage and the formation of bile stones. Clonorchiasis has been associated with cholangiocarcinoma.

Paragonimus westermani is the lung fluke. It is capable of causing a pulmonary disease characterized by lung lesions at the sites of larval encystment.

Schistosoma mansoni and the other etiologic agents of schistosomiasis are commonly known as blood flukes. They are spread throughout the body by hematogenous dissemination.

The fish tapeworm *Diphyllobothrium latum* inhabits the small intestines of the human host.

136–141. The answers are: 136-E, 137-A, 138-C, 139-A, 140-D, 141-B. [*Chapter 13 I A, B, C, D; II E 4; III A*] *Borrelia recurrentis* is the etiologic agent of relapsing fever in humans. The disease occurs worldwide and is characterized by a febrile bacteremia. The disease name is derived from the fact that 3–10 recurrences can occur, apparently from the original infection. The disease is transmitted to humans from infected animals by ticks and from human to human by lice.

Bejel is caused by a variant of *Treponema pallidum* designated endemic syphilis. The disease usually is seen in children in the Middle East and Africa. Transmission appears to be through the shared use of drinking and eating utensils; bejel is not transmitted sexually. The disease develops in primary, secondary, and tertiary stages.

Pinta is a tropical disease caused by *Treponema carateum*. It occurs primarily in Central and South America, where it appears to be spread by person-to-person contact. Unlike other treponemal diseases, the lesions of pinta remain localized in the skin.

The etiologic agent of syphilis is *T. pallidum*. Humans are the only natural host of the spirochete, and venereal transmission is the most common means of acquiring the infection. Congenital syphilis occurs when the fetus is infected transplacentally and survives to delivery. Accidental laboratory infections occasionally occur.

Fort Bragg fever is a localized name of pretibial fever caused by *Leptospira interrogans* var. fort bragg. The disease is characterized by a rash on the shins. Humans probably acquire the infection by contact with the urine of infected animals.

Treponema pertenue is the etiologic agent of yaws. The disease occurs primarily in children in tropical regions, where it appears to be transmitted by direct contact or by vectors such as flies. This potentially disfiguring disease has primary, secondary, and tertiary stages.

142–146. The answers are: 142-D, 143-D, 144-B, 145-E, 146-A. [*Chapter 4 II A 5, B 1, C 3, 4, D 4*] Erythromycin inhibits protein synthesis by binding to peptidyltransferase in the 50S ribosomal subunit. The inactivation of the enzyme inhibits the formation of peptide bonds between the amino acids forming the polypeptide chain.

Chloramphenicol, like erythromycin, inhibits protein synthesis by binding to peptidyltransferase in the 50S subunit of bacterial ribosomes. It inhibits the formation of peptide bonds between amino acids of the elongating polypeptide chain. Chloramphenicol has the advantage of crossing the blood-brain barrier in sufficient quantities to treat life-threatening meningitis, but the antibiotic may cause severe aplastic anemia.

Polymyxin antibiotics act as cationic detergents and disrupt the integrity of the cell membrane. They are highly toxic to mammalian cells; only polymyxin B and E are used clinically as topical antibiotics. Polymyxin E is also known as colistin.

Metronidazole is a nitroimidazole compound that has specific toxicity for anaerobic bacteria and anaerobic protozoa, probably by reducing the nitro group to a nitrosohydroxyl amino group. The metabolized drug inhibits DNA synthesis, probably by causing strand breakage.

Cycloserine inhibits bacterial cell wall synthesis by preventing the complete synthesis of peptidoglycan. The antibiotic interferes with the production of alanine and its incorporation into the transpeptide bridges of peptidoglycan.

147–150. The answers are: 147-D, 148-C, 149-A, 150-E. [*Chapter 2 III B 1 b (4), c (1), 2 a (1), 3*] Energy used by bacteria primarily is produced by fermentative, respiratory, and autotrophic metabolism.

In autotrophic metabolism, a variety of inorganic energy sources are used. A unique form of chemoautotrophy used by bacteria occurs anaerobically. Whereas some chemoautotrophs are aerobic and use oxygen as the ultimate electron acceptor, many anaerobes use inorganic salts (e.g., nitrates) as terminal electron acceptors.

Respiratory metabolism usually uses oxygen as the terminal electron acceptor; respiratory metabolism may be aerobic or anaerobic. In complete aerobic oxidation, glucose is converted to carbon dioxide and water via terminal respiration through the tricarboxylic acid cycle. This process yields 38 mol of adenosine triphosphate (ATP) molecules per mol of glucose used.

Fermentative metabolism uses organic compounds as both the electron donors and electron acceptors. A variety of fermentation processes may be used. In mixed acid fermentation—which is common among members of the Enterobacteriaceae family—lactate, acetate, and formate are produced from pyruvate. The Entner-Doudoroff pathway is the major hexose-degrading pathway in pseudomonads.

Index